T0074271

For further volumes:
http://www.springer.com/series/8578

Monica Bianchini, Marco Maggini,
and Lakhmi C. Jain (Eds.)

Handbook on Neural Information Processing

 Springer

Editors
Professor Monica Bianchini
Dipartimento di Ingegneria dell'Informazione
Università degli Studi di Siena
Siena
Italy

Dr. Lakhmi C. Jain
Adjunct Professor
University of Canberra ACT 2601
Australia

Professor Marco Maggini
Dipartimento di Ingegneria dell'Informazione
Università degli Studi di Siena
Siena
Italy

ISSN 1868-4394
ISBN 978-3-642-36656-7
DOI 10.1007/978-3-642-36657-4
Springer Heidelberg New York Dordrecht London

ISSN 1868-4408   (electronic)
ISBN 978-3-642-36657-4   (eBook)

Library of Congress Control Number: 2013932221

Printed on acid-free paper

Springer is part of Springer Science+Business Media (www.springer.com)

Foreword

Neural information processing is an interdisciplinary field of research with contributions from statistics, computer science, neuroscience, electrical engineering, cognitive science, mathematics, and physics. Researchers in the field share a common desire to develop and understand computational and statistical strategies for information processing, in the brain or in artifacts thereof.

The field of neural information processing systems has seen a tremendous growth in the last couple of decades, both in volume and, more importantly, in reputation. It helps that there is now a quite clear separation between computational neuroscience (modeling the brain) and machine learning (developing artifacts that can "learn", possibly but not necessarily inspired by the way the brain works). This handbook focuses on the latter, and does so with great success.

Although the editors do not claim to be fully comprehensive, the handbook has an amazing coverage of the important paradigms in machine learning, by some of the best experts available. It contains chapters going all the way from relevant mathematical theory, to state-of-the-art architectures and learning algorithms, to actual applications.

The nice thing about the chapters is that the authors can take a bit more space to introduce and explain the relevant concepts than they are normally allowed or expected to do in journal and conference papers. You can tell that they enjoyed writing their contributions, which makes the chapters a pleasure to read.

In short, the editors and authors are to be congratulated and thanked for a very interesting and readable contemporary perspective on the state-of-the-art in the machine learning part of neural information processing systems. This handbook will be a valuable asset for graduate students, research, as well as practitioners for the next years to come, until there will hopefully be another edition. I can hardly wait!

Prof.dr. Tom Heskes
Institute for Computing and Information Sciences
Radboud University Nijmegen

Preface

This handbook is inspired by two fundamental questions: "Can intelligent learning machines be built?" and "Can they be applied to face problems otherwise unsolvable?". A simple unique answer certainly does not exist to the first question. Instead in the last three decades, the great amount of research in machine learning has succeeded in answering many related, but far more specific, questions. In other words, many automatic tools able to learn in particular environments have been proposed. They do not show an "intelligent" behavior, in the human sense of the term, but certainly they can help in addressing problems that involve a deep perceptual understanding of such environments. Therefore, the answer to the second question is also partial and not fully satisfactory, even if a lot of challenging problems (computationally too hard to be faced in the classic algorithmic framework) can actually be tackled with machine learning techniques.

In this view, the handbook collects both well-established and new models in connectionism, together with their learning paradigms, and proposes a deep inspection of theoretical properties and advanced applications using a plain language, particularly tailored to non experts. Not pretending to be exhaustive, this chapter and the whole book delineate an evolving picture of connectionism, in which neural information systems are moving towards approaches that try to keep most of the information unaltered and to specialize themselves, sometimes based on biological inspiration, to cope expertly with difficult real–world applications.

Chapters Included in the Book

Chapters composing this book can be logically grouped into three sections:

- Neural Network Architectures
- Learning Paradigms
- Reasoning and Applications

Neural Network Architectures

Chapter 1, by Yoshua Bengio and Aaron Courville, gives an overview on deep learning algorithms, that are able to better capture invariances, and discover non–local structure in data distributions. Actually, deep architectures are obtained by composing several levels of distributed representations, as in neural networks with many layers and in the mammalian cortex. Unsupervised learning of representations, applied in deep architectures, has been found useful in many applications and benefits from several advantages, e.g. when there are many unlabeled examples and few labeled ones (semi-supervised learning), or when the examples come from a distribution different but related to the one of interest (self–taught learning, multi–task learning, and domain adaptation). Learning algorithms for deep architectures have recently be proven to be important in modeling rich data, with the kind of complexity exhibited in images, video, natural language and other AI tasks.

In Chapter 2, by Sajid A. Marhon, Chistopher J. F. Cameron, and Stefan C. Kremer, the readers are introduced to the basic principles of recurrent networks, a variant of feedforward networks which additionally includes recurrent connections (loops) between processing elements. Recurrence forces a cyclic temporal dependency between the activation values of the processing elements, that injects a temporal dynamics in their evaluation. The ability to evolve the activation values over time makes recurrent networks capable of processing temporal input patterns (as opposed to the spatial patterns that limit conventional models), generating temporal output patterns, and performing internal, iterative computations. Computational capabilities and learning abilities of recurrent networks are assessed, and real–world state–of–the–art applications explored.

In the supervised framework, Graph Neural Networks (GNNs) are a powerful tool for processing structured data in the form of general graphs. In Chapter 3, by Monica Bianchini and Marco Maggini, a unified framework is proposed for learning in neural networks, based on the BackPropagation algorithm, that starts from GNNs and graphs, and views each other type of architecture/learning environment (f.i., recursive networks/trees, recurrent networks/sequences, static networks/arrays) as a particular and simpler subcase of the BackPropagation Through Graph algorithm. The derivation of the learning algorithm is described in details, showing how the adaptation is driven by an error diffusion process on the graph structure.

Chapter 4, by Liviu Goraş, Ion Vornicu, and Paul Ungureanu, introduces the Cellular Neural Network (CNN) paradigm, and several analog parallel architectures inspired by CNNs. CNNs are massive parallel computing architectures defined in discrete N–dimensional spaces, and represented by regular arrays of elements (cells). Cells are multiple input–single output processors, described by one or few parametric functionals. The spatio–temporal dynamics of CNNs, possibly associated with image sensors, can be used for high speed 1D and 2D signal processing, including linear and nonlinear filtering, and feature extraction. The VLSI implementability of CNNs in standard CMOS technology is also explored.

In Chapter 5, by Paul C. Kainen, Věra Kůrková, and Marcello Sanguineti, tools from nonlinear approximation theory are used to derive useful estimates of network complexity, which depends on the type of computational units and the input dimension. Major tools are described for estimating the rates of decrease of approximation errors with increasing model complexity. Properties of best approximation are also discussed. Finally, recent results concerning the dependence of model complexity on the input dimension are described, together with examples of multivariable functions that can be tractably approximated.

A neural network utilizes data to find a function consistent with the data and with further conceptual desiderata, such as desired smoothness, boundedness or integrability. The weights and the functions embodied in the hidden units can be thought of as determining a finite sum that approximates some function. This finite sum is a kind of quadrature for an integral formula that would represent the function exactly. In the last chapter of this section (Chapter 6), authored by Paul C. Kainen and Andrew Vogt, Bochner integration is applied to neural networks, in order to understand their approximation capabilities and to make effective choices of the network parameters. By modifying the traditional focus on the data to the construction of a family of functions able to approximate such data, gives not only a deep theoretical insight, but also helps in constructing networks capable of performing more sophisticated tasks.

Learning Paradigms

In traditional supervised learning, labeled data are used to build a model. However, labeling training data for real–world applications is difficult, expensive and time consuming, as it requires the efforts of humans with a specific domain experience. This is especially true for applications that involve learning with a large number of classes, and sometimes with similarities among them. Semi–supervised learning addresses this inherent bottleneck by allowing the model to combine labeled and unlabeled data to boost the performance. In Chapter 7, by Mohamed Farouk Abdel Hady and Friedhelm Schwenker, an overview of recent research advances in semi–supervised learning is provided, whereas its combination with active learning techniques is explored.

Relational learning refers to learning from data that have a complex structure. This structure may be either internal (a data instance may be complex) or external (relationships among data). Statistical relational learning refers to the use of statistical learning methods in a relational learning context. In Chapter 8, by Hendrik Blockeel, statistical relational learning is presented in its concrete forms (learning from graphs, learning from logical interpretations, learning from relational databases) and state–of–the–art methods in this framework — such as inductive logic programming, relational neural networks, and probabilistic logical models — are extensively reviewed.

Kernel methods are a class of non–parametric learning techniques relying on kernels. A kernel generalizes dot products to arbitrary domains and can thus be seen as a similarity measure between data points with complex structures. Key to the success of any kernel method is the definition of an appropriate kernel for the data at hand. A well–designed kernel should capture the aspects characterizing similar instances, while being computationally efficient. Kernels have been designed for sequences, trees and graphs, as well as arbitrary relational data represented in first or higher order logic. In Chapter 9, by Andrea Passerini, the basic principles underlying kernel machines are revised, together with some of the most popular approaches that have recently been developed. Finally, kernel methods for predicting structures are also presented. These algorithms deal with structured–output prediction, a learning framework in which the output is itself a structure that has to be predicted from the input one.

In Chapter 10, by Francesco Gargiulo, Claudio Mazzariello, and Carlo Sansone, a survey on multiple classifier systems is presented. In the area of Pattern Recognition, this idea has been proposed based on the rationale that the consensus of a set of classifiers may compensate for the weakness of a single one, while each classifier preserves its own strength. Classifiers could actually be constructed following a variety of classification methodologies, and they could achieve different rates of correctly classified samples. Moreover, different combining rules and strategies, independent of the adopted classification model, have been proposed in literature. Some of the currently available tools (following different approaches) for implementing multiple classifier systems are reviewed in this chapter, together with a taxonomy of real–world applications.

Modal learning in neural computing refers to the strategic combinations of modes of adaptation and learning within a single artificial neural network architecture. Modes, in this context, are learning methods, and two or more modes may proceed in parallel in different parts of the neural computing structure (layers and neurons) or, alternatively, can be applied to the same part of the structure with a mechanism for allowing the network to switch between modes. Chapter 11, by Dominic Palmer–Brown and Chrisina Draganova, presents an introduction to modal learning techniques, with a particular focus on modal self–organization methods, applied to grouping learners' responses to multiple choice questions, natural language phrase recognition and pattern classification. Algorithms, dataset descriptions, pseudocode and Matlab code are included.

Reasoning and Applications

Bayesian networks represent a widely accepted model for reasoning with uncertainty. In Chapter 12, by Wim Wiegerink, Willem Burgers, and Bert Kappen, Bayesian networks are approached from a practical point of view, putting emphasis on modelling and practical applications. A short theoretic introduction to Bayesian networks is provided and some of their typical usages are reported, e.g. for reasoning and diagnostics. Furthermore, peculiar network behaviors are described, such as the *explaining away phenomenon*. Finally, practical applications to victim identification by kinship analysis based on DNA profiles, and to the construction of a petrophysical decision support system are provided, in order to illustrate the data modelling process in detail.

In content based image retrieval (CBIR), relevance feedback is an interactive process, that builds a bridge between users and a search engine. In fact, it leads to a much improved retrieval performance by updating queries and similarity measures according to a user's preference. Chapter 13, by Jing Li and Nigel M. Allinson, introduces the basic elements of CBIR (from low–level feature extraction to classification methods) and reviews recently proposed CBIR techniques, with a particular focus on long–term learning, that allows the system to record and collect feedback knowledge from different users over a variety of query sessions, and it is particularly tailored for multimedia information searching. Some representative short–term learning techniques are also presented, in order to provide the reader with a comprehensive reference source for CBIR.

In Chapter 14, authored by Ah Chung Tsoi, Markus Hagenbuchner, Milly Kc, and ShiJua Zhang, an application of learning in structured domains is presented, namely classification/regression problems on text documents from large document collections. In fact, while text documents are often processed as unstructured data, the performance and problem solving capability of machine learning methods can be enhanced through the use of suitable structured representations. In particular, for classification problems, the incorporation of the relatedness information, as expressed by Concept Link Graphs, allows the development of tighter clusters with respect to the bag–of–words representation, whereas such graphs can also be useful in the regression framework, in order to rank the items in a large text corpus.

Finally, Chapter 15, by Masood Zamani and Stefan C. Kremer, constitutes a survey on the application of connectionist models to bioinformatics. Actually, many problems in bioinformatics involve predicting later stages in the information flow from earlier ones. Bioinformatics methods capable of such predictions can often eliminate costly, difficult or time–consuming tasks in important biological research. For example, predicting protein structure and function based on the amino acid sequence is an essential component of modern drug design, and can replace expensive wet–lab work. This chapter is intended to provide an introduction to the predominant research areas and some neural network approaches used within bioinformatics.

Conclusions

This chapter has presented an introduction to recent research in neural information processing paradigms, starting from connectionist models recently developed to learn graphs — and showing how they represent a powerful tool to address all those problems where the information is naturally organized in entities and relationships among entities — to classical paradigms, theoretically inspected in order to establish new significant properties, and rivisited for guaranteeing better performances, and to advanced applications, derived from real–world problems and strongly biologically inspired. Not having the presumption to exhaust a so extensive and evolving argument, if ever something should be concluded on the universe of connectionism is that neural information paradigms are moving towards approaching a vast variety of problems, derived from every–day life and, in so doing, they both try to keep most of the information unaltered and to specialize themselves, to attack each problem with the weapons of an expert in the particular field. This book collects just some suggestions on which directions connectionism is following/will follow in view of the future challenges to be faced.

Aknowledgements

The Editors hope that this handbook will prove an interesting and valuable tool to researchers/practitioners, so as to graduate students, in the vast and partially unfathomed area of learning in connectionism. Being conscious of its incopleteness, we are, however, confident in its utility as a sourcebook, in that it draws the trends of current reasearch.

We have been really fortunate in attracting contributions from top class scientists and wish to offer our deep gratitude for their support in this project. We also aknowledge the reviewers for their precious expertise and time and Prof. Franco Scarselli for his fundamental help in realizing this book. Finally, we thanks Springer–Verlag for their support.

Contents

Chapter 1
Deep Learning of Representations

Yoshua Bengio and Aaron Courville

Abstract. Unsupervised learning of representations has been found useful in many applications and benefits from several advantages, e.g., where there are many un-labeled examples and few labeled ones (semi-supervised learning), or where the unlabeled or labeled examples are from a distribution different but related to the one of interest (self-taught learning, multi-task learning, and domain adaptation). Some of these algorithms have successfully been used to learn a hierarchy of features, i.e., to build a deep architecture, either as initialization for a supervised predictor, or as a generative model. Deep learning algorithms can yield representations that are more abstract and better disentangle the hidden factors of variation underlying the unknown generating distribution, i.e., to capture invariances and discover non-local structure in that distribution. This chapter reviews the main motivations and ideas behind deep learning algorithms and their representation-learning components, as well as recent results in this area, and proposes a vision of challenges and hopes on the road ahead, focusing on the questions of invariance and disentangling.

1   Introduction

This chapter is a review of recent work in the area of Deep Learning, which is mostly based on unsupervised learning of representations. It follows up on Bengio (2009) and presents our vision of the main challenges ahead for this type of learning algorithms.

Why Learn Representations?

Machine learning is about capturing dependencies between random variables and discovering the salient but hidden structure in an unknown distribution (conditional

Yoshua Bengio · Aaron Courville
Dept. IRO, Université de Montréal

M. Bianchini et al. (Eds.): *Handbook on Neural Information Processing*, ISRL 49, pp. 1–28.
DOI: 10.1007/978-3-642-36657-4_1   © Springer-Verlag Berlin Heidelberg 2013

or not), from examples. Current machine learning algorithms tend to be very dependent on the choice of data representation on which they are applied: with the
right representation, almost every learning problem becomes very easy. This is particularly true of non-parametric and kernel-based algorithms, which have been very
successful in recent years (Schölkopf and Smola, 2002), but it is also true of the
numerous and equally successful systems based on linear predictors. It is therefore
very tempting to ask the questions: can we learn better representations? what makes
a good representation? The probabilistic and geometric points of view also help us
to see the importance of a good representation. The data density may concentrate in
a complicated region in raw input space, but a linear or non-linear change in coordinates could make the learning task much easier, e.g., extracting uncorrelated (PCA)
or independent (ICA) factors may yield much more compact descriptions of the raw
input space density, and this is even more true if we consider the kind of non-linear
transformations captured by manifold learning algorithms (Saul and Roweis, 2002;
Bengio *et al.*, 2006; Lee and Verleysen, 2007).

Imagine a probabilistic graphical model (Jordan, 1998) in which we introduce latent variables which correspond to the true explanatory factors of the observed data.
It is very likely that answering questions and learning dependencies in that space
is going to be much easier. To make things concrete, imagine a graphical model
for images that, given an image of a car breaking at a red light, would have latent
variables turning "ON", whose values would directly correspond to these events
(e.g., a variable could encode the fact that a car is located at a particular position in
a particular pose, others that there is a traffic light at a particular position and orientation, that it is red, and a higher-level one that detects that the traffic light's and
car's orientation mean that the former applies to the latter, etc). Starting from this
representation, *even if the semantics of these latent variables is not determined a
priori (because they have been automatically discovered)*, learning to answer questions about the scene would be very easy indeed. A simple linear classifier trained
from one or just a few examples would probably do the job. Indeed, this is the kind
of surprisingly fast supervised learning that humans perform *in settings that are familiar to them*, i.e., in which they had the chance to collect many examples (without
a label associated with the task on which they are finally tested, i.e., this is really the
transfer learning setting in machine learning). In fact, with linguistic associations of
these latent variables to words and sentences, humans can answer questions about a
new task with zero labeled training examples for the new task, i.e., they are simply
doing inference: the new task is specified in a language that is already well connected to these latent variables, and one can thus generalize to new tasks with zero
examples (Larochelle *et al.*, 2008), if one has set or learned a good representation
for tasks themselves.

So that is the objective we are setting: **learning representations that capture
the explanatory factors of variation, and help to disentangle them.**

Why Distributed Representations?

A distributed representation is one which has many components or attributes, and such that many of them can be independently active or varied simultaneously. Hence the number of possible objects that can be represented can grow up to exponentially with the number of attributes that can be simultaneously active. In a dense distributed representation, all of the components can be active simultaneously, whereas in a *sparse representation*, only a few can be (while the others are 0). The other end of the spectrum is a *local representation*, such as a hard clustering, in which each input object is represented by the activation of a single cluster (the one to which the object belongs), i.e., an integer (ranging from 1 to the number of clusters). It is clear that even with a discrete representation, if there are many factors of variation, representing the input object by a set of N attributes (each taking one out of several values, at least 2) provides for a much richer representation, with which up to 2^N different objects can be distinguished. With a sparse representation with k active attributes, the number of objects that can be distinguished is on the order of N choose k, which also grows faster than exponential in k (when $k = \alpha N$ and α a fixed small fraction). An advantage of sparse representations (besides the inspiration from the brain) is that it allows to represent a variable-dimension (generally small-dimensional) object within a (large) fixed-dimension vectors, by having a few of the dimensions active for any particular input. We also hypothesize that sparsity may help to disentangle factors of variation, as observed in Goodfellow *et al.* (2009).

Why Deep?

How should our learned representation by parametrized? A **deep architecture** is one in which there are *multiple levels of representation*, with higher levels built on top of lower levels (the lowest being the original raw input), and higher levels representing more abstract concepts defined in terms of less abstract ones from lower levels. There are several motivations for deep architectures:

- *Brain inspiration*: several areas of the brain, especially those better understood such as visual cortex and auditory cortex, are organized as a deep architecture, with each brain area associated with a level of representation (Serre *et al.*, 2007).
- *Computational complexity*: as discussed in (Bengio and LeCun, 2007), some computational complexity results suggest that some functions which can be represented compactly with a deep architecture would require an exponential number of components if represented with a shallow (e.g. 2-level) architecture. For example, a family of positive-weight formal-neuron deep networks can be exponentially more expensive to represent with depth $k - 1$ than with depth k (Håstad and Goldmann, 1991), and similarly for logic gates (Håstad, 1986). More recently Braverman (2011) showed that if all marginals of an input distribution involving at most k variables are uniform, more depth makes it exponentially easier to distinguish the joint from the uniform. In addition, Bengio and Delalleau (2011) showed two families of sum-product networks

(i.e. to represent polynomials) in which a 2-layer architecture required exponentially more units (i.e., computations and parameters) than a suffcently deep architecture. All these theoretical results illustrate that depending on the task and the types of computation performed at each layer, the sufficient depth will vary. It is therefore important to have algorithms that can accommodate different depths and choose depth empirically.

- *Statistical efficiency and sharing of statistical strength.* First, if deeper architectures can be more efficient in terms of number of computational units (to represent the same function), that in principle means that the number of parameters to estimate is smaller, which gives rise to greater statistical efficiency. Another way to see this is to consider the sharing of statistical strength that occurs when different components of an architecture are *re-used* for different purposes (e.g., in the computation for different outputs, or different tasks, or in the computation for different intermediate features). In the deep architecture, the lower level features can be shared in the definition of the higher-level features, which thus provides more sharing of statistical strength then in a shallow architecture. In addition the higher-level shared features can be more complicated functions (whereas for example the hidden units of a single hidden-layer network are very restricted in their expressive power, e.g. hidden=sigmoid(affine(input))). Since the parameters of a component are used for different purposes, they share statistical strength among the different examples (or parts of examples) that rely on these parameters. This is similar and related to the sharing of statistical strength that occurs in distributed representations. For example, if the parameters of one hidden unit of an RBM are "used" for many examples (because that unit turns on for many examples), then there is more information available to estimate those parameters. When a new configuration of the input is presented, it may not correspond to any of those seen in the training set, but its "components" (possibly represented at a higher level of abstraction in intermediate representations) may have been seen previously. See Bengio (2009) for a longer discussion of the statistical advantages of deep and distributed architectures to fight the so-called "curse of dimensionality". First of all, it is not so much dimensionality as much as the amount of variations to be captured that matters (even a one-dimensional function can be difficult to learn if it is not sufficiently smooth). Previous theoretical work on the use of shallow neural networks to bypass the curse of dimensionality (Barron, 1993; Kurkova and Sanguineti, 2008) rely on the assumption of *smoothness* of the target function in order to avoid dependence on dimensionality. As argued in Bengio (2009) the smoothness assumption (also exploited in most non-parametric statistical models) is insufficient because the kinds of functions we want to learn for AI are not smooth enough, have too much complexity and variations. Deep architectures, with their ability to generalize non-locally, can generalize even to variations never seen in the training set. They can represent a function that has lots of variation (e.g. a very high-order polynomial with sharp non-linearities in many places) with a comparatively small number of parameters, compared to the number of variations one can capture.

- *Cognitive arguments and engineering arguments for compositionality.* Humans very often organize ideas and concepts in a modular way, and at multiple levels. Concepts at one level of abstraction are defined in terms of lower-level concepts. For example it is possible to write a dictionary whose definitions depend only of a very small number of core words. Note that this gives rise to a graph which is generally not a tree (like in ontologies) because each concept can be re-used in defining many other concepts. Similarly, that is also how problems are solved and systems built by engineers: they typically construct a chain or a graph of processing modules, with the output of one feeding the inputs of another. The inputs and outputs of these modules are intermediate representations of the raw signals that enter the system. Software engineering is also about decomposing computation into re-usable modules, and the word *factorization* is often used to describe that process which enables more powerful compositionality. Contrast this with a computer program written as a single main function without any calls to sub-routines. This would be a very shallow program and is not the way software engineers like to design software. Such hand-crafted modules are designed thanks to human ingenuity. We would like to add to human ingenuity the option to *learn such decompositions and intermediate representations*. In both the dictionary example and the engineering example, the power stems from compositionality.
- *Sharing of statistical strength for multi-task learning, semi-supervised learning, self-taught learning, and out-of-domain generalization.* Sharing of statistical strength is a core idea behind many advances in machine learning. Components and parameters are shared across tasks in the case of multi-task learning, and deep architectures are particularly well suited for multi-task learning (Collobert and Weston, 2008). Similarly semi-supervised learning exploits statistical sharing between the tasks of learning the input distribution $P(X)$ and learning the conditional distribution $P(Y|X)$. Because deep learning algorithms often rely heavily on unsupervised learning, they are well suited to exploit this particular form of statistical sharing. A very related form of sharing occurs in self-taught learning (Raina *et al.*, 2007), whereby we consider unlabeled training data from $P(X|Y)$ for a set of classes Y's but really care about generalizing to tasks $P(Y|X)$ for a different set of Y's. Recent work showed that deep learners benefit more from the self-taught learning and multi-task learning frameworks than shallow learners (Bengio *et al.*, 2010). This is also a form of out-of-domain generalization, for which deep learners are well suited, as shown in (Bengio *et al.*, 2010) for pattern recognition.

Why Semi-supervised or Unsupervised Learning?

An important prior exploited in many deep learning algorithms (such as those based on greedy layer-wise pre-training, detailed below, sec. 2.1) is the following: *representations that are useful for capturing $P(X)$ can be useful (at least in part, or as initialization) for capturing $P(Y|X)$*. As discussed above, this is beneficial as a statistical sharing strategy, and especially so because X is usually very high-dimensional and rich, compared to Y, i.e., it can contain very detailed structure that can be relevant to predicting Y given X. See (Erhan *et al.*, 2010b) for a discussion of the advantages brought by this prior, and a comprehensive set of experiments showing how it helps not only as a regularizer, but also to find better training error when the training set is large, i.e., *to find better local minima of the generalization error (as a function of the parameters)*. This is an important consideration because deep learners have a highly non-convex training criterion (whether it be supervised or unsupervised) and the greedy layer-wise initialization strategy, based on unsupervised pre-training, can make a huge difference.

More generally, the very task of learning representations with the objective of sharing statistical strengths across tasks, domains, etc. begs for unsupervised learning, modeling for all the variables observed (including the Y's) and good representations for their joint distribution. Consider an agent immersed in a stream of observations of X's and Y's, where the *set* of values that Y can take is non-stationary, i.e., new classes can appear on which we will later want to make predictions. Unsupervised learning is a way to collect as much information as possible ahead of time about all those observations, so as to later be in the best possible position in order to respond to new requests, possibly from very few labels associated with a new class Y. Unsupervised learning is what we ultimately need if we consider multi-task / semi-supervised / self-taught learning and the number of possible tasks or classes becomes very large or unbounded.

2 Deep Learning of Representations: A Review and Recent Trends

2.1 *Greedy Layerwise Pre-training*

The following basic recipe was introduced in 2006 (Hinton and Salakhutdinov, 2006; Hinton *et al.*, 2006; Ranzato *et al.*, 2007a; Bengio *et al.*, 2007):

1. Let $h_0(x) = x$ be the lowest-level representation of the data, given by the observed raw input x.
2. For $\ell = 1$ to L

   > Train an unsupervised learning model taking as observed data the training examples $h_{\ell-1}(x)$ represented at level $\ell - 1$, and producing after training representations $h_\ell(x) = R_\ell(h_{\ell-1}(x))$ at the next level.

From this point on, several variants have been explored in the literature. For supervised learning with fine-tuning, which is the most common variant (Hinton *et al.*, 2006; Ranzato *et al.*, 2007b; Bengio *et al.*, 2007):

3. Initialize a supervised predictor whose first stage is the parametrized representation function $h_L(x)$, followed by a linear or non-linear predictor as the second stage (i.e., taking $h_L(x)$ as input).

4. Fine-tune the supervised predictor with respect to a supervised training criterion, based on a labeled training set of (x, y) pairs, and optimizing the parameters in both the representation stage and the predictor stage.

Another supervised variant involves using all the levels of representation as input to the predictor, keeping the representation stage fixed, and optimizing only the predictor parameters (Lee *et al.*, 2009a,b):

3. Train a supervised learner taking as input $(h_k(x), h_{k+1}(x), \ldots, h_L(x))$ for some choice of $0 \leq k \leq L$, using a labeled training set of (x, y) pairs.

Finally, there is a common unsupervised variant, e.g. for training deep auto-encoders (Hinton and Salakhutdinov, 2006) or a Deep Boltzmann Machine (Salakhutdinov and Hinton, 2009):

3. Initialize an unsupervised model of x based on the parameters of all the stages.
4. Fine-tune the unsupervised model with respect to a global (all-levels) training criterion, based on the training set of examples x.

2.2 Undirected Graphical Models and Boltzmann Machines

The first unsupervised learning algorithm (Hinton and Salakhutdinov, 2006; Hinton *et al.*, 2006) that has been proposed for training each level of the above algorithm (step 2) is based on a Restricted Boltzmann Machine (Smolensky, 1986), which is an undirected graphical model that is a particular form of Boltzmann Machine (Hinton *et al.*, 1984). An undirected graphical model for observed variable x based on latent variable h is specified by an *energy function* Energy(x, h):

$$P(x, h) = \frac{e^{-\text{Energy}(x,h)}}{Z}$$

where Z is a normalization constant called the partition function. A Boltzmann machine is one where Energy(x, h) is a second-order polynomial in (x, h), e.g.,

$$\text{Energy}(x, h) = h'Wx + h'Uh + x'Vx + b'h + c'x$$

and in general both x and h are considered to be binary vectors, which makes Z intractable except when both x and h have very few components. The coefficients $\theta = (W, U, V, b, c)$ of that second-order polynomial are the parameters of the model. Given an observed x, the inference $P(h|x)$ is generally intractable but can be estimated by sampling from a Monte-Carlo Markov Chain (MCMC), e.g. by Gibbs sampling, or using loopy belief, variational or mean-field approximations. Even though computing the energy is easy, marginalizing over h in order to compute the likelihood $P(x)$ is generally intractable, so that the exact log-likelihood gradient is also intractable. However, several algorithms have been proposed in recent years to estimate the gradient, most of them based on the following decomposition into the so-called "positive phase part" (x is fixed to the observed value, the gradient term tends to decrease the associated energies) and "negative phase part" (both x and h are sampled according to P, and the gradient term tends to increase their energy):

$$\frac{\partial}{\partial\theta}(-\log P(x)) = E_h\left[\frac{\partial\text{Energy}(x,h)}{\partial\theta}|x\right] - E_{x,h}\left[\frac{\partial\text{Energy}(x,h)}{\partial\theta}\right].$$

Even though a Boltzmann Machine is a parametric model when we consider the dimensionality n_h of h to be fixed, in practice one allows n_h to vary, making it a non-parametric model. With n_h large enough, one can model any discrete distribution: Le Roux and Bengio (2008) shows that RBMs are universal approximators, and since RBMs are special cases of Boltzmann Machines, Boltzmann Machines also are universal approximators. On the other hand with $n_h > 0$ the log-likelihood is not anymore convex in the parameters, and training can potentially get stuck in one of many local minima.

2.3   The Restricted Boltzmann Machine

The Restricted Boltzmann Machine (RBM) is one without lateral interactions, i.e., $U = 0$ and $V = 0$. In turns out that the positive phase part of the gradient can be computed exactly and tractably in the easier special case of the RBM, because $P(h|x)$ factorizes into $\prod_i P(h_i|x)$. Similarly $P(x|h)$ factorizes into $\prod_j P(x_j|h)$, which makes it possible to apply blocked Gibbs sampling (sampling h given x, then x given h, again h given x, etc.). For a trained RBM, the learned representation $R(x)$ of its input x is usually taken to be $E[h|x]$, as a heuristic.

RBMs are typically trained by stochastic gradient descent, using a noisy (and generally biased) estimator of the above log-likelihood gradient. The first gradient estimator that was proposed for RBMs is the Contrastive Divergence estimator (Hinton, 1999; Hinton et al., 2006), and it has a particularly simple form: the negative phase gradient is obtained by starting a very short chain (usually just one step) at the observed x and replacing the above expectations by the corresponding samples. In practice, it has worked very well for unsupervised pre-training meant to initialize each layer of a deep supervised (Hinton et al., 2006; Bengio et al., 2007; Erhan et al., 2010b) or unsupervised (Hinton and Salakhutdinov, 2006) neural network.

Another common way to train RBMs is based on the Stochastic Maximum Likelihood (SML) estimator (Younes, 1999) of the gradient, also called Persistent Contrastive Divergence (PCD) (Tieleman, 2008) when it was introduced for RBMs. The idea is simply to keep sampling negative phase x's (e.g. by blocked Gibbs sampling) even though the parameters are updated once in a while, i.e., without restarting a new chain each time an update is done. It turned out that SML yields RBMs with much better likelihood, whereas CD updates sometimes give rise to worsening likelihood and suffers from other issues (Desjardins et al., 2010). Theory suggests (Younes, 1999) this is a good estimator if the parameter changes are small, but practice revealed (Tieleman, 2008) that it worked even for large updates, in fact giving rise to faster mixing (Tieleman and Hinton, 2009; Breuleux et al., 2011). This is happening because learning actually interacts with sampling in a useful way, pushing the MCMC out of the states it just visited. This principle may also explain some of the fast mixing observed in a related approach called Herding (Welling, 2009; Breuleux *et al.*, 2011).

RBMs can be stacked to form a Deep Belief Network (DBN), a hybrid of directed and undirected graphical model components, which has an RBM to characterize the interactions between its top two layers, and then generates the input through a directed belief network. See Bengio (2009) for a deeper treatment of Boltzmann Machines, RBMs, and Deep Belief Networks.

2.4 The Zoo: Auto-Encoders, Sparse Coding, Predictive Sparse Decomposition, Denoising Auto-Encoders, Score Matching, and More

Auto-encoders are neural networks which are trained to reconstruct their input (Rumelhart *et al.*, 1986; Bourlard and Kamp, 1988; Hinton and Zemel, 1994). A one-hidden layer auto encoder is very similar to an RBM and its reconstruction error gradient can be seen as an approximation of the RBM log-likelihood gradient (Bengio and Delalleau, 2009). Both RBMs and auto-encoders can be used as one-layer unsupervised learning algorithms that give rise to a new representation of the input or of the previous layer. In the same year that RBMs were successfully proposed for unsupervised pre-training of deep neural networks, auto-encoders were also shown to help initialize deep neural networks much better than random initialization (Bengio *et al.*, 2007). However, ordinary auto-encoders generally performed worse than RBMs, and were unsatisfying because they could potentially learn a useless identity transformation when the representation size was larger than the input (the so-called "overcomplete" case).

Sparse coding was introduced in computational neuroscience (Olshausen and Field, 1997) and produced filters very similar to those observed in cortex visual area V1 (before similar filters were achieved with RBMs, sparse predictive decomposition, and denoising auto-encoders, below). They correspond to a linear directed graphical model with a continuous-valued latent variable associated with a sparsity prior (Student or Laplace, the latter corresponding to an L1 penalty on the value of the latent variable). This is like an auto-encoder, but without a parametric encoder, only a parametric decoder. The "encoding" corresponds to inference (finding the most likely hidden code associated with observed visible input) and involves solving a lengthy but convex optimization problem and much work has been devoted to speeding it up. A very interesting way to do so is with **Predictive Sparse Decomposition** (Kavukcuoglu *et al.*, 2008), in which one learns a parametric encoder that approximates the result of the sparse coding inference (and in fact changes the solution so that both approximate encoding and decoding work well). Such models based on approximate inference were the first successful examples of stacking a sparse encoding (Ranzato *et al.*, 2007a; Jarrett *et al.*, 2009) into a deep architecture (fine-tuned for supervised classification afterwards, as per the above greedy-layerwise recipe).

Score Matching is an alternative statistical estimation principle (Hyvärinen, 2005) when the maximum likelihood framework is not tractable. It can be applied to models of continuous-valued data when the probability function can be computed tractably up to its normalization constant (which is the case for RBMs), i.e., it has a tractable energy function The *score* of the model is the partial derivative of the log-likelihood with respect to the input, and indicates in which direction the likelihood would increase the most, from a particular input x. Score matching is based on minimizing the squared difference between the score of the model and a target score. The latter is in general unknown but the score match can nonetheless be rewritten in terms of the expectation (under the data generating process) of first and (diagonal) second derivatives of the energy with respect to the input, which correspond to a tractable computation.

Denoising Auto-Encoders were first introduced (Vincent *et al.*, 2008) to bypass the frustrating limitations of auto-encoders mentioned above. Auto-encoders are only meant to learn a "bottleneck", a reduced-dimension representation. The idea of Denoising Auto-Encoders (DAE) is simple: feed the encoder/decoder system with a *stochastically corrupted input*, but ask it to *reconstruct the clean input* (as one would typically do to train any denoising system). This small change turned out to systematically yield better results than those obtained with ordinary auto-encoders, and similar or better than those obtained with RBMs on a benchmark of several image classification tasks (Vincent *et al.*, 2010). Interestingly, the denoising error can be linked in several ways to the likelihood of a generative model of the distribution of the uncorrupted examples (Vincent *et al.*, 2008; Vincent, 2011), and in particular through the Score Matching proxy for log-likelihood (Vincent, 2011): the denoising error corresponds to a form of *regularized* score matching criterion (Kingma and LeCun, 2010). The link also sheds light on why a denoising auto-encoder captures the input distribution. The difference vector between the reconstruction and the corrupted input is the model's guess as to the direction of greatest increase in the likelihood (starting from a corrupted example), whereas the difference vector between the corrupted input and the clean original is nature's hint of a direction of greatest increase in likelihood (since a noisy version of a training example is very likely to have a much lower probability under the data generating distribution than the original). The difference of these two differences is just the denoising reconstruction error residue.

Noise-Contrastive Estimation is another estimation principle which can be applied when the energy function can be computed but not the partition function (Gutmann and Hyvarinen, 2010). It is based on training not only from samples of the target distribution but also from samples of an auxiliary "background" distribution (e.g. a flat Gaussian). The partition function is considered like a free parameter (along with the other parameters) in a kind of logistic regression trained to predict the probability that a sample belongs to the target distribution vs the background distribution.

Semi-supervised Embedding is an interesting and different way to use unlabeled data to learn a representation (e.g., in the hidden layers of a deep neural network), based on a hint about *pairs of examples* (Weston *et al.*, 2008). If some pairs are expected to have a similar semantic, then their representation should be encouraged to be similar, whereas otherwise their representation should be at least some distance away. This idea was used in unsupervised and semi-supervised contexts (Chopra *et al.*, 2005; Hadsell *et al.*, 2006; Weston *et al.*, 2008), and originates in the much older idea of *siamese networks* (Bromley *et al.*, 1993).

3 Convolutional Architectures

When there are directions of variation around input examples that are known to be irrelevant to the task of interest (e.g. translating an image), it is often very useful to exploit such *invariances* in a learning machine. We would like features that characterize the presence of objects in sequences and images to have some form of *translation equivariance*: if the object is translated (temporally or spatially), we would like the associated feature detectors to also be translated. This is the basic insight behind the convolutional network (LeCun *et al.*, 1989, 1998), which has been very influential and regained importance along with new algorithms for deep architectures.

Recently, convolutional architectures have grown in popularity (Jain and Seung, 2008; Lee *et al.*, 2009a,b; Kavukcuoglu *et al.*, 2010; Turaga *et al.*, 2010; Taylor *et al.*, 2010; Krizhevsky, 2010; Le *et al.*, 2010). As researchers attempt to move toward ever larger, more realistic and more challenging datasets, the convolutional architecture has been widely recognized for computational scalability and ease of learning. These properties are due to the basic architectural features that distinguish convolutional networks from other network models: weight sharing and feature pooling.

3.1 *Local Receptive Fields and Weight Sharing*

The classical way of obtaining some form of translation equivariance is the use of weight sharing in a network whose units have a local receptive field, i.e., only receiving as input a subset (e.g. subimage) of the overall input. The same feature detector is applied at different positions or time steps[1], thus yielding a sequence or a "map" containing the detector output for different positions or time steps. The idea of local receptive fields goes back to the Perceptron and to Hubel and Wiesel's discoveries (Hubel and Wiesel, 1959) in the cat's visual cortex.

---

[1] Or, equivalently, one can say that many feature detectors, each associated with a different receptive field, *share the same weights*, but applied to a different portion of the input.

3.2 Feature Pooling

Aside from weight sharing, the other signature architectural trait of the convolutional network is feature pooling. The standard implementation of feature pooling groups together locally translated copies of a single feature detector (single weight vector) as input into a single pooling layer unit. The output of the pooled feature detectors are either summed, or more commonly their maximal value is taken as the output of the pooling feature. The use of local pooling adds robustness to the representation in giving the pooling layer units a degree of translational invariance – the unit activates irrespective of where the image feature is located inside its pooling region. Empirically the use of pooling seems to contribute significantly to improved classification accuracy in object classification tasks (LeCun et al., 1998).

4   Learning Invariant Feature Sets

For many AI tasks, the data is derived from a complex interaction of factors that act as sources of variability. When combined together, these factors give rise to the rich structure characteristic of AI-related domains. For instance, in the case of natural images, the factors can include the identity of objects in a scene, the orientation and position of each object as well as the ambient illumination conditions. In the case of speech recognition, the semantic content of the speech, the speaker identity and acoustical effects due to the environment are all sources of variability that give rise to speech data. In each case, factors that are relevant to a particular task combine with irrelevant factors to render the task much more challenging.

As a concrete example consider the task of face recognition. Two images of the same individual with different poses (e.g., one image is in a full frontal orientation, while the other image is of the individual in profile) may result in images that are well separated in pixel space. On the other hand images of two distinct individuals with identical poses may well be positioned very close together in pixel space. In this example, there are two factors of variation at play: (1) the identity of the individual in the image, and (2) the person's pose with respect to the image plane. One of these factors (the pose) is irrelevant to the face recognition task and yet of the two factors it could well dominate the representation of the image in pixel space. As a result, pixel space-based face recognition systems are destined to suffer from poor performance due to sensitivity to pose, but things really get interesting when many factors are involved.

The key to understanding the significance of the impact that the combination of factors has on the difficulty of the task is to understand that these factors typically do not combine as simple superpositions that can be easily separated by, for example, choosing the correct lower-dimensional projection of the data. Rather, as our face recognition example illustrates, these factors often appear tightly entangled in the raw data. The challenge for deep learning methods is to construct representations of the data that somehow attempt to cope with the reality of entangled factors that account for the wide variability and complexity of data in AI domains.

4.1 Dealing with Factors of Variation: Invariant Features

In an effort to alleviate the problems that arise when dealing with this sort of richly structured data, there has recently been a very broad based movement in machine learning toward building feature sets that are *invariant* to common perturbations of the datasets. The recent trend in computer vision toward representations based on large scale histograming of low-level features is one particularly effective example (Wang *et al.*, 2009).

To a certain degree, simply training a deep model – whether it be by stacking a series of RBMs as in the DBN (in section 2.3) or by the joint training of the layers of a Deep Boltzmann Machine (discussed in section 6) – should engender an increasing amount of invariance in increasingly higher-level representations. However with our current set of models and algorithms, it appears as though depth alone is insufficient to foster a sufficient degree of invariance at all levels of the representation and that an explicit modification of the inductive bias of the models is warranted.

Within the context of deep learning, the problem of learning invariant feature sets has long been considered an important goal. As discussed in some detail in section 3, one of the key innovations of the convolutional network architecture is the inclusion of max-pooling layers. These layers pool together locally shifted versions of the filters represented in the layer below. The result is a set of features that are invariant to local translations of objects and object parts within the image.

More generally, invariant features are designed to be insensitive to variations in the data that are uninformative to the target task while remaining selective to relevant aspects of the data. The result is a more stable representation that is well suited to be used as an input to a classifier. With irrelevant sources of variance removed, the resulting feature space has the property that distances between data points represented in this space are a more meaningful indicator of their true similarity. In classification tasks, this property naturally simplifies the discrimination of examples associated with different class labels.

Thus far, we have considered only invariant features, such as those found in the convolutional network, whose invariant properties were hand-engineered by specifying the filter outputs to be pooled. This approach, while clearly effective in constructing features invariant to factors of variation such as translation, are fundamentally limited to expressing types of invariance that can be imposed upon them by human intervention. Ideally we would like our learning algorithms to automatically discover appropriate sets of invariant features. Feature sets that *learn* to be invariant to certain factors of variation have the potential advantage of discovering patterns of invariance that are either difficult to hand-engineer (e.g. in-plane object rotations) or simply *a priori* not known to be useful. In the remainder of this section we review some of the recent progress in techniques for learning invariant features and invariant feature hierarchies.

4.2 Invariance via Sparsity

Learning sparse feature sets has long been popular both in the context of learning feature hierarchies as a deep learning strategy and as a means of learning effective shallow representations of the data. In the context of an object recognition task, Raina *et al.* (2007) established that using a sparse representations learned from image data as input to an SVM classifier led to better classification performance than using either the raw pixel data or a non-sparse PCA representation of the data.

Recently, Goodfellow *et al.* (2009) showed that sparsity can also lead to more invariant feature representations. In the context of auto-encoder networks trained on natural images, they showed that when adding a sparsity penalty to the hidden unit activations, the resulting features are more invariant to specific transformations such as translations, rotations normal to the image plane as well as in-plane rotations of the objects.

At this point it is not clear by what mechanism sparsity promotes the learning of invariant features. It is certainly true that sparsity tends to cause the learned feature detectors to be more localized: in training on natural images, the learned features form Gabor-like edge detectors (Olshausen and Field, 1997); in training on natural sounds, the learned features are wavelet-like in that they tend to be localized in both time and frequency (Smith and Lewicki, 2006). It is also true that localized filters are naturally invariant to variations outside their local region of interest or *receptive field*. It is not clear that feature locality is sufficient to entirely account for the invariance observed by Goodfellow *et al.* (2009). Nor is it clear that the often observed superior classification performance of sparse representations (Raina *et al.*, 2007; Manzagol *et al.*, 2008; Bagnell and Bradley, 2009) may be attributed to this property of generating more invariant features. What is clear is that these issues merit further study.

4.3 Teasing Apart Explanatory Factors via Slow Features Analysis

We perceive the world around us through a temporally structured stream of perceptions (e.g. a video). As one moves their eyes and head, the identity of the surrounding objects generally does not change. More generally, a plausible hypothesis is that many of the most interesting high-level explanatory factors for individual perceptions have some form *temporal stability*. The principle of identifying slowly moving/changing factors in temporal/spatial data has been investigated by many (Becker and Hinton, 1993; Wiskott and Sejnowski, 2002; Hurri and Hyvärinen, 2003; Körding *et al.*, 2004; Cadieu and Olshausen, 2009) as a principle for finding useful representations of images, and as an explanation for why V1 simple and complex cells behave the way they do. This kind of analysis is often called *slow feature analysis*. A good overview can be found in Berkes and Wiskott (2005). Note that it is easy to obtain features that change slowly when they are computed through averaging (e.g. a moving average of current and past observations). Instead, with slow feature analysis, one learns features of an instantaneous perception (e.g. a single image) such that consecutive feature values are similar to each other. For this to be of any use, it is also required that these features capture as much as possible of the input variations (e.g. constant features would be very stable but quite useless). A very interesting recent theoretical contribution (Klampfl and Maass, 2009) shows that *if there exist categories and these categories are temporally stable, then slow feature analysis can discover them even in the complete absence of labeled examples.*

Temporal coherence is therefore a prior that could be used by learning systems to discover categories. Going further in that direction, we hypothesize that more structure about the underlying explanatory factors could be extracted from temporal coherence, still without using any labeled examples. First, different explanatory factors will tend to operate at *different time scales*. With such a prior, we can not only separate the stable features from the instantaneous variations, but we could disentangle different concepts that belong to different time scales. Second, instead of the usual squared error penalty (over the time change of each feature), one could use priors that favor the kind of time course we actually find around us. Typically, a factor is either present or not (there is a notion of sparsity there), and when it is present, it would tend to change slowly. This corresponds to a form of sparsity not only of the feature, but also of its change (either no change, or small change). Third, an explanatory factor is rarely represented by a single scalar. For example, camera geometry in 3 dimensions is characterized by 6 degrees of freedom. Typically, these factors would either not change, or change together. This could be characterized by a group sparsity prior (Jenatton *et al.*, 2009).

4.4 Learning to Pool Features

Most unsupervised learning methods that are used as building blocks for Deep Learning are based on a distributed or factorial representation. Each unit is associated with a filter that is compared to the observation via a simple linear operation – most commonly the dot product – and activates the unit accordingly, typically through a non-linearity. This is the case for Autoencoder Networks, Denoising Autoencoder Networks and Restricted Boltzmann Machines. With this activation mechanism, the comparison between the filter and the observed vector is maximal when the two vectors are co-linear and decreases smoothly as the observation vector is perturbed in any direction. With regard to the activation of the feature, the only relevant aspects of the observation vector is the angle it makes with the filter and its norm. Features built this way are not differentially sensitive to variations in the observed vector. In other words, they do not, by themselves, possess directions of invariance. Even when sparsity or a temporal coherence penalty is used, the units activations may be more robust to small perturbations of the observations – particularly along the length of an edge, but there is no mechanism to allow them to exhibit robustness to significant perturbations in arbitrary directions while remaining sensitive to perturbations in other directions.

In section 3, we encountered feature pooling as a way to ensure that the representation in the pooling layers of the convolutional network were invariant to small translations of the image. Feature pooling works by combining filters together to form a single feature that activates with the activation of any feature in the pool. With this simple mechanism, the pooling feature has the capacity to represent directions of invariance. Consider the way pooling is used in the convolutional network. The pooling features take as their input units possessing filters that are locally shifted copies of each other. When an input pattern is presented to the network that is "close" (in the sense of cosine distance) to one of the pooled filters, the pooling feature is activated. The result is a pooling feature that is invariant to translation of the target input pattern but remains sensitive to other transformations or distortions of the pattern.

The principle that invariance to various factors of the data can be induced through the use of pooling together of a set of simple filter responses is a powerful one – exploited to good effect in the convolutional network. However, the convolutional network's use of the pooling is restricted to pooling of features that are shifted copies of each other – limiting the kind of invariance captured by the convolutional network to translational invariance. It is possible to conceive of generating other forms of invariant features by pooling together variously-transformed copies of filters. For instance, in the case of natural images, we might like some features to be invariant to small amounts of rotation of a particular image pattern while being highly sensitive to changes in the morphology of the image pattern. Ultimately this approach suffers from the disadvantage of limiting the set of invariances represented in the representation to hand-crafted transformations that are well-studied and understood to be useful in the domain of application. Consider the case of audio sequences, where we want a set of features to be invariant to changes in environmental properties such as reverberation while remaining highly sensitive to the source pitch and timbre. The set of transformations that render a feature set invariant to environmental conditions is not straightforward to encode. In this section we explore methods that seek to exploit the basic feature pooling architecture to *learn* the set of filters to be pooled together and thereby learn the invariance reflected in the pooling feature.

The ASSOM Model. The principle that invariant features can actually *emerge* from the organization of features into pools was first established by Kohonen (1996) in the ASSOM model. Synthesizing the self-organizing map (SOM) (Kohonen, 1990) with the learning subspace method (Kohonen *et al.*, 1979), Kohonen (1996) generalized the notion of a filter to a filter subspace. Filter subspaces are defined by the space spanned by the set of filters in the pool. The principle states that one may consider an invariant feature as a linear subspace in the feature space. According to the ASSOM model, the value of the higher-order pooling feature is given by the square of the norm of the projection of the observation on the feature subspace.

Using a competitive learning strategy with an image sequence of colored noise, Kohonen (1996) showed that learning feature subspaces can lead to features that are invariant to the geometric transformations (including translation, rotation and scale) reflected in the structure of the colored noise sequence.

Subspace and Topological ICA. Hyvärinen and Hoyer (2000) later integrated the principle of feature subspaces (Kohonen, 1996) within the independent component analysis (ICA) paradigm via Multidimensional ICA (Cardoso, 1998) to form the Subspace ICA model. Multidimensional independent component analysis loosens the standard ICA requirement of strictly independent components to allow dependencies between groups or pools of components and only enforces the independence constraint across the component pools. By allowing dependencies within the feature subspaces defined by the pools and enforcing the ICA independence constraint across feature pools, Hyvärinen and Hoyer (2000) were able to demonstrate the emergence of higher-order edge-like feature detectors that display some amount of invariance to edge orientation, phase, or location.

In their Topographic ICA (TICA) model, Hyvärinen *et al.* (2001) generalized the Subspace-ICA model by allowing for overlapping pools, arranged topographically such that neighbouring pools shared the linear filters along their shared border. Training on a natural image dataset, Hyvärinen *et al.* (2001) demonstrated that the topographic arrangement of the feature pools led to the emergence of pinwheel patterns in the topographically arranged filter set. The pinwheel pattern of filters is highly reminiscent of the pinwheel-like structure of orientation preference observed in primary visual cortex (Bartfeld and Grinvald, 1992).

Tiled-Convolutional Models. More recently, Le *et al.* (2010) used a training algorithm based on Topographic ICA to learn the filters in a tiled-convolutional arrangement. Unlike a standard convolution model, where the features pooled together share translated copies of the weights, the tiled convolution model pools together features that do not have the weights tied. However, at a distance away from that pool, another non-overlapping pool may share the same weights, allowing such models to handle variable-size images, like convolutional nets. In essence, this model can be seen as developing a convolutional version of the topographic ICA model that combines the major advantages of both approaches. On the one hand, the model preserves TICA's capability to learn potentially complex invariant feature subspaces. On the other hand, the use of convolutional weight sharing means that the model possesses a relatively small number of parameters which can dramatically improve scalability and ease of learning.

Invariant Predictive Sparse Decomposition. Another recent approach to learn invariant features subspaces via a topographic pooling of low-level features is the Invariant Predictive Sparse Decomposition (IPSD) (Kavukcuoglu *et al.*, 2009). The IPSD is a variant of Predictive Sparse Decomposition (discussed in section 2.4) that replaces the L_1 penalty on the coefficients z_j with a group sparsity term: $\lambda \sum_{i=1}^{K} v_i$ where K is the number of pooling features and $v_i = \sqrt{\sum_{j \in P_i} w_j z_j^2}$ is the value of the pooling feature. Here, P_i is the set of low-level features associated with pooling unit i, w_j reflects the relative weight of coefficient z_j in the pool, and λ is a parameter that adjusts the relative strength of the sparsity penalty in the loss function. This modification to the PSD loss function has the effect of encouraging units within a pool to be nonzero together – forming the feature subspace that defines the invariance properties of the pooling feature v_i.

Mean and Covariance Restricted Boltzmann Machine. Within the RBM context, Ranzato and Hinton (2010) follow the same basic strategy of learning invariant feature subspaces with their mean and covariance RBM (mcRBM). Using an energy function with elements that are very similar to elements of the objective functions of both Topographic-ICA and IPSD, Ranzato and Hinton (2010) demonstrated the model's capacity to learn invariant feature subspaces by pooling filters. A very similar type of RBM-based pooling arrangement was presented by Courville *et al.* (2011).

4.5   Beyond Learning Invariant Features

As is evident from the above paragraphs, the feature subspace approach has been very popular recently as a means of learning features that are invariant to learned transformations of the data (those spanned by the feature subspace), while maintaining selectivity in all other directions in feature space. The ability to learn invariant features is obviously a crucially important step toward developing effective and robust representations of data. However, we feel it is doubtful that these strategies for creating invariant features are, by themselves, sufficient to encode the rich interactions extant in the kind of data we would like to model and represent. We would like to learn features capable of *disentangling* the web of interacting factors that make up practically all data associated with AI-related tasks. In the next section, we explore what we mean by *disentangling* and how one can think about beginning to approach this ambitious goal.

5   Disentangling Factors of Variation

Complex data arise from the rich interaction of many sources. These factors interact in a complex web that can complicate AI-related tasks such as object classification. For example, an image is composed of the interaction between one or more light sources, the object shapes and the material properties of the various surfaces present in the image. Shadows from objects in the scene can fall on each other in complex patterns, creating the illusion of object boundaries where there are none and dramatically effect the perceived object shape. How can we cope with these complex interactions? How can we *disentangle* the objects and their shadows? Ultimately, we believe the approach we adopt for overcoming these challenges must leverage the data itself, using vast quantities of unlabeled examples, to learn representations that separate the various explanatory sources. Doing so should give rise to a representation significantly more robust to the complex and richly structured variations extant in natural data sources for AI-related tasks.

It is important to distinguish between the related but distinct goals of learning invariant features and learning to disentangle explanatory factors. The central difference is the preservation of information. Invariant features, by definition, have reduced sensitivity in the direction of invariance. This is the goal of building invariant features and fully desirable if the directions of invariance all reflect sources of variance in the data that are uninformative to the task at hand. However it is often the case that the goal of feature extraction is the *disentangling* or separation of many distinct but informative factors in the data. In this situation, the methods of generating invariant features – namely, the feature subspace method – may be inadequate.

Roughly speaking, the feature subspace method can be seen as consisting of two steps (often performed together). First, a set of low-level features are recovered that account for the data. Second, subsets of these low level features are pooled together to form higher level invariant features. With this arrangement, the invariant representation formed by the pooling features offers a somewhat incomplete window on the data as the detailed representation of the lower-level features is abstracted away in the pooling procedure. While we would like higher level features to be more abstract and exhibit greater invariance, we have little control over what information is lost through feature subspace pooling. For example, consider higher level features made invariant to the color of its target stimulus by forming a subspace of low-level features that represent the target stimulus in various colors (forming a basis for the subspace). If this is the only higher level feature that is associated with these low-level colored features then the color information of the stimulus is lost to the higher-level pooling feature and every layer above. This loss of information becomes a problem when the information that is lost is necessary to successfully complete the task at hand such as object classification. In the above example, color is often a very discriminative feature in object classification tasks. Losing color information through feature pooling would result in significantly poorer classification performance.

Obviously, what we really would like is for a particular feature set to be invariant to the irrelevant features and disentangle the relevant features. Unfortunately, it is often difficult to determine *a priori* which set of features will ultimately be relevant to the task at hand. Further, as is often the case in the context of deep learning methods (Collobert and Weston, 2008), the feature set being trained may be destined to be used in multiple tasks that may have distinct subsets of relevant features. Considerations such as these lead us to the conclusion that the most robust approach to feature learning is *to disentangle as many factors as possible, discarding as little information about the data as is practical*. If some form of dimensionality reduction is desirable, then we hypothesize that the local directions of variation least represented in the training data should be first to be pruned out (as in PCA, for example, which does it globally instead of around each example).

One solution to the problem of information loss that would fit within the feature subspace paradigm, is to consider many overlapping pools of features based on the same low-level feature set. Such a structure would have the potential to learn a redundant set of invariant features that may not cause significant loss of information. However it is not obvious what learning principle could be applied that can ensure that the features are invariant while maintaining as much information as possible. While a Deep Belief Network or a Deep Boltzmann Machine (as discussed in sections 2.3 and 6 respectively) with 2 hidden layers would, in principle, be able to preserve information into the "pooling" second hidden layer, there is no guarantee that the second layer features are more invariant than the "low-level" first layer features. However, there is some empirical evidence that the second layer of the DBN tends to display more invariance than the first layer (Erhan *et al.*, 2010a). A second issue with this approach is that it could nullify one of the major motivations for pooling features: to reduce the size of the representation. A pooling arrangement with a large number of overlapping pools could lead to as many pooling features as low-level features – a situation that is both computationally and statistically undesirable.

A more principled approach, from the perspective of ensuring a more robust compact feature representation, can be conceived by reconsidering the disentangling of features through the lens of its generative equivalent – feature composition. Since most (if not all) of our unsupervised learning algorithms have a generative interpretation, the generative perspective can provide insight into how to think about disentangling factors. The majority of the models currently used to construct invariant features have the interpretation that their low-level features linearly combine to construct the data.[2] This is a fairly rudimentary form of feature composition with significant limitations. For example, it is not possible to linearly combine a feature with a generic transformation (such as translation) to generate a transformed version of the feature. Nor can we even consider a generic color feature being linearly combined with a gray-scale stimulus pattern to generate a colored pattern. It would seem that if we are to take the notion of disentangling seriously we require a richer interaction of features than that offered by simple linear combinations.

Disentangling via Multilinear Models. While there are presently few examples of work dedicated to the task of learning features that disentangle the factors of variation in the data, one promising direction follows the development of bilinear (Tenenbaum and Freeman, 2000; Grimes and Rao, 2005) and multilinear (Vasilescu and Terzopoulos, 2005) models. Bilinear models are essentially linear models where the latent state is factored into the product of two variables. Formally, the elements of observation x are given by $\forall k, x_k = \sum_i \sum_j W_{ijk} y_i z_j$, where y_i and z_j are elements of the two latent factors (y and z) representing the observation and W_{ijk} is an element of the tensor of model parameters (Tenenbaum and Freeman, 2000). The tensor W can be thought of as a generalization of the typical weight matrix found in most unsupervised models we have considered above.

Tenenbaum and Freeman (2000) developed an EM-based algorithm to learn the model parameters and demonstrated, using images of letters from a set of distinct fonts, that the model could disentangle the style (font characteristics) from content

---

[2] As an aside, if we are given only the values of the higher-level pooling features, we cannot accurately recover the data because we do not know how to apportion credit for the pooling feature values to the lower-level features. This is simply the generative version of the consequences of the loss of information caused by pooling.

(letter identity). Grimes and Rao (2005) later developed a bilinear sparse coding model of a similar form as described above but included additional terms to the objective function to render the elements of both y and z sparse. They used the model to develop transformation invariant features of natural images.

Multilinear models are simply a generalization of the bilinear model where the number of factors that can be composed together is 2 or more. Vasilescu and Terzopoulos (2005) develop a multilinear ICA model, which they use to model images of faces, to disentangle factors of variation such as illumination, views (orientation of the image plane relative to the face) and identities of the people.

Neuroscientific Basis for Disentangling. Interestingly there is even some evidence from neuroscience that bilinear feature composition may play a role in the brain (Olshausen and Field, 2005). Olshausen *et al.* (1993) proposed a model of visual attention for forming position- and scale-invariant representations of objects. The model relies on a set of control neurons that dynamically and selectively routes information from lower level cortical areas (V1) to higher cortical areas. If we interpret the bilinear model as a standard linear model, by folding one of the latent state factors (say, y) into the parametrization of a dynamic weight matrix, then we can interpret these control neurons as playing the role of y, routing information from x to z.

6   On the Importance of Top-Down Connections

Whereas RBMs and other layer-wise representation learning algorithms are interesting as a means of initializing deeper models, we believe that it is important to develop algorithms that can *coordinate the representations at multiple levels of a deep architecture*. For example, to properly interpret a particular patch as an eye and estimate its characteristics (e.g. its pose and color) in the features computed at a particular level of a hierarchy, it helps to consider the top-down influence of higher-level face features (for the face to which this eye hypothetically belongs), to make sure that both eyes are generally in agreement about pose and color (this agreement being mediated by pose and eye-color features at the level of the face). If we imagine that the value of a particular intermediate-level feature (maybe with a receptive field that only covers a part of the whole input) is associated with a belief about some factor of variation (within that receptive field), consider how this belief could be improved by taking into account the higher-level abstractions that are captured at higher levels of the hierarchy. Observations of the mechanisms and role of attention in brains clearly suggest that top-down connections are used to bias and "clean up" the activations of lower-level detectors. In the case of images, where feature detectors are associated with a receptive field, the top-down connections allow one to *take context into account* when interpreting what is going on in a particular patch of the image. Top-down connections allow the "conclusions" reached in the higher levels to influence those obtained in the lower levels, but since the latter also influence the former, some kind of inference mechanism is needed (to get a handle on appropriate values for the features associated with all the layers, given the network input).

In the Deep Learning literature, this approach is currently solely seen in the work on Deep Boltzmann Machines (Salakhutdinov and Hinton, 2009; Salakhutdinov, 2010; Salakhutdinov and Larochelle, 2010). A Deep Boltzmann Machine is just a Boltzmann machine whose hidden units are organized in layers, and with connections mostly between consecutive layers. A Deep Boltzmann Machine differs from a Deep Belief Network in that proper inference (that involves top-down connections) can be done, albeit at the price of running an MCMC or a variational approximation each time an input is seen. Interestingly, whereas a mean-field variational approximation (which gives a bound on the quantity of interest) can be used in the positive phase of Boltzmann machine gradient computation, it would hurt to do it in the negative phase, hence Salakhutdinov and Hinton (2009); Salakhutdinov (2010) recommend mean-field approximations for the positive phase and Gibbs sampling for the negative phase. Because of the overhead involved in both phases, an interesting thread of research is the exploration of heuristics aiming to speed up inference (Salakhutdinov and Larochelle, 2010), for example by training a feedforward approximation of the posterior which is a good initialization for the iterative inference procedure. This idea is already present in some form in Predictive Sparse Decomposition (Kavukcuoglu et al., 2008) (see section 2.4).

We hypothesize that deep architectures with top-down connections used to coordinate representations at all levels can give rise to "better representations", that can be *more invariant* to many of the changes in input. Indeed, invariance is achieved by composing non-linearities in the appropriate way, and the outcome of inference involves many iterations back and forth through the deep architecture. When unfolded in the time course of these iterations, this corresponds to an even deeper network, but with shared parameters (across time), like in a recurrent neural network. If an intermediate-level feature is informed by higher-level interpretations of the input, then it can be "cleaned-up" to better reflect the prior captured by higher levels, making it more robust to changes in the input associated with "noise" (i.e. variations in the directions not very present in the input distribution, and hence not well represented in the higher levels).

7 Conclusion

Unsupervised learning of representations promises to be a key ingredient in learning hierarchies of features and abstractions from data. This chapter asked: "why learn representations?", "why deep ones?", and especially "*what makes a good representation?*". We have provided some partial answers, but since many algorithms and training criteria have already been proposed for unsupervised learning of representations, the right answer is not clear, and we would like to see algorithms more directly motivated by this last question. In particular, the chapter focuses on the question of *invariance*, and what architectures and criteria could help discover features that are invariant to major factors of variation in the data. We argue that a more appropriate objective is to discover representations that *disentangle the factors of variation*: keeping all of them in the learned representation but making each feature invariant to most of the factors of variation.

References

Bagnell, J.A., Bradley, D.M.: Differentiable sparse coding. In: Koller, D., Schuurmans, D., Bengio, Y., Bottou, L. (eds.) Advances in Neural Information Processing Systems 21 (NIPS 2008), pp. 113–120 (2009)

Barron, A.E.: Universal approximation bounds for superpositions of a sigmoidal function. IEEE Trans. on Information Theory 39, 930–945 (1993)

Bartfeld, E., Grinvald, A.: Relationships between orientation-preference pinwheels, cytochrome oxidase blobs, and ocular-dominance columns in primate striate cortex. Proc. Nati. Acad. Sci. USA 89, 11905–11909 (1992)

Becker, S., Hinton, G.E.: Learning mixture models of spatial coherence. Neural Computation 5, 267–277 (1993)

Bengio, Y.: Learning deep architectures for AI. Foundations and Trends in Machine Learning 2(1), 1 127 (2009); also published as a book. Now Publishers (2009)

Bengio, Y., Delalleau, O.: Justifying and generalizing contrastive divergence. Neural Computation 21(6), 1601–1621 (2009)

Bengio, Y., Delalleau, O.: Shallow versus deep sum-product networks. In: The Learning Workshop, Fort Lauderdale, Florida (2011)

Bengio, Y., LeCun, Y.: Scaling learning algorithms towards AI. In: Bottou, L., Chapelle, O., DeCoste, D., Weston, J. (eds.) Large Scale Kernel Machines. MIT Press (2007)

Bengio, Y., Delalleau, O., Le Roux, N., Paiement, J.-F., Vincent, P., Ouimet, M.: Spectral Dimensionality Reduction. In: Guyon, I., Gunn, S., Nikravesh, M., Zadeh, L. (eds.) Feature Extraction, Foundations and Applications, vol. 207, pp. 519–550. Springer, Heidelberg (2006)

Bengio, Y., Lamblin, P., Popovici, D., Larochelle, H.: Greedy layer-wise training of deep networks. In: Schölkopf, B., Platt, J., Hoffman, T. (eds.) Advances in Neural Information Processing Systems 19 (NIPS 2006), pp. 153–160. MIT Press (2007)

Bengio, Y., Bastien, F., Bergeron, A., Boulanger-Lewandowski, N., Chherawala, Y., Cisse, M., Côté, M., Erhan, D., Eustache, J., Glorot, X., Muller, X., Pannetier-Lebeuf, S., Pascanu, R., Savard, F., Sicard, G.: Deep self-taught learning for handwritten character recognition. In: NIPS*2010 Deep Learning and Unsupervised Feature Learning Workshop (2010)

Berkes, P., Wiskott, L.: Slow feature analysis yields a rich repertoire of complex cell properties. Journal of Vision 5(6), 579–602 (2005)

Bourlard, H., Kamp, Y.: Auto-association by multilayer perceptrons and singular value decomposition. Biological Cybernetics 59, 291–294 (1988)

Braverman, M.: Poly-logarithmic independence fools bounded-depth boolean circuits. Communications of the ACM 54(4), 108–115 (2011)

Breuleux, O., Bengio, Y., Vincent, P.: Quickly generating representative samples from an RBM-derived process. Neural Computation 23(8), 2058–2073 (2011)

Bromley, J., Benz, J., Bottou, L., Guyon, I., Jackel, L., LeCun, Y., Moore, C., Sackinger, E., Shah, R.: Signature verification using a siamese time delay neural network. In: Advances in Pattern Recognition Systems using Neural Network Technologies, pp. 669–687. World Scientific, Singapore (1993)

Cadieu, C., Olshausen, B.: Learning transformational invariants from natural movies. In: Koller, D., Schuurmans, D., Bengio, Y., Bottou, L. (eds.) Advances in Neural Information Processing Systems 21, pp. 209–216. MIT Press (2009)

Cardoso, J.-F.: Multidimensional independent component analysis. In: Proceedings of the 1998 IEEE International Conference on Acoustics, Speech and Signal Processing, vol. 4, pp. 1941–1944 (1998)

Chopra, S., Hadsell, R., LeCun, Y.: Learning a similarity metric discriminatively, with application to face verification. In: Proceedings of the Computer Vision and Pattern Recognition Conference (CVPR 2005). IEEE Press (2005)

Collobert, R., Weston, J.: A unified architecture for natural language processing: Deep neural networks with multitask learning. In: Cohen, W.W., McCallum, A., Roweis, S.T. (eds.) Proceedings of the Twenty-Fifth International Conference on Machine Learning (ICML 2008), pp. 160–167. ACM (2008)

Courville, A., Bergstra, J., Bengio, Y.: A spike and slab restricted Boltzmann machine. In: Proceedings of the Fourteenth International Conference on Artificial Intelligence and Statistics (AISTATS 2011) (2011)

Desjardins, G., Courville, A., Bengio, Y., Vincent, P., Delalleau, O.: Tempered Markov chain Monte-Carlo for training of restricted Boltzmann machine. In: Proceedings of the Thirteenth International Conference on Artificial Intelligence and Statistics (AISTATS 2010), pp. 145–152 (2010)

Erhan, D., Courville, A., Bengio, Y.: Understanding representations learned in deep architectures. Technical Report 1355, Université de Montréal/DIRO (2010a)

Erhan, D., Bengio, Y., Courville, A., Manzagol, P.-A., Vincent, P., Bengio, S.: Why does unsupervised pre-training help deep learning? Journal of Machine Learning Research 11, 625–660 (2010b)

Goodfellow, I., Le, Q., Saxe, A., Ng, A.: Measuring invariances in deep networks. In: Bengio, Y., Schuurmans, D., Williams, C., Lafferty, J., Culotta, A. (eds.) Advances in Neural Information Processing Systems 22 (NIPS 2009), pp. 646–654 (2009)

Grimes, D.B., Rao, R.P.: Bilinear sparse coding for invariant vision. Neural Computation 17(1), 47–73 (2005)

Gutmann, M., Hyvarinen, A.: Noise-contrastive estimation: A new estimation principle for unnormalized statistical models. In: Proceedings of the Thirteenth International Conference on Artificial Intelligence and Statistics, AISTATS 2010 (2010)

Hadsell, R., Chopra, S., LeCun, Y.: Dimensionality reduction by learning an invariant mapping. In: Proceedings of the Computer Vision and Pattern Recognition Conference (CVPR 2006), pp. 1735–1742. IEEE Press (2006)

Håstad, J.: Almost optimal lower bounds for small depth circuits. In: Proceedings of the 18th Annual ACM Symposium on Theory of Computing, Berkeley, California, pp. 6–20. ACM Press (1986)

Håstad, J., Goldmann, M.: On the power of small-depth threshold circuits. Computational Complexity 1, 113–129 (1991)

Hinton, G.E.: Products of experts. In: Proceedings of the Ninth International Conference on Artificial Neural Networks (ICANN), Edinburgh, Scotland, vol. 1, pp. 1–6. IEE (1999)

Hinton, G.E., Salakhutdinov, R.: Reducing the dimensionality of data with neural networks. Science 313(5786), 504–507 (2006)

Hinton, G.E., Zemel, R.S.: Autoencoders, minimum description length, and Helmholtz free energy. In: Cowan, D., Tesauro, G., Alspector, J. (eds.) Advances in Neural Information Processing Systems 6 (NIPS 1993), pp. 3–10. Morgan Kaufmann Publishers, Inc. (1994)

Hinton, G.E., Sejnowski, T.J., Ackley, D.H.: Boltzmann machines: Constraint satisfaction networks that learn. Technical Report TR-CMU-CS-84-119, Carnegie-Mellon University, Dept. of Computer Science (1984)

Hinton, G.E., Osindero, S., Teh, Y.: A fast learning algorithm for deep belief nets. Neural Computation 18, 1527–1554 (2006)

Hubel, D.H., Wiesel, T.N.: Receptive fields of single neurons in the cat's striate cortex. Journal of Physiology 148, 574–591 (1959)

Hurri, J., Hyvärinen, A.: Temporal coherence, natural image sequences, and the visual cortex. In: Advances in Neural Information Processing Systems 15 (NIPS 2002), pp. 141–148 (2003)

Hyvärinen, A.: Estimation of non-normalized statistical models using score matching. Journal of Machine Learning Research 6, 695–709 (2005)

Hyvärinen, A., Hoyer, P.: Emergence of phase and shift invariant features by decomposition of natural images into independent feature subspaces. Neural Computation 12(7), 1705–1720 (2000)

Hyvärinen, A., Hoyer, P.O., Inki, M.O.: Topographic independent component analysis. Neural Computation 13(7), 1527–1558 (2001)

Jain, V., Seung, S.H.: Natural image denoising with convolutional networks. In: Koller, D., Schuurmans, D., Bengio, Y., Bottou, L. (eds.) Advances in Neural Information Processing Systems 21 (NIPS 2008), pp. 769–776 (2008)

Jarrett, K., Kavukcuoglu, K., Ranzato, M., LeCun, Y.: What is the best multi-stage architecture for object recognition? In: Proc. International Conference on Computer Vision (ICCV 2009), pp. 2146–2153. IEEE (2009)

Jenatton, R., Audibert, J.-Y., Bach, F.: Structured variable selection with sparsity-inducing norms. Technical report, arXiv:0904.3523 (2009)

Jordan, M.I.: Learning in Graphical Models. Kluwer, Dordrecht (1998)

Kavukcuoglu, K., Ranzato, M., LeCun, Y.: Fast inference in sparse coding algorithms with applications to object recognition. Technical report, Computational and Biological Learning Lab, Courant Institute, NYU. Tech Report CBLL-TR-2008-12-01 (2008)

Kavukcuoglu, K., Ranzato, M., Fergus, R., LeCun, Y.: Learning invariant features through topographic filter maps. In: Proceedings of the Computer Vision and Pattern Recognition Conference (CVPR 2009), pp. 1605–1612. IEEE (2009)

Kavukcuoglu, K., Sermanet, P., Boureau, Y.-L., Gregor, K., Mathieu, M., LeCun, Y.: Learning convolutional feature hierarchies for visual recognition. In: Advances in Neural Information Processing Systems 23 (NIPS 2010), pp. 1090–1098 (2010)

Kingma, D., LeCun, Y.: Regularized estimation of image statistics by score matching. In: Lafferty, J., Williams, C.K.I., Shawe-Taylor, J., Zemel, R., Culotta, A. (eds.) Advances in Neural Information Processing Systems 23, pp. 1126–1134 (2010)

Klampfl, S., Maass, W.: Replacing supervised classification learning by slow feature analysis in spiking neural networks. In: Bengio, Y., Schuurmans, D., Williams, C., Lafferty, J., Culotta, A. (eds.) Advances in Neural Information Processing Systems 22 (NIPS 2009), pp. 988–996 (2009)

Kohonen, T.: The self-organizing map. Proceedings of the IEEE 78(9), 1464–1480 (1990)

Kohonen, T.: Emergence of invariant-feature detectors in the adaptive-subspace self-organizing map. Biological Cybernetics 75, 281–291 (1996), doi:10.1007/s004220050295

Kohonen, T., Nemeth, G., Bry, K.-J., Jalanko, M., Riittinen, H.: Spectral classification of phonemes by learning subspaces. In: IEEE International Conference on Acoustics, Speech, and Signal Processing, ICASSP 1979, vol. 4, pp. 97–100 (1979)

Körding, K.P., Kayser, C., Einhäuser, W., König, P.: How are complex cell properties adapted to the statistics of natural stimuli? Journal of Neurophysiology 91, 206–212 (2004)

Krizhevsky, A.: Convolutional deep belief networks on cifar-10 (2010) (unpublished manuscript)
http://www.cs.utoronto.ca/~kriz/conv-cifar10-aug2010.pdf

Kurkova, V., Sanguineti, M.: Geometric upper bounds on rates of variable-basis approxima-tion. IEEE Trans. on Information Theory 54, 5681–5688 (2008)

Larochelle, H., Erhan, D., Bengio, Y.: Zero-data learning of new tasks. In: Proceedings of the 23rd National Conference on Artificial Intelligence, vol. 2, pp. 646–651. AAAI Press (2008)

Le, Q., Ngiam, J., Chen, Z., Hao Chia, D.J., Koh, P.W., Ng, A.: Tiled convolutional neu-ral networks. In: Lafferty, J., Williams, C.K.I., Shawe-Taylor, J., Zemel, R., Culotta, A. (eds.) Advances in Neural Information Processing Systems 23 (NIPS 2010), pp. 1279–1287 (2010)

Le Roux, N., Bengio, Y.: Representational power of restricted Boltzmann machines and deep belief networks. Neural Computation 20(6), 1631–1649 (2008)

LeCun, Y., Boser, B., Denker, J.S., Henderson, D., Howard, R.E., Hubbard, W., Jackel, L.D.: Backpropagation applied to handwritten zip code recognition. Neural Computation 1(4), 541–551 (1989)

LeCun, Y., Bottou, L., Bengio, Y., Haffner, P.: Gradient-based learning applied to document recognition. Proceedings of the IEEE 86(11), 2278–2324 (1998)

Lee, H., Grosse, R., Ranganath, R., Ng, A.Y.: Convolutional deep belief networks for scalable unsupervised learning of hierarchical representations. In: Bottou, L., Littman, M. (eds.) Proceedings of the Twenty-Sixth International Conference on Machine Learning (ICML 2009). ACM, Montreal (2009a)

Lee, H., Pham, P., Largman, Y., Ng, A.: Unsupervised feature learning for audio classification using convolutional deep belief networks. In: Bengio, Y., Schuurmans, D., Williams, C., Lafferty, J., Culotta, A. (eds.) Advances in Neural Information Processing Systems 22 (NIPS 2009), pp. 1096–1104 (2009b)

Lee, J.A., Verleysen, M.: Nonlinear dimensionality reduction. Springer (2007)

Manzagol, P.-A., Bertin-Mahieux, T., Eck, D.: On the use of sparse time-relative auditory codes for music. In: Proceedings of the 9th International Conference on Music Information Retrieval (ISMIR 2008), pp. 603–608 (2008)

Olshausen, B., Field, D.J.: How close are we to understanding V1? Neural Computation 17, 1665–1699 (2005)

Olshausen, B.A., Field, D.J.: Sparse coding with an overcomplete basis set: a strategy em-ployed by V1? Vision Research 37, 3311–3325 (1997)

Olshausen, B.A., Anderson, C.H., Van Essen, D.C.: A neurobiological model of visual atten-tion and invariant pattern recognition based on dynamic routing of information. J. Neu-rosci. 13(11), 4700–4719 (1993)

Raina, R., Battle, A., Lee, H., Packer, B., Ng, A.Y.: Self-taught learning: transfer learning from unlabeled data. In: Ghahramani, Z. (ed.) Proceedings of the Twenty-Fourth Interna-tional Conference on Machine Learning (ICML 2007), pp. 759–766. ACM (2007)

Ranzato, M., Hinton, G.H.: Modeling pixel means and covariances using factorized third-order Boltzmann machines. In: Proceedings of the Computer Vision and Pattern Recogni-tion Conference (CVPR 2010), pp. 2551–2558. IEEE Press (2010)

Ranzato, M., Poultney, C., Chopra, S., LeCun, Y.: Efficient learning of sparse representations with an energy-based model. In: Schölkopf, B., Platt, J., Hoffman, T. (eds.) Advances in Neural Information Processing Systems 19 (NIPS 2006), pp. 1137–1144. MIT Press (2007a)

Ranzato, M., Poultney, C., Chopra, S., LeCun, Y.: Efficient learning of sparse representations with an energy-based model. In: NIPS 2006 (2007b)

Rumelhart, D.E., Hinton, G.E., Williams, R.J.: Learning representations by back-propagating errors. Nature 323, 533–536 (1986)

Salakhutdinov, R.: Learning deep Boltzmann machines using adaptive MCMC. In: Bottou, L., Littman, M. (eds.) Proceedings of the Twenty-Seventh International Conference on Machine Learning (ICML 2010), vol. 1, pp. 943–950. ACM (2010)

Salakhutdinov, R., Hinton, G.E.: Deep Boltzmann machines. In: Proceedings of the Twelfth International Conference on Artificial Intelligence and Statistics (AISTATS 2009), vol. 5, pp. 448–455 (2009)

Salakhutdinov, R., Larochelle, H.: Efficient learning of deep Boltzmann machines. In: Proceedings of the Thirteenth International Conference on Artificial Intelligence and Statistics (AISTATS 2010), JMLR W&CP, vol. 9, pp. 693–700 (2010)

Saul, L., Roweis, S.: Think globally, fit locally: unsupervised learning of low dimensional manifolds. Journal of Machine Learning Research 4, 119–155 (2002)

Schölkopf, B., Smola, A.J.: Learning with Kernels: Support Vector Machines, Regularization, Optimization and Beyond. MIT Press, Cambridge (2002)

Serre, T., Kreiman, G., Kouh, M., Cadieu, C., Knoblich, U., Poggio, T.: A quantitative theory of immediate visual recognition. Progress in Brain Research, Computational Neuroscience: Theoretical Insights into Brain Function 165, 33–56 (2007)

Smith, E.C., Lewicki, M.S.: Efficient auditory coding. Nature 439(7079), 978–982 (2006)

Smolensky, P.: Information processing in dynamical systems: Foundations of harmony theory. In: Rumelhart, D.E., McClelland, J.L. (eds.) Parallel Distributed Processing, ch. 6, vol. 1, pp. 194–281. MIT Press, Cambridge (1986)

Taylor, G.W., Fergus, R., LeCun, Y., Bregler, C.: Convolutional Learning of Spatio-temporal Features. In: Daniilidis, K., Maragos, P., Paragios, N. (eds.) ECCV 2010, Part VI. LNCS, vol. 6316, pp. 140–153. Springer, Heidelberg (2010)

Tenenbaum, J.B., Freeman, W.T.: Separating Style and Content with Bilinear Models. Neural Computation 12(6), 1247–1283 (2000)

Tieleman, T.: Training restricted Boltzmann machines using approximations to the likelihood gradient. In: Cohen, W.W., McCallum, A., Roweis, S.T. (eds.) Proceedings of the Twenty-Fifth International Conference on Machine Learning (ICML 2008), pp. 1064–1071. ACM (2008)

Tieleman, T., Hinton, G.: Using fast weights to improve persistent contrastive divergence. In: Bottou, L., Littman, M. (eds.) Proceedings of the Twenty-Sixth International Conference on Machine Learning (ICML 2009), pp. 1033–1040. ACM (2009)

Turaga, S.C., Murray, J.F., Jain, V., Roth, F., Helmstaedter, M., Briggman, K., Denk, W., Seung, H.S.: Convolutional networks can learn to generate affinity graphs for image segmentation. Neural Computation 22, 511–538 (2010)

Vasilescu, M.A.O., Terzopoulos, D.: Multilinear independent components analysis. In: IEEE Computer Society Conference on Computer Vision and Pattern Recognition (CVPR 2005), vol. 1, pp. 547–553 (2005)

Vincent, P.: A connection between score matching and denoising autoencoders. Neural Computation 23(7), 1661–1674 (2011)

Vincent, P., Larochelle, H., Bengio, Y., Manzagol, P.-A.: Extracting and composing robust features with denoising autoencoders. In: Cohen, W.W., McCallum, A., Roweis, S.T. (eds.) Proceedings of the Twenty-Fifth International Conference on Machine Learning (ICML 2008), pp. 1096–1103. ACM (2008)

Vincent, P., Larochelle, H., Lajoie, I., Bengio, Y., Manzagol, P.-A.: Stacked denoising autoencoders: Learning useful representations in a deep network with a local denoising criterion. Journal of Machine Learning Research 11, 3371–3408 (2010)

Wang, H., Ullah, M.M., Kläser, A., Laptev, I., Schmid, C.: Evaluation of local spatio-temporal features for action recognition. In: British Machine Vision Conference (BMVC), London, UK, p. 127 (2009)

Welling, M.: Herding dynamic weights for partially observed random field models. In: Proceedings of the 25th Conference in Uncertainty in Artificial Intelligence (UAI 2009). Morgan Kaufmann (2009)

Weston, J., Ratle, F., Collobert, R.: Deep learning via semi-supervised embedding. In: Cohen, W.W., McCallum, A., Roweis, S.T. (eds.) Proceedings of the Twenty-Fifth International Conference on Machine Learning (ICML 2008), pp. 1168–1175. ACM, New York (2008)

Wiskott, L., Sejnowski, T.: Slow feature analysis: Unsupervised learning of invariances. Neural Computation 14(4), 715–770 (2002)

Younes, L.: On the convergence of Markovian stochastic algorithms with rapidly decreasing ergodicity rates. Stochastics and Stochastic Reports 65(3), 177–228 (1999)

Chapter 2
Recurrent Neural Networks

Sajid A. Marhon, Christopher J.F. Cameron, and Stefan C. Kremer

1 Introduction

This chapter presents an introduction to recurrent neural networks for readers familiar with artificial neural networks in general, and multi-layer perceptrons trained with gradient descent algorithms (back-propagation) in particular. A recurrent neural network (RNN) is an artificial neural network with internal loops. These internal loops induce recursive dynamics in the networks and thus introduce delayed activation dependencies across the processing elements (PEs) in the network.

While most neural networks use distributed representations, where information is encoded across the activation values of multiple PEs, in recurrent networks a second kind of distributed representation is possible. In RNNs, it is also possible to represent information in the time varying activations of one or more PEs. Since information can be encoded spatially, across different PEs, and also temporally, these networks are sometimes also called Spatio-Temporal Networks [29].

In a RNN, because time is continuous, the PE activations form a dynamical system which can be described by a system of integral equations; and PE activations can be determined by integrating the contributions of earlier activations over a temporal kernel. If the activation of PE j at time t is denoted by $y_j(t)$, the temporal kernel by $k(\cdot)$; and, f is a transformation function (typically a sigmoidal non-linearity), then

$$ y_j(t) = f\left(\sum_i \int_{t'=0}^{t} k_{ji}(t'-t) \cdot y_i(t') \partial t' \right). \tag{1} $$

This formulation defines a system of integral equations in which the variables $y_j(t)$, for different values of j, depend temporally on each other. If we place no restrictions on the indices, i and j, in the kernels, k, then these temporal dependencies can

Sajid A. Marhon · Christopher J.F. Cameron · Stefan C. Kremer
The School of Computer Science at the University of Guelph,
Guelph, Ontario
e-mail: {smarhon,ccameron,skremer}@uoguelph.ca

M. Bianchini et al. (Eds.): *Handbook on Neural Information Processing*, ISRL 49, pp. 29–65.
DOI: 10.1007/978-3-642-36657-4_2   © Springer-Verlag Berlin Heidelberg 2013

contain loops. If we do place dependencies on the indices then we can restrict the graphical topologies of the dependencies to have no loops. In this case, the only temporal behaviour is that encoded directly in the kernel, k, that filters the inputs to the network. In both cases, but particularly in the latter, the kernel, k, is defined by a design choice to model a specific type of temporal dependency.

For the purposes of simulating these systems on digital computers, it is convenient to describe the evolution of the activation values in the models at fixed, regular time intervals. The frequency of the simulation required to maintain fidelity with the original model is governed by Nyquist's Theorem [45]. In this case, the dynamics of the model can be expressed without the necessity of an explicit integration over a temporal kernel. In fact, many models are described only in terms of activations that are defined on the activations at a previous time-step. If w_{ji} is used to represent the impact of the previous time-step's value of PE i on PE j, then

$$y_j(t) = f(p_j(t)), \qquad (2)$$

where,

$$p_j(t) = \sum_i w_{ji} \cdot y_i(t-1). \qquad (3)$$

There is a number of operational paradigms in which RNNs can be applied:

- **Vector to Vector mapping** — Conventional multi-layer perceptrons (MLPs) map vectors of inputs into vectors of outputs (possibly using some intermediate hidden layer activations in the process). MLPs can be viewed as special cases of RNNs in which there either are no recurrent connections, or all recurrent connections have weights of zero. Additionally, winner take all, Hopfield networks and other similar networks which receive a static input and are left to converge before an output is withdrawn, fall into this category.
- **Sequence to Vector mapping** — RNNs are capable of processing a sequence of input activation vectors over time, and finally rendering an output vector as a result. If the output vector is then post-processed, to map it to one of a finite number of categories, these networks can be used to classify input sequences. In a degenerate case there may be a single time-varying input, resulting in a device that maps a single input signal to a category; but in general, a number of input values that vary in time can be used.
- **Vector to Sequence mapping** — Less common are RNNs used to generate an output sequence in response to a single input pattern. These are generative models for sequences, in which a dynamical system is allowed to evolve under a constant input signal.
- **Sequence to Sequence mapping** — The final case is the one where both input and output are vector sequences that vary over time. This is the case of sequence transduction and is the most general of the 4 cases. In general, this approach assumes a synchronization between the sequences in the sense that there is a one-to-one mapping between activations values at all points in time among the input and output sequences (for an asynchronous exception see [11]).

The most obvious application of these networks is, of course, to problems where input signals arrive over time, or outputs are required to be generated over time. But, this is not the only way in which these systems are used. Sometimes it is desirable to convert a problem that is not temporal in nature into one that is. A good example of such a problem is one in which input sequences of varying lengths must be processed.

If input sequences vary in length, the use of MLPs may impose an unnatural encoding of the input. One can either: 1) use a network with an input vector large enough to accommodate the longest sequence and pad shorter inputs with zeros, or 2) compress the input sequence into a smaller, fixed-length vector. The first option is impractical for long sequences or when the maximum sequence length is unbounded, and generally leads to non-parsimonious solutions. The second option is very effective when prior knowledge about the problem exists to formulate a compressed representation that does not lose any information that is germaine to the problem at hand. Some approaches to reducing input sequences into vectors use a temporal window which only reveals some parts (a finite vector) of the input sequence to the network, or to use a signal processing technique and feature reduction to compress the information. But, in many cases, such apriori knowledge is not available. Clearly, reducing the amount of information available to the network can be disastrous if information relevant to the intended solution is lost in the process.

Another scenario is one in which input sequences are very long. A long input sequence necessitates a large number of inputs which in turn implies a large number of parameters, corresponding to the connection weights from these inputs to subsequent PEs. If a problem is very simple, then having a large number of parameters typically leads to overfitting. Moreover, if such a network is trained using sequences with some given maximum length, and then evaluated on longer sequences, the network will not only be incapable of correctly generalizing reasonable outputs for these longer patterns, it will not even be able to process a pattern that is longer than its maximum input size at all. By contrast a RNN with only a single input and a moderate number of other PEs could very well solve the problem. The reduced number of parameters in the latter system not only reduces the chance of overfitting, but also simplifies the training process.

Thus, RNNs represent a useful and sometimes essential alternative to conventional networks that every neural network practitioner should be aware of.

This chapter is organized as follows. We begin with a section on "Architecture" in which we present a very general RNN architecture which subsumes all other RNN and MLP models. Next we explore some topologies and some very specific models. We then present a section on "Memory" in which we describe different ways in which RNNs incorporate information about the past. Again we begin with a general model and then present specific examples. The third section of the chapter is about "Learning". In it, we describe the methods for updating the parameters of RNNs to improve performance on training datasets (and hopefully unseen test-data). We also describe an important limitation of all gradient based approaches to adapting the parameters of RNNs. In the section titled "Modeling", we compare these systems to more traditional computational models. We focus on a comparison between

RNNs and finite state automata, but also mention other computational models. The "Applications" section describes some example applications to give the reader some insight into how these networks can be applied to real-world problems. Finally, the "Conclusion" presents a summary, and some remarks on the future of the field.

2 Architecture

In this section, we discuss RNN architectures. As mentioned before, RNNs are a generalization of MLPs, so it is appropriate to begin our discussion with this more familiar architecture. A typical MLP architecture is shown in Figure 1. In this figure, the mid-sized, white circles represent the input units of the network, solid arrows are connections between input and/or PEs, and large circles are PEs. The network is organized into a number of node layers; each layer is fully connected with its preceding and subsequent layers (except of course the original input and terminal output layers). Note that there are no recurrent connections in this network (e.g. amongst PEs in the same layer, or from PEs in subsequent layers to PEs in previous layers). The activation value of each PE y_j in the first PE-layer of this network can be computed as

$$y_j = f(p_j), \tag{4}$$

where $f(p_j)$ is generally a non-linear squashing function, such as the sigmoid function:

$$f(p_j) = \frac{1}{1 + e^{-p_j}}, \tag{5}$$

$$p_j = \sum_i w_{ji} \cdot x_i + b_j, \tag{6}$$

where w_{ji} represents the weight of the connection from input i to PE j, p_j is the weighted sum of the inputs to PE j, b_j is a bias term associated with PE j, and x_i is the i-th component of the input vector. For subsequent layers, the formula for the weighted summation (Equation 6) is changed to:

$$p_j = \sum_i w_{ji} \cdot y_i + b_j, \tag{7}$$

where, this time p_j is the weighted sum of the activation values of the PEs in the previous layer.

In a MLP network, this formulation assumes that the weights w_{ji} for any units j and i not in immediately subsequent layers are zero. We can relax this restriction somewhat and still maintain a feedforward neural network (FNN) by requiring only that the weights w_{ji} for $j \leq i$ be zero. This results in a cascade [10] architecture in which every input and every lower indexed unit can connect to every higher index unit. This generalization also allows us to identify an extra input $i = 0$ whose value is

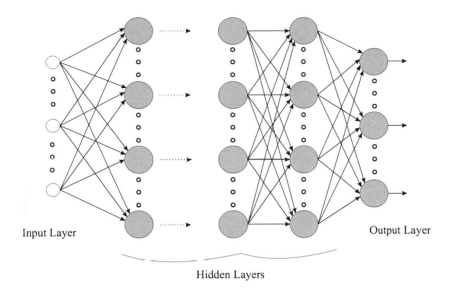

Fig. 1 A multi-layer perceptron neural network

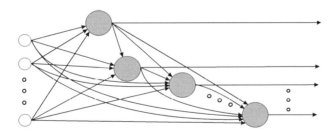

Fig. 2 A cascade architecture of FNN

permanently fixed at $y_0 - 1$ to act as a biasing influence via the connections $w_{j0} = b_j$ to all other PEs j. An example of a cascade architecture is shown in Figure 2.

We now go one step further by removing all restrictions on weights to form a fully connected recurrent network (FCRN) in which every PE is connected to every other PE (including itself). At this stage, it becomes necessary to introduce a temporal index to our notation in order to disambiguate the activation values and unsatisfiable equalities. So, we can now formulate the weighted summations of the PEs as

$$p_j(t) = \sum_h w_{jh} \cdot y_h(t-1) + \sum_i w_{ji} \cdot x_i(t), \qquad (8)$$

where the index h is used to sum over the PEs, and the index i sums over the inputs. We assume that the weight values do not change over time (at least not at the same

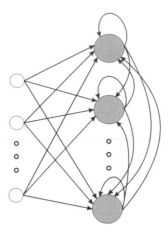

Fig. 3 A fully-connected recurrent network

scale as the activation values). A FCRN is shown in Figure 3. Note that these figures become very difficult to draw as the number of PEs are increased, since there are not topological restrictions to organize elements into layers. However, it is possible to produce a simplified illustration that matches Equation 8 as shown in Figure 4. Note that in this figure time is being represented spatially by the appearance of an additional, virtual layer of PEs that are in fact the same PEs as already shown, but in a previous time-step. This virtual layer has been called a "context" layer [8]. We add a series of additional connections labeled z^{-1} from the PEs to the context units in order to indicate the temporally delayed copy of activation values implied in the system. Figure 5 shows the RNN architecture in Figure 4 as a block diagram. In this figure and all the following figures that present block diagrams in this chapter, the thick arrows indicate full connectivity (many-to-many) between the units in the linked sets. However, the thin bold arrows indicate the unit-to-unit connectivity (one-to-one).

There are a few important things to note about the FCRN architecture presented. First, it is the most general. The MLP and FNN architectures can be implemented within the FCRN paradigm by restricting some of the weights to zero values as indicated in the first two paragraphs of this section. Second, any discrete time RNN can be represented as a FNN. This includes the specific layered recurrent networks discussed in the following sub-sections. Third, the networks introduce an additional parameter set whose values need to be determined. Namely, the initial values of the PE activations $y_j(0)$. Finally, the illustration of activation value calculation in Figure 4 can be solved recursively all the way to the base-case at $t' = 0$, as shown in Figure 6.

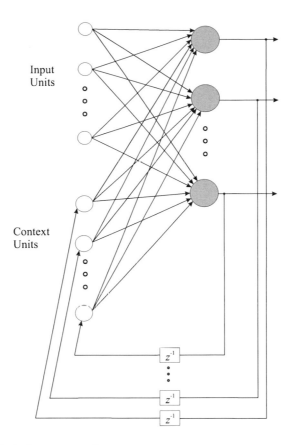

Fig. 4 A RNN including a context layer

2.1 *Connectionist Network Topologies*

While the general FCRN described in the previous subsection is often used, many other RNNs are structured in layers. A RNN includes an input layer, output layer and typically one or more hidden layers. Each layer consists of a set of PEs. The feedback connections, which are specific to RNNs, can exist within or between any of the network layers. Typically, the inputs to the PE, in a RNN, are from other PEs in a preceding layer and delayed feedback from the PE itself or from other PEs in the same layer or in a successive layer. The sum of the inputs is presented as an activation to a nonlinear function to produce the activation value of the PE.

In RNNs, the topology of the feedforward connections is similar to MLPs. However, the topology of feedback connections, which is limited to RNNs, can be classified into locally recurrent, non-local recurrent and globally recurrent connections. In locally recurrent connections, a feedback connection originates from the output of a PE and feeds back the PE itself. In non-local recurrent connections, a feedback connection links the output of a PE to the input of another PE in the same layer.

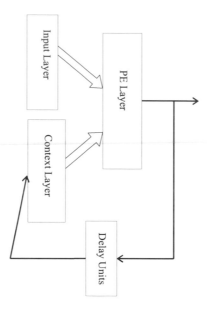

Fig. 5 The block diagram of the RNN architecture in Figure 4

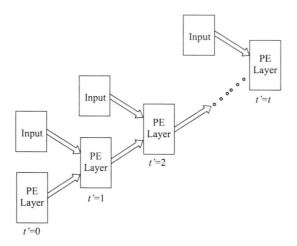

Fig. 6 The unrolled architecture

In globally recurrent connections, the feedback connection is between two PEs in different layers. If we extend this terminology to feedforward connections, all MLPs are considered as global feedforward networks. The non-local recurrent connection class is a special case of the globally recurrent connection class. Based on the feedback topologies, the architecture of RNNs can take different forms as follows:

2.1.1 Locally Recurrent Globally Feedforward (LRGF) Networks

In this class of recurrent networks, recurrent connections can occur in a hidden layer or the output layer. All feedback connections are within the intra PE level. There are no feedback connections among different PEs [51]. When the feedback connection is in the first PE layer of the network, the activation value of PE j is computed as follows:

$$y_j(t) = f\left(w_{jj} \cdot y_j(t-1) + \sum_i w_{ji} \cdot x_i(t)\right) \tag{9}$$

where w_{jj} is the intensity factor at the local feedback connection of PE j, the index i sums over the inputs, and $f(\cdot)$ is a nonlinear function, usually a sigmoid function as denoted in Equation 5. For subsequent layers, Equation 9 is changed to:

$$y_j(t) = f\left(w_{jj} \cdot y_j(t-1) + \sum_i w_{ji} \cdot y_i(t)\right) \tag{10}$$

where $y_i(t)$ are the activation values of the PEs in the preceding PE layer.

There are three different models of LRGF networks depending on the localization of the feedback.

Local Activation Feedback — In this model, the feedback can be a delayed version of the activation of the PE. The local activation feedback model was studied by [12]. This model can be described by the following equations:

$$y_j(t) = f\left(p_j(t)\right), \tag{11}$$

$$p_j(t) = \sum_{t'=1}^{m} w_{jj}^{t'} \cdot p_j(t-t') + \sum_i w_{ji} \cdot x_i(t), \tag{12}$$

where $p_j(t)$ is the activation at the time step t, t' is a summation index over the number of delays in the system, the index i sums over the system inputs, and $w_{jj}^{t'}$ is the weight of activation feedback of $p_j(t-t')$. Figure 7 illustrates the architecture of this model.

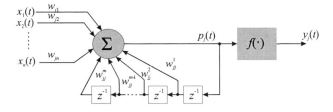

Fig. 7 A PE with local activation feedback

Local Output Feedback — In this model, the feedback is a delayed version of the activation output of the PE. This model was introduced by Gori *et al.* [19]. This feedback model can be illustrated as in Figure 8a, and its mathematical formulation can be given as follows:

$$y_j(t) = f\Big(w_{jj} \cdot y_j(t-1) + \sum_i w_{ji} \cdot x_i(t)\Big).$$ (13)

Their model can be generalized by taking the feedback after a series of delay units and the feedback is fed back to the input of the PE as illustrated in Figure 8b. The mathematical formulation can be given as follows:

$$y_j(t) = f\Big(\sum_{t'=1}^{m} w_{jj}^{t'} \cdot y_j(t-t') + \sum_i w_{ji} \cdot x_i(t)\Big),$$ (14)

where $w_{jj}^{t'}$ is the intensity factor of the output feedback at time delay $z^{-t'}$, and the index i sums over the input units. From Equation 14, it can be noticed that the output of the PE is filtered by a finite impulse response (FIR) filter.

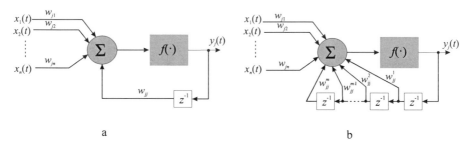

a   b

Fig. 8 A PE with local output feedback. a) With one delay unit. b) With a series of delay units.

Local Synapse Feedback — In this model, each synapse may include a feedback structure, and all feedback synapses are summed to produce the activation of the PE. Local activation feedback model is a special case of the local synapse feedback model since each synapse represents an individual local activation feedback structure. The local synapse feedback model represents FIR filter or infinite impulse response (IIR) filter [3]. A network of this model is called a FIR MLP or IIR MLP when the network incorporates FIR synapses or IIR synapses respectively since the globally feedforward nature of this class of networks makes it identical to MLP networks [3, 32]. Complex structures can be designed to incorporate combination of both FIR synapses and IIR synapses [3].

In this model, a linear transfer function with poles and zeros is introduced with each synapse instead of a constant synapse weight. Figure 9 illustrates a PE architecture of this model. The mathematical description of the PE can be formulated as follows:

$$y_j(t) = f\left(\sum_i G_i(z^{-1}) \cdot x_i(t)\right), \tag{15}$$

$$G_i(z^{-1}) = \frac{\sum_{l=0}^{q} b_l z^{-l}}{\sum_{l=0}^{r} a_l z^{-l}}, \tag{16}$$

where $G_i(z^{-1})$ is a linear transfer function, and b_l ($l = 0, 1, 2, \cdots, q$) and a_l ($l = 0, 1, 2, \cdots, r$) are its zeros' and poles' coefficients respectively.

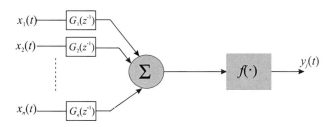

Fig. 9 A PE with local synapse feedback

2.1.2 Non-local Recurrent Globally Feedforward (NLRGF) Networks

In this class of RNNs, the feedback connections to a particular PE are allowed to originate from the PE itself (like LRGF networks) and from other PEs in the same layer. Some researchers classify this type of feedback connections (non-local) as global feedback connections [37]. Based on the description of NLRGF networks, Elman [8] and Williams-Zipser [54] architectures are considered examples of this class of networks [8, 26, 54]. Non-local feedback connections can appear in the hidden or output layers. The mathematical description of PE j that has non-local connections in the first PE layer can be given as follows:

$$y_j(t) = f\left(\sum_h w_j^h \cdot y_h(t-1) + \sum_i w_{ji} \cdot x_i(t)\right), \tag{17}$$

where w_j^h is the weight of the feedback connections including the non-local connections from PE h to PE j ($j \neq h$) and the local connection from the PE itself ($j = h$), the index h sums over the PEs, and the index i sums over the inputs. For subsequent layers, Equation 17 is changed to:

$$y_j(t) = f\left(\sum_h w_j^h \cdot y_h(t-1) + \sum_i w_{ji} \cdot y_i(t)\right), \tag{18}$$

where the index i of the summation sums over the PEs in the preceding PE layer.

2.1.3   Globally Recurrent Globally Feedforward (GFGR) Networks

In this class of recurrent networks, the feedback connections are in the inter layer
level. The feedback connections originate from PEs in a certain layer and feed back
PEs in a preceding layer. It can be from the output layer to a hidden layer, or it
can be from a hidden layer to a preceding hidden layer. Like the MLP, feedforward
connections are global connections from a particular layer to the successive layer.
The difference between this class and the non-local recurrent network class is that
in the latter the feedback connections are in the intra layer level, while in the former
the feedback connections are in the inter layer level. When the first PE layer in
the network receives global recurrent connections from a successive PE layer, the
activation value of PE j in the layer that receives this feedback can be computed as
follows:

$$y_j(t) = f\left(\sum_h w_j^h \cdot y_h(t-1) + \sum_i w_{ji} \cdot x_i(t)\right), \tag{19}$$

where $y_h(t)$ is the output of PE h in a successive layer which the feedback connections originate from, and the index i sums over the inputs of PE j. For subsequent
layers Equation 19 is changed to:

$$y_j(t) = f\left(\sum_h w_j^h \cdot y_h(t-1) + \sum_i w_{ji} \cdot y_i(t)\right). \tag{20}$$

Jordan's second architecture is one of the models of this class of recurrent networks
[25]. In this architecture, the feedback connections from the output layer are fed
back to the hidden layer through a context layer. A unit in the context layer serves
as an intermediate state in the model. Figure 10 illustrates the block diagram of the
Jordan's second architecture.

There are other classes of RNN topologies which incorporate two of the previously mentioned topologies. For example, the locally recurrent globally recurrent
(LRGR) class represents models that include globally recurrent connections as well

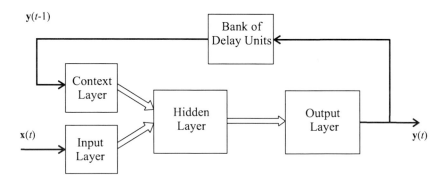

Fig. 10 A block diagram of the Jordan's 2nd architecture

as locally recurrent connections. Another class of RNNs includes networks incorporating non-local recurrent connections and globally recurrent connections at the same time [41].

2.2 Specific Architectures

In this section, we present some common RNN architectures which have been proposed in the literature. These architectures can be related to the classes mentioned in Subsection 2.1.

2.2.1 Time Delay Neural Networks (TDNN)

This architecture was proposed by Sejnowski and Rosenberg [44] and applied by Waibel et al. to phoneme recognition [52], and it is a variation of the MLP. It incorporates delay units at the inputs of the PEs. Each input to the PE is delayed with a series of time delays z^{-i}, $(i = 0, 1, 2, \cdots, N)$. The delayed signals are multiplied by weight factors. The sum of the weighted signals is presented to a nonlinear function which is usually a sigmoid function [52]. The mathematical description of the TDNN PE can be formulated as in Equation 21. The architecture of a basic TDNN PE is illustrated in Figure 11.

$$y_j(t) = f\left(\sum_{l=1}^{M} \sum_{i=0}^{N} w_{jli} \cdot x_l(t-i) \right) \tag{21}$$

The TDNN network can relate and compare the current input to the history of that input. In the mentioned phoneme recognition application, this architecture was shown to have the ability to learn the dynamic structure of the acoustic signal. Moreover, it has the property of translation invariance which means the features learned by this model are not affected by time shifts.

2.2.2 Williams-Zipser Recurrent Networks

This model has been introduced in this section as the most general form of RNNs. The architecture was proposed by Williams and Zipser [54]. It is called a real-time recurrent network since it was proposed for real-time applications. The network consists of a single layer of PEs. Each PE receives feedback from all other PEs as well as from itself via time delay units. Thus, a fully-connected network is obtained. In addition, the PEs receive external inputs. Each recurrent connection has a unique, adjustable weight to control the intensity of the delayed signal. The diagram of this model is illustrated in Figure 12. The activation of a PE in this network is same as that in Equation 19. This model can be classified to incorporate both local and non-local recurrent connections.

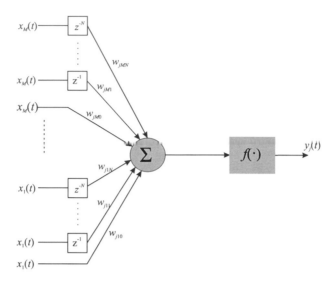

Fig. 11 A typical TDNN PE architecture. The PE has M input signals. Each input signal is delayed by $z^{-0}, \cdots, z^{-N}$. The delayed version of the input as well as the current input (without delay) are multiplied by weights. The weighted sum of all the input signals is computed and presented to a nonlinear function $f(\cdot)$.

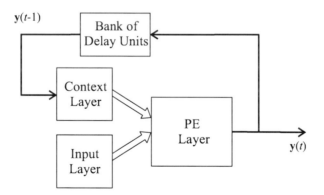

Fig. 12 The architecture of the Williams-Zipser model. Every PE gets feedback from other PEs as well as from itself.

2.2.3 Partially-Connected Recurrent Networks

Unlike the FCRNs, the partially-connected recurrent networks (PCRNs) are based on the MLP. The most obvious example of this architecture is the Elman's [8] and Jordan's [25] models. The Elman model consists of three layers which are the input layer, hidden layer and output layer in addition to a context layer. The context layer receives feedback from the hidden layer, so its units memorize the output of the hidden layer. At a particular time step, the output of the PEs in the hidden layer depends on the current input and the output of the hidden PEs in the previous time step. The mathematical formulation of the output of a hidden PE in this model is given in Equation 17 considering that $y_j(t)$ is the output of hidden PE j. This model can be classified as a non-local recurrent model since each hidden PE receives feedback from itself and from other PEs in the hidden layer. The block scheme of the Elman's model is illustrated in Figure 13.

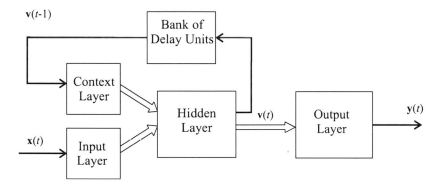

Fig. 13 The architecture of the Elman's model. $\mathbf{y}(t)$ is the network output vector. $\mathbf{x}(t)$ is the input vector. $\mathbf{v}(t)$ is the output vector of the hidden PEs (states).

In contrast to the Elman's model, the Jordan's 2nd model incorporates feedback connections from the output layer to feed back the PEs in the hidden layer via memory unit delays. A context layer receives feedback from the output layer and feeds it to the PEs in the hidden layer. This model is classified as a globally recurrent network since the feedback connections are global between the output and the hidden layers [25]. Figure 10 illustrates the block diagram of the Jordan's 2nd architecture. The mathematical formulation of a hidden PE in the Jordan's 2nd architecture is similar to what has been given in Equation 19.

2.2.4 State-Space Recurrent Networks

This model can include one or more hidden layers. The hidden PEs in a specific layer determine the states of the model. The output of this hidden layer is fed back to that layer via a bank of delay units. The feedback topology of this model can be classified as a non-local recurrent connection class. The number of hidden PEs (states) that feed back the layer can be variant and determine the order of the model [57]. Figure 14 describes the block diagram of the state-space model. The mathematical description of the model is given by the following two equations:

$$\mathbf{v}(t) = \mathbf{f}\big(\mathbf{v}(t-1), \quad \mathbf{x}(t-1)\big), \tag{22}$$

$$\mathbf{y}(t) = \mathbf{C}\mathbf{v}(t), \tag{23}$$

where $\mathbf{f}(\cdot)$ is a vector of nonlinear functions, $\mathbf{C}$ is the matrix of the weights between the hidden and output layers, $\mathbf{v}(t)$ is the vector of the hidden PE activations, $\mathbf{x}(t)$ is a vector of source inputs, and $\mathbf{y}(t)$ is the vector of output PE activations.

By comparing the diagrams in Figures 13 and 14, it can be noticed that the architecture of the state-space model is similar to the Elman's (PCRN) architecture except that the Elman's model uses a nonlinear function in the output layer, and there are no delay units in the output of the network. One of the most important features of the state-space model is that it can approximate many nonlinear dynamic functions [57]. There are two other advantages of the state-space model. First, the number of states (the model order) can be selected independently by the user. The other advantage of the model is that the states are accessible from the outside environment which makes the measurement of the states possible at specific time instances [37].

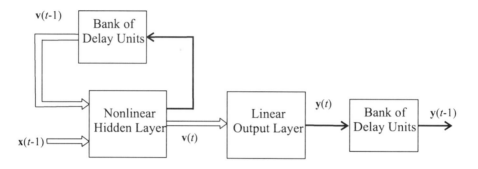

Fig. 14 The block diagram of the state-space model

2.2.5   Second-Order Recurrent Networks

This model was proposed by Giles *et al.* [16]. It incorporates a single layer of PEs. It was developed to learn grammars. The PEs in this model are referred to as second-order PEs since the activation of the next state is computed as the multiplication of the previous state with the input signal. The output of each PE is fed back via a time delay unit and multiplied by each input signal. If the network has N feedback states and M input signals, $N \times M$ multipliers are used to multiply every single feedback state by every single input signal [16]. Thus, the activation value $y_j(t)$ of PE j can be computed as follows:

$$y_j(t) = f\left(\sum_i \sum_l w_{jil} y_i(t-1) x_l(t-1) \right),\qquad(24)$$

where the weight w_{jil} is applied to the multiplication of the activation value $y_i(t-1)$ and the input $x_l(t-1)$. Figure 15 shows the diagram of a second-order recurrent network.

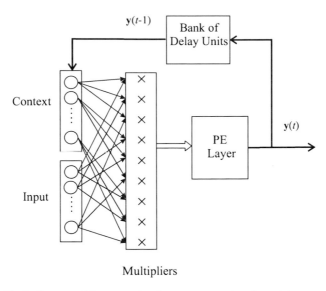

Fig. 15 The block diagram of the second-order recurrent network model

2.2.6 Nonlinear Autoregressive Model with Exogenous Inputs (NARX) Recurrent Networks

In this class of neural networks, the memory elements are incorporated in the input and output layers. The topology of NARX networks is similar to that of the finite memory machines, and this has made them good representative of finite state machines. The model can include input, hidden and output layers. The input to the network is fed via a series of delay units. The output is also fed back to the hidden layer via delay units [6]. The model has been successful in time series and control applications. The architecture of a NARX network with three hidden units is shown in Figure 16. The mathematical description of the model can be given as follows:

$$y(t) = f\left(\sum_{i=1}^{N} a_i y(t-i) + \sum_{i=1}^{M} b_i x(t-i)\right), \tag{25}$$

where $x(t)$ is the source input; $y(t)$ is the output of the network; N and M are constants; and a_i and b_i are constants.

In this section, we reviewed the possible topologies of the RNN architectures and the common proposed architectures. The RNN architectures mentioned above have been proposed to tackle different applications. Some models have been proposed for grammatical inference and other models have been proposed for identification and control of dynamic systems. In addition, there is a coordination between the network architecture and the learning algorithm used for training.

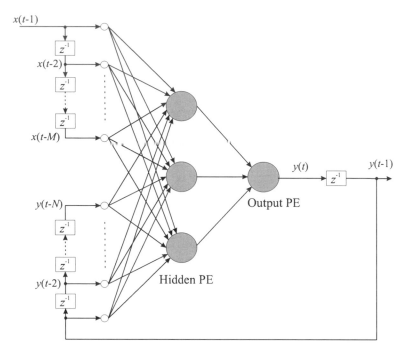

Fig. 16 The architecture of a NARX network

3 Memory

This section of the chapter contains a more descriptive understanding of the importance of the memory for RNNs, how the memory works, and different types of memory.

3.1 Delayed Activations as Memory

Every recurrent network in which activation values from the past are used to compute future activation values, incorporates an implicit memory. This implicit memory can stretch back in time over the entire past history of processing. Consider Figure 5 which depicts a network with recurrent connections within the *PE Layer*. In this illustration, the block labeled *Context Layer* represents a *virtual* set of elements that contain a copy of the *PE Layer*'s elements at the previous time step. Figure 6, shows the same network over a number of time intervals. In this figure, the *PE Layer*'s elements can be seen to depend on the entire history of inputs all the way back to $t' = 0$.

3.2 Short-Term Memory and Generic Predictor

In this subsection, neural networks without global feedback paths, but intra PE level, will be considered. These RNNs consist of two subsystems: a short-term memory and a generic predictor. Short-term memory will be assumed to be of a linear, time invariant, and causal nature. While a generic predictor is considered to be a feed-forward neural network predictor, this predictor consists of nonlinear elements (i.e. a sigmoid function) with associated weights and zero, or more, hidden layers. In the case of this discussion, the generic predictor consists of constant parameters (i.e. weights), nonlinear elements, and it is time invariant.

This structure consisting of the short-term memory and the generic predictor will be considered and referenced as the time invariant nonlinear short-term memory architecture (TINSTMA); the TINSTMA is alternatively known as a memory kernel [34]. The simple structure of the memory kernel is shown in Figure 17.

Fig. 17 Example of a memory kernel, the class of RNNs being observed in this section

When developing a TINSTMA, there are 3 issues that must be taken into consideration: architecture, training, and representation. The architecture refers to the internal structural form of a memory kernel, which involves the PE consideration of the number of layers and PEs within the network, the connections formed between these PEs/layers, and the activation function associated to these PEs. Training (as discussed in the next section of the chapter) involves taking into consideration how the kernel (in this case, the internal parameters will be weights) will adapt to the introduction of a set of input patterns to match the targets associated with the input patterns. Representation refers to retention of an input pattern within the short-term memory (i.e. how should it be stored?); nature and quantity of information to be retained is domain dependent. These three issues are views on the same problem, and thus related. The desired representation for a TINSTMA may or may not alter the architecture of a kernel and, consequently, has the possibility of affecting the design of network training. Alternatively a given training design may influence the structure and possible representation for a TINSTMA. In the following subsection, possible types of kernels will be discussed [29].

3.3 Types of Memory Kernels

Memory kernels can typically be considered one of two forms: each modular component consists of the same structural form, but with different parameters (i.e. weights) OR each modular component consists of the same structural form with identical weights.

3.3.1   Modular Components with Different Parameters

In this subsection, cases where each modular component in a memory kernel has the same structural form, but allows different parameters in each node, will be discussed.

Tapped Delay Lines. Considered to be the simplest of memory kernels, a tapped delay line (TDL) is a delay line (a series of nodes) with at least one tap. A delay-line tap extracts a signal output from somewhere within the delay line, optionally scales it, and usually sums it with other taps (if existing) to form an output signal (shown in Figure 18). A tap may be either interpolating or non-interpolating. A non-interpolating tap extracts the signal at some fixed integer delay relative to the input. Tapped delay lines efficiently simulate multiple echoes from the same source signal. Thus, a tap implements a shorter delay line within a larger one. As a result, they are extensively used in the field of artificial reverberation [49].

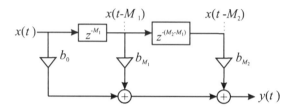

Fig. 18 Example of a TDL network. a TDL consists of an internal tap located at M_1, a total delay length of M_2 samples. Each node may be considered a layer of a multilayer neural network. The output signal of the TDL is a linear combination of the input signal $x(t)$, the delay-line output $x(t - M_2)$, and the tap signal $x(t - M_1)$.

Therefore, the filter output corresponding to Figure 18:

$$y(t) = b_0 x(t) + b_{M_1} x(t - M_1) + b_{M_2} x(t - M_2) \tag{26}$$

Laguerre Filter. Another popular memory kernel applies a filter based on Laguerre polynomials [55], formulated based on Figure 19 as follows:

$$L_i(z^{-1}, u) = \sqrt{1 - u^2} \frac{(z^{-1} - u)^i}{(1 - uz^{-1})^{i+1}}, \quad i \geq 0, \tag{27}$$

$$y_k(t, u) = \sum_{i=0}^{k} w_{k,i}(u) x_i(t, u), \tag{28}$$

where

$$x_i(t, u) = L_i(z^{-1}, u) x(t). \tag{29}$$

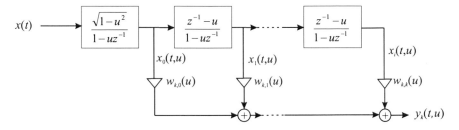

Fig. 19 Example of a Laguerre filter of size k. This filter is stable only if $|u| < 1$. When $u = 0$ the filter degenerates into the familiar transversal filter [48].

FIR/IIR Filters. FIR filters are considered to be nonrecursive since they do not provide feedback ($a_i = 0$, for $i > 0$), compared to IIR or 'recursive' filters which provide feedback ($a_i \neq 0$, for $i > 0$). An example of the format of the filter is shown in the following equations based on Figure 20:

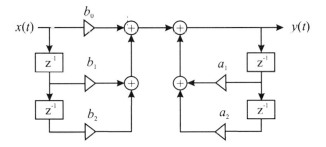

Fig. 20 Example of a second-order IIR filter

$$y(t) = b_0 x(t) + b_1 x(t-1) + \cdots + b_M x(t-M) - a_1 y(t-1) - \cdots - a_N y(t-N) \quad (30)$$

$$y(t) = \sum_{i=0}^{M} b_i x(t-i) - \sum_{j=1}^{N} a_j y(t-j) \quad (31)$$

Gamma Filter. The Gamma filter is a special class of IIR filters where the recursion is kept locally, proving effective in identification of systems with long impulse responses [30]. The structure of the Gamma filter is similar to a TDL (refer above). The format of the Gamma filter is shown in the following equations [39]:

$$y(t) = \sum_{k=0}^{K} w_k x_k(t) \quad (32)$$

$$x_k(t) = G(z^{-1}) x_{k-1(t)}, \quad k = 1, ..., K, \quad (33)$$

where $y(t)$ is the output signal, $x(t)$ is an input signal, and $G(z^{-1})$ is the generalized delay operator (tap-to-tap transfer function, either recursive or non-recursive).

Moving Average Filter. A Moving Average filter (MAF) is a simplified FIR, alternatively called the exponential filter; and, as the name suggests, it works by averaging a number of points from the input signal to produce each point in the output signal. This filter is considered the moving average, since at any moment, a moving window is calculated using M values of the data sequence [50]. Since the MAF places equal emphasis on all input values, two areas of concern for the MAF is the need for n measurements to be made before a reliable output can be computed and for this computation to occur there needs to be storage of n values [50]. The simplified equation of the MAF is:

$$y(t) = \frac{1}{M} \sum_{j=0}^{M-1} x(t - j), \tag{34}$$

where $y(t)$ is the output signal, $x(t)$ is the input signal, and M is the number of points used in the moving average.

3.3.2   Modular Components with Identical Parameters

In this subsection, cases where each modular component in a memory kernel has the same structural form and node parameter(s) will be discussed. With the possibility of each node's weight being the same, parameter estimation is kept to minimum and each node may be fabricated (i.e. in gate array technology). Note, TDLs may be considered within this section as well.

Modular Recurrent Form. A modular recurrent form is a series of independent nodes organized by some intermediary. Each node provides inputs to the intermediary, which the intermediary processes to produce an output for the network as a whole. The intermediary only accepts nodes outputs, no response (i.e. signal) may be reported back to the node. Nodes do not typically interact with each other.

4   Learning

The most desirable aspect of artificial neural networks is their learning capability. When provided input and output values, a function approximation between these values can be approximated by the network. In this section, an understanding of how a RNN can learn through various learning methods to determine this relationship will be provided.

4.1   Recurrent Back-Propagation: Learning with Fixed Points

Fixed point learning algorithms assume that the network will converge to a stable fixed point. This type of learning is useful for computation tasks such as constraint satisfaction and associative memory tasks. Such problems are provided to the network through an initial input signal or through a continuous external signal, and the solution is provided as the state of the network when a fixed point has been reached.

A problem with fixed point learning is whether or not a fixed point will be reached, since RNNs do not always reach a fixed point [29]. There are several ways to guarantee that a fixed point will be reached with certain special cases:

Weight Symmetry. Linear conditions on weights such as zero-diagonal symmetry ($w_{ij} = w_{ji}, w_{ii} = 0$) guarantee that the Lyapunov function (Equation 35) will decrease until a fixed point has been reached [7]. If weights are considered to be Bayesian constraints, as in Boltzmann Machines, the weight symmetry condition will arise [21].

$$L = -\sum_{j,i} w_{ji} y_i y_j + \sum_i \left(y_i \log(y_i) + (1 - y_i) \log(1 - y_z) \right) \tag{35}$$

Setting Weight Boundaries. If $\sum_{ji} w_{j,i}^2 < \max_x\{f'(x)\}$ where $\max_x\{f'(x)\}$ is the maximal value of $f'(x)$ for any x [2], and $f'(\cdot)$ is the derivative of $f(\cdot)$, convergence to a fixed point will occur. In practice it has been shown that much weaker bounds on weights have an effect [40].

Asymptotic Fixed point Behavior. Some studies have shown that applying the fixed point learning to a network causes the network to exhibit asymptotic fixed point behavior [1, 13]. There is no theoretical explanation of this behavior as of yet, nor replication on larger networks.

Even with the guarantee of a network reaching a fixed point, the fixed point learning algorithm can still have problems reaching that state. As the learning algorithm moves the fixed point location by changing the weights, there is the possibility of the error jumping suddenly due to discontinuity. This occurs no matter how gradually weights are manipulated as the underlying mechanisms are a dynamical system subject to bifurcations, and even chaos [29].

4.1.1 Traditional Back-Propagation Algorithm

The traditional back-propagation algorithm [42, 53] involves the computation of an error gradient with respect to network weights by means of a three-phase process: (1) activation forward propagation, (2) error gradient backward propagation, and (3) weight update. The concept is based on the premise that by making changes in weights proportional to the negation of the error gradient, and the error will be reduced until a locally minimal error can be found. The same premise can be applied to recurrent networks. However, the process of back-propagation becomes complex as soon as recurrent connections are introduced. Also, as we will discover at the end of this section, some complications arise when applying the gradient descent method in recurrent networks.

4.2 Back-Propagation through Time: Learning with Non-fixed Points

The traditional back-propagation algorithm cannot directly be applied to RNNs since the error backpropagation pass assumes that the connections between PEs induce a cycle-free ordering. The solution to the back-propagation through time (BPTT) application is to "unroll" the RNN in time. This "unrolling" involves the stacking of identical copies of the RNN (displayed in Figure 6) and redirecting connections within the network to obtain connections between subsequent copies. The end result of this process provides a feedforward network, which is able to have the backpropagation algorithm applied by back-propagating the error gradient through previous time-steps to $t' = 0$. Note that in Figure 6 the arrows from each of the "Input" rectangles to each of the "PE Layer" rectangles represent different "incarnations" of the exact same set of weights during different time-steps. Similarly, the arrows from each of the "PE Layer" rectangles to their subsequent "PE Layer" rectangles also represent the exact same set of weights in different temporal incarnations. This implies that all the changes to be made to all of the incarnations of a particular weights can be cumulatively applied to that weight. This implies that the traditional equation for updating the weights in a network,

$$\Delta w_{ji} = -\eta a_i \delta_j = -\eta a_i \frac{\partial E}{\partial p_j}, \tag{36}$$

is replaced by a temporal accumulation:

$$\Delta w_{ji} = -\eta \sum_{t'} a_i(t') \delta_j(t'+1) = -\eta \sum_{t'} a_i(t') \frac{\partial E}{\partial p_j(t')}. \tag{37}$$

With the "unrolled" network, a forward propagation begins from the initial copy propagation through the stack updating each connection. For each copy or time t: the input ($u(t)$) is read in, the internal state ($y(t)$) is computed from the input and the previous internal state ($y(t-1)$), and then output ($y(t)$) is calculated. The error (E) to be minimized is:

$$E = \sum_{t=1,\dots,T} \|d(t) - y(t)\|^2 = \sum_{t=1,\dots,T} E(t), \tag{38}$$

where T represents the total length of time, $d(t)$ is the target output vector, and $y(t)$ is the output vector.

The problem with the slow convergence for traditional backpropagation is carried on in BPTT. The computation complexity of an epoch for the network is $O(T N^2)$, N is equal to the number of internal/hidden PEs. As with traditional backpropagation, there tends to be the need for multiple epochs (on the scale of 1000s) for a convergence to be reached. It does require manipulation with the network (and processing time) before a desired result can be achieved. This hindrance of the BPTT algorithm tends to lead to smaller networks being used with this design (3-20 PEs), where larger networks tend to go for many hours before convergence.

4.2.1 Real-Time Recurrent Learning

Real-time recurrent learning (RTRL) [54] allows for computation of the partial error gradients at every time step within the BPTT algorithm in a forward-propagated fashion eliminating the need for a temporal back-propagation step (thus, making it very useful for online learning). Rather than computing and recording only the partial derivative of each net value with respect to the total error,

$$\delta_j = \frac{\partial E}{\partial p_j}, \tag{39}$$

it is noted that the total gradient error in a recurrent network is given by:

$$\frac{\partial E}{\partial w_{ji}} = \sum_k (d_k(t) - y_k(t)) \frac{\partial y_k(t)}{\partial w_{ji}}, \tag{40}$$

and that

$$\frac{\partial y_k(t+1)}{\partial w_{ji}} = f'(p_k(t)) \cdot \left(\sum_l w_{lk} \frac{\partial y_l(t)}{\partial w_{ji}} + y_j(t) \right). \tag{41}$$

Note that in this equation, the partial derivatives

$$\frac{\partial y_k(t+1)}{\partial w_{ji}} \tag{42}$$

depend on the previous (not future) values of the same partial derivatives

$$\frac{\partial y_l(t)}{\partial w_{ji}}. \tag{43}$$

This implies that by storing the values

$$\frac{\partial y_k(t+1)}{\partial w_{ji}} \tag{44}$$

at each time-step, the next steps values can be computed until at the final time-step, where a complete error derivative can be computed.

Since w_{kl} has been assumed to be constant, the learning rate or η must be kept small. RTRL has a computational cost of $O(N+L)^4$ for each update step in time ($(N+L)$ dimensional system solved each time), which means that networks employing RTRL must be kept small.

4.2.2 Schmidhuber's Algorithm

Schmidhuber [43] has developed a hybrid algorithm which applies the two previous approaches in a clever, alternating fashion and is able to manage the superior time performance of BPTT while not requiring unbounded memory as sequence length increases.

4.2.3 Time Constants and Time Delays

One advantage of temporal continuous networks is the addition of parameters to control the temporal behavior of the network in ways known to be associated with natural tasks. Time constants represent an example of these additional parameters for networks as shown in [23, 33, 34, 35]. For time delays, consider that a network's connections take a finite period of time between PEs, such that:

$$y_j(t) = \sum_i w_{ji} y_i(t - \tau_{ji}), \tag{45}$$

where τ_{ji} represents the time delay from PE i to PE j.

Using a modification of RTRL, parameters like τ can be learned by a gradient descent approach.

4.3 Long-Term Dependencies

RNNs provide an explicit modeling of time and memory, allowing, in principle, the modelling of any type of open nonlinear dynamical systems. The dynamical system is described by the RNN with a set of real numbers represented by a point in an appropriate state space. There is often an understanding of negativity towards RNNs due to their inability to identify and learn long-term dependencies over time; authors have stated no more than ten steps [4]. Long-term dependencies, in this case, can be defined as a desired output at time T which is dependent on inputs from times t less than T. This difficulty with RNN understanding of long term dependencies has been noted by Mozer, where that RNNs were able to learn short-term musical structure with gradient learning based methods, but had difficulty with a global behavior [34]. Therefore the goal of a network should be to robustly latch information (i.e. a network should be able to store previous inputs with the presence of noise for a long period of time).

As a system robustly latches information, the fraction of the gradient due to information t-steps in the past approaches zero as t becomes large. This is known as the 'problem of vanishing gradients'. Bengio *et al.* [4] and Hochreiter [22] have both noted that the problem of vanishing gradients is the reason why a network trained by gradient-descent methods is unable to distinguish a relationship between target outputs and inputs that occur at a much earlier time. This problem has been termed 'long-term dependencies'.

One way to counteract long-term dependencies is TDL, such as BPTT with a NARX network. BPTT occurs through the process of unrolling the network in time and then back propagating the error through the 'unrolled network' (Figure 6). From this, the output delays will occur as jump ahead connections in the 'unrolled network'. These jump ahead connections provide a shorter path for propagation of the gradient information, and this decreases the sensitivity of a network to long-term dependencies.

Another solution is presented in [23].

5 Modeling

RNNs, which have feedback in their architectures, have the potential to represent and learn discrete state processes. The existence of feedback makes it possible to apply RNN models to solve problems in control, speech processing, time series prediction, natural language processing, and so on [15]. In addition, apriori knowledge can be encoded into these networks which enhances the performance of the applied network models. This section presents how RNNs can represent and model theoretical models such as finite state automata (FSA) and Turing machines (TM). The understanding of the theoretical aspect of these networks in relation to formal models such as FSA is important to select the most appropriate model to solve a given problem. In this section, we will review RNNs for representing FSA and Turing machines.

5.1 Finite State Automata

FSA have a finite number of input and output symbols, and a finite number of internal states. Large portion of discrete processes can be modeled by deterministic finite-state automata (DFA). The mathematical formulation of the FSA M is a 6-tuple and can be defined by:

$$M = \{Q, q_0, \Sigma, \Delta, \delta, \varphi\} \tag{46}$$

where Q is a set of finite number of state symbols: $\{q_1, q_2, \cdots, q_n\}$, n is the number of states, $q_0 \in Q$ is the initial state, Σ is the set of input symbols: $\{u_1, u_2, \cdots, u_m\}$, m is the number of input symbols, Δ is the set of output symbols: $\{o_1, o_2, \cdots, o_r\}$, r is the number of output symbols, $\delta: Q \times \Sigma \to Q$ is the state transition function, and φ is the output function. The class of FSA is basically divided into Mealy and Moore models. Both of the models are denoted by the 6-tuple formulation defined in Equation 46. However, the only difference between the two models is the formulation of the output function φ. In the case of the Mealy model, the output function is $\varphi: Q \times \Sigma \to \Delta$, while, in the Moore model, it is: $\varphi: Q \to \Delta$ [17, 24]. In general, the FSA M is described by the following dynamic model:

$$M: \begin{cases} q(t+1) = \delta(q(t), u(t)), & q(0) = q_0 \\ o(t) = \varphi(q(t), u(t)) \end{cases} \tag{47}$$

Many RNN models have been proven to model FSA. The existence of such equivalence between RNNs and FSA has given the confidence to apply RNN models in solving problems with dynamical systems. Omlin and Giles [36] proposed an algorithm to construct DFA in second-order networks. The architecture of the network is same as the architecture illustrated in Figure 15. The outputs of the PEs in the output layer are the states of the DFA. The network accepts a temporal sequence of inputs and evolves with the dynamic determined by the activation function of the network defined by Equation 24, so the state PEs are computed according to that equation. One of the PEs o_0 is a special PE. This PE represents the output of the network after a string of input is presented to the network. The possible value of this state PE is accept/reject. Therefore, if the modeled DFA has n states and m input symbols, the construction of the network includes $n+1$ state PEs and m inputs. The proposed algorithm is composed of two parts. The first part is to adjust the weights to determine the DFA state transitions. The second part is to adjust the output of the PE for each DFA state.

Kuroe [31] introduced a recurrent network called a hybrid RNN to represent the FSA. In this model, the state symbols q_i ($i = 1, 2, \cdots, n$), input symbols u_i ($i = 1, 2, \cdots, m$) and the output symbols o_i ($i = 1, 2, \cdots, r$) are encoded as binary numbers. $q(t)$, $u(t)$ and $o(t)$ will be expressed as follows:

$$\begin{aligned}
q(t) &= (s_1(t), s_2(t), \cdots, s_A(t)) & (s_i(t) \in \{0, 1\}) \\
u(t) &= (x_1(t), x_2(t), \cdots, x_B(t)) & (x_i(t) \in \{0, 1\}) \\
o(t) &= (y_1(t), y_2(t), \cdots, y_D(t)) & (y_i(t) \in \{0, 1\})
\end{aligned} \tag{48}$$

where A, B and D are integer numbers and their values are selected in a way such that $n \leq 2^A$, $m \leq 2^B$ and $r \leq 2^D$ respectively. In this way, there is no need to increase the number of state PEs in the network according to the 1 against 1 basis as the number of states in the FSA increases. The architecture of the hybrid network consists of two types of PEs, which are static and dynamic PEs. The difference between them is that the dynamic PE feedback originates from its output, while the static PE does not [31]. Figure 21 illustrates the diagram of an arbitrary hybrid neural network representing a FSA. The dynamic PEs are represented by darkly-shaded circles, while the static PEs are represented by lightly-shaded circles.

Won et al. [55] proposed a new recurrent neural architecture to identify discrete-time dynamical systems (DTDS). The model is composed of two MLPs of five layers. The first MLP is composed of layers 0,1 and 2. The second MPL is composed of layers 3 and 4. The first set of layers $\{0, 1, 2\}$ approximates the states of the FSA, while the second set of layers $\{3, 4\}$ approximates the output of the FSA. Figure 22 shows the architecture of the network proposed by [55].

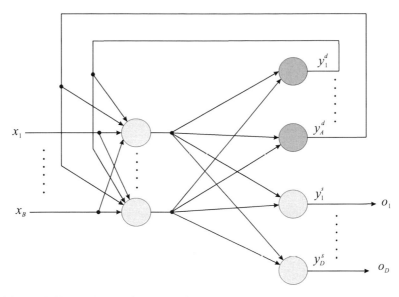

Fig. 21 A hybrid neural network representing a FSA

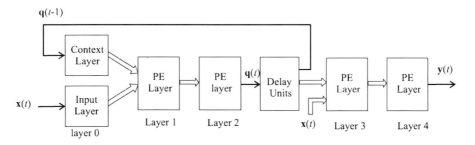

Fig. 22 A RNN composed of two MLPs to identify and extract FSA

5.2 Beyond Finite State Automata

The ability of RNNs to model FSA should inspire great confidence in these systems as FSA represent the computational limits of digital computers with finite memory resources. I.e. anything that a digital computer with finite memory can compute, can also be computed by a FSA, and by a RNN.

Sometimes it is interesting to study even greater levels of computational power by asking questions about what a device could do without memory limitations. This has led to the study, within theoretical computer science of devices such as pushdown automata and Turing machines. It is interesting to note that RNNs measure up to these devices as well. The reader is referred to [28, 38, 47].

6 Applications

The application of RNNs has proved itself in various fields. The spectrum of RNN applications is so wide, and it touched various aspects. Various architectures and learning algorithms have been developed to be applied in solving problems in various fields. The spectrum of application is ranging from natural language processing, financial forecasting, plant modeling, robot control, and dynamic system identification and control. In this section, we review two case studies. One of these cases will be on grammatical inference and the other will be on control.

6.1 Natural Language Processing

In the last years, there has been a lot of efforts and progress in developing RNN architectures for natural language processing. Harigopal and Chen [20] proposed a network to recognize strings which are much longer that the ones which the network was trained on. They used a second-order recurrent network model for the problem of grammatical inference. Zheng et al. [59] proposed a discrete recurrent network for learning deterministic context-free grammar. Elman [9] addressed three challenges of natural language processing. One challenge is the nature of the linguistic representations. Second, is the representation of the complex structural relationships. The other challenge is the ability of a fixed resource system to accommodate the open-ended nature of a language.

Grammatical inference is the problem of extracting the grammar from the strings of a language. There exists a FSA that generates and recognizes that grammar. In order to give the reader a clear idea about the application of RNNs in grammatical inference, we will review a method proposed by Chen et al. [5] to design a RNN for grammatical inference. Chen et al. [5] proposed an adaptive RNN to learn a regular grammar and extract the underlying grammatical rules. They called their model as adaptive discrete recurrent neural network finite state automata (ADNNFSA). The model is based on two recurrent network models, which are the neural network finite state automata (NNFSA) proposed by Giles et al. [18] and the discrete neural network finite state automata (DNNFSA) proposed by Zeng et al. [58].

Figure 23 shows the network architecture for both NNFSA and DNNFSA which was also used in the ADNNFSA model. The network consists of two layers. The first layer consists of N units (context units) that receive feedback from the next layer (state layer) and M input units that receive input signals. The outputs of this layer is connected to the inputs of the second layer via second-order weight connections. The second layer consists of N PEs. The state PEs are denoted by the vector $\mathbf{s}(t-1)$ and the input units are denoted by the vector $\mathbf{x}(t-1)$. In the second layer, $\mathbf{s}(t)$ is the current-state output vector and $\mathbf{h}(t)$ is the current-state activation vector. The activation of the second layer can be computed as follows:

$$h_j(t) = f\left(\sum_i \sum_n w_{jin} s_i(t-1) x_n(t-1)\right), \tag{49}$$

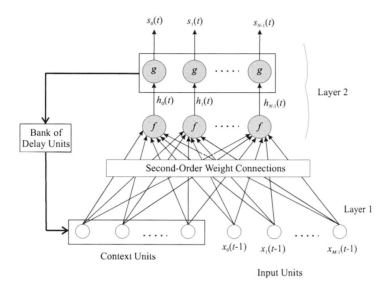

Fig. 23 The general architecture for NNFSA, DNNFSA, and ADNNFSA models

$$s_j(t) = g\big(h_j(t)\big). \tag{50}$$

In the implementation of the NNFSA model, $f(\cdot)$ is a sigmoid function and $g(\cdot)$ is an identity function, while in the implementation of DNNFSA, $g(\cdot)$ is a discrete hard-limiter as follows:

$$g(a) = \begin{cases} 0.8 & \text{if } a > 0.5 \\ 0.2 & \text{if } a < 0.5 \end{cases} \tag{51}$$

The NNFSA model applies the true-gradient descent real-time recurrent learning (RTRL) algorithm, which had a good performance. In the NNFSA model, Giles *et al.* [18] used the analog nature of the network, which does not match with the discrete behavior of a FSA. Therefore, Zeng *et al.* [58], in DNNFSA, discretized the analog NNFSA by using the function in Equation 51. Therefore, all the states are discretized, and the RTRL algorithm is no longer applicable. The pseudo-gradient algorithm was used, and it hinders training because it is an approximation of the true gradient. Therefore, Chen *et al.* [5] used analog internal states at the beginning of the training, and as the training progresses, the model changes gradually to the discrete mode of the internal states. Thus, the current-state activation output $h_j(t)$ is computed same as in Equation 49, and current-state output $s_j(t)$ is computed as follows:

$$s_j(t) = \begin{cases} h_j(t) & \text{if } j \in \text{analog mode} \\ g\big(h_j(t)\big) & \text{if } j \in \text{discrete mode} \end{cases} \tag{52}$$

To decide whether the mode of a state PE has to be switched to the discrete mode, a quantization threshold parameter β is used. If the output of the sate PE j, $s_j(t) < \beta$ or $s_j(t) > 1.0 - \beta$ for all the training strings, the mode of this state PE is switched to the discrete phase. This recurrent network model adapts the training from the initial analog phase, which has a good training performance, to the discrete phase, which fits properly with the nature of the FSA, through the progress of the training for automatic rule extraction.

6.2 Identification and Control of Dynamical Systems

In dynamic systems, different time-variant variables interact to produce outputs. To control such systems, a dynamic model is required to tackle the unpredictable variations in such systems. Adaptive control systems can be used to control dynamic systems since these control systems use a control scheme that have the feature to modify its behavior in response to the variations in dynamic systems. However, most of the adaptive control techniques require the availability of an explicit dynamic structure of the system, and this is impossible for most nonlinear systems which have poorly known dynamics. In addition, these conventional adaptive control techniques lack the ability of learning. This means that such adaptive control systems cannot use the knowledge available from the past and apply it in similar situations in the present or future. Therefore, a class of intelligent control has been applied which is based on neural modeling and learning [27, 46].

Neural networks can deal with nonlinearity, perform parallel computing and process noisy data. These features have made neural networks good tools for controlling nonlinear and time-variant systems. Since dynamic systems involve model states at different time steps, it has been important for the neural network model to memorize the previous states of the system and deal with the feedback signals. This is impossible with the conventional feedforward neural networks. To solve the problem, it has been important to make neural networks have a dynamic behavior by incorporating delay units and feedback links. In other words, RNNs with time delay units and feedback connections can tackle the dynamic behavior of nonlinear time-variant systems. RNNs have proved successfulness in the identification and control of systems which are dynamic in nature. We will review a model that was proposed by Ge et al. [14] for the identification and control of nonlinear systems.

Yan and Zhang [56] presented two main characteristics to measure the dynamic memory performance of neural networks. They called them the "depth" and "resolution ratio". Depth refers to how far information can be memorized by the model, while resolution ratio refers to the amount of information of the input to the model that can be retained. Time-delay recurrent neural networks (TDRNN) can retain much information about the input and memorize information of a short time period. Thus, it has a good resolution ratio and poor depth. On the other hand, most recurrent networks such as Elman networks have a good depth and poor resolution ratio [14]. Ge et al. [14] proposed a model that has a memory of better depth and

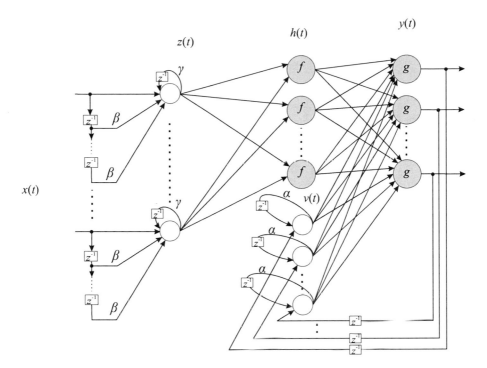

Fig. 24 The architecture of the TDRNN for identification and control of dynamic systems

resolution ratio to identify and control dynamic systems. Figure 24 shows the architecture of the TDRNN proposed by Ge *et al.* [14].

The model incorporates memory units in the input layer with local feedback gain
γ ($0 \leq \gamma \leq 1$) to enhance the resolution ratio. The architecture includes input layer,
hidden layer, output layer, and context layer. In this model, the input and context
units are different from the traditional recurrent networks since, in this model, the
units in the input and context layers are PEs with linear transfer functions. Therefore,
in this instance only, we will call them input PEs and context PEs for consistency.
The context PEs memorize the activations of the output PEs; in addition, there are
local feedback links with constant coefficient α in the context PEs. The output of
the context PE can be given as follows:

$$v_j(t) = \alpha v_j(t-1) + y_j(t-1), \quad j = 1, 2, \cdots, N \qquad (53)$$

where $v_j(t)$ and $y_j(t)$ are the outputs of the jth context PE and the jth output PE
respectively, and N is the number of the context PEs and the output PEs.

The mathematical description of the output PEs, hidden PEs, and input PEs are
respectively described by the following three equations:

$$y_j(t) = g\left(\sum_{i=1}^{M} w_{ji}^2 h_i(t) + \sum_{i=1}^{N} w_{ji}^3 v_i(t)\right), \qquad (54)$$

$$h_j(t) = f\left(\sum_{i=1}^{K} w_{ji}^1 z_i(t)\right), \tag{55}$$

$$z_j(t) = x_j(t) + \beta \sum_{i=1}^{r} x_j(t-i) + \gamma z_j(t-1), \tag{56}$$

where w^2 is the weight between the hidden and the output PEs, w^3 is the weight between the context and the output PEs, $h_j(t)$ is the output of the jth hidden PE, M is the number of the hidden PEs, w^1 is the weight between the input and hidden PEs, $z_j(t)$ is the output of the jth input PE, K is the number of the input PEs, $x_j(t)$ is the jth external input, r is the number of the unit time delays, and $0 \leq \alpha, \beta, \gamma \leq 1$ and $\beta + \gamma = 1$. The activation function $f(\cdot)$ is a sigmoid function, and the function $g(\cdot)$ is a linear function.

The network was learned by a dynamic recurrent backpropagation learning algorithm which was developed based on the gradient descent method. The model shows good effectiveness in the identification and control of nonlinear systems.

7   Conclusion

In this chapter, we presented an introduction to RNNs. We began by describing the basic paradigm of this extension of MLPs, and the need for networks that can process sequences of varying lengths. Then we classified the architectures used and identified the prevailing models and topologies. Subsequently we tacked the implementation of memory in these systems by describing different approaches for maintaining state in such devices. Then we described the prevailing learning methods and an important limitation that all gradient based approaches to learning in RNNs face. In the next section, we described the relationship between the ability of RNNs to process symbolic input sequences and more familiar computational models that can handle the same types of data. This provided confidence in the model as a general purpose computational tool. The final section presented two sample applications to elucidate the applicability of these networks on real-world problems.

We hope that this chapter has motivated the use of RNNs and whetted the reader's appetite to explore this rich paradigm further. A good place to begin a deeper exploration is [29], which presents the most important results in the field in a collection of mutually consistent papers by the primary researchers in these areas.

References

1. Allen, R.B., Alspector, J.: Learning of stable states in stochastic asymmetric networks. Technical Report TM-ARH-015240, Bell Communications Research, Morristown, NJ (1989)
2. Atiya, A.F.: Learning on a general network. In: Neural Information Processing Systems, New York, pp. 22–30 (1988)

3. Back, A.D., Tsoi, A.C.: FIR and IIR synapses, a new neural network architecture for time series modeling. Neural Computation 3, 375–385 (1991)
4. Bengio, Y., Simard, P., Frasconi, P.: Learning long-term dependencies with gadient descent is difficult. IEEE Transactions on Neural Networks 5, 157–166 (1994)
5. Chen, L., Chua, H., Tan, P.: Grammatical inference using an adaptive recurrent neural network. Neural Processing Letters 8, 211–219 (1998)
6. Chen, S., Billings, S., Grant, P.: Nonlinear system identification using neural networks. International Journal of Control 51(6), 1191–1214 (1990)
7. Cohen, M.A., Grossberg, S.: Stability of global pattern formation and parallel memory storage by competitive neural networks. IEEE Transactions on Systems, Man and Cybernetics 13, 815–826 (1983)
8. Elman, J.L.: Finding structure in time. Cognitive Science 14, 179–211 (1990)
9. Elman, J.L.: Distributed representations, simple recurrent networks and grammatical structure. Machine Learning 7, 195–225 (1991)
10. Fahlman, S.E., Lebiere, C.: The cascade-correlation learning architecture. Technical Report CMU-CS-90-100, School of Computer Science, Carnegie Mellon University, Pittsburgh, PA (February 1990)
11. Forcada, M.L., Ñeco, R.P.: Recursive Hetero-Associative Memories for Translation. In: Mira, J., Moreno-Díaz, R., Cabestany, J. (eds.) IWANN 1997. LNCS, vol. 1240, pp. 453–462. Springer, Heidelberg (1997)
12. Frasconi, P., Gori, M., Soda, G.: Local feedback multilayered networks. Neural Computation 4, 120–130 (1992)
13. Galland, C.C., Hinton, G.E.: Deterministic Boltzman learning in networks with asymmetric connectivity. Technical Report CRG-TR-89-6, University of Toronto Department of Computer Science (1989)
14. Ge, H., Du, W., Qian, F., Liang, Y.: Identification and control of nonlinear systems by a time-delay recurrent neural network. Neurocomputing 72, 2857–2864 (2009)
15. Giles, C., Kuhn, G., Williams, R.: Dynamic recurrent neural networks: theory and applications. IEEE Trans. Neural Netw. 5(2), 153–156 (1994)
16. Giles, C.L., Chen, D., Miller, C.B., Chen, H.H., Sun, G.Z., Lee, Y.C.: Second-order recurrent neural networks for grammatical inference. In: 1991 IEEE INNS International Joint Conference on Neural Networks, Seattle, Piscataway, NJ, vol. 2, pp. 271–281. IEEE Press (1991)
17. Giles, C.L., Horne, B.G., Lin, T.: Learning a class of large finite state machines with a recurrent neural network. Neural Networks 8, 1359–1365 (1995)
18. Giles, C.L., Miller, C.B., Chen, D., Chen, H.H., Sun, G.Z., Lee, Y.C.: Learning and extracting finite state automata with second-order recurrent neural networks. Neural Computation 4, 395–405 (1992)
19. Gori, M., Bengio, Y., Mori, R.D.: Bps: A learning algorithm for capturing the dynamic nature of speech. In: International Joint Conference on Neural Networks, vol. II, pp. 417–423 (1989)
20. Harigopal, U., Chen, H.C.: Grammatical inference using higher order recurrent neural networks. In: Proceedings of the Twenty-Fifth Southeastern Symposium on System Theory, SSST 1993, pp. 338–342 (1993)
21. Hinton, T.J., abd Sejnowski, G.E.: Optimal perceptual inference. In: Proceedines of the IEEE Conference on Computer Vision and Pattern Recognition, pp. 448–453. IEEE Computer Society (1983)
22. Hochreiter, S.: Untersuchungen zu dynamischen neuronalen Netzen. Diploma thesis (1991)

23. Hochreiter, S., Schmidhuber, J.: Long short-term memory. Neural Computation 9, 1735–1780 (1997)
24. Hopcroft, J.E., Ullman, J.D.: Introduction to Automata Theory, Languages, and Computation. Addison-Wesley (1979)
25. Jordan, M.I.: Supervised learning and systems with excess degrees of freedom. Technical Report COINS Technical Report 88–27, Massachusetts Institute of Technology (1988)
26. Karakasoglu, A., Sudharsanan, S., Sundareshan, M.K.: Identification and decentralized adaptive control using dynamic neural networks with application to robotic manipulators. IEEE Trans. Neural Networks 4, 919–930 (1993)
27. Karray, F.O., Silva, C.: Soft Computing and Intelligent Systems Design. Addison Wesley (2004)
28. Kilian, J., Siegelmann, H.T.: On the power of sigmoid neural networks. In: Proceedings of the Sixth ACM Workshop on Computational Learning Theory, pp. 137–143. ACM Press (1993)
29. Kolen, J.F., Kremer, S.C. (eds.): A Field Guide to Dynamical Recurrent Networks. Wiley-IEEE Press (2001)
30. Kuo, J., Celebi, S.: Adaptation of memory depth in the gamma filter. In: Acoustics, Speech and Signal Processing IEEE Conference, pp. 1–4 (1994)
31. Kuroe, Y.: Representation and Identification of Finite State Automata by Recurrent Neural Networks. In: Pal, N.R., Kasabov, N., Mudi, R.K., Pal, S., Parui, S.K. (eds.) ICONIP 2004. LNCS, vol. 3316, pp. 261–268. Springer, Heidelberg (2004)
32. Lippmann, R.P.: An introduction to computing with neural nets. IEEE ASSP Magazine 4, 4–22 (1987)
33. Mozer, M.: A focused background algorithm for temporal pattern recognition. Complex Systems 3 (1989)
34. Mozer, M.C.: Induction of multiscale temporal structure. In: Advances in Neural Information Processing Systems 4, pp. 275–282. Morgan Kaufmann (1992)
35. Nguyen, M., Cottrell, G.: A technique for adapting to speech rate. In: Kamm, C., Kuhn, G., Yoon, B., Chellapa, R., Kung, S. (eds.) Neural Networks for Signal Processing 3. IEEE Press (1993)
36. Omlin, C.W., Giles, C.L.: Constructing deterministic finite-state automata in recurrent neural networks. Journal of the ACM 43(6), 937–972 (1996)
37. Patan, K.: Locally Recurrent Neural Networks. In: Patan, K. (ed.) Artificial. Neural Net. for the Model. & Fault Diagnosis. LNCIS, vol. 377, pp. 29–63. Springer, Heidelberg (2008)
38. Pollack, J.B.: On Connectionist Models of Natural Language Processing. PhD thesis, Computer Science Department of the University of Illinois at Urbana-Champaign, Urbana, Illinois, Available as TR MCCS-87-100, Computing Research Laboratory, New Mexico State University, Las Cruces, NM (1987)
39. Principe, J.C., de Vries, B., de Oliveira, P.G.: The gamma filter - a new class of adaptive IIR filter with restricted feedback. IEEE Transactions on Signal Processing 41, 649–656 (1993)
40. Renals, S., Rohwer, R.: A study of network dynamics. Journal of Statistical Physics 58, 825–848 (1990)
41. Robinson, A.J.: Dynamic Error Propagation Networks. Ph.d., Cambridge University Engineering Department (1989)
42. Rumelhart, D.E., Hinton, G.E., Williams, R.J.: Learning internal representations by error propagation. In: Parallel Distributed Processing. MIT Press, Cambridge (1986)
43. Schmidhuber, J.H.: A fixed size storage $o(n^3)$ time complexity learning algorithm for fully recurrent continually running networks. Neural Computation 4(2), 243–248 (1992)

44. Sejnowski, T.J., Rosenberg, C.R.: Parallel networks that learn to pronounce english text. Complex Syst. I, 145–168 (1987)
45. Shannon, C.E.: Communication in the presence of noise. Proc. Institute of Radio Engineers 37(1), 10–21 (1949); reprinted as classic paper in: Proc. IEEE 86(2) (February 1998)
46. Shearer, J.L., Murphy, A.T., Richardson, H.H.: Introduction to System Dynamics. Addison-Wesley, Reading (1971)
47. Siegelmann, H.T., Sontag, E.D.: Turing computability with neural nets. Applied Mathematics Letters 4(6), 77–80 (1991)
48. Silva, T.O.: Laguerre filters - an introduction. Revista do Detua 1(3) (1995)
49. Smith, J.O.: Delay lines. Physical Audio Signal Processing (2010), http://ccrma.stanford.edu/ jos/pasp/ Tapped_Delay_Line_TDL.htm (cited November 28, 2010)
50. Smith, S.W.: The scientist and engineer's guide to digital signal processing. California Technical Publishing (2006), http://www.dspguide.com/ch15.htm (cited November 29, 2010)
51. Tsoi, A.C., Back, A.D.: Locally recurrent globally feedforward networks: A critical review of architectures. IEEE Transactions on Neural Networks 5, 229–239 (1994)
52. Waibel, A., Hanazawa, T., Hinton, G., Shikano, K., Lang, L.: Phonemic recognition using time delay neural networks. IEEE Trans. Acoustic Speech and Signal Processing 37(3), 328–339 (1989)
53. Werbos, P.: Beyond Regression: New Tools for Prediction and Analysis in the Behavioural Sciences. Phd thesis, Harvard University (1974)
54. Williams, R.J., Zipser, D.: A learning algorithm for continually running fully recurrent neural networks. Neural Computation 1, 270–289 (1989)
55. Won, S.H., Song, I., Lee, S.Y., Park, C.H.: Identification of finite state automata with a class of recurrent neural networks. IEEE Transactions on Neural Networks 21(9), 1408–1421 (2010)
56. Yan, P.F., Zhang, C.S.: Artificial Neural Network and Simulated Evolutionary Computation. Thinghua University Press, Beijing (2000)
57. Zamarreno, J.M., Vega, P.: State space neural network. Properties and application. Neural Networks 11, 1099–1112 (1998)
58. Zeng, Z., Goodman, R.M., Smyth, P.: Learning finite state machines with self-clustering recurrent networks. Neural Computation 5(6), 977–990 (1993)
59. Zeng, Z., Goodman, R.M., Smyth, P.: Discrete recurrent neural networks for grammatical inference. IEEE Transactions on Neural Networks 5(2), 320–330 (1994)

Chapter 3
Supervised Neural Network Models for Processing Graphs

Monica Bianchini and Marco Maggini

An intelligent agent interacts with the environment where it lives taking its decisions on the basis of sensory data that describe the specific context in which the agent is currently operating. These measurements compose the environment representation that is processed by the agent's decision algorithms and, hence, they should provide sufficient information to yield the correct actions to support the agent's life. In general, the developed environment description is redundant to provide robustness with respect to noise and eventual missing data. On the other hand, a proper organization of the input data can ease the development of successful processing and decisional schemes.

Therefore, input data encoding is a crucial step in the design of an artificial intelligent agent. This is particularly true for agents featuring also learning capabilities, since an appropriate data representation, that faithfully preserves the environment properties, can make the extraction of general rules from the available examples simpler. The most natural organization of the input data depends on the specific task. In some cases, the decision is taken on a single event that is described by a certain number measurements, encoded as a *vector* of real numbers. For instance, if the agent should decide if a piece of food is good, a vector collecting measurements related to its color, temperature, and chemical properties as detected by olfactory and/or taste sensors is likely to provide the relevant information for the decision. When the event is a sound and the agent should interpret if it can be the hint of a possible threat, or the presence of a prey, or a message from another agent, the most appropriate input data model is a *temporal sequence* of measurements, like, for instance, the signal spectrogram. When considering simple visual tasks, as recognizing the shape of a potential predator, a two dimensional spatial sensory input can be processed as encoded by a bidimensional *matrix*, whose entries measure the luminance or color of a given point in the space projected by the agent visual system

M. Bianchini · M. Maggini
Dipartimento di Ingegneria dell'Informazione e Scienze Matematiche,
Università degli Studi di Siena, Italy
e-mail: {monica,maggini}@dii.unisi.it

M. Bianchini et al. (Eds.): *Handbook on Neural Information Processing*, ISRL 49, pp. 67–96.
DOI: 10.1007/978-3-642-36657-4_3   © Springer-Verlag Berlin Heidelberg 2013

onto the retina. However, when more complex tasks, such as scene interpretation, are considered, a successful approach may require to exploit structured data, that provide a more balanced split of the intrinsic complexity between the task of data representation and that of data processing. In these cases, a *graph* can describe more precisely the environment, by modeling it through a set of entities and relationships among them, that are encoded by the graph nodes and arcs, respectively. For instance, in a vision task, the nodes may represent parts of objects (as regions having homogenous visual features), whereas the arcs may encode the spatial relationships among pairs of adjacent regions.

Hence, the design of an intelligent learning agent requires to jointly devise an optimal feature representation of the task inputs and the related processing/learning algorithm. In fact, even if it is possible to obtain simpler representations from complex data structures, this process always implies the loss of information, an inaccurate modeling of the problem at hand or a higher complexity in the input data distribution. For instance, given an input representation encoded as a tree, it is possible to map it to a vector by performing a visit of its nodes. However, such a representation would not be suitable for trees with a varying number of nodes that would yield vectors with different dimensions, unless a maximum number of nodes is assumed and a padding technique is used to encode the missing nodes. Moreover, the relationships among the nodes in the tree would be encoded by their position in the vector, making it more difficult to exploit the structural regularities by an algorithm designed to process generic vectors.

In this chapter, we will show how an agent based on artificial neural networks (ANNs) can be designed in order to naturally process structured input data encoded as graphs. Graph Neural Networks (GNNs) [23] are an extension of classical MultiLayer Perceptrons (MLPs) that accept input data encoded as general undirected/directed labeled graphs. GNNs are provided with a supervised learning algorithm that, beside the classical input-output data fitting measure, incorporates a criterion aimed at the development of a contractive dynamics, in order to properly process the cycles in the input graph. A GNN processes a graph in input and it can be naturally employed to compute an output for each node in the graph (*node–focused* computation). The training examples are provided as graphs for which supervisions are given as output target values for a subset of their nodes. This processing scheme can be adapted to perform a *graph–based* computation in which only one output is computed for the whole graph.

For instance, an object detection problem may be faced as a node–focused task (see figure 1-a). The input image can be represented by a Region Adjacency Graph (RAG) [5, 13], where each homogenous region, extracted by an image segmentation algorithm, is associated to a node and the edge between two nodes encodes the adjacency relationship among them. The label assigned to the node represents visual features of the associated region (color, shape, etc.), whereas the edge labels may encode parameters related to the mutual spatial position of the two adjacent regions (distance of the barycenters, rotation angles, etc.). The GNN computes a label for each node in the graph, stating if the corresponding region is a part of the object to be detected. A post-processing can be applied to the GNN output to find the node sets that match the target object profile, providing also the localization of the object in the input image.

As an example of a graph–based problem, we can consider the prediction of some chemical property of a molecule (see figure 1-b). In this case, the graph nodes model the atoms or the atomic groups in the molecule, whereas the edges represent the chemical bonds between them. Each node (edge) can store a label that describes the physicochemical properties of the atom (bond). The GNN processes an input graph, modeling a specific molecule, yielding a prediction for the presence of the target chemical property. The training is performed by providing a learning set containing graphs (molecules) for which the value of the considered chemical property is known a priori.

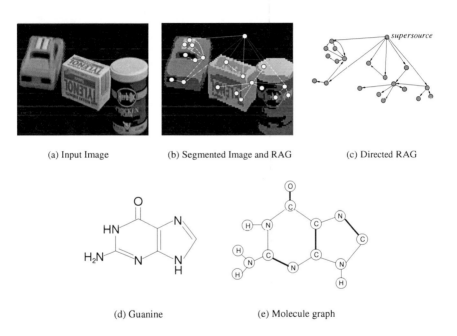

(a) Input Image   (b) Segmented Image and RAG   (c) Directed RAG

(d) Guanine   (e) Molecule graph

Fig. 1 Examples of structured data encodings. (a-c) An object detection and localization problem. The input image is encoded by a Region Adjacency Graph. (d-e) A chemical compound and its graph encoding (nodes are described by the features of the corresponding atom, edges are enriched with features representing the bond type — thicker lines correspond to double bonds).

The chapter describes in details the GNN model, showing how simpler models can be derived from this more general computational scheme. In particular, when considering acyclic graphs, such as *Directed Acyclic Graphs with Labeled Edges* (DAGs-LE), *Directed Positional Acyclic Graphs* (DPAGs), or generic *Positional Trees* (PTs), the model reduces to that of *Recursive Neural Networks* (RNNs) [6, 11]. Also Recurrent Neural Networks, that are designed to process temporal input sequences [1, 18, 19, 29] can be viewed as a specific case of GNNs, since sequences can be represented as linked lists of nodes, where arcs encode the *followed-by* relationship. Finally, MLPs processing vectors can be viewed as a GNN model processing graphs containing a single node.

The chapter is organized as follows. The next section describes the input representation as graphs and introduces the required notation. Section 2 details the neural network architectures for processing structured information. In particular, the Graph Neural Network model is introduced for processing generic labelled graphs, together with simpler models that can be used with less general cases, such as acyclic graphs. Then, Section 3 presents the supervised learning algorithm for GNNs, showing how it can be viewed as a generalization of the classical BackPropagation algorithm for MLPs. Finally, Section 4 summarizes the main concepts presented in this chapter, reviewing some applications reported in the literature.

1 Graphs

A graph is a data model that allows the representation of complex structures as a set of elements and binary relationships among them. Nodes represent the basic entities that compose the atomic parts of the information, whereas an edge encodes a property that links a pair of entities. In general, different types of relations can be defined on the pairs of nodes, such that the same two nodes may be connected by a set of edges each representing a different relationship. For instance, if two nodes stand for two regions in an image, they can be connected by an edge encoding the *adjacent-to* relationship and an arc for the *on-top-of* property. As shown in this example, the relationships encoded by the graph edges may be *symmetrical* (*adjacent-to*), or *asymmetrical* (*on-top-of*). The two cases would correspond to an *undirected* edge and a *directed* arc between the two nodes, respectively. In the following, we formally introduce the graph data model, considering specific choices to provide a simple and uniform representation able to deal with multiple types of relationships, that can be both symmetrical or asymmetrical.

A *directed unlabeled graph* is defined by the pair $G_U = (V, E)$, where V is the finite set of nodes and $E \subseteq V \times V$ collects the arcs. An arc from node u to node v is a directed link represented by the ordered pair $(u, v) \in E$, $u, v \in V$. It is clear that an arc can encode an asymmetrical relationship between the two nodes, since each node can be associated to a specific role by its position in the pair. For instance, if the arc encodes the *on-top-of* property, the first node u will represent the region that is on the top of the region corresponding to node v. An undirected graph can be conveniently represented as a directed graph by substituting each undirected edge with a pair of directed arcs: an edge between nodes u and v will correspond to the two directed arcs (u, v) and (v, u). Hence, if we need to model symmetrical relationships, we will need to explicitly encode the symmetry by adding the two directed arcs. For instance, the *adjacent-to* relationship will be represented by two arcs stating that the region of node u is adjacent to that of node v and viceversa. Even if this choice does not provide a natural representation of symmetric relationships, the resulting model allows a uniform and simple encoding of both symmetric and asymmetric relationships.

The pair (V, E) specifies the underlying structure of the data organization through the topology of the connections among the nodes. This model can be enriched to incorporate attributes or measures that may be available for the entities represented by the graph nodes and/or for the relationships encoded by the arcs. In particular, attributes on the arcs may allow us to encode the presence of different types of relationships between two nodes, as it is detailed in the following.

First, a label can be attached to each node in the graph, encoding the values of a set of properties that are measured for the entity represented by the node. For instance, if a node is a model of a region in an image, features describing perceptual and geometrical properties can be stored in the node to describe same visual characteristics of the region (e.g. average color, shape descriptors, area of the region, perimeter of the region contour, etc). In general, a different set of attributes can be attached to each node, but in the following we will assume that the labels are chosen from the same label space L. Since we will deal with connectionist models that are devised to process information encoded by real numbers, in the following we will consider labels modeled as vectors of reals, i.e. $L \subset \mathbb{R}^m$. Given a node label space, a *directed labeled graph* is a triple $G_L = (V, E, \mathscr{L})$, where V and E are the set of nodes and arcs respectively, and $\mathscr{L} : V \to L$ is a node labeling function which defines the label $\mathscr{L}(v) \in L$ for each node v in the graph. The label of node v will be compactly denoted by $\mathbf{l}_v \in \mathbb{R}^m$.

Moreover, a label can also be associated to each arc (u, v) in the graph, to enrich its semantics. A graph with labeled edges can encode also attributes attached to the relationships between pairs of nodes. For instance, we can associate a label to the arc encoding the *adjacent-to* relationship, in order to specify a set of features that describe the mutual position of the two regions (e.g. distance of the region barycenters, angles between the region principal axes, etc.). In particular, the label can also be exploited to encode the types of relationships existing between two nodes, thus providing a uniform method to model the presence of different kinds of relationships in the data. In the previous example concerning image representation, if we would like to encode both the relationships *adjacent-to* and *on-top-of* between pairs of regions, we need just to add two boolean attributes to the arc stating if the associated type of relationship is applicable. In general, we will assume that the labels for the arcs belong to a given edge label space L_e. A *directed labeled graph with labeled edges* is defined by a quadruple $G_{LE} = (V, E, \mathscr{L}, \mathscr{E})$, where the edge labeling function $\mathscr{E} : E \to L_e$ attaches a label $\mathscr{E}((u, v)) \in L_e$ to the arc $(u, v) \in E$. More compactly, the label attached to the arc (u, v) will be denoted as $\mathbf{l}_{(u,v)}$. Notice that, in general, the two directed arcs (u, v) and (v, u) can have different labels. When dealing with Artificial Neural Networks the attributes must be encoded with real valued vectors, such that $L_e \subset \mathbb{R}^d$. In this case, if T different types of relationships can be present between a pair of nodes, T entries of the vector may be allocated to encode the actual types that are applicable for the pair (u, v). Each entry will be associated to a specific type of relationship, such that its value is 1 if it is present and 0 otherwise. For instance, given the two relationships *adjacent-to* and *on-top-of*, the configuration of the label entries $[1,0]$ will encode that the region u is adjacent to region v but not on its top, whereas $[1,1]$ will represent the fact that the two regions are adjacent and region u is on the top of region v. The configuration $[0,0]$ would make no sense if *adjacent-to* and *on-top-of* are the only modeled relationships between pairs of regions, as, in this case, a correct modeling would require to omit the arc between the two nodes. Finally, the entries $[0,1]$ may be used if we suppose that the relation *on-top-of* does not need the regions to be adjacent.

Properties deriving from the graph topology are defined as follows. Given a node $v \in V$, the set of its *parents* is defined as $pa[v] = \{w \in V | (w,v) \in E\}$, whereas the set of its *children* is $ch[v] = \{w \in V | (v,w) \in E\}$. The *outdegree* of v is the cardinality of $ch[v]$, $od[v] = |ch[v]|$, and $o = \max_v od[v]$ is the maximum outdegree in the graph. The *indegree* of node v, is the cardinality of $pa[v]$ ($|pa[v]|$). Nodes having no parents (i.e. $|pa[v]| = 0$) are called *sources*, whereas nodes having no children (i.e. $|ch[v]| = 0$) are referred to as *leaves*. The class of graphs with maximum indegree i and maximum outdegree o is denoted as $\#^{(i,o)}$. Moreover, the class of graphs with bounded indegree and outdegree (but unspecified) is indicated as $\#$. The set of the *neighbors* of node v is defined as $ne[v] = ch[v] \bigcup pa[v]$, i.e. the neighborhood of v contains all the nodes connected to it by an arc independently of its direction. Given a labeled graph G_L, the structure obtained by ignoring the node and/or edge labels will be referred to as the *skeleton* of G_L, denoted as $ske(G_L)$. Finally, the class of all the data structures defined over the domain of the labeling function $\mathscr{L}$ and skeleton in $\#^{(i,o)}$ will be denoted as $\mathscr{L}^{\#^{(i,o)}}$ and will be referred to as a structured space.

A *directed path* from node u to node v in a graph G is a sequence of nodes $(w_1, w_2, \ldots, w_p)$ such that $w_1 = u$, $w_p = v$ and the arcs $(w_i, w_{i+1}) \in E$, $i = 1, \ldots, p - 1$. The *path length* is $p - 1$. If there exists at least one path such that $w_1 = w_p$, the graph is *cyclic*. If the graph is *acyclic*, we can define a partial ordering, referred to as *topological order*, on the set of nodes V, such that $u \prec v$ if u is connected to v by a directed path. The set of the *descendants* of a node u, $desc(u) = \{v \in V | u \prec v\}$, contains all the nodes that follow v in the partial ordering. The *ancestors* of a node u, $anc(u) = \{v \in V | v \prec u\}$, contains all the nodes that precede v in the partial ordering.

The class of *Directed Acyclic Graphs* (DAGs) is particularly interesting since it can be processed by employing simpler computational schemes. In particular, *Directed Positional Acyclic Graphs* (DPAGs) form a subclass of DAGs for which an injective function $o_v : ch[v] \to [1,o]$ assigns a position $o_v(c)$ to each child c of a node v. Therefore, a DPAG is represented by the tuple $(V, E, \mathscr{L}, \mathscr{O})$, where $\mathscr{O} = \{o_1, \ldots, o_{|V|}\}$ is the set of functions defining the position of the children for each node. Since the codomain for each function $o_v(c)$ is the interval $[1,o]$, if the outdegree of v is less than the graph maximum outdegree o, i.e. $|ch[v]| < o$, some positions will be empty. The missing links will be encoded by using *null point-ers* (NIL). Hence, in a DPAG the children of each node v can be organized in a fixed size vector $ch[v] = [ch_1[v], \ldots, ch_o[v]]$, where $ch_k[v] \in V \bigcup \{NIL\}$, $k = 1, \ldots, o$. Finally, we denote with PTREEs the subset of DPAGs that contains graphs which are trees, i.e. there is only one node with no parents, the *root* node r ($|pa[r]| = 0$), and any other node v has just one parent ($|pa[v]| = 1$). Instead, *Directed Acyclic Graphs with Labeled Edges* (DAGs–LE) represent the subclass of DAGs for which an edge labeling function $\mathscr{E}$ is defined. In this case it is not required to define an ordering among the children of a given node, since an eventual order can be encoded in the arc labels. Finally, we denote with TREEs–LE the subset of DAGs–LE that contains graphs which are trees.

For *graph–focused* processing tasks, the graph G may be required to possess a *supersource*, that is a node $s \in V$ such that any other node in G can be reached by a directed path starting from s. Note that, if the graph does not have a supersource, it is still possible to define a convention for adding an extra node s with a minimal number of outgoing edges, such that s is a supersource for the expanded graph [26].

2 Neural Models for Graph Processing

The Graph Neural Network (GNN) model defines a computational scheme that allows the processing of an input graph G_{LE} in order to implement a *node–focused* function. Formally, a GNN realizes a function $\tau(G_{LE}, v)$ that maps the input graph G_{LE} and one of its nodes v to a real valued vector in $\mathbb{R}^r$. The computational scheme at the basis of the GNN model is that of information diffusion. A computational unit is associated to each node in the input graph, and the units are interconnected to each other following the link topology of the input graph. In order to define the computation of the GNN model we introduce the concepts of state and state transition functions.

2.1 The Graph Neural Network Model

The information needed to perform the overall computation is locally stored in a *state variable* $\mathbf{x}_v$ for each node v. The set of the state variables, available for all the computational units attached to the graph nodes, represents the current state of the computation. The state of each node $\mathbf{x}_v$ is an s–dimensional vector that is supposed to encode the information relevant for node v and its context. Hence, the state $\mathbf{x}_v \in \mathbb{R}^s$ should depend on the information contained in the neighborhood of v, defined by the nodes that link or arc linked by v.

Following the idea that the global computation emerges from the diffusion of the information embedded into the states of the graph nodes, the model assumes that the state variables can be computed locally at each node depending on the states of its neighbors and on the node's local information and connectivity (see Fig. 2). Basically, this framework requires the definition of a *state transition function f* that models the dependence of $\mathbf{x}_v$ with respect to the context of node v. The function will depend on the label $\mathbf{l}_v$ of node v, the labels of the incoming arcs $\mathbf{l}_{(u,v)}$, those of the outgoing arcs $\mathbf{l}_{(v,u)}$, and on the states of the node's neighbors $\mathbf{x}_{ne[v]}$. In a learning framework, the transition function will also depend on a set of learnable parameters $\theta_f \in \mathbb{R}^p$. Hence, in general, the state computation will exploit a *state transition function* defined as

$$\mathbf{x}_v = f(\mathbf{x}_{ne[v]}, \mathbf{l}_{(v,ch[v])}, \mathbf{l}_{(pa[v],v)}, \mathbf{l}_v | \theta_f) . \tag{1}$$

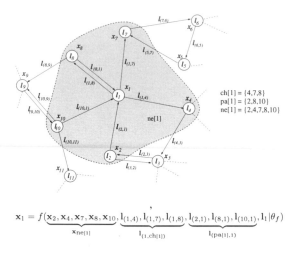

$$x_1 = f(\underbrace{x_2, x_4, x_7, x_8, x_{10}}_{x_{ne[1]}}, \underbrace{l_{(1,4)}, l_{(1,7)}, l_{(1,8)}}_{l_{(1,ch[1])}}, \underbrace{l_{(2,1)}, l_{(8,1)}, l_{(10,1)}}_{l_{(pa[1],1)}}, l_1 | \theta_f)$$

Fig. 2 The neighborhood of a node exploited for the state computation when processing a directed graph in input

This model can be made more general by exploiting other definitions of neighborhood that imply the dependence of the function f from additional variables. For instance, the proposed formulation assumes that the labels of the nodes in ne[v] are properly encoded by the corresponding states $x_{ne[v]}$, but we can rewrite f such that this dependence is made explicit. Moreover, the concept of neighborhood exploited in eq. (1) just includes the nodes one link away from the current unit v. In general, we can redefine the context by including the nodes two or more links away from the current node. Anyway, such a model would not be minimal since we can assume that the information of larger contexts is properly encoded by the states in ne[v]. Finally, the model of eq. (1) assumes that the same computational unit is applied to any node in the graph, since both the function model and the parameters θ_f are the same for all the nodes in V. We can relax this requirement provided that there is an explicit rule that allows us to choose a specific function and the associated parameters given some node's properties (e.g. the node in–degree and/or out–degree). In general, the function f can be implemented by a neural network and the set of trainable parameters θ_f will represent the neural network connection weights. Apart from the constraints on the number and type of inputs and outputs of the neural network, there are no other assumptions on its architecture (type of neurons, number of layers, etc.). We will give more details on the possible implementations of f in the following.

The result of the state computation for a given input graph G is not completely defined until we do not devise an appropriate global scheme for applying the local state transition function f. In fact, due to the presence of cycles in the input graph there is a circular dependence of the variables x_v on their own values. This is evident just by the definition of neighborhood, since the value of x_v depends on the values of the neighbors' states $x_{ne[v]}$ whose computation depends on x_v itself. The state variables for the nodes in a given input graph can be stacked into a vector $\mathbf{x} \in \mathbb{R}^{s|V|}$. Similarly, the labels defined on the input graph nodes and arcs can be collected into the vector $\mathbf{l} \in \mathbb{R}^{m|V|+d|E|}$. With this notation, the collective state computation can be written as the solution of the following vectorial equation

$$\mathbf{x} = F(\mathbf{x}, \mathbf{l} | \theta_f) , \tag{2}$$

where F is the *global transition function*, whose components are obtained by concatenating the single functions f. The actual parameters for each function f in F are easily obtained by projecting the needed components of $\mathbf{x}$ and $\mathbf{l}$, as required by eq. (1). This equation shows that the state resulting from the computation is the solution of a system of non–linear equations in the variables $\mathbf{x}$. In fact, the state $\mathbf{x}$ is both on the left and the right side of eq. (2). This is due to the fact that the state for each node v depends on the states of its neighbors in ne[v] that, on turn, depend on the state of v. This cyclic dependence is independent on the actual presence of explicit cycles in the graph, since the considered model for the transition function exploits the states of the neighbor nodes despite the direction of the arcs.

In general, eq. (2) may have multiple solutions that can be single points or a manifold in the state space. However, an useful computation requires that the final state is uniquely defined. This requirement can be met by satisfying the conditions of the Banach fixed point theorem [14]. The theorem states that, if F is a *contraction map* with respect to the state variables $\mathbf{x}$, then the system in eq. (2) has a unique solution. The global transition function is a contraction map if there exists μ, $\mu \in [0,1)$ such that $\|F(\mathbf{x}, \mathbf{l} | \theta_f) - F(\mathbf{y}, \mathbf{l} | \theta_f)\| \leq \mu \|\mathbf{x} - \mathbf{y}\|$, for any $\mathbf{x}$ and $\mathbf{y}$ in the state space.

The Banach's fixed point theorem provides also a method to compute the solution of the system in eq. (2). Given that F is a contraction map, the unique solution can be found by employing an iterative state update scheme, defined by the equation

$$\mathbf{x}(t+1) = F(\mathbf{x}(t), \mathbf{l} | \theta_f) , \tag{3}$$

where $\mathbf{x}(t)$ is the state at iteration t. The theorem guarantees that the sequence $\mathbf{x}(t)$ converges exponentially to the solution of the non-linear system (2). This computational scheme motivates the term *state transition function* for the function f; in fact, given an input graph G, the states at its nodes are updated by the attached computational units as

$$\mathbf{x}_v(t+1) = f(\mathbf{x}_{\text{ne}[v]}(t), \mathbf{l}_{(v,\text{ch}[v])}, \mathbf{l}_{(\text{pa}[v],v)}, \mathbf{l}_v | \theta_f) , \tag{4}$$

until all the state vectors $\mathbf{x}_v(t)$ converge to the fixed point. It should be noted that the final result of the computation depends on the input graph, since the function F depends both on the graph topology and on the node and arc labels. Whereas the dependence on the labels is clear, since they are explicit parameters of the transition function, the contribution of the graph topology is not directly evident, being due to the computation scheme based on the information diffusion among the neighboring nodes. This scheme gives rise to a computational structure made up of

computational units that are interconnected following the link topology of the input
graph. When processing a different graph, the units will be the same but their computation will be arranged in a different way. The resulting computational structure
is a network that is referred to as *encoding network* (see Fig. 3), since it encodes
relevant information on the input graph into the state vectors. The encoding network is built by placing a computational unit f in each node in the graph, and by
interconnecting them following its arcs. By applying the same transition function
to different input graphs, we obtain different encoding networks, featuring the same
building block but assembled with a different structure. The current states $\mathbf{x}_v(t)$,
$v \in V$, are stored in each computational unit that calculates the new state $\mathbf{x}_v(t+1)$
using the information provided by both the node label and its neighborhood.

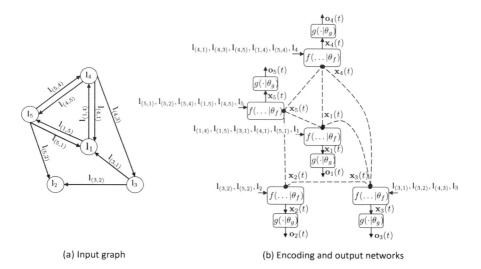

(a) Input graph   (b) Encoding and output networks

Fig. 3 The *encoding* and the *output* networks (b) associated to an input graph (a). The *graph
neural network*, implementing the functions $f(\ldots, \theta_f)$ and $g(\cdot, \theta_g)$, is replicated following
the structure of the graph and its weights, θ_f and θ_g, are shared among all the replicas.

Hence, the output of the state propagation is a graph having the same skeleton
of the input graph G. The states $\mathbf{x}_v$ are essentially new labels attached to the nodes
of G. As a final remark, it should be noted that all the computational units in the
encoding network share the same weights θ_f.

The GNN model considers also a *local output function g*, that maps the states to
the actual output of the computation for each node. This function will be applied
at each node after the states have converged to the fixed point and will, in general,
depend on a set of parameters θ_g, that can be tuned by the learning algorithm. Hence,
the output $\mathbf{o}_v$, for each node $v \in V$, will be computed as

$$\mathbf{o}_v = g(\mathbf{x}_v | \theta_g) , \tag{5}$$

where $\mathbf{o}_v$ belongs to the output space. In the following we will consider $\mathbf{o}_v \in \mathbb{R}^r$. The function g can be computed for each node in the input graph G, thus yielding an output graph with the same skeleton of G and the nodes labeled with the values $\mathbf{o}_v$. In this case the GNN network realizes a transduction from a graph G to a graph G', such that $skel(G) = skel(G')$. Otherwise, the output can be computed only for a given node in the input graph realizing the node–focused function $\tau(G,v)$, from the space of labeled graphs to $\mathbb{R}^r$, defined as $\tau(G,v) = g(\mathbf{x}_v|\theta_g)$. Figure 3 shows how the output function is combined with the state encoding process to yield the final result of the computation. A graph–focused computation can be obtained by properly combining the outputs $\mathbf{o}_v$ for all the nodes $v \in V$ (f.i. we can use the average operator), or by choosing a predefined node with specific properties, usually the supersource s, such that the final output is $\mathbf{o}_s$.

As shown in Figure 3, the functions f and g define the *Graph Neural Network*. In particular, the connections among the units computing the function f define the dependences among the variables in the connected nodes. In fact, the recursive connections define the topology of the encoding network, stating the modality of combination of the states of the neighbors of each node. The parameters θ_f and θ_g are the trainable connection weights of the graph neural network, being θ_f and θ_g independent of node v. The parametric representations of f and g can be implemented by a variety of neural network models as detailed in the following.

The function g can be implemented by any MultiLayer Perceptron (MLP) with s inputs and r outputs. The function g does not need to meet any other particular requirement, and, hence, there will be no restrictions on the MLP implementing it. The parameter vector θ_g will collect the MLP connection weights. Instead, the function f should be selected to guarantee that the global transition function F is a contraction map with respect to the state $\mathbf{x}$. Moreover, the implementation of f should take into account that the number of input arguments may vary with the processed node v. In fact, the size of the neighborhood $\text{ne}[v]$ and the number of incoming and outgoing arcs can be different among the nodes in the graph.

Let us first consider the case of *positional graphs* with bounded indegree and outdegree, i.e. the set $\#^{(i,o)}$ as defined in section 1. In this case the set $\text{ne}[v]$ will contain at most $i + o$ nodes, that can be ordered on the basis of a position assigned to each node in $\text{ne}[v]$. Basically, the state transition function will process at most i states of nodes in $\text{pa}[v]$, and the related arc labels $\mathbf{l}_{(\text{pa}[v],v)}$, and a maximum of o states for the nodes in $\text{ch}[v]$, with the attached labels $\mathbf{l}_{(v,\text{ch}[v])}$. Hence, the local transition function can be rewritten as

$$\mathbf{x}_v(t+1) = f\big(\mathbf{x}_{\text{ch}_1[v]}(t),\dots,\mathbf{x}_{\text{ch}_o[v]}(t),\mathbf{x}_{\text{pa}_1[v]}(t),\dots,\mathbf{x}_{\text{pa}_i[v]}(t),$$
$$\mathbf{l}_{(v,\text{ch}_1[v])},\dots,\mathbf{l}_{(v,\text{ch}_o[v])},\mathbf{l}_{(\text{pa}_1[v],v)},\dots,\mathbf{l}_{(\text{pa}_i[v],v)},\mathbf{l}_v|\theta_f\big)\,, \tag{6}$$

where the function arguments are mapped to the associated graph variables, taking into account the position map defined for the node v. When processing a node having an indegree (outdegree) less than the maximum value i (o), the missing positions will be padded using specific constants for the state ($\mathbf{x}_{nil} \in \mathbb{R}^s$) and for the arc labels ($\mathbf{l}_{nil} \in \mathbb{R}^d$). The transition function of eq. (6) has a predefined number of arguments, and can be implemented by an MLP with $(s+d) \cdot (i+o) + m$ inputs and s outputs. When processing undirected graphs, there is no distinction between parent and child nodes. In this case, it is preferable to avoid to transform the undirected graph into a directed one, that would just lead to a duplication of the function arguments and make the implementation more complex, without adding any useful information in the computation.

The positional implementation is meaningful when a significant positional map is actually available for any graph to be processed by the GNN, and the node indegrees/outdegrees do not vary too much among the nodes to be processed. In fact, when the number of null pointers in the input padding is highly variable, the learning process may be hindered if we exploit only one positional state transition function. In this specific case, we can use a different function f (i.e. a different MLP) for processing nodes featuring the indegree and the outdegree in a given range. The number of functions needed for this implementation depends on the actual distribution of the in degrees/outdegrees. For instance, if only two configurations are possible, i.e. (i_1, o_1) and (i_2, o_2), then we may exploit two different implementations $f_{(i_1, o_1)}$ with $(s+d) \cdot (i_1 + o_1) + m$ inputs and $f_{(i_2, o_2)}$ with $(s+d) \cdot (i_2 + o_2) + m$ inputs, respectively. The two implementations will depend on two different sets of parameters $\theta_{f_{(i_1, o_1)}}$ and $\theta_{f_{(i_2, o_2)}}$. Clearly, when building the encoding network, the proper implementation will be selected for each node based on its indegree/outdegree configuration. This simple case can be extended to more general configurations involving ranges for the indegree/outdegree.

On the other hand, when the graph is not positional, the state transition function can be realized by a scheme that is independent on the neighborhood size. A simple solution is to use a state transition function implemented as

$$\mathbf{x}_v(t+1) = \sum_{u \in \text{ne}[v]} h(\mathbf{x}_u(t), \mathbf{l}_{(v,u)}, \mathbf{l}_{(u,v)}, \mathbf{l}_v | \theta_h) \,, \tag{7}$$

where the arc label is set to $\mathbf{l}_{nil}$ if the ingoing or the outgoing link is missing. In the case of undirected graphs, we can avoid to duplicate the arc label and simplify the function by removing one argument.

In the case of a *linear (non–positional) GNN*, the non–positional transition function h is implemented as

$$h(\mathbf{x}_u(t), \mathbf{l}_{(v,u)}, \mathbf{l}_{(u,v)}, \mathbf{l}_v | \theta_h) = \mathbf{A}_{v,u} \mathbf{x}_u + \mathbf{b}_v \tag{8}$$

where the matrix $\mathbf{A}_{v,u} \in \mathbb{R}^{s \times s}$ and the vector $\mathbf{b}_v \in \mathbb{R}^s$ are the output of two MLPs, whose connection weights correspond to the parameters of the GNN θ_h. In particular, the *transition neural network* computes the elements of the matrix $\mathbf{A}_{v,u}$ given the labels of the arcs between the nodes v and u, i.e. $\mathbf{l}_{(v,u)}$ and $\mathbf{l}_{(u,v)}$, and the label of the current node $\mathbf{l}_v$. The transition MLP will have $2d + m$ inputs and s^2 outputs, implementing a function $\phi : \mathbb{R}^{2d+m} \rightarrow \mathbb{R}^{s^2}$ that depends on a set of connection weights collected into the parameter vector θ_ϕ. The s^2 values computed by the function $\phi(\mathbf{l}_{(v,u)}, \mathbf{l}_{(u,v)}, \mathbf{l}_v | \theta_\phi)$ can be arranged into a $s \times s$ square matrix $\Phi_{v,u}$ such that the

(i, j) entry of $\mathbf{\Phi}_{v,u}$ is the component $i + j \cdot s$ of the vector computed by the function ϕ. Finally, the *transition matrix* is computed as

$$\mathbf{A}_{v,u} = \frac{\mu}{s|\text{ne}[u]|} \mathbf{\Phi}_{v,u} , \qquad (9)$$

where $\mu \in (0,1)$. This particular expression to compute the transition matrix guarantees that the resulting global transition map is a contraction, provided that the MLP implementing the function ϕ has a bounded activation function in the output neurons (e.g. a sigmoid ora a hyperbolic tangent) [23]. The *forcing term* $\mathbf{b}_v$ can be computed by a *forcing neural network*, that is an MLP that implements a function $\rho : \mathbb{R}^m \to \mathbb{R}^s$, such that

$$\mathbf{b}_v = \rho(\mathbf{l}_v|\theta_\rho) , \qquad (10)$$

where θ_ρ is the parameter vector collecting the connection weights of the MLP implementing the function ρ.

In a *non–linear (non–positional) GNN* the function h is implemented by an MLP having $s + 2d + m$ inputs and s outputs. In particular, three–layer MLPs are universal approximators [24] and, hence, a three–layer MLP can be employed to approximate any function h provided that a sufficient number of neurons is used. However, this solution does not necessarily implements a contraction map for any value of the MLP connection weights θ_h. Hence, when adopting this solution, the learning objective needs to include a cost term that penalizes the development of non–contractive mappings (see section 3 for the details).

2.2 Processing DAGs with Recursive Neural Networks

When dealing with directed acyclic graphs, we can devise a simpler processing scheme that does not need a relaxation procedure to compute the states $\mathbf{x}_v$ for the nodes in the input graph. The setting is a particular case of Graph Neural Networks for which the local state is updated using the transition function

$$\mathbf{x}_v = f(\mathbf{x}_{\text{ch}[v]}, \mathbf{l}_{(v,\text{ch}[v])}, \mathbf{l}_v|\theta_f) . \qquad (11)$$

In this model, the state information flows from the children to their parents following backwards the links between the nodes. Since there are no cycles, we can order the graph nodes using their *inverse topological order*. In fact, in the topological order, any node precedes all its descendants, i.e. all the nodes that can be reached from it following a direct path (see section 1). Hence, if the node u is an ancestor of node v, $u \prec v$ in the topological order. In particular, the parent always precedes its children. If, for a given input DAG G, we apply the transition function eq. (11) recursively following the inverse topological order of the nodes, it is guaranteed that when we process node v the states of its children have already been computed. In particular, the first nodes to be processed are the leaves, whose state only depends on the local label $\mathbf{l}_v$ (the order in which the leaves are processed is irrelevant).

Then, the computation is propagated to the upper levels of the graph until the source nodes are reached (or the supersource if there is only one source node).

Hence, in the case of DAGs the simplified version of the local state transition function allows us to define a processing scheme that does not require to iterate the state update equations to converge to a fixed point. Following the inverse topological order, the fixed point of the state computation is reached just after applying the local update equations only once. Given this general procedure for calculating the states for a given input graph, specific models are devised assuming some hypotheses on the structure of the function f and on the type of DAGs in input.

For graph–focused tasks, the output is computed at the supersource. Hence, the recursive neural network implements a function from the set of DAGs to $\mathbb{R}^r$, ψ : DAGs $\rightarrow \mathbb{R}^r$, where $\psi(G) = \mathbf{o}_s$. Formally, $\psi = g \circ \tilde{f}$, where $\tilde{f}$, recursively defined as

$$\tilde{f}(G) = \begin{cases} \mathbf{x}_{nil} & \text{if } G \text{ is empty,} \\ f(\tilde{f}(G_1), \dots, \tilde{f}(G_o), \mathbf{l}_v) & \text{otherwise,} \end{cases}$$

denotes the process that maps a graph to the state at its supersource, $\tilde{f}(G) = \mathbf{x}_s$. The encoding function $\tilde{f}$ yields a result that depends both on the topology and on the labels of the input DAG.

2.2.1 DPAGs

In DPAGs the children of a given node, $\text{ch}[v]$, are ordered and can be organized in a vector using their position as index. Hence, the first argument $\mathbf{x}_{\text{ch}[v]}$ of the state transition function in eq. (11) is a vector in $\mathbb{R}^{o \cdot s}$, such that

$$\mathbf{x}_{\text{ch}[v]} = \left[\mathbf{x}'_{\text{ch}_1[v]}, \dots, \mathbf{x}'_{\text{ch}_o[v]} \right]', \quad o = \max_{v \in V} \{ \text{od}[v] \},$$

where $\mathbf{x}_{\text{ch}_i[v]}$ is equal to the *null state* $\mathbf{x}_{nil}$, if node v has no child in position i. In particular, when processing a leaf node, the state only depends on its label $\mathbf{l}_v$ and on the null state $\mathbf{x}_{nil}$, since $\mathbf{x}_{leaf} = f([\mathbf{x}'_{nil}, \dots, \mathbf{x}'_{nil}]', \mathbf{l}_{leaf} | \theta_f)$.

The function f may be implemented by a three–layer perceptron, with sigmoidal activation functions in the q hidden units and linear activation functions in the output units, such that the state for each node v is calculated according to

$$\mathbf{x}_v = \mathbf{V} \cdot \sigma \left(\sum_{k=1}^{o} \mathbf{A}_k \cdot \mathbf{x}_{\text{ch}_k[v]} + \mathbf{B} \cdot \mathbf{l}_v + \mathbf{C} \right) + \mathbf{D}, \tag{12}$$

where σ is a vectorial sigmoid and the parameters θ_f of the local transition function (11) collect the *pointer matrices* $\mathbf{A}_k \in \mathbb{R}^{q \times s}$, $k = 1, \dots, o$, the local label connections $\mathbf{B} \in \mathbb{R}^{q \times m}$, the hidden layer biases $\mathbf{C} \in \mathbb{R}^q$, the output layer biases $\mathbf{D} \in \mathbb{R}^s$, and hidden to output connection weights $\mathbf{V} \in \mathbb{R}^{s \times q}$.

The state transition function (12) exploits a different pointer matrix $\mathbf{A}_k$ for each child position k. This choice allows to implement a dependence of the output value on the child position, but has some limitations when the node outdegree has a high variability and the empty positions are just added to pad the child vectors to have the same size. In this case, the pointer matrices associated to the last positions in the child vector are often used to propagate the null pointer value $\mathbf{x}_{nil}$, that is introduced by the padding rule. Since padding is not a natural property of the environment and it is introduced just to remove a limitation of the model, this choice may cause artificial results and a sub–optimal use of the network parameters. In this cases, different transition functions can be employed to process nodes having very different values for the outdegree. In practice, all the exploited functions will have the same structure of eq. (12) but with different sets of parameters. By using this approach, we avoid to introduce noisy information due to the padding of empty positions with the null state, otherwise needed for nodes having outdegrees lower than the maximum.

Similarly the output function g can be implemented by a three–layer MLP with q' hidden neurons with a sigmoidal activation function, such that

$$\mathbf{o}_v = \mathbf{W} \cdot \sigma\left(\mathbf{E} \cdot \mathbf{x}_v + \mathbf{F}\right) + \mathbf{G}, \tag{13}$$

where the parameters of the output function θ_g are $\mathbf{E} \in \mathbb{R}^{q' \times s}$, $\mathbf{F} \in \mathbb{R}^{q'}$, $\mathbf{G} \in \mathbb{R}^{r}$, $\mathbf{W} \in R^{r \times q'}$.

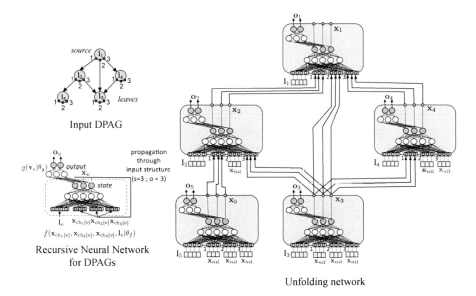

Fig. 4 Transition function implemented by a three–layer MLP. The resulting recursive neural network can process graphs with a maximum outdegree $o = 3$.

Figure 4 shows an example of a neural network architecture implementing the transition function of eq. (12). The recursive connections link the output of the state neurons to the network inputs for each position in the child vector. This notation describes the template that is exploited to assemble the encoding network. Apart from the recursive connections, the transition function is implemented by a classical feedforward MLP with a layer of hidden units. The network has $s \cdot o + m$ inputs corresponding to the o states of the children (s components each) and to the node label (m components).

2.2.2   DAGs–LE

In many applications, the DPAG–based model adds unnecessary constraints in the data encoding. First, the assignment of a position to each child of a node may not be naturally defined, and the chosen criterion may be arbitrary. Second, as also noted previously, the need to have a uniform outdegree or to bound its maximum value in the graphs can cause the loss of important information. For example, when considering the image representation based on the Region Adjacency Graph, the order of the adjacent regions may be significant, since they can be listed following the region contour in a given direction, but to map them to a specific position we need to define a criterion to choose the starting point on the contour and the contour direction. Unless there is a natural and uniform criterion we may end up with arbitrary encodings that fail to preserve natural properties of the modeled environment. Moreover, the maximum graph outdegree may be bounded by defining a criterion to prune some arcs when the number of the node children exceeds a predefined value. Anyway, the pruning process may lead to a loss of useful information and may not completely eliminate the need to arbitrarily pad the last positions in the child vector with the null state $\mathbf{x}_{nil}$, for each node v having $|ch[v]| < o$. In fact, the need to have exactly o children is a limitation of the model and not a feature of the problem.

These limitations may be overcome by modeling the data with DAGs–LE [12]. In fact, the arc label $\mathbf{l}_{(u,v)}$ can encode attributes of the relationship represented by the arcs, allowing us to devise a model able to distinguish the role of each child without the need to encode it by its position. For DAGs–LE, we can define a transition function f that has not a predefined number of arguments and that does not depend on their order, following eq. (7). The general form for the state update function for DAGs–LE is

$$\mathbf{x}_v = \sum_{u \in ch[v]} h(\mathbf{x}_u, \mathbf{l}_{(v,u)}, \mathbf{l}_v | \theta_h) \,, \tag{14}$$

that can be applied for any internal node, i.e. the nodes with at least one child. For the leaf nodes the set $ch[leaf]$ is empty and the previous update equation is undefined. In this case the state will depend on the node label $\mathbf{l}_{leaf}$, such that $\mathbf{x}_{leaf} = h'(\mathbf{l}_{leaf} | \theta_{h'})$.

However, there are other solutions to implement a non–positional transition function, such as that proposed in [12]. In this model, first the contributions of the children are summed to obtain an intermediate node state $\mathbf{y}_v \in \mathbb{R}^p$, then this state is non–linearly combined with the node label to yield the node state $\mathbf{x}_v$. The state update function is then defined as

$$\begin{aligned}
\mathbf{y}_v &= \sum_{u \in ch[v]} \phi(\mathbf{x}_u, \mathbf{l}_{(v,u)} | \theta_\phi) \,, \\
\mathbf{x}_v &= \tilde{f}(\mathbf{y}_v, \mathbf{l}_v | \theta_{\tilde{f}}) \,,
\end{aligned} \tag{15}$$

where the *edge–weighting function* $\phi : \mathbb{R}^{(s+d)} \to \mathbb{R}^p$ is a non–linear function parameterized by θ_ϕ, and $\tilde{f} : \mathbb{R}^{(p+m)} \to \mathbb{R}^s$ depends on the trainable parameters $\theta_{\tilde{f}}$. The state update function defined by (15) can be applied to nodes with any number of children and it is also independent of the order of the children, similarly to the update rule of eq. (14). For leaf nodes we need to define $\mathbf{y}_v$. A straightforward solution is to choose a constant $\mathbf{y}_{leaf}$ that is used for all the leaf nodes.

The functions ϕ and $\tilde{f}$ can be implemented by MLPs. For instance, ϕ can be computed by a three–layer perceptron with q sigmoidal hidden units and p linear outputs, as

$$\phi(\mathbf{x}_u, \mathbf{l}_{(v,u)} | \theta_\phi) = \mathbf{V}\sigma(\mathbf{A}\mathbf{x}_u + \mathbf{B}\mathbf{l}_{(v,u)} + \mathbf{C}) + \mathbf{D},$$

where θ_ϕ collects the state–to–hidden weights $\mathbf{A} \in \mathbb{R}^{q \times s}$, the arc–label–to–hidden weights $\mathbf{B} \in \mathbb{R}^{q \times d}$, the hidden unit biases $\mathbf{C} \in \mathbb{R}^q$, the output biases $\mathbf{D} \in \mathbb{R}^p$, and the hidden–to–output weights $\mathbf{V} \in \mathbb{R}^{p \times q}$.

On the other hand, the function ϕ can be also realized by an *ad hoc* model. For instance, the solution described in [12], exploits an implementation of ϕ as

$$\phi(\mathbf{x}_u, \mathbf{l}_{(v,u)} | \theta_\phi) = \left(\sum_{j=1}^{d} \mathbf{H}_j \mathbf{l}_{(v,u),j} \right) \mathbf{x}_u, \tag{16}$$

being $\mathbf{H} \in \mathbb{R}^{p \times s \times d}$ the *arc–weight matrix*. In particular, $\mathbf{H}_j \in \mathbb{R}^{p \times s}$ is the j–th layer of the matrix $\mathbf{H}$ and $\mathbf{l}_{(v,u),j}$ is the j–th component of the arc label.

3   Supervised Learning for Graph Neural Networks

In the following, we consider a supervised learning framework for graph neural networks. This setting requires to choose a specific network architecture (type and structure of both the state transition function and the output function), a learning set $\mathscr{L}$, that collects an ensemble of supervised examples, and a cost function, to evaluate how well the network responses approximate the target values provided by the supervisor for the examples in $\mathscr{L}$. Given the learning set, the cost function depends only on the free parameters of the graph neural network, i.e. the vectors θ_f and θ_g. All the network parameters can be collected into an unique vector, i.e. $\theta = [\theta_f' \theta_g']'$, such that the cost function to be optimized during the learning process can be written as $E = E(\theta | \mathscr{L})$. Therefore, the learning task is formulated as the optimization of a multivariate function.

3.1   Learning Objective

The learning set collects supervised examples consisting in graphs for which a supervisor provided target values for the network outputs at given nodes. More formally, the learning set is defined as

$$\mathscr{L} = \big\{ (G_p, v_{p,j}, \mathbf{t}_{p,j}) \mid G_p \in \mathscr{G}, p = 1, \ldots, P, \\ v_{p,j} \in V_{G_p}, \mathbf{t}_{p,j} \in \mathbb{R}^r, j = 1, \ldots, q_p \big\} \qquad (17)$$

where for a given input graph G_p we can provide a set of target values $\mathbf{t}_{p,j}$, $j = 1, \ldots, q_p$ for a subset of its nodes $S_{G_p} = \{ v_{p,j} \in V_{G_p}, j = 1, \ldots, q_p \}$. For graph-focused classification or regression tasks, the supervision is only provided at the graph supersource, thus only one target is specified for each graph in the training set (i.e. $q_p = 1, p = 1, \ldots, P$). Usually, the learning objective is defined by a quadratic cost function

$$E(\theta | \mathscr{L}) = \frac{1}{P} \sum_{p=1}^{P} e(\theta | G_p) = \frac{1}{2P} \sum_{p=1}^{P} \sum_{v_{p,j} \in S_{G_p}} \| \mathbf{t}_{p,j} - \mathbf{o}_{v_{p,j}}(\theta | G_p) \|_2^2. \qquad (18)$$

This cost function measures the approximation of each target value by the Euclidean norm and linearly combines the contributions for each node and each graph. In general, other loss functions can be employed to measure the fitting of the target values. If the graph neural network functions, f and g, are differentiable with respect to the parameters θ, the cost function $E(\theta | \mathscr{L})$ is a continuous differentiable function and can be optimized by using a gradient descent technique. This property holds because of the assumption that the dynamical system defined by the graph neural network is based on a contractive mapping. In fact, when the global transition function $F(\mathbf{x}, \mathbf{l} | \theta_f)$ is a contraction, the cost function depends on its unique fixed point whose dependence on the parameters θ_f can be proven to be continuous and continuously differentiable (see Theorem 1 in [23]). Instead, for a general dynamical system a slight change in the parameters can cause abrupt changes in the fixed point, due to the peculiar properties of non–linear dynamical systems, such as the presence of bifurcations in the parameter space [8]. For generic non–linear dynamical systems, there can also be more than one fixed point or other asymptotic behaviors, like limit cycles or chaotic trajectories. In these cases the computation would have an undefined result and the learning procedure could not be applied.

When the implementation of the global transition function F does not guarantee that it is a contraction mapping, a penalty term can be added to the learning objective to favor the development of this property. However, the penalty approach does not assure that the requirement is actually satisfied. Given a target contraction constant μ, the penalty term is

$$p(\theta_f) = \beta L \left(\left\| \frac{\partial F(\mathbf{x}, \mathbf{l} | \theta_f)}{\partial \mathbf{x}} \right\| \right), \qquad (19)$$

where β is the penalty weight to balance the contribution of this term with respect to the data fitting term of eq. (18), and $L(y)$ may be the hinge-like loss function $\max(0, y - \mu)$, that penalizes values of y beyond μ. More details on the implementation of the penalty function, and on the computations needed to calculate its gradient with respect to the GNN parameters θ_f, can be found in [23].

3.2 Learning Procedure for GNNs

For each graph $G_p \in \mathcal{L}$, the learning procedure is based on the following steps.

1. Given an initial state $\mathbf{x}(0|G_p)$, the state update function

$$\mathbf{x}(t+1|G_p) = F(\mathbf{x}(t|G_p), \mathbf{l}_{G_p}|\theta_f)$$

   is iterated until, at time T, the state approaches the fixed point, i.e. $\mathbf{x}(t|G_p) \approx const$ for $t \geq T$. In practice the stopping condition is determined by the condition $\|\mathbf{x}(T|G_p) - \mathbf{x}(T-1|G_p)\| < \varepsilon$, where $\varepsilon > 0$ is a predefined threshold.

2. The outputs at the supervised nodes $v_{p,j} \in S_{G_p}$ are computed as

$$\mathbf{o}_{v_{p,j}}(\theta|G_p) = g(\mathbf{x}_{v_{p,j}}(T|G_p)|\theta_g)$$

   and the partial contribution $e(\theta|G_p)$ to the cost function of eq. (18) is accumulated.

3. The partial gradient $\nabla_\theta e(\theta|G_p)$ is computed as detailed in the following, and its contribution is accumulated to compute the global gradient, since $\nabla_\theta E(\theta|\mathcal{L}) = \frac{1}{P} \sum_{p-1}^{P} \nabla_\theta e(\theta|G_p)$.

Once all the graphs in $\mathcal{L}$ have been processed, the gradient $\nabla_\theta E(\theta|\mathcal{L})$ is available and a gradient descent rule can be applied to update the GNN weights in order to optimize the cost function $E(\theta|\mathcal{L})$. Given the value of the weight parameters θ_k at the iteration (*learning epoch*) k, the simplest update rule that can be applied is

$$\theta_{k+1} = \theta_k - \eta_{k+1} \nabla_\theta E(\theta|\mathcal{L})|_{\theta=\theta_k} , \qquad (20)$$

where the gradient of $E(\theta|\mathcal{L})$ is computed for $\theta = \theta_k$ and η_{k+1} is the *learning rate*, whose value may be eventually adapted during the learning process. The initial parameters θ_0 are usually initialized with small random values (for instance, sampled from a uniform distribution in an interval around 0). Different stopping criteria are usually applied to end the iteration of the gradient descent procedure. The learning algorithm can be stopped when the value of the cost function is below a predefined threshold, when a maximum number of iterations (epochs) is reached, or when the norm of the gradient is smaller than a given value. Unfortunately, the training algorithm based on this procedure is not guaranteed to halt yielding a global optimum. In particular, the learning process can be trapped in a local minimum of the function $E(\theta|\mathcal{L})$, that may represent a sub–optimal solution to the problem. In some cases, the obtained sub–optimum can be acceptable, but, when the performances are not satisfactory, a new learning procedure should be run, starting from a different initialization for the network parameters. More sophisticated gradient–based optimization techniques can be applied to increase the speed of convergence, such as *resilient propagation* [21], that exploits the changes in the sign of the gradient components during the update iterations to obtain an adapted learning rate for each weight in θ, or second–order methods based also on the curvature as described by the Hessian matrix.

3.2.1 Gradient Computation for Graph Neural Networks

The partial gradient of the cost function $\nabla_\theta e(\theta | G_p)$ can be efficiently computed by
a diffusion process that derives from the iterative procedure followed to compute
the approximation of the fixed point $\mathbf{x}(T | G_p)$. In fact, the relaxation process, that
is applied to the state $\mathbf{x}(t)$ for T steps until the fixed point is approached, yields a
layered unfolding network in which each of the T layers corresponds to a replica of
the state transition function $F(\mathbf{x}(t | G_p), \mathbf{l}_{G_p} | \theta_f)$ at each time step $t = 1, \ldots, T$. The
units in layers t and $t + 1$ are connected through the state variables, being the output
of layer t, $\mathbf{x}(t)$, the input for the following layer that computes $\mathbf{x}(t + 1)$. As shown in
Figure 5, the resulting unfolding network has a multilayered structure, in which the
topology of the connections is determined by the connectivity of the input graph.
The same weight vector θ_f is shared among all the layers. Finally, the state units of
the layer T are connected to the output networks $g(\mathbf{x}_v(T | G_p) | \theta_g)$ to yield the GNN
outputs (actually we connect the output networks only to the supervised nodes, i.e.
$v \in S_{G_p}$).

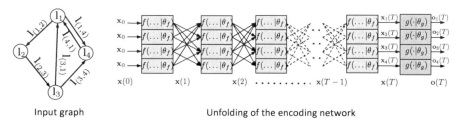

Input graph   Unfolding of the encoding network

Fig. 5 The unfolding network resulting from the state computation for an input graph G_p

The unfolding network provides the schema for computing efficiently the gradient of the cost function with respect to the GNN parameters θ. In fact, the original back–propagation algorithm for feed–forward neural networks can be applied
straightforwardly on the layered unfolding network. In particular, the gradient computation can be performed in two steps as in the back–propagation algorithm: in the
forward pass the GNN states for $t = 1, \ldots, T$ and the outputs at T are computed and
stored building the encoding network; in the *backward pass* the errors are computed
at the nodes where the targets are provided and they are back–propagated through
the unfolding network to compute the gradient components. The procedure requires
to store all the state sequence $\mathbf{x}(t | G_p)$, $t = 0, \ldots, T$ generated in the forward step,
since these values are used in the back propagation process. For graphs having a
large number of nodes and for long relaxation sequences, the memory requirements
may become unfeasible. To overcome this problem, it is possible to assume that the
state sequence has already converged to the fixed point $\mathbf{x}(G_p) \approx \mathbf{x}_v(T | G_p)$, such
that for all the considered layers in the unfolding network $\mathbf{x}(t | G_p) = \mathbf{x}(G_p)$. This
assumption is motivated by the fact that the state transition function is a contraction
map and there is a unique fixed point that can be approached starting from any given
initial state $\mathbf{x}(0 | G_p)$. In particular, if an oracle allows us to set $\mathbf{x}(0 | G_p) = \mathbf{x}(G_p)$, we

actually obtain that the state is constant in the relaxation process. Hence, we don't need to store all the different state vectors, but only the fixed point approximation $\mathbf{x}(T|G_p)$, that can replace all the other states needed in the backward propagation.

Output network

First, let us consider a generic weight $\vartheta_g \in \theta_g$ of the output function in the GNN. The cost function depends on ϑ_g through the outputs computed on the supervised nodes $\mathbf{o}_{v_{p,j}}(\theta|G_p)$. Hence, the partial derivative with respect to ϑ_g can be obtained by applying the chain rule

$$\frac{\partial e(\theta|G_p)}{\partial \vartheta_g} = \sum_{v_{p,j} \in S_p} \sum_{k=1}^{r} \frac{\partial e(\theta|G_p)}{\partial \mathbf{o}_{v_{p,j}}(\theta|G_p)[k]} \frac{\partial \mathbf{o}_{v_{p,j}}(\theta|G_p)[k]}{\partial \vartheta_g} , \qquad (21)$$

where $\mathbf{o}_{v_{p,j}}(\theta|G_p)[k]$ is the k–th component of the GNN output vector computed at the supervised node $v_{p,j} \in S_p$. When using the quadratic cost function of eq. (18), the first factor in the sum is the output–target error

$$\frac{\partial e(\theta|G_p)}{\partial \mathbf{o}_{v_{p,j}}(\theta|G_p)[k]} = \mathbf{o}_{v_{p,j}}(\theta|G_p)[k] - \mathbf{t}_{p,j}[k] , \qquad (22)$$

whereas the second factor is

$$\frac{\partial \mathbf{o}_{v_{p,j}}(\theta|G_p)[k]}{\partial \vartheta_g} = \frac{\partial g(\mathbf{x}_{v_{p,j}}(G_p)|\theta_g)[k]}{\partial \vartheta_g} , \qquad (23)$$

that depends on the gradient of k–th component of the output function g computed at the state fixed point $\mathbf{x}_{v_{p,j}}(G_p)$. The actual computation of this gradient depends on the specific implementation of g.

State transition function

On the other hand, the computation of partial derivatives of the cost function with respect to a weight $\vartheta_f \in \theta_f$ is more complex, since it must consider the iterative procedure that defines the fixed point and yields the encoding network. In fact, the value of the parameters θ_f affects the value of the state fixed point $\mathbf{x}(G_p)$, that in turn is exploited to compute the GNN outputs on the supervised nodes. Theorem 2 in [23] derives a method to efficiently compute the gradient, also giving a formal proof based on the implicit function theorem (the fixed point is the solution of equation $\mathbf{x}(G_p) - F(\mathbf{x}(G_p), \mathbf{l}_{G_p}|\theta_f) = 0$). Instead, we derive the equations needed to compute the gradient by following an intuitive approach based on the encoding network. As it is done in the derivation of the Back–Propagation Through Time (BPTT) algorithm for Recurrent Neural Networks processing time sequences [29], we consider that, for each iteration t of the fixed point computation, a different weight vector $\theta(t)$ is used. Under this assumption we can compute the derivative of the partial cost function $e(\theta|G_p)$ with respect to the *unfolded* parameters $\theta(t)$. More precisely, we consider the cost function as $e(\theta(1), \ldots, \theta(T)|G_p)$. Since the replicas of the global state transition function in the iterations actually share the same set of weights, i.e. $\theta(1) = \ldots = \theta(T) = \theta$, by applying the rule for the derivative of a composition of functions, we obtain the following expression for each weight ϑ_f

$$\frac{\partial e(\theta|G_p)}{\partial \vartheta_f} = \sum_{t=1}^{T} \frac{\partial e(\theta|G_p)}{\partial \vartheta_f(t)} . \tag{24}$$

Once the weights are decoupled among the different iterations, we can describe the propagation of a change in $\vartheta_f(t)$ through the state $\mathbf{x}(t|G_p)$, that in turn affects the objective function $e(\theta|G_p)$, due to the future evolution it causes. Using again the chain rule for function composition, we can write

$$\frac{\partial e(\theta|G_p)}{\partial \vartheta_f(t)} = \sum_{v \in V_p} \sum_{h=1}^{s} \frac{\partial e(\theta|G_p)}{\partial \mathbf{x}_v(t|G_p)[h]} \frac{\partial \mathbf{x}_v(t|G_p)[h]}{\partial \vartheta_f(t)} . \tag{25}$$

As in the classical back–propagation algorithm, we introduce the *generalized errors* as

$$\delta_v(t|G_p)[h] = \frac{\partial e(\theta|G_p)}{\partial \mathbf{x}_v(t|G_p)[h]} , \tag{26}$$

such that eq. (25) can be rewritten as

$$\frac{\partial e(\theta|G_p)}{\partial \vartheta_f(t)} = \sum_{v \in V_p} \sum_{h=1}^{s} \delta_v(t|G_p)[h] \frac{\partial f(\mathbf{x}_{\mathrm{ne}[v]}(t-1|G_p), \mathbf{l}_{(v,\mathrm{ch}[v])}, \mathbf{l}_{(\mathrm{pa}[v],v)}, \mathbf{l}_v | \theta_f)[h]}{\partial \vartheta_f} , \tag{27}$$

where in the last term we exploited the local state transition function of eq. (1), that defines the explicit dependence of $\mathbf{x}_v(t|G_p)$ with respect to the weight at a given iteration t. In this last derivative, we removed the explicit dependence of ϑ_f on t, since the iteration number affects only the value of the first argument of f. The computation of this derivative depends on the specific implementation of f. Actually, since we consider the fixed point equation at its convergence, we need to compute this derivative only once at the fixed point state $\mathbf{x}(G_p)$, removing completely the dependence of this term on t. Hence, by merging eq. (27) with eq. (24) we obtain

$$\frac{\partial e(\theta|G_p)}{\partial \vartheta_f} = \sum_{v \in V_p} \sum_{h=1}^{s} \frac{\partial f(\mathbf{x}_{\mathrm{ne}[v]}(G_p), \mathbf{l}_{(v,\mathrm{ch}[v])}, \mathbf{l}_{(\mathrm{pa}[v],v)}, \mathbf{l}_v | \theta_f)[h]}{\partial \vartheta_f} \sum_{t=1}^{T} \delta_v(t|G_p)[h] . \tag{28}$$

The sequence of the generalized errors $\delta(t|G_p)$ can be computed backwards starting from $\delta(T|G_p)$. In fact, given that the GNN outputs are computed from $\mathbf{x}_v(T|G_p)$, that approximates the fixed point, we can directly compute the error derivative by applying the chain rule

$$\delta_v(T|G_p)[h] = \sum_{k=1}^{r} \frac{\partial e(\theta|G_p)}{\partial \mathbf{o}_v(T|G_p)[k]} \frac{\partial \mathbf{o}_v(T|G_p)[k]}{\partial \mathbf{x}_v(T|G_p)[h]} \tag{29}$$

where

$$\frac{\partial e(\theta|G_p)}{\partial \mathbf{o}_v(\theta|G_p)[k]} = \begin{cases} \mathbf{o}_v(\theta|G_p)[k] - \mathbf{t}_v[k] & \text{if } v \in S_p \\ 0 & \text{if } v \notin S_p \end{cases}, \tag{30}$$

in the case of the quadratic cost function of eq. (18), and

$$\frac{\partial \mathbf{o}_v(T|G_p)[k]}{\partial \mathbf{x}_v(T|G_p)[h]} = \frac{\partial g(\mathbf{x}_v(G_p)|\theta_g)[k]}{\partial \mathbf{x}_v(G_p)[h]} \tag{31}$$

is the derivative of the output function g with respect to the state computed at the fixed point, $\mathbf{x}_v(G_p)$. Finally, the equations to propagate the $\delta(t|G_p)$ backward in time can be obtained by considering the dependence of $\mathbf{x}(t|G_p)$ from $\mathbf{x}(t-1|G_p)$ through the global state transition function F. In particular, we can apply the chain rule deriving the following expansion

$$
\begin{aligned}
\delta_v(t|G_p)[h] &= \sum_{u \in V_p} \sum_{k=1}^{s} \frac{\partial e(\theta|G_p)}{\partial \mathbf{x}_u(t+1|G_p)[k]} \frac{\partial \mathbf{x}_u(t+1|G_p)[k]}{\partial \mathbf{x}_v(t|G_p)[h]} \\
&= \sum_{u \in V_p} \sum_{k=1}^{s} \delta_u(t+1|G_p)[k] \frac{\partial f(\mathbf{x}_{\text{ne}[u]}(G_p), \mathbf{l}_{(u,\text{ch}[u])}, \mathbf{l}_{(\text{pa}[u],u)}, \mathbf{l}_u|\theta_f)}{\partial \mathbf{x}_v(G_p)[h]}
\end{aligned}
\tag{32}
$$

where, in the last term, we exploited the fact that at convergence we can compute the derivatives at the fixed point. Clearly, the last term is null if $v \notin \text{ne}[u]$ and the generalized errors are only propagated to the node's neighbors. The actual form of this term depends on the specific implementation of the local state transition function. If we arrange the variables $\delta_v(t|G_p)[h]$ into a row vector $\delta(t|G_p) \in \mathbb{R}^{s|V_p|}$ the back propagation equation can be rewritten in matrix form as

$$\delta(t|G_p) = \delta(t+1|G_p) \cdot \frac{\partial F(\mathbf{x}(G_p), \mathbf{l}_{G_p}|\theta_f)}{\partial \mathbf{x}(G_p)}, \tag{33}$$

where the last term is the gradient of the global state transition function (it is a $s|V_p| \times s|V_p|$ matrix). Since F is a contraction, $\left\| \frac{\partial F(\mathbf{x}, \mathbf{l}|\theta_f)}{\partial \mathbf{x}} \right\| \le \mu < 1$ holds and, hence, the equation

$$\|\delta(t|G_p)\| \le \mu \|\delta(t+1|G_p)\| \le \mu^{t-T} \|\delta(T|G_p)\| \tag{34}$$

shows that $\delta(t|G_p)$ decays exponentially to the null vector while going backward in time. As a consequence, the sum $\sum_{t=1}^{T} \delta_v(t|G_p)[h]$ converges to $\delta_v(G_p)[h]$, and it can be proven that the computation of eq. (28) yields the correct gradient of the implicit function (see Theorem 2 in [23]). The equations (28), (30), and (32) define the algorithm to compute the gradient with respect to parameters of the state transition function. In particular the backward phase for the gradient computation requires the following steps.

1. Initialize the generalized errors $\delta_v(T, G_p)[h]$ using eq. (22).
2. Iterate the backward propagation of eq. (32) until $\|\delta(t|G_p)\| < \varepsilon_\delta$, updating at each iteration t the variables $\mathbf{z}_v(t|G_p)[h] = \mathbf{z}_v(t+1|G_p)[h] + \delta_v(t|G_p)[h]$, that accumulate the generalized errors as required by eq. (28).
3. Compute the gradient using eq. (28) where $\mathbf{z}_v(t_0|G_p)[h] = \sum_{t=t_0}^{T} \delta_v(t|G_p)[h]$ and t_0 is the iteration at which the stopping criterion is verified.

This procedure defines the *backward step* of the learning algorithm. Hence, there is not the need to predefine a number of backward propagation steps, but the stopping criterion is based on the threshold ε_δ on the norm of the generalized errors.

3.3 Learning Procedure for Recursive Neural Networks

In the case of Recursive Neural Networks the training procedure can be optimized with respect to the more general Graph Neural Network model, due to the fact that the input graph has no cycles. In section 2.2 we have already shown that the forward step does not need an undefined number of iterations on the global state transition function F to approach the fixed point. By rearranging the node updates using the inverse topological order, just one step is needed to compute the final state for each node. The lack of cycles allows also the optimization of the backward step needed to backpropagate the generalized errors and to compute the gradient of the objective function. In this case, the errors are propagated backwards through the node links, starting from the graph supersources to the leaf nodes. Hence, also the gradient computation can be completed with just one iteration on the input graph nodes.

In the case of RNNs the learning set of eq. (17) will contain only directed acyclic graphs (i.e. $\mathscr{G} \subseteq \text{DAG}$), and the objective function will be defined as in eq. (18). The optimization of the cost function is based on a gradient descent technique as, for instance, the simple weight update rule of eq. (20).

Let us suppose that for each DAG G_p in $\mathscr{L}$, the set of its nodes is ordered following their topological order, i.e. $V_p = (v_1, \ldots, v_n)$ such that $v_i \prec v_j \rightarrow i < j$. As pointed out in Section 1, the order is partial since there may be pairs of nodes between which the $\prec$ relationship is not defined. This happens for instance for sibling nodes, such as the children of a given node in a tree. In this case, unless the graph is positional, the mutual position of these nodes is not defined and many permutations of the node set are compatible with the ordering criterion. However, this is not an issue for defining the RNN computation, that only requires that the ancestors are processed after their descendants in the forward step and viceversa in the backward step. Hence, the ordering of the sibling nodes is irrelevant for the RNN computation and we do not need any particular assumption on the actual permutation of these nodes in V_p. The gradient contribution $\nabla_\theta e(\theta|G_p)$, for each graph G_p in the learning set, is computed in two steps as it happens for GNNs.

1. *Forward step*
   Following the node order in V_p, the local state transition function is applied and the states are computed and stored in the state vector $\mathbf{x}(G_p)$. In this case the sub-vector $\mathbf{x}_{v_j}(G_p)$ will correspond to the component in $\mathbf{x}(G_p)$ starting from

index $s \cdot (j-1)$, $j = 1, \ldots, n$. In the case of RNNs, we do not need to iterate the global state transition function in time, but we can organize the computation such that, for each iteration, we just update only those nodes for which the local state transition function yields the final result. This strategy allows us to avoid the computation for the nodes that have already reached their final state and those for which the fixed point of the child states is not available yet. Therefore, we can use the index of the nodes in V_p as the iteration step. The state of node v_j, $j = 1, \ldots, n$, will be computed as

$$\mathbf{x}_{v_j} = f(\mathbf{x}_{\mathrm{ch}[v_j]}, \mathbf{l}_{(v_j, \mathrm{ch}[v_j])}, \mathbf{l}_{v_j} | \theta_f) ,$$

where the states $\mathbf{x}_{\mathrm{ch}[v_j]}$ have already been computed, since the children of v_j precede it in the ordered node list V_p. In the case of leaf nodes (that are in the first positions of the list V_p) or when dealing with positional DAGs with missing links, a given initial state $\mathbf{x}_0$ is possibly used to complete the argument list of f. In any case, the inserted null links are not stored into the state vector $\mathbf{x}(G_p)$. Once all the states have been computed, the network outputs are evaluated and stored for the supervised nodes $v_j \in S_p$ as

$$\mathbf{o}_{v_j} = g(\mathbf{x}_{v_j} | \theta_g) .$$

2. *Backward step*

In the optimized organization of the state updates in the forward step, we dropped the iteration index t and we used instead the position in the ordered node list V_p. Hence, we can follow the same approach to derive the RNN backward step, by replacing the backward iterations in t with iterations on the nodes v_j, $j = n, \ldots, 1$. First, as in eq. (24), we assume that the weights exploited at each node v_j by the RNN are independent from each other, i.e. we consider the $|V_p|$ weight vectors $\theta_f(v_j)$. Since the weight values are actually shared, the constraint $\theta_f(v_1) = \ldots = \theta_f(v_n) = \theta_f$ requires that, for any weight $\vartheta_f \in \theta_f$,

$$\frac{\partial e(\theta | G_p)}{\partial \vartheta_f} = \sum_{j=1}^{n} \frac{\partial e(\theta | G_p)}{\partial \vartheta_f(v_j)} .$$

The local derivatives $\frac{\partial e(\theta | G_p)}{\partial \vartheta_f(v_j)}$ can be computed using the chain rule by considering the state variables that are directly affected in the forward computation by the value of $\vartheta_f(v_j)$. Since this value is exploited by the state transition function when applied at node v_j, similarly to eq. (25) for GNNs, we can expand the derivative as

$$\begin{aligned}
\frac{\partial e(\theta | G_p)}{\partial \vartheta_f(v_j)} &= \sum_{h=1}^{s} \frac{\partial e(\theta | G_p)}{\partial \mathbf{x}_{v_j}(G_p)[h]} \frac{\partial \mathbf{x}_{v_j}(G_p)[h]}{\partial \vartheta_f(v_j)} \\
&= \sum_{h=1}^{s} \delta_{v_j}(G_p)[h] \frac{\partial f(\mathbf{x}_{\mathrm{ch}[v_j]}, \mathbf{l}_{(v_j, \mathrm{ch}[v_j])}, \mathbf{l}_{v_j} | \theta_f)[h]}{\partial \vartheta_f(v_j)} ,
\end{aligned}$$

where $\delta_{v_j}(G_p) = \frac{\partial e(\theta|G_p)}{\partial \mathbf{x}_{v_j}(G_p)}$ is the vector of the generalized errors at node v_j, and the last term depends on the specific implementation of the function f. We can derive the relation for the backward propagation of the generalized errors as in eq. (32) for GNNs, with the difference that the propagation is more explicitly expressed as being performed through the graph structure. In fact, given a node v_j, a variation in its state affects the states of its parents $\mathrm{pa}[v_j]$, such that we can apply the chain rule as

$$
\begin{aligned}
\delta_{v_j}(G_p)[h] &= \sum_{u \in \mathrm{pa}[v_j]} \sum_{k=1}^{s} \frac{\partial e(\theta|G_p)}{\partial \mathbf{x}_u(G_p)[k]} \frac{\partial \mathbf{x}_u(G_p)[k]}{\partial \mathbf{x}_{v_j}(G_p)[h]} \\
&= \sum_{u \in \mathrm{pa}[v_j]} \sum_{k=1}^{s} \delta_u(G_p)[k] \frac{\partial f(\mathbf{x}_{\mathrm{ch}[u]}, \mathbf{l}_{(u,\mathrm{ch}[u])}, \mathbf{l}_u | \theta_f)[k]}{\partial \mathbf{x}_{v_j}(G_p)[h]} .
\end{aligned}
$$

The inner sum in the above formula implements the propagation of the generalized errors at node $u \in \mathrm{pa}[v_j]$ towards the input components corresponding to the state of the child v_j, for the module implementing the function f. If the function f is realized by a module (f.i. an MLP function), the implementation will embed the backward propagation of the generalized errors from its outputs to its inputs. The specific implementation, that takes into account the structure exploited to realize the function f, may also perform the backward step efficiently. For instance, if f is implemented by an MLP, the backpropagation schema can be employed internally instead of computing directly the derivatives of f with respect to each input component, as in the last term of the equation. Hence, if we model the function f as a black box, that provides both a *forward method* to compute the function f given its inputs, and a *backward method* to propagate backwards the generalized errors available at its outputs to its input units, we can rewrite the propagation equation from parents to children as

$$
\delta_{v_j}(G_p) = \sum_{u \in \mathrm{pa}[v_j]} \delta_{(u,v_j)}(G_p)
$$

where by $\delta_{(u,v_j)}(G_p)$ we indicate the generalized error vector that is computed by propagating $\delta_u(G_p)$, through the module implementing f, to the input corresponding to the graph link between the node u and its child v_j. This equation describes the BackPropagation Through Structure (BPTS) schema, showing how the error information flows through the graph topology.

If we consider the output function g, the generalized errors for its output variables $\mathbf{o}_{v_j}(G_p)[k]$ at node v_j can be computed directly from the cost function. In fact, if there is no supervision at node v_j, then $\delta_{v_j}(\theta|G_p) = 0$ holds; otherwise, in the case of the quadratic cost function, $\delta_{v_j}(G_p) = \mathbf{o}_{v_j}(\theta|G_p) - \mathbf{t}_{p,j}$. These generalized errors can be backpropagated through the output network g to yield the generalized errors for its inputs $\delta_{v_j}^g = \frac{\partial e(\theta|G_p)}{\partial \mathbf{x}_{v_j}(G_p)}$, corresponding to the state variables at node v_j. This term is an additional contribution to the generalized error $\delta_{v_j}(G_p)$ and, thus, the complete equation for computing the generalized errors is

$$\delta_{v_j}(G_p) = \sum_{u \in pa[v_j]} \delta_{(u,v_j)}(G_p) + \delta_{v_j}^g(G_p) \; .$$

The generalized errors can be computed by processing the nodes from the last positions in the list V_p, thus guaranteeing that the generalized errors of the parents have already been accumulated in the children before the propagation proceeds with the following layer of nodes.

Once the generalized errors $\delta_{v_j}(G_p)$ are computed, the backpropagation procedure for the replica of the transition function at node v_j is executed, yielding both the partial gradients $\frac{\partial e(\theta|G_p)}{\partial \vartheta_f(v_j)}$ and the generalized errors at the network inputs, $\delta_{(v_j,\text{ch}[v_j])}(G_p)$.

4 Summary

Graph Neural Networks are a powerful tool for processing graphs, that represent a natural way to collect information coming from several areas of science and engineering – e.g. data mining, computer vision, molecular chemistry, molecular biology, pattern recognition –, where data are intrinsically organized in entities and relationships among entities. Moreover, GNNs have been proved to be universal approximators, in the sense that they can approximate, in probability and up to any prescribed degree of precision, a wide class of functions on graphs, namely those preserving the unfolding equivalence. Actually, such class includes most of the practically useful functions on graphs [22]. Therefore, GNNs have recently been used in many cutting–edge applications. In particular, in the bioinformatic field, GNNs were employed for QSAR problems, i.e., for the prediction of the mutagenicity and the biodegradability properties of some molecules [27, 28, 2] and for the secondary protein structure prediction [2]. In computer vision, GNNs were applied to the problem of object localization in images [9, 16, 20], whereas, with respect to Internet applications, they were successfully exploited for many different problems, such that web page ranking [25, 10, 13], network security [15], sentence extraction [17], and web document classification and clustering [30, 7]. Just as an example, let us briefly deepen how GNNs can represent a very useful tool for biological data processing. Since the protein expression is mostly due to post–translational processes, biological networks, such as gene interaction networks or metabolic networks, play a fundamental role to understand cell processes. On the other hand, the continuous development of high quality biotechnology, e.g. micro–array techniques and mass spectrometry, provides complex patterns for the characterization of cell processes. In this way, challenging problems from clinical proteomics, drug design, or design of species can be tackled. However, in all these problems, a variety of different biological information is collected, which can be appropriately represented by structured data. This puts new challenges not only on appropriate data storage, visualization, and retrieval of heterogeneous information, but also on machine learning tools used in this context, which must adequately process and integrate heterogeneous information into a global picture. The very immediate way to integrate data which are distinct in nature, but related to each other, is to represent them by graphs, which can be positional or not, directed or not, cyclic or acyclic, but can always be appropriately processed by GNNs.

Apart from their vast applicability, nonetheless, all recurrent/recursive models suffer from the long–term dependency pathology, i.e. they are unable to properly process deep structures. This is due to the very local nature of the learning procedure, that prevents both the state calculation and the weight updating to be influenced from weights, labels, states related to far nodes. In other words, practical difficulties are easily verifiable when dynamic neural network models are trained on tasks where the temporal contingiencies present in the input/output data span long intervals. These ideas have been clearly formalized in [4] for recurrent networks, where it is proven that if the system is to latch information robustly, then the fraction of the gradient due to information t time steps in the past approaches zero as t becomes large. Interestingly, the long–term dependency problem conflicts with the idea, which has recently been explored by several researchers, that deep architectures, composed by numerous layers of neurons, are necessary in order to cope with complex applications. In fact, it has been proven that deep networks can implement, with a smaller number of neurons, functions that cannot be implemented with shallow architectures [3]. To cope with this problem, a composite architecture, called Layered GNN (LGNN), was recently proposed in [2]. LGNN collects a cascade of GNNs, each of which is fed with the original data (coming from the problem to be faced) and with the information calculated by the previous GNN in the cascade. LGNNs are trained layer by layer, using the original targets and forcing each network in the cascade to improve the solution of the previous one. Intuitively, this allows each GNN to solve a subproblem, related only to those patterns which were misclassified by the previous GNNs. This approach is able to realize a sort of incremental learning that is progressively enriched by taking into account far and far information.

Finally, it is worth mentioning that the possibility of dealing with complex data structures gives rise to several new topics of research. Actually, while it is usually assumed that the learning domain is static, it may happen that the input graphs change in time. In this case, at least two interesting issues can be considered: first, GNNs must be extended to cope with a dynamic learning environment; and second, no method exists to model its evolution. The solution of the latter problem, for instance, may allow us to describe the way in which the Web and, more generally, social networks alter themselves. Furthermore, an open research problem regards how to deal with domains where the relationships, not known in advance, must be inferred. In this last case, in fact, the input is constituted by flat data and must automatically be transformed into a set of graphs in order to discover possible hidden relationships.

References

1. Almeida, L.B.: Backpropagation in perceptrons with feedback. In: Eckmiller, R., Von der Malsburg, C. (eds.) Neural Computers, pp. 199–208. Springer (1988)
2. Bandinelli, N., Bianchini, M., Scarselli, F.: Learning long–term dependencies using layered graph neural networks. In: Proceedings of the IEEE International Joint Conference on Neural Networks, pp. 1–8 (2010)
3. Bengio, Y.: Learning deep architectures for AI. Foundations and Trends in Machine Learning 2(1), 1–127 (2009)
4. Bengio, Y., Simard, P., Frasconi, P.: Learning long–term dependencies with gradient descent is difficult. IEEE Transactions on Neural Networks 5(2), 157–166 (1994)
5. Bianchini, M., Maggini, M., Sarti, L.: Object localization using input/output recursive neural networks. In: Proceedings of the 18th International Conference on Pattern Recognition, vol. 3, pp. 95–98 (2006)
6. Bianchini, M., Maggini, M., Sarti, L., Scarselli, F.: Recursive neural networks for processing graphs with labelled edges: Theory and applications. Neural Networks 18(8), 1040–1050 (2005)
7. Chau, R., Tsoi, A.-C., Hagenbuchner, M., Lee, V.: A conceptlink graph for text structure mining. In: Proceedings of the Thirty-Second Australasian Conference on Computer Science, ACSC 2009, vol. 91, pp. 141–150. Australian Computer Society, Inc. (2009)
8. Crawford, J.D.: Introduction to bifurcation theory. Reviews of Modern Physics 63, 991–1037 (1991)
9. Di Massa, V., Monfardini, G., Sarti, L., Scarselli, F., Maggini, M., Gori, M.: A comparison between recursive neural networks and graph neural networks. In: Proceedings of the IEEE International Joint Conference on Neural Networks, pp. 778–785 (2010)
10. Di Noi, L., Hagenbuchner, M., Scarselli, F., Tsoi, A.-C.: Web spam detection by probability mapping graphSOMs and graph neural networks. In: Proceedings of the International Conference on Artificial Neural Networks, pp. 372–381. Springer (2010)
11. Frasconi, P., Gori, M., Sperduti, A.: A general framework for adaptive processing of data structures. IEEE Transactions on Neural Networks 9(5), 768–786 (1998)
12. Gori, M., Maggini, M., Sarti, L.: A recursive neural network model for processing directed acyclic graphs with labeled edges. In: Proceedings of the International Joint Conference on Neural Networks, vol. 2, pp. 1351–1355 (2003)
13. Gori, M., Monfardini, G., Scarselli, F.: A new model for learning in graph domains. In: Proceedings of the International Joint Conference on Neural Networks, vol. 2, pp. 729–734 (2005)
14. Khamsi, M.A.: An Introduction to Metric Spaces and Fixed Point Theory. Wiley, New York (2001)
15. Lu, L., Safavi-Naini, R., Hagenbuchner, M., Susilo, W., Horton, J., Yong, S.L., Tsoi, A.-C.: Ranking Attack Graphs with Graph Neural Networks. In: Bao, F., Li, H., Wang, G. (eds.) ISPEC 2009. LNCS, vol. 5451, pp. 345–359. Springer, Heidelberg (2009)
16. Monfardini, G., Di Massa, V., Scarselli, F., Gori, M.: Graph neural networks for object localization. In: Proceedings of the Conference on Artificial Intelligence, pp. 665–669. IOS Press (2006)
17. Muratore, D., Hagenbuchner, M., Scarselli, F., Tsoi, A.-C.: Sentence Extraction by Graph Neural Networks. In: Diamantaras, K., Duch, W., Iliadis, L.S. (eds.) ICANN 2010, Part III. LNCS, vol. 6354, pp. 237–246. Springer, Heidelberg (2010)
18. Pearlmutter, B.A.: Learning state space trajectories in recurrent neural networks. Neural Computation 1, 263–269 (1989)

19. Pineda, F.J.: Generalization of Back–Propagation to recurrent neural networks. Physical Review Letters 59, 2229–2232 (1987)
20. Quek, A., Wang, Z., Zhang, J., Feng, D.: Structural image classification with graph neural networks. In: Proceedings of IEEE International Conference on Digital Image Computing Techniques and Applications, pp. 416–421 (2011)
21. Riedmiller, M., Braun, H.: A direct adaptive method for faster Backpropagation learning: The RPROP algorithm. In: Proceedings of the IEEE International Conference on Neural Networks, pp. 586–591 (1993)
22. Scarselli, F., Gori, M., Tsoi, A.-C., Hagenbuchner, M., Monfardini, G.: Computational capabilities of graph neural networks. IEEE Transactions on Neural Networks 20(1), 81–102 (2009)
23. Scarselli, F., Gori, M., Tsoi, A.-C., Hagenbuchner, M., Monfardini, G.: The graph neural network model. IEEE Transactions on Neural Networks 20(1), 61–80 (2009)
24. Scarselli, F., Tsoi, A.-C.: Universal approximation using feedforward neural networks: A survey of some existing methods, and some new results. Neural Networks 11(1), 15–37 (1998)
25. Scarselli, F., Yong, S.L., Gori, M., Hagenbuchner, M., Tsoi, A.-C., Maggini, M.: Graph Neural Networks for ranking web pages. In: Proceedings of IEEE/WIC/ACM Conference on Web Intelligence, pp. 666–672. IEEE Computer Society (2005)
26. Sperduti, A., Starita, A.: Supervised neural networks for the classification of structures. IEEE Transactions on Neural Networks 8, 429–459 (1997)
27. Uwents, W., Monfardini, G., Blockeel, H., Gori, M., Scarselli, F.: Neural networks for relational learning: An experimental comparison. Machine Learning 82(3), 315–349 (2011)
28. Uwents, W., Monfardini, G., Blockeel, H., Scarselli, F., Gori, M.: Two connectionist models for graph processing: An experimental comparison on relational data. In: Proceedings of the International Workshop on Mining and Learning with Graphs, ECML 2006, pp. 211–220 (2006)
29. Williams, R.J., Zipser, D.: A learning algorithm for continually running fully recurrent neural networks. Neural Computation 1, 270–280 (1989)
30. Yong, S.L., Hagenbuchner, M., Tsoi, A.-C., Scarselli, F., Gori, M.: Document Mining Using Graph Neural Networks. In: Fuhr, N., Lalmas, M., Trotman, A. (eds.) INEX 2006. LNCS, vol. 4518, pp. 458–472. Springer, Heidelberg (2007)

Chapter 4
Topics on Cellular Neural Networks

Liviu Goraş, Ion Vornicu, and Paul Ungureanu

In this chapter we present the CNN paradigm introduced by Chua and Yang and several analog parallel architectures inspired by it as well as aspects regarding their spatio-temporal dynamics and applications.

The first part of the chapter is an extremely concise description of the standard CNN philosophy related to the idea of parallel computing. The equations of the standard architecture and influence of the template parameters on the dynamics are briefly discussed. The concept of the CNN universal computing machine is mentioned as well. A few examples of image processing and feature extraction are given.

The second part is devoted to the presentation of a generalization of the original CNN architecture consisting of an array of identical linear cells coupled by voltage controlled current sources which can be studied using the decoupling technique tailored to the boundary conditions and mode competition concept. It is shown that when such architectures exhibit unstable spatial modes, they can be used for linear spatial filtering purposes and for feature extraction by freezing the dynamics before any nonlinearity has been reached. The influence of the cells and coupling on the spatial frequency behavior is investigated and examples are given. Another aspect shortly discussed refers to VLSI implementation possibilities in standard CMOS technology mentioning the linear Operational Transconductance Amplifier and log-domain realizations. Several considerations regarding the effect of cells and template parameters non-homogeneities are made as well.

The last part of the chapter is devoted to the study of the spatio-temporal dynamics of the reaction-diffusion homogeneous two-grid coupling architecture with piecewise linear characteristic cells which has been shown to be able to produce Turing patterns based on spatial mode competition dynamics.

Liviu Goraş · Ion Vornicu · Paul Ungureanu
"Gheorghe Asachi" Technical University, Iasi, Romania

Liviu Goraş
Institute of Computer Science,
Romanian Academy, Iasi Branch

M. Bianchini et al. (Eds.): *Handbook on Neural Information Processing*, ISRL 49, pp. 97–141.
DOI: 10.1007/ 978-3-642-36657-4_4   © Springer-Verlag Berlin Heidelberg 2013

The influence of the cells and grid parameters on the dispersion curve is presented. It is shown that the architecture can be useful for circular spatial filtering and feature extractions based on the fulfillment of Turing conditions.

1 The CNN Concept

1.1 The Architecture

Cellular Neural Networks (CNN's) are parallel computing systems characterized by an architecture consisting of cells connected only with their neighbors [1]-[5]. This feature makes the difference between CNN's and general neural networks whose cells are totally and unhomogeneously interconnected [6]. The spatio-temporal dynamics of such analog architectures possibly associated with image sensors [7]-[12] can be used for high speed 1D and 2D signal processing including linear [13], [14] or nonlinear filtering [15]-[22] and feature extraction [23]. CNN's as they have been defined in the seminal papers by Chua and Yang [1],[2] consisted of an array of identical and identically coupled cells as suggested by the sketch of a 2D 4×4 CNN shown in Fig. 1a. Each cell ij in an M×N array is connected with the cells kl within a neighborhood $N_r(i,j)$ of order r:

$$N_r(i,j) = \left\{ C(k,l) \mid \max\left\{ |k-i|, |l-j| \right\} \leq r, 1 \leq k \leq M, 1 \leq l \leq N \right\} \quad (1)$$

as shown in Fig. 1b

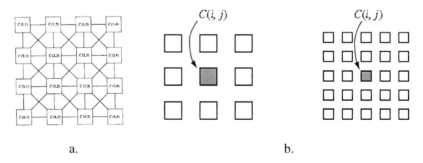

a.   b.

Fig. 1 a. Sketch of a 4×4 CNN; b. Neighborhoods N_r of order r=1 and 2 [1]

The schematic of a cell ij in an M×N array (Fig. 2a) consists of a parallel RC linear circuit, a nonlinear voltage controlled current source with the characteristic shown in Fig. 2b loaded by a linear resistor, an input voltage source E_{ij} and a biasing current source I. The state of the cell ij i.e., the voltage v_{xij} on the cell capacitor is determined by the input and output voltages E_{kl} and v_{ykl} of its r×r neighboring cells (including itself) through the voltage controlled current sources $I_{xu}(i,j,k,l)$ and $I_{xy}(i,j,k,l)$ respectively.

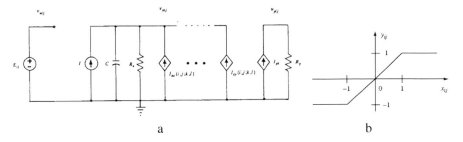

Fig. 2 a. Schematic of a cell; b. Characteristic of the nonlinear voltage controlled current source [1]

1.2   Mathematical Description

The spatio-temporal dynamics of a homogeneous CNN is determined, besides the cells, by two connection matrices i.e., the so-called A and B templates, which represent the weights with which the voltage outputs and respectively the inputs of the neighboring cells influence the state of the cells by means of the voltage controlled current sources discussed above. The processing capabilities of a CNN can be greatly extended if the template elements are nonlinear operators [24], [25]. The equations that describe a CNN are [1]:

State equation:

$$C\frac{dv_{xij}(t)}{dt} = -\frac{1}{R_x}v_{xij}(t) + \sum_{C(k,l)\in N_r(i,j)} A(i,j;k,l)v_{ykl}(t) + \sum_{C(k,l)\in N_r(i,j)} B(i,j;k,l)v_{ukl}(t) + I, \quad 1\le i\le M; 1\le j\le N \quad (2a)$$

$$\text{Output equation: } v_{yij}(t) = \frac{1}{2}\left(\left|v_{xij}(t)+1\right|-\left|v_{yij}(t)-1\right|\right), \quad 1\le i\le M; 1\le j\le N \quad (2b)$$

$$\text{Input equation: } v_{uij}(t) = E_{ij}, \quad 1\le i\le M; 1\le j\le N \quad (2c)$$

$$\text{Constraint conditions: } \left|v_{yij}(0)\right|\le 1, \quad 1\le i\le M; 1\le j\le N$$
$$\left|v_{uij}\right|\le 1, \quad 1\le i\le M; 1\le j\le N \quad (2d)$$

Basically, the CNN's applications in image processing and feature extraction consists in loading images as inputs and/or states of the array and getting the processed image as the outputs of the cells after a transient. In other words, the output image is the equilibrium point of the array. The great majority of CNN's applications of are based on fundamental results concerning the stability of the equilibrium points under various constraints for the template parameters [26]-[40] since oscillatory or chaotic dynamics can be obtained as well.

Since the aim of this Chapter is mainly to present several special CNN architectures, in what follows we give only two application examples.

Two Application Examples

The above described architecture has been thoroughly studied with respect to various templates designed such that a large variety of processing of black and white as well as gray-scale and colored images could be obtained. A plethora of templates useful for various processing tasks can be found in [4] and [24], [25] from which two examples of templates and processing tasks are given below.

1. Convex Corner Detection

$$A = \begin{bmatrix} 0\,0\,0 \\ 0\,1\,0 \\ 0\,0\,0 \end{bmatrix} \qquad B = \begin{bmatrix} -1\,-1\,-1 \\ -1\,\ 4\,-1 \\ -1\,-1\,-1 \end{bmatrix} \qquad I = [-5]$$

1.a. Global Task
Given: static binary image P;
Input: U(t)=P;
Initial State: X(0)=Arbitrary (in the example $x_{ij} = 0$);

Boundary conditions: Fixed type, $u_{ij} = 0$ for all virtual cells, denoted by [u]=0;
Output: Y(t)=Y(∞) = Binary image where black pixels represent the convex corners of objects in P

Remark: Black pixels having at least 5 white neighbors are considered to be convex of the objects.

1.b. Example: image name: chinese.bmp, image size: 16x16; template name: corner.tem

input   output

2. Grayscale Contour Detector

$$A = \begin{bmatrix} 0\,0\,0 \\ 0\,2\,0 \\ 0\,0\,0 \end{bmatrix} \qquad B = \begin{bmatrix} a\,a\,a \\ a\,0\,a \\ a\,a\,a \end{bmatrix} \qquad I = [0.7]$$

where a is defined by the following nonlinear function:

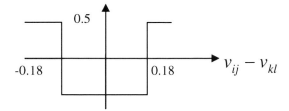

2.a. Global task

Given: static grayscale image P
Input: U(t)=P
Initial State: X(0)=P

Boundary conditions: Fixed type, $u_{ij} = 0$ for all virtual cells, denoted by [u]=0;

Output: Y(t)–Y(∞) – Binary image where black pixels represent the contours of the objects in P

Remark: The template extracts contours which resemble edges (resulting from big changes in gray level intensities) from grayscale images.

2.b. Example: image name: Madonna.bmp, image size: 59x59; template name: contour.tem

input   contour

1.3   Other Tasks CNN's Can Accomplish – The CNN Universal Machine

The main goal CNN's have been designed for was high speed image processing; currently they can process up to 50,000 frames/s which makes them more suitable for tasks like missile tracking, flash detection and spark-plug diagnostics than conventional supercomputers. The major advantage of CNN's is the high processing speed due to the hardware implementation of processor intensive operations. Even though, compared to digital implementations, the precision is lower there are applications like *feature extraction, level and gain adjustments, color constancy detection, contrast enhancement, deconvolution, image compression, motion estimation, image encoding, image decoding, image segmentation, orientation preference maps, pattern learning/recognition, multi-target tracking, image stabilization, resolution enhancement, image deformations and mapping, image inpainting, optical flow, contouring, moving object detection, axis of symmetry detection, and image fusion* [5] which have been proved to work very successfully. Such tasks have been realized using a highly flexible architecture proposed and implemented under the name of **CNN Universal Machine** [15]-[22] which, based on local memory and template programming, speculates the possibility of changing the templates and thus performing various as well as successive processing tasks. There exist specialized programming languages and algorithms that facilitate the implementation of global tasks.

Before proceeding we make several considerations concerning the criteria CNN's can be classified from the point of view of circuit and system theory. In all cases the cells or neurons are dynamical systems described as shown above by state and output equations (2a, b, c). From the point of view of linearity, the cells as well as the interconnections can be linear or nonlinear. The dynamics of the array is *continuous* in *time* (even though discrete time array have also been reported) and *discrete in space*. From the point of view of spatial homogeneity, the array can be homogeneous or non-homogeneous due to the cells and/or to the templates. Obviously, the nonlinear and non-homogeneous character complicates the analysis and design problems. This is why most approaches assume linear or nonlinear but homogeneous arrays.

2 A Particular Architecture

In what follows we are going to describe and analyze an architecture inspired from the standard CNN for operation in the central linear part of the cell characteristic [41]-[43]. As it will be shown shortly, the hypothesis of linearity and spatial homogeneity allows the use of the powerful decoupling mode technique to analyze the spatio-temporal dynamics of the array.

2.1 The Architecture and the Equations

To simplify things, in the following we consider a 1D array with the architecture shown in Fig. 3; generalizations to 2D are straightforward and will be presented later. The array consists of linear (or piecewise linear working in the central linear part) cells represented by admittances denoted by Y(s) and coupled between them as well as with the inputs, using voltage controlled current sources over a neighborhood of radius r, N_r.

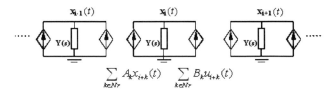

Fig. 3 1D array architecture

The 1D template elements have been denoted by A_k for inter-cell connections and by B_k for the sources connections – their physical dimension is (trans)conductance. The neighborhood dimension has been chosen the same for both cases, a fact that does not restrict the generality as any of the template coefficients might be zero.

The CNN output is the voltage $x_i(t)$ across the admittance $Y(s)$ which is a linear integro-differential operator of the form:

$$Y(s) = \frac{Q(s)}{P(s)} = \frac{\displaystyle\sum_{l=0}^{q} q_l s^l}{\displaystyle\sum_{n=0}^{p} p_n s^n} \tag{3}$$

in particular a real-positive function where $Q(s)$ and $P(s)$ are polynomials in the variable $s \leftrightarrow d/dt$.

The equations that formally describe the network are:

$$Y(s)x(t) = \sum_{k \in Nr} A_k x_{i+k}(t) + \sum_{k \in Nr} B_k u_{i+k}(t) \tag{4}$$

or

$$\sum_{l=0}^{q} q_l \frac{d^l}{dt^l} x_i(t) = \sum_{k \in Nr} \sum_{n=0}^{p} A_k p_n \frac{d^n}{dt^n} x_{i+k}(t) + \sum_{k \in Nr} \sum_{n=0}^{p} B_k p_n \frac{d^n}{dt^n} u_{i+k}(t), \quad i = 0,1,...,M-1 \tag{5}$$

2.2 The Decoupling Technique

The above relations represent a set of coupled integro-differential equations which can be elegantly solved by the decoupling technique [43] which basically consists of the change of variables

$$x_i(t) = \sum_{m=0}^{M-1} \Phi_M(i,m)\hat{x}_m(t) \tag{6}$$

$$u_i(t) = \sum_{m=0}^{M-1} \Phi_M(i,m)\hat{u}_m(t) \tag{7}$$

where the M functions $\Phi_M(i,m)$ depend on the boundary conditions and are orthogonal with respect to the scalar product in C^M, i.e.,

$$< \Phi_M(i,m), \Phi_M(i,n) > = \sum_{i=0}^{M-1} \Phi_M^*(i,m)\Phi_M(i,n) = \sum_{i=0}^{M-1} \Phi_M(m,i)\Phi_M(i,n) = \delta_{mn} \tag{8}$$

so that $\hat{x}_m$ and $\hat{u}_m$ can be expressed, by means of the inversion formulas:

$$\hat{x}_m(t) = \sum_{i=0}^{M-1} \Phi_M^*(m,i)x_i(t)$$
$$\hat{u}_m(t) = \sum_{i=0}^{M-1} \Phi_M^*(m,i)u_i(t)$$
$$m = 0,...,M-1 \tag{9}$$

where

$$\Phi_M(m,i) = \Phi_M^*(i,m) \tag{10}$$

For ring boundary conditions $\Phi_M(m,i)$ have the form $\Phi_M(i,m) = e^{j2\pi mi/M} = e^{j\omega_m i} = e^{j\alpha_m i}$ where $\omega_0 = 2\pi/M$ and $\omega_m = 2\pi m/M = \omega_0 m$ so that $\Phi_M(i+k,m) = e^{j2\pi mk/M}\Phi_M(i,m)$. For other types of boundary conditions suitable orthogonal functions have been reported in [41] and will be presented in the third part of this Chapter. Thus the action of the spatial operators A and B on $\Phi_M(i,m)$ gives

$$\sum_{k \in Nr} A_k \Phi_M(i+k,m) = K_A(m)\Phi_M(i,m) \tag{11}$$

$$\sum_{k \in Nr} B_k \Phi_M(i+k,m) = K_B(m)\Phi_M(i,m) \tag{12}$$

where

$$K_A(m) = \sum_{k=-r}^{r} A_k e^{j2\pi mk/M} = A_0 + \sum_{k=1}^{r}(A_k + A_{-k})\cos 2\pi mk / M + j\sum_{k=1}^{r}(A_k - A_{-k})\sin 2\pi mk / M \tag{13}$$

and $K_B(m)$ has a similar form.

Therefore $\Phi_M(i,m)$ are eigenfunctions of the spatial operators represented by the A and B templates and $K_A(m)$ and $K_B(m)$ are the corresponding spatial eigenvalues which are complex in general and depend on the parameters of the template and on the spatial mode number m. For symmetric templates, $A_{-k} = A_k$, the spatial eigenvalues are real:

$$K_A(m) = A_0 + 2\sum_{k=1}^{r} A_k \cos \frac{2\pi mk}{M} \tag{14}$$

$$K_B(m) = B_0 + 2\sum_{k=1}^{r} B_k \cos \frac{2\pi mk}{M} \tag{15}$$

and, in particular, for r=1 and r =2 they are

$$K_A(m) = A_0 + 2A_1 \cos \omega(m) = A_0 + 2A_1 \cos \frac{2\pi}{M} m \tag{15}$$

and

$$K_A(m) = A_0 + 2A_1 \cos \omega(m) + 2A_2 \cos 2\omega(m) = A_0 + 2A_1 \cos \frac{2\pi}{M}m + 2A_2 \cos \frac{4\pi}{M}m \quad (16)$$

respectively, with similar expressions for $K_B(m)$.

Using the above change of variable and the properties of $\Phi_M(i,m)$, equations (5) become successively

$$\sum_{l=0}^{q} q_l \frac{d^l}{dt^l} \sum_{m=0}^{M-1} \Phi_M(m,i)\hat{x}_m(t) = \sum_{k \in Nr} \sum_{n=0}^{p} A_k P_n \frac{d^n}{dt^n} \sum_{m=0}^{M-1} \Phi_M(m,i+k)\hat{x}_m(t) + \sum_{k \in Nr} \sum_{n=0}^{p} B_k P_n \frac{d^n}{dt^n} \sum_{m=0}^{M-1} \Phi_M(m,i+k)\hat{u}_m(t), \quad i=0,1,...,M-1$$

$$\sum_{m=0}^{M-1} \Phi_M(m,i) \sum_{l=0}^{q} q_l \frac{d^l}{dt^l} \hat{x}_m(t) = \sum_{m=0}^{M-1} \sum_{k \in Nr} A_k \Phi_M(m,i+k) \sum_{n=0}^{p} P_n \frac{d^n}{dt^n} \hat{x}_m(t) + \sum_{m=0}^{M-1} \sum_{k \in Nr} B_k \Phi_M(m,i+k) \sum_{n=0}^{p} P_n \frac{d^n}{dt^n} \hat{u}_m(t), \quad i=0,1,...,M-1$$

$$\sum_{m=0}^{M-1} \Phi_M(m,i) \sum_{l=0}^{q} q_l \frac{d^l}{dt^l} \hat{x}_m(t) = \sum_{m=0}^{M-1} K_A(m)\Phi_M(m,i) \sum_{n=0}^{p} P_n \frac{d^n}{dt^n} \hat{x}_m(t) + \sum_{m=0}^{M-1} K_B(m)\Phi_M(m,i) \sum_{n=0}^{p} P_n \frac{d^n}{dt^n} \hat{u}_m(t), \quad i=0,1,...,M-1$$

If we take the scalar product of both sides of the last equations with $\Phi_M(n,i)$ and then replace the index n with m, we have

$$\sum_{l=0}^{q} q_l \frac{d^l}{dt^l} \hat{x}_m(t) = K_A(m) \sum_{n=0}^{p} P_n \frac{d^n}{dt^n} \hat{x}_m(t) + K_B(m) \sum_{n=0}^{p} P_n \frac{d^n}{dt^n} \hat{u}_m(t), \quad m=0,1,...,M-1 \quad (17)$$

which can be written symbolically either

$$Y(s)\hat{x}_m(t) = K_A(m)\hat{x}_m(t) + K_B(m)\hat{u}_m(t) \quad (18)$$

or

$$Q(s)\hat{x}_m(t) = K_A(m)P(s)\hat{x}_m(t) + K_B(m)P(s)\hat{u}_m(t) \quad (19)$$

The above equation is the differential equation satisfied by the amplitude of the m-th spatial mode. Thus, the initial set of equations has been decoupled, the new variables being the amplitudes of the spatial modes of the cell signals with respect to $\Phi_M(i,m) = e^{j2\pi mi/M}$.

It easy to observe that a transfer functions valid for each spatial mode can be defined as

$$H_m(s) = \frac{\hat{x}_m(s)}{\hat{u}_m(s)} = \frac{K_B(m)}{Y(s) - K_A(m)} = \frac{K_B(m)Z(s)}{1 - K_A(m)Z(s)} = \frac{K_B(m)P(s)/Q(s)}{1 - K_A(m)P(s)/Q(s)}, \quad m=0,1,...M \quad (20)$$

where $\hat{x}_m(s)$ and $\hat{u}_m(s)$ are the Laplace transforms of $\hat{x}_m(t)$ and $\hat{u}_m(t)$ respectively.

The transfer functions correspond to a feedback system for each of the modes as depicted in Fig. 4 where $Z(s)=1/Y(s)$.

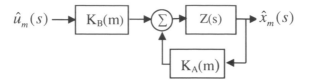

Fig. 4 Feedback schematic for the spatial mode m

The characteristic polynomial of the m-th mode is

$$R(s) = Q(s) - K_A(m)P(s) \qquad (21)$$

Thus, the stability and dynamics of the spatial modes will depend both on the A-template eigenvalues and on the cell admittance $Y(s)=Q(s)/P(s)$. The spatio-temporal dynamics of the array can be studied using classical methods from feedback/control theory such as the root locus and Nyquist criterion and, of course, the Routh-Hurwitz test, all valid for each of the spatial modes, conveniently modified for non-symmetric templates [41]. Depending on $K_A(m)$ the differential equation describing the evolution of the amplitude of the m-th spatial mode can have roots either in the left complex half plane or in the right one – a case when the mode is unstable and will increase in time until the dynamics will be frozen by means of switches. The stability or instability of the spatial modes is thus determined by the sign of $K_A(m)$, the so-called called *dispersion curve*.

2.3 Particular Cases

For the particular case $Y(s)=Cs$ and $B_k=0$ we obtain the following set of decoupled autonomous first order differential equations:

$$C\frac{d\hat{x}_m(t)}{dt} = K_A(m)\hat{x}_m(t) \qquad m = 0,...,M-1 \qquad (22)$$

The roots of the characteristic equation are $s=K_A(m)/C$, so that the dispersion curve is a straight line with respect to $K_A(m)$.

For $Y(s)=Cs+G$ we obtain a similar equation (the -G constant can be always absorbed by A_0 in $K_A(m)$:

$$C\frac{d\hat{x}_m(t)}{dt} = \left(-G + K_A(m)\right)\hat{x}_m(t) \qquad m = 0,...,M-1 \qquad (23)$$

The roots of the m-th characteristic equation are thus $s_m=-G+K_A(m)$ so that, again, the dispersion curve is a straight line with respect to $K_A(m)$ but not with m where $K_A(m)$ is given by formula (16) for second order neighborhood. In this last case, the roots of $K_A(m)=0$ where A_0 contains all constant terms are:

$$m_1 = \frac{M}{2\pi}\left(\arccos\left(\frac{-A_1 + \sqrt{A_1^2 - 4A_0A_2 + 8A_2^2}}{4A_2}\right)\right) \quad (24)$$

$$m_2 = \frac{M}{2\pi}\left(\pi - \arccos\left(\frac{A_1 + \sqrt{A_1^2 - 4A_0A_2 + 8A_2^2}}{4A_2}\right)\right) \quad (25)$$

When real, the above roots represent the intersection of the dispersion curve with the abscissa axis in the hypothesis of a continuous variation of m.

Thus, if the quantity under the square root is positive and the module of the argument of the *arcos* function is less than unity, the system might have poles on the positive real axis of the complex plane on the condition that at least one integer m is placed in the domain of unstable modes, i.e., for which $K_A(m)$ is positive. In such a case the unstable spatial modes will increase according to their weight in the initial conditions and/or input signal and the value of the (positive) temporal eigenvalues, while stable modes will decrease and finally vanish. Using appropriate switches the dynamics of the array can be stopped/frozen before any nonlinearity has been reached. In this way a linear time dependent spatial filter is obtained. By freezing the spatio-temporal dynamics at a certain moment of time, a filtered version of the input signal can be obtained. By increasing the final time the filter selectivity will increase as well since the "more unstable" spatial modes will increase much more than the "less unstable" ones.

In Fig. 5 the dispersion curves for a high-pass and a band-pass spatial filter are presented for an M=50 cells 1D array. The bands of unstable modes, i.e., with real positive roots are marked with black lines. Note that the dispersion curve is significant for values of m less than M/2 as the complex exponential for m>M/2 combine to those with m<M/2 to give cosine functions according to Euler formulas.

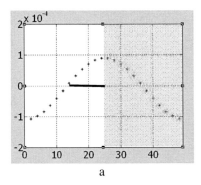

a

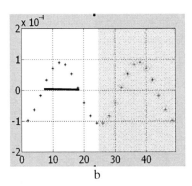

b

Fig. 5 $K_A(m)$ for a high pass (a) and a band-pass (b) spatial filter (M=50)

2.4 Implementation Issues

2.4.1 Using OTA's

The implementation of the above discussed architectures is significantly determined by the type of voltage controlled current sources used. Their implementation can be based either on standard OTA's or on log-domain structures.

In the standard approach the implementation of the voltage controlled current sources implies the design of transconductors whose g_m's are the coefficients of the template matrix. For large scale integration, which means a better spatial filter resolution, it is advisable that the current amplifiers have a simple structure, in order to occupy a smaller silicon area, and have low power consumption. In all cases a compromise should be made between *bandwidth, linearity, output swing and power consumption*. For our simulations the basic transconductor amplifier structure presented in Fig. 6 has been chosen.

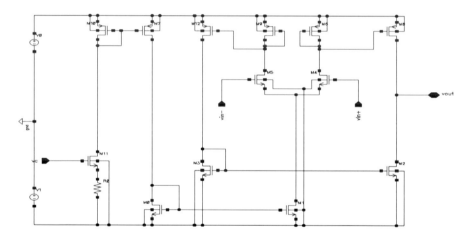

Fig. 6 Basic voltage controlled OTA's structure

The simulations were realized in Spectre - Cadence, the circuitry being sourced by a +/-1.65V differential voltage. A Monte Carlo analysis showed small g_m dispersion as seen in Fig. 7a. The dependence of the filter selectivity on A_0 for the same value of the freezing time is shown in Fig. 7b. The purpose of this type of analysis was to see the influence of process variations on the dispersion curve. As it can be seen in Figure 7a, the high-pass filter characteristics, resulted from the Monte Carlo simulation, are very close, fact that shows network *robustness* with technology process variations.

As we will further show, transistor level circuit simulations match very well the ideal ones. The main disadvantage, consisting in pixel level high power consumption, makes this solution to be rather unpractical for high resolution applications. The cell capacity can not be made too small in order to increase the processing speed, since the time constant should be large enough to ensure data sampling with a good enough accuracy. The system dynamic must be "frozen" before any transconductor reaches the saturation region, otherwise the system will be no more linear. From the *"modes competition"* point of view, the dynamics of the cell voltages shows a remarkable robustness of the filter implemented with transconductances designed without special constraints on the dynamic range.

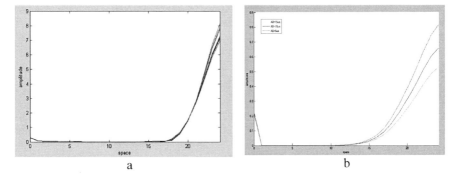

a   b

Fig. 7 a. High-pass filter characteristics frozen at the same time and affected by process variations; b. High pass-filter characteristics for A_0=5us, 10us, 15us

Even though keeping the same freezing time and changing the value of A_0 or changing it and keeping the same value of A_0 seems to give the same dynamic, there is a significant difference between these two cases: in the first one the dispersion curve is different thus changing the bandwidth of the filter and keeping the same selectivity, while in the second one, the selectivity of the filter varies with the freezing time, corresponding to the same dispersion curve. For large enough variations of A_0, this difference is highlighted better.

2.4.2 Using Log-Domain Techniques

Another possibility of implementation of the discussed architectures is based on *log-domain techniques.*

As it is well known [44], [45], the dynamic translinear circuit principle is based on the fact that the derivative of a current can be written as a product of two currents. Such circuits compand the input (by making the voltage logarithm dependent on a current), which is an advantage from the dynamic range point of view and can be easily implemented using class AB circuitry (offering a large dynamic range with a low mean power consumption). Moreover, having a low voltage range, they exhibit a relatively large bandwidth, dynamic translinear circuits being tunable for a wide range of frequencies, quality factor and gain, making them attractive to be used as standard cells or programmable building blocks.

In the following we briefly show the conversion of the CNN equations (1D for simplicity) into the log-domain and the implementation based on CMOS transistors in weak inversion [44].

Thus, using the change of variable

$$x(i) = I_S e^{\alpha v_x(i)} - I_S \tag{26}$$

for a symmetric second order neighborhood, state equation "i",
$C \dfrac{x_i(t)}{dt} = \sum\limits_{k \in Nr} A_k x_{i+k}(t)$ can be written in the form

$$CI_S \alpha \frac{v_x(i)}{dt} e^{\alpha v_x(i)} = A_2 I_S e^{\alpha v_x(i-2)} + A_1 I_S e^{\alpha v_x(i-1)} - A_0 I_S e^{\alpha v_x(i)} + A_1 I_S e^{\alpha v_x(i+1)} + A_2 I_S e^{\alpha v_x(i+2)} \tag{27}$$
$$- I_S (2A_1 + 2A_2 - A_0)$$

Dividing equation (27) by $e^{\alpha v_x(i)}$ and using the notations:

$$C_x = CI_S \alpha, \ x_{offset} = I_S(2A_1 + 2A_2 - A_0) = I_{x_0}$$
$$A_1 I_S = I_{A_1}, \ A_2 I_S = I_{A_2}, \ A_0 I_S = I_{A_0} \tag{28}$$

the new state equation "i" can be rewritten as:

$$C_x \dot{v}_x(i) = CT - I_{x_0} e^{-\alpha v_x(i)} \tag{29}$$

where CT represent the coupling terms

$$CT = I_{A_2} e^{\alpha(v_x(i-2)-v_x(i))} + I_{A_1} e^{\alpha(v_x(i-1)-v_x(i))} + I_{A_2} e^{\alpha(v_x(i+2)-v_x(i))} + I_{A_1} e^{\alpha(v_x(i+1)-v_x(i))} - I_{A_0} \tag{30}$$

Equation (27) can be interpreted as representing the Kirchhoff current law for cell i: the current through the capacitor C_x equals the sum of currents injected by the current sources nonlinearly controlled by the voltages of the neighboring cells while CT represents the contribution of neighboring cells over the i-th cell. Each term within CT represents the nonlinear equivalent in the log-domain of the voltage controlled current sources and $I_{x_0} e^{-\alpha v_x(i)}$ defines a nonlinear conductance.

2.4.3 Log-Domain Transistor Level Simulations

As it has been previously shown (29) represents the state equation associated to cell i of a first order high-pass spatial filter translated into the log-domain. The structure of a pixel obtained by the nonlinear mapping of the state equations of a linear autonomous system is composed of a nonlinear conductance, four current sources exponentially controlled by the voltages of the neighboring cells, and a dc current source.

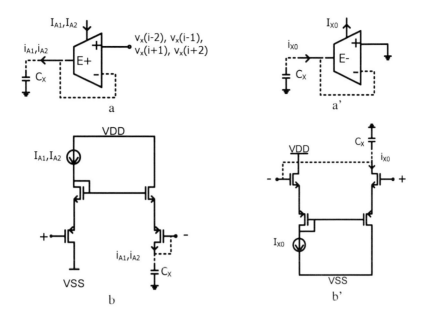

Fig. 8 a) Current source exponentially controlled by the voltages of neighboring cells (injects current in cell capacitor); a') current source exponentially controlled by the voltage of the current cell (sinks current from cell capacitor), b) transistor implementation of a), b') transistor implementation of b)

The basic building blocks used for the implementation of the log-domain equations are shown in Fig. 8. The nonlinear controlled sources in Fig. 8 a, b, respectively 8 a', b' have three inputs: a current input (I_{A1}, I_{A2} or I_{X0}), and two voltage inputs (v_+, v_-). The scaling factor of the sources can be controlled by an external biasing current. In this way the template coefficients and thus the spatial filtering characteristics can be modified.

According to the above considerations, the schematic of a logarithmic cell is presented in Fig. 9 and corresponds to equation (29).

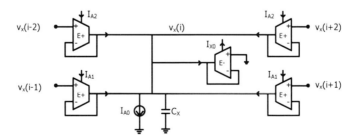

Fig. 9 Cell schematic for a low-pass/stop-band filter structure

The above structure has been used to simulate the implementation of a low-pass filter ($A_1 \neq 0$, $A_2 = 0$) and a stop-band filter ($A_1 = 0$, $A_2 \neq 0$).

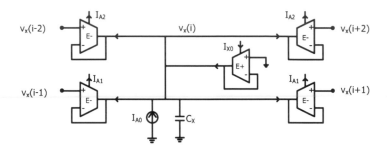

Fig. 10 Cell schematic for a high-pass/band-pass filter structure

Multiplying equation (27) by minus one, the cell structure for a high-pass/band-pass spatial filter shown in Fig. 10 can be easily obtained.

2.4.4 Comparison between 1D OTA and Log-Domain Implementations

In what follows we present the spatial frequency characteristics of a 50 cell 1D architecture configured for high-pass, low-pass, band-pass and stop-band simulated with linear OTA implementations and at log-domain transistor level. For simplicity, the term A_0 has been considered zero.

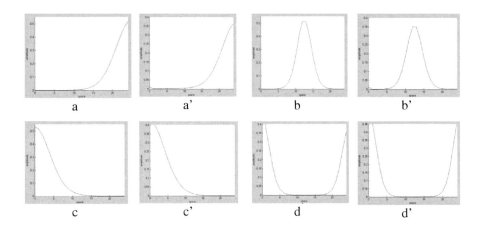

Fig. 11 High pass, band-pass, low pass respectively stop-band spatial frequency characteristics for OTA (a, b, c, d) and log-domain transistor level implementation (a', b', c', d')

The characteristics have been obtained by seeding only one cell of the network with a nonzero initial condition of 10mV and freezing the transient at the same time moment. In this way the initial condition contained all spatial frequencies with equal weights and at the freezing moment the *DFT of the output represents the frequency characteristics of the array*. The system level and log-domain filter parameters are shown in the table below:

Table 1 Filters parameters

Cell Implementation	Filter type	Freq. characteristic	Implementation type					
			Log-domain transistor filter				OTA filter	
			Filter coefficients					
			I_{X0}	I_{A0}	I_{A1}	I_{A2}	A_1	A_2
Fig 9	HPF	Fig. 4a/a'	2nA	0	1nA	0	1nS	0
	BPF	Fig. 4b/b'	2nA	0	0	1nA	0	1nS
Fig. 10	LPF	Fig. 4c/c'	2nA	0	1nA	0	-1nS	0
	SBF	Fig. 4d/d'	2nA	0	0	1nA	0	-1nS

For different "freezing" times various spatial frequency selectivity can be obtained. In Fig. 12 the time evolution of the spatial frequency characteristic of high-pass and band-pass architectures for the OTA transistor level and log domain transistor level are presented.

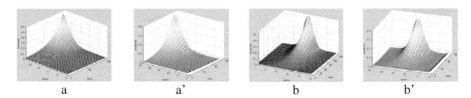

| a | a' | b | b' |

Fig. 12 Time evolution of the spatial frequency characteristics for a high-pass (a, a') and band-pass (b, b') CNN filter at OTA and log-domain transistor level respectively

2.5 A "Toy" Application: 1D "Edge" Detection

An example of a 1D "edge" detection using a high-pass filter realized with linear OTA and log-domain architecture is shown in Figure 13, the input state being a 10 mV amplitude square spatial signal; the transient has been stopped in both cases after 53us. It is apparent that the responses of the two implementations are almost identical.

HP or BP ideal filter characteristics are obtained taking a_0=10us, a_1=50us, a_2=0 respectively A_0=10us, A_1=0, A_2=50us, and for LP or SB they can be obtained setting the coefficients as follows: A_0=10us, A_1=-50us, A_2=0 and A_0=10us, A_1=0, A_2=-50us respectively; in the transistor level case, the HP or BP frequency characteristic result setting the voltages control as: V_{c1}=1.65V, V_{c2}=-1.65V and V_{c1}=-1.65, V_{c2}=1.65 respectively, while the LP or SB frequency characteristic are obtained with the same voltages control by changing the polarity of the transconductors inputs.

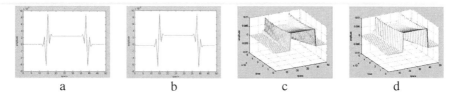

a   b   c   d

Fig. 13 1D "edge" detection using linear OTA (a) and log-domain, (b) implementations, respective time evolution (c, d) for a square spatial input signal

Among the advantages of the log-domain realization are the *small power consumption* and *silicon area* while the main disadvantage is the *high technological dispersion* in the implementation. However, some recent results show a *certain robustness* of the spatio-temporal dynamics even for rather high deviations of the parameters from their nominal values. Last but not least, a significant aspect related to the design of parallel architectures using exponentially voltage controlled current sources is the need of keeping the transistors width as low as possible in order to *minimize parasitic capacitance* since the cell capacitance should be much higher than the parasitic capacitance of each source.

2.5.1 A 2D Log-Domain Architecture

Many implementations of image segmentation can be found in the literature. Among them, interesting performances have been obtained with CNN's based on a non-linear resistive grid. The nonlinear resistor used in such networks, so called "resistive fuse", can be implemented in different ways. Several circuit solutions are based on Chua's negative resistor [46], on transmission gate controlled by a combinational digital circuit [47], on pulse-modulation techniques [48], or on switch capacitors [49]

In the following segmentation results obtained using the above discussed spatial-temporal filter [49] will be presented and compared with those reported in the literature. Moreover, a modified log-domain architecture that includes selective filtering techniques in order to preserve edges has been tested and simulated at transistor level. The array is designed such that the initial images are loaded from an external memory or an embedded CMOS imager and the output data can be read out line by line after the temporal dynamic of the network is frozen. The following basic image processing operations can be performed: *edge detection, smoothing, noise cancellation, contrast enhancement* and *segmentation*.

The implementation of several operations for a bi-dimensional image, using the proposed analog parallel architecture [42] will be briefly presented as well.

In the following, a 2D array of MxM identically coupled identical cells as shown in Fig. 14 is considered. Each cell is connected with its neighboring cells belonging to a maximum second order neighborhood. The boundary conditions are ring, but can be zero-flux as well as it will be shown for segmentation using the selective filtering technique.

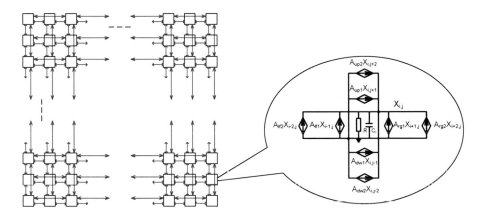

Fig. 14 2D array architecture/ System level 2D pixel structure

The system level structure of a single cell is presented in Fig. 14 as well. The voltage controlled current sources are implemented with OTA's using an adaptive biasing technique in order to decrease the DC power consumption.

The linear differential equation valid for the (i,j) node, taking into account ring boundary conditions for homogeneous networks ($A_{lf1}=A_{rg1}=A_{up1}=A_{dw1}=A_1$; $A_{lf2}=A_{rg2}=A_{up2}=A_{dw2}=A_2$), has the following expression:

$$C\frac{dx_{i,j}(t)}{dt} = A_2(x_{i-2,j} + x_{i+2,j} + x_{i,j-2} + x_{i,j+2}) + A_1(x_{i-1,j} + x_{i+1,j} + x_{i,j-1} + x_{i,j+1}) - A_0 x_{i,j}, \quad \forall i,j=0..M\text{-}1 \ (31)$$

where "$x_{i,j}(t)$" is the (i, j) node voltage.

The above linear differential equations can be translated into the log-domain as in the 1D case by using the change of variable (26).

Denoting $x_{i,j}(t) = x(i,j)$ and $v_{x,ij}(t) = v_x(i,j)$, the linear differential equation (31) becomes:

$$CI_S \alpha \dot{v}_x(i,j)e^{\alpha v_x(i,j)} = A_2 I_S(e^{\alpha v_x(i-2,j)} + e^{\alpha v_x(i+2,j)} + e^{\alpha v_x(i,j-2)} + e^{\alpha v_x(i,j+2)}) + A_1 I_S(e^{\alpha v_x(i-1,j)} +$$
$$e^{\alpha v_x(i+1,j)} + e^{\alpha v_x(i,j-1)} + e^{\alpha v_x(i,j+1)}) - A_0 I_S e^{\alpha v_x(i,j)} - I_S(4A_1 + 4A_2 - A_0) \quad (32)$$

Dividing the nodal state equation (32) by $e^{\alpha v_x(i,j)}$ and using the same notations (28) except for:

$$x_{offset} = I_S(4A_1 + 4A_1 - A_0) = I_{x_0}$$

the new (i,j) state equation takes the form :

$$C_x \overset{\bullet}{v}_x(i,j) = CT - I_{x_0} e^{-\alpha v_x(i,j)}$$

$$CT = I_{A_2}(e^{\alpha(v_x(i-2,j)-v_x(i,j))} + e^{\alpha(v_x(i+2,j)-v_x(i,j))} + e^{\alpha(v_x(i,j-2)-v_x(i,j))} + e^{\alpha(v_x(i,j+2)-v_x(i,j))}) + \quad (33)$$

$$I_{A_1}(e^{\alpha(v_x(i-1,j)-v_x(i,j))} + e^{\alpha(v_x(i+1,j)-v_x(i,j))} + e^{\alpha(v_x(i,j-1)-v_x(i,j))} + e^{\alpha(v_x(i,j+1)-v_x(i,j))}) - I_{A_0}$$

The above equation can describe a low-pass/ stop-band filter for positive coupling coefficients and a high-pass/ band-pass one for negative coefficients, providing that a band of unstable modes has been ensured [43]. Based on the above equations, the log-domain implementation of a 2D pixel is given in Fig. 15.

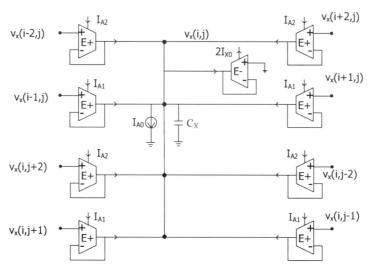

Fig. 15 Log-domain implementation of a 2D pixel structure

2.5.2 Comparison between 2D OTA and Log-Domain Implementations for Applications Using 64x64 Log-Domain Architectures

Edge detection and contrast enhancement on one side and smoothing, i.e., noise cancellation or details blurring on the other side can be performed using the same network configured to have a high-pass or low-pass spatial frequency characteristics respectively.

Fig. 16 presents the obtained results for a chessboard type input image.

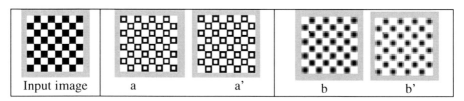

Fig. 16 a, a') edges extraction obtained with OTA (a) and log-domain (a') filters; b, b') smoothed image obtained with OTA (b) and log-domain (b') filters. Freezing time =435ns

Other results obtained for gray-scale images are presented in Figure 17.

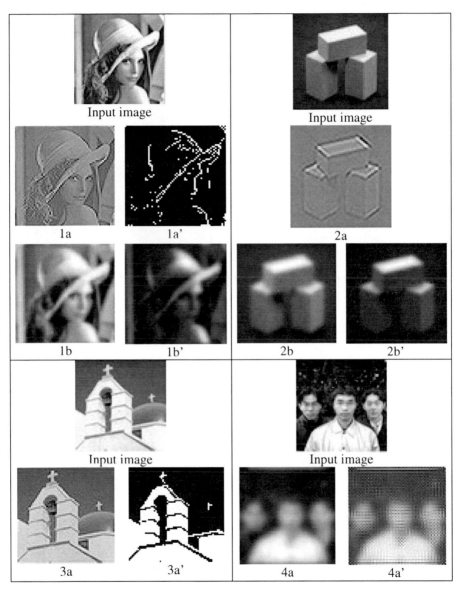

Fig. 17 1a) high-pass filtering after 55us; 1a') high-pass filtered version with a 300mV threshold; 1b) low-pass filtering after 105us; 1b')low-pass filtering after 78us, performed by log-domain implementation; 2a) high-pass filtering after 71us; 2b) low-pass filtering after 160us, with the OTA network; 2b') low-pass filtering after 71us, using a log-domain filter; 3a,a') high-pass filtering with 300mV threshold for edge detection, after 28us 4a) stop-band filtering using asymmetric spatial frequency characteristic with higher low-band amplification; 4b') stop-band filtering using asymmetric spatial frequency characteristic with higher high-band amplification

2.5.3 Image Segmentation

Image segmentation aims at objects recognition, borders positions estimation for moving objects and image compression. It is known that linear resistive grids can be used only for smoothing operations. However, if the so called "resistive fuses" are used instead of linear resistors, the basic network is fragmented into zones that have the same spatial contrast that do not surpass a given threshold.

The analog parallel network described above can be easily modified according to the principle of the nonlinear resistive grid - the pixel structure is slightly changed compared with the basic scheme shown in Fig. 14 and a circuit that calculates the module of the difference between the voltages of two consecutive pixels is introduces. This value will be compared with a given threshold and stored in order to control the gain of the voltage controlled current sources. In this way, a connection between two neighbors can be kept or cut off so that fragmentations in the compact network can be achieved.

Thus every fragmented sub-network has the difference between any two neighbors under a given threshold, meaning that the voltage map of this sub-network is a rather uniform surface.

Each sub-network can be analyzed by the decoupling technique valid for the homogeneous architecture. All sub-network features are kept only if all parameters of each active pixel remain unchanged.

The interconnectivity map between network cells can be set from the beginning and kept during the filtering process, continuously updated or updated only at a given moment.

On the other hand, the compact analog architecture is able to perform segmentation without the above described fragmentation techniques borrowed from the nonlinear resistive grid. This behavior is obtained by programming the network in a low-pass configuration. Since the initial differences between similar contrast level are also amplified by the low-pass filter, it somehow compensate the unwanted edges filtering, finally resulting that the need for fragmentation is not a must in order to obtain a segmentation effect. Even though it is hard to appreciate the behavior of the segmented network compared to the counterpart compact filter in what concerns the temporal evolution of the spectral components as it can be observed from Fig. 18.3a, b, c, the fragmented network has a different dynamic than the compact one (Fig. 18.1a, 2a, 2c). In order to see the differences between the two implementations before the filter reaches nonlinearities, the unfragmented architecture has to be setup with a large selectivity (close to the instability limit) to slow down the unstable behavior. Another remark regarding these two types of implementation refers to the processing speed per frame: the fragmented filter performs the same results like the compact one, but in a shorter time.

The advantage of the so-called fragmented implementation becomes significant when different regions from an image have to be filtered in different ways. This is possible only because each sub-network exhibit an independent dynamics compared to the others.

Also, a selective kind of decoupling technique is useful for a nonlinear processing by disconnecting the saturated region from the rest of the network, thus the nonlinear part of the filter does not affect the linear one.

In the following we present several results for image segmentation, obtained using a 2D network, based on OTA or log-domain simulated at transistor level.

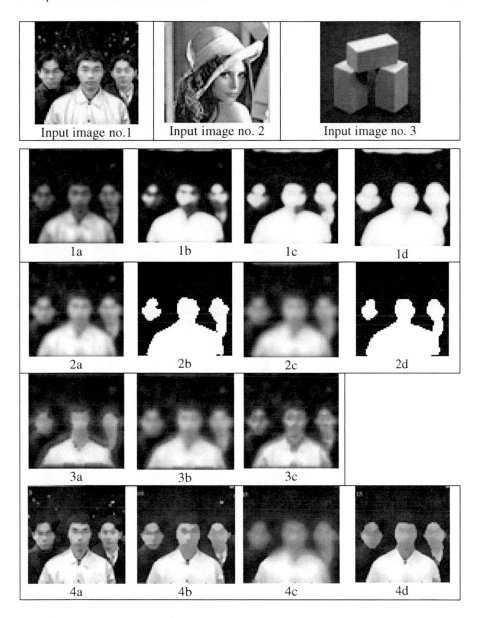

Fig. 18 Image segmentation using different techniques compared with the results obtained with the non-linear network implemented with pulse-modulation techniques. 1a, b, c, d – low-pass filtered images after 71, 96, 104, 113us respectively 2 a, b – low-pass filtering with 1.65 threshold after 160us and 306us(2c, 2d) respectively 3a – low-pass filter using the segmented filter, after 300us and 240us (3b) respectively; 3c – low-pass filter using the segmented filter, with the reloading of the network interconnectivity configuration after 140us; 4a, b, c, d – segmented versions obtained with a nonlinear resistive network implemented using pulse-modulation technique [48].

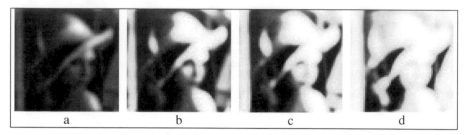

Fig. 19 Image segmentation using log-domain filter after reaching saturation. a, b, c, d – snapshot of the dynamics of the log-domain filter frozen after 78, 95, 103 and 110us

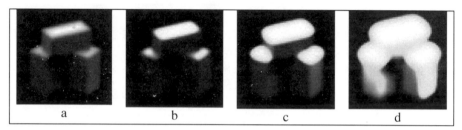

Fig. 20 Image segmentation using log-domain filter after reaching saturation a, b, c, d – snapshot of the dynamics of the log-domain filter frozen after 71, 98, 107 and 126us

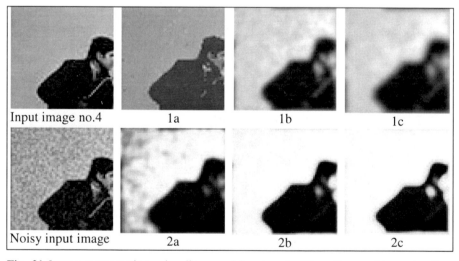

Fig. 21 Image segmentation using linear and log-domain filter after reaching saturation, compared with the non-linear resistive grid 1a – segmentation obtained with the nonlinear resistive grid [46], 1b, 1c – low-pass filtering with the linear filter after 192, 352us respectively, 2a, 2b, 2c – segmentation using the log-domain filter frozen after 98, 108, 116us respectively

Moreover, we make a comparison between the fragmented filter, log-domain filter frozen after reaching saturation and the nonlinear resistive grid implemented by pulse-modulation technique [48].

Figures 19 and 20 show several relevant snapshots taken from the log-domain filter at different times, this time allowing the voltage cells reaching saturation.

Thus the *nonlinear* log-domain filter can be used for image segmentation as well. Fig. 21 confirms the usefulness of linear/nonlinear low-pass filtering for image segmentation as seen from the comparison of the simulations performed with the compact linear filter, nonlinear log-domain filter and nonlinear resistive grid implemented using another circuit solution [46].

From the above it follows that fragmenting the network with zero-flux boundary conditions in applications like *edge detection, smoothing, image segmentation* presents certain advantages. Since any cell that reaches saturation affects the linear behavior of the rest of the network, the decoupling technique of some parts of an imager can be useful in nonlinear processing if the saturated part is cutoff from the filter, and the linear parts have independent evolution. The time constant is another significant difference between compact filter and the fragmented one; the fragmented network is faster than the other one with the same nodal capacitance. Thus, this technique might be useful for increasing the processing speed.

3 Two-Grid Coupled CNN's

An interesting phenomenon, which has been shown to appear in CNN's, is that of pattern formation - a property that, perhaps, has not been yet enough exploited. Pattern will be the name for any stable equilibrium point.

Various connections with phenomena from other domains including biology have been made so far and interesting analogies have been established. Among them, pattern formation based on a mechanism similar to that proposed by Turing [50] to explain morphogenesis has been reported in two-grid coupled second order cell CNN's [51]-[53]. In the following, several results on pattern formation and spatial filtering in two-grid coupled CNN's will be presented. The two-grid coupled CNN is a rather special case of homogeneous parallel architecture, derived from the reaction-diffusion model proposed by Turing.

The previously introduced concepts of mode decoupling, dispersion curves, band of unstable modes will be used again. We will refer to cells, interconnections, equations, mechanism of pattern formation and the influence of various factors in the process of pattern formation.

Turing patterns have been shown to appear in an architecture consisting of two-port second order identical cells sandwiched between two homogeneous resistive grids [52] which simulate the activation-inhibition mechanism [50]. The specific feature of Turing patterns is that the isolated cells are stable while the dynamics of the array can exhibit unstable spatial modes. If the cells are piecewise linear, a powerful method of investigation is the decoupling technique which, as shown, basically consists of a change of variable chosen according to the boundary conditions. The transformed differential equations corresponding to each spatial mode are decoupled, part of them having unstable solutions – a *necessary condition for pattern formation*. The competition of the unstable spatial modes leads to a pattern which depends on the shape of the dispersion curve, initial conditions and on the nonlinearity of the cells characteristics. Of course, the method is valid only for the central linear part of the cell characteristics but offers useful insight on the shape of final pattern obtained after the nonlinearity has been reached.

In fact, two mechanisms can be used to limit the pattern evolution: either through the nonlinearity of the cell characteristics or by "freezing" the transient typically before any nonlinearity has been reached, as shown in the previous section. If the emerging pattern is frozen before the signals leave the central linear part of the cell characteristics, the CNN behaves again as a spatial time variable filter, the spatial frequency response being dependent on the moment the transient has been stopped exactly in the same way that has been described in the previous section. Various aspects regarding the spatio-temporal dynamics of two-grid coupled CNN's and their applications as texture classification have been reported in [14].

In the following we present the two-grid CNN architecture capable to produce patterns and in particular, Turing patterns. The architecture is based on second order two-port cells coupled by means of two resistive grids.

3.1 The Architecture and the Equations

3.1.1 The Cells

In general, a cell consists of a *nonlinear resistive two port* characterized by the relations

$$i_1 = f(u,v)$$
$$i_2 = \tilde{g}(u,v)$$
$$\text{(34)}$$

where u and v are the port voltages and *two capacitors* as shown in Fig. 22.

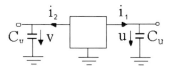

Fig. 22 Two-port cell

The analysis is greatly simplified if the nonlinearity is piecewise linear. A cell consisting of four linear elements including a voltage controlled current source and a nonlinear resistor [52] is represented in Fig. 23 and is described by the equations:

$$i_1 = f(u,v) = -Gu - f(u) + Gv$$
$$i_2 = \tilde{g}(u,v)i_1 = (G-g)u - Gv$$

(35)

where f(u) is the piecewise linear characteristic of the nonlinear resistor.

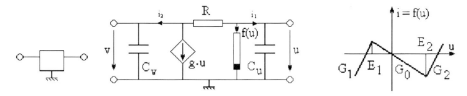

Fig. 23 Two-port cell and i-v characteristic of the piecewise nonlinear resistor

3.1.2 The Interconnections

The CNN based on the above cell is built by connecting the cells using two resistive grids. The architecture is like a sandwich of cells between grids, in the sense that each grid connects similar ports.

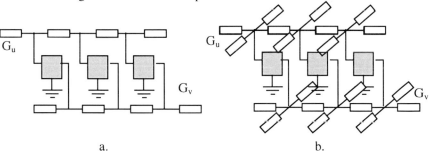

a.   b.

Fig. 24 Sketch of a 1D two-grid coupled CNN architecture (a) and the way towards a 2D array (b)

3.1.3 The Equations

In the general case, the behavior of a 2D CNN composed of M×N cells is described by the following system of equations

$$C_u \frac{du_{ij}(t)}{dt} = f(u_{ij}, v_{ij}) + G_u \nabla^2 u_{ij}$$
$$C_v \frac{dv_{ij}(t)}{dt} = \tilde{g}(u_{ij}, v_{ij}) + G_v \nabla^2 v_{ij}$$

$$i = 0,...,M-1, j = 0,...,N-1$$

(36)

where $\nabla^2 x_{ij} = x_{(i+1)j} + x_{(i-1)j} + x_{i(j+1)} + x_{i(j-1)} - 4x_{ij}$ is the Laplacean (which, for the 1-D case, has the form $\nabla^2 x_i = x_{i+1} + x_{i-1} - 2x_i$).

With the notations

$$\gamma = \frac{1}{C_u}; D_u = \frac{G_u}{C_u}; D_v = \frac{G_v}{C_v}; g(u_{ij}, v_{ij}) = \frac{C_u}{C_v} \tilde{g}(u_{ij}, v_{ij}) \tag{37}$$

the equations become

$$\frac{du_{ij}(t)}{dt} = \gamma f(u_{ij}, v_{ij}) + D_u \nabla^2 u_{ij}$$
$$\frac{dv_{ij}(t)}{dt} = \gamma g(u_{ij}, v_{ij}) + D_v \nabla^2 v_{ij}$$
$$\quad i = 0,...,M-1, j = 0,...,N-1 \tag{38}$$

Linearization of these equations gives

$$\frac{du_{ij}(t)}{dt} = \gamma(f_u u_{ij} + f_v v_{ij}) + D_u \nabla^2 u_{ij}$$
$$\frac{dv_{ij}(t)}{dt} = \gamma(g_u u_{ij} + g_v v_{ij}) + D_v \nabla^2 v_{ij}$$
$$\quad i = 0,...,M-1, j = 0,...,N-1 \tag{39}$$

where f_u, f_v, g_u, g_v are the elements of the Jacobian matrix of $f(u,v)$ and $g(u,v)$, D_u and D_v are the diffusion coefficients and γ is a scaling coefficient. For the cell in Fig. 23 the above equation are valid for u-voltages within the interval $[E_1, E_2]$. In this case, the relations between the Jacobian parameters and the circuit elements are

$$f_u = -(G + G_0), f_v = G, \ g_u = \frac{C_u}{C_v}(G - g), \ g_v = -\frac{C_u}{C_v}G \tag{40}$$

The saturation type piecewise linear shape for the cell nonlinearity is convenient from the implementation point of view as well as for the theoretical tractability. Thus, in the case when all cell voltages are in the central linear part of the nonlinear characteristics the analysis simplifies considerably due to linearity and symmetry (which is valid until at least one cell reaches saturation).

3.2 The Decoupling Technique

In the following we analyze a CNN made of piecewise nonlinear cells as shown in Fig. 23 and suppose that all voltages are within the central linear part of the cells characteristics. For the sake of simplicity we consider the 1D version of equations (39):

$$\frac{du_i(t)}{dt} = \gamma(f_u u_i + f_v v_i) + D_u \nabla^2 u_i$$
$$\hspace{4cm} i = 0,...,M-1 \hspace{2cm} (41)$$
$$\frac{dv_i(t)}{dt} = \gamma(g_u u_i + g_v v_i) + D_v \nabla^2 v_i$$

Using the notations from [52] we transform the system of equations by means of the change of variable

$$u_i(t) = \sum_{m=0}^{M-1} \Phi_M(i,m)\hat{u}_m(t)$$
$$\hspace{4cm} i = 0,...,M-1 \hspace{2cm} (42)$$
$$v_i(t) = \sum_{m=0}^{M-1} \Phi_M(i,m)\hat{v}_m(t)$$

where $\Phi_M(i,m)$ are eigenfunctions (dependent on the boundary conditions) of the 1D Laplacean i.e., $\nabla^2 \Phi_M(i,m) = -k_m^2 \Phi_M(i,m)$ and $-k_m^2$ are the eigenvalues, proportional to the square (or sum of squares) of sine functions.

If the set $\Phi_M(i,m)$ of M functions are orthogonal with respect to the scalar product in C^M, i.e.,

$$\sum_{i=0}^{M-1} \Phi_M^*(m,i)\Phi_M(i,n) = \delta_{mn} \hspace{2cm} (43)$$

$\hat{u}_m$ and $\hat{v}_m$ can be expressed, by means of the inversion formulas:

$$\hat{u}_m(t) = \sum_{i=0}^{M-1} \Phi_M^*(m,i)u_i(t)$$
$$\hspace{4cm} m = 0,...,M-1 \hspace{2cm} (44)$$
$$\hat{v}_m(t) = \sum_{i=0}^{M-1} \Phi_M^*(m,i)v_i(t)$$

where

$$\Phi_M^*(m,i) = \Phi_M(i,m) \hspace{2cm} (45)$$

Making the change of variable and taking the scalar product of both sides of the equations, the dynamics of the 1D CNN is described by the following set of pairs of *decoupled* linear equations

$$\begin{bmatrix} \dot{\hat{u}}_m \\ \dot{\hat{v}}_m \end{bmatrix} = \left(\gamma \begin{bmatrix} f_u & f_v \\ g_u & g_v \end{bmatrix} - k_m^2 \begin{bmatrix} D_u & 0 \\ 0 & D_v \end{bmatrix}\right)\begin{bmatrix} \hat{u}_m \\ \hat{v}_m \end{bmatrix} \hspace{0.5cm} m = 0,...,M-1 \hspace{1cm} (46)$$

Thus, the set of 2×M coupled differential equations in the u and v variables transforms into M sets of pairs of second order differential equations in the new variables - the amplitudes of the spatial components of the voltages.

The natural frequencies, λ_{m1} and λ_{m2} are the roots of the characteristic polynomials

$$\lambda_m^2 + \lambda_m[k_m^2(D_u + D_v) - \gamma(f_u + g_v)] + D_u D_v k_m^4 \quad m = 0,...,M-1$$
$$- \gamma(D_v f_u + D_u g_v)k_m^2 + (f_u g_v - f_v g_u) = 0 \tag{47}$$

The solution of the 1-D CNN equations is thus

$$u_i(t) = \sum_{m=0}^{M-1}(a_m e^{\lambda_{m1}t} + b_m e^{\lambda_{m2}t})\Phi_M(i,m)$$
$$\qquad\qquad\qquad\qquad\qquad\qquad i = 0,...,M-1 \tag{48}$$
$$v_i(t) = \sum_{m=0}^{M-1}(c_m e^{\lambda_{m1}t} + d_m e^{\lambda_{m2}t})\Phi_M(i,m)$$

The integration constants satisfy the constraints

$$c_m = p_m a_m; \ d_m = q_m d_m \tag{49}$$

where

$$p_m = \frac{\lambda_{m1} - \gamma f_u + D_u k_m^2}{\gamma f_v}; q_m = \frac{\lambda_{m2} - \gamma f_u + D_u k_m^2}{\gamma f_v} \tag{50}$$

and can be expressed in terms of the initial conditions of the voltages in the two "layers" of the CNN by means of the formulas

$$a_m = \frac{\hat{v}_m(0) - q_m \hat{u}_m(0)}{p_m - q_m}; \ b_m = \frac{\hat{v}_m(0) - p_m \hat{u}_m(0)}{q_m - p_m} \tag{51}$$

Thus, the complete response of the CNN in terms of the spectrum of the initial conditions with respect to the corresponding boundary conditions (which will be discussed soon) can be expressed easily in terms of

$$\hat{u}_m(t) = \frac{\hat{v}_m(0) - q_m \hat{u}_m(0)}{p_m - q_m}e^{\lambda_{m1}t} + \frac{\hat{v}_m(0) - p_m \hat{u}_m(0)}{q_m - p_m}e^{\lambda_{m2}t}$$
$$\hat{v}_m(t) = p_m\frac{\hat{v}_m(0) - q_m \hat{u}_m(0)}{p_m - q_m}e^{\lambda_{m1}t} + q_m\frac{\hat{v}_m(0) - p_m \hat{u}_m(0)}{q_m - p_m}e^{\lambda_{m2}t} \tag{52}$$

When biasing current sources are used at the u ports of the cells, the equations become

$$\begin{bmatrix} \dot{\hat{u}}_m(t) \\ \dot{\hat{v}}_m(t) \end{bmatrix} = \left(\gamma \begin{bmatrix} f_u & f_v \\ g_u & g_v \end{bmatrix} - k_m^2 \begin{bmatrix} D_u & 0 \\ 0 & D_v \end{bmatrix} \right) \begin{bmatrix} \hat{u}_m(t) \\ \hat{v}_m(t) \end{bmatrix} + \gamma \begin{bmatrix} \hat{J}_m \\ 0 \end{bmatrix} \tag{53}$$

where $\hat{J}_m$ is the amplitude of the m-th spatial spectral component of the biasing source. In this case the general form of the transient expressed in terms of the decoupled variables is

$$\begin{cases} \hat{u}_m(t) = a_m(\hat{J}_m,\hat{u}_m(0),\hat{v}_m(0))e^{\lambda_{m1}t} + b_m(\hat{J}_m,\hat{u}_m(0),\hat{v}_m(0))e^{\lambda_{m2}t} + f_1\hat{J}_m \\ \hat{v}_m(t) = c_m(\hat{J}_m,\hat{u}_m(0),\hat{v}_m(0))e^{\lambda_{m1}t} + d_m(\hat{J}_m,\hat{u}_m(0),\hat{v}_m(0))e^{\lambda_{m2}t} + f_2\hat{J}_m \end{cases} \quad (54)$$

where f_1 and f_2 are:

$$\begin{cases} f_1 = \dfrac{-\gamma(\gamma s_v - k_m^2 D_v)}{(\gamma s_u - k_m^2 D_u)(\gamma s_v - k_m^2 D_v) - \gamma^2 f_v g_u} \\ f_2 = \dfrac{\gamma^2 g_u}{(\gamma s_u - k_m^2 D_u)(\gamma s_v - k_m^2 D_v) - \gamma^2 f_v g_u} \end{cases} \quad (55)$$

The above results extend easily to the 2D case with the change of variable

$$u_{ij}(t) = \sum_{m=0}^{M-1}\sum_{n=0}^{N-1} \Phi_{MN}(i,j,m,n)\hat{u}_{mn}(t)$$

$$v_{ij}(t) = \sum_{m=0}^{M-1}\sum_{n=0}^{N-1} \Phi_{MN}(i,j,m,n)\hat{v}_{mn}(t)$$

$$(56)$$

where $<\Phi_{MN}(i,j,m,n),\Phi_{MN}(i,j,p,q)> = \sum_{i=0}^{M-1}\sum_{j=0}^{N-1}\Phi*_{MN}(m,n,i,j)\Phi_{MN}(i,j,p,q) = \delta_{mnpq}$.

3.3 Boundary Conditions (BC's) and Their Influence on Pattern Formation

BC's reflect the way the cells on the edges of the array are connected and influence the behavior of the CNN. This influence will be stronger for smaller arrays going up to the aspect of allowing or not the development of a pattern. The spatial operator represented by the connection template (in our case the Laplacean) should be specified for the edge cells and the virtual cells around the array within the neighborhood radius.

The periodic and the zero-flux BC's are common in physics, biology, chemistry etc. but also in CNN implementation where they are considered as "natural". However, in the case of CNN's many other BC's may be imagined, some of them having no counterpart in physical, biological, chemical etc. problems but allowing analytical tractability [54], [55].

In the table below we list the definition, the eigenvectors and eigenvalues for various BC's in the 1-D case. The results are useful to decouple the system of linear differential equations describing the CNN behavior in the central linear part.

Table 2 Eigenvectors and eigenvalues for various BC's

Left/right	Boundary conditions	Eigenvectors	Eigenvalues
ring	$u(-1)=u(M-1)$ $u(M)=u(0)$	$e^{j\frac{2\pi}{M}mi}$	$-4\sin^2\frac{m\pi}{M}$
zero-flux zero-flux	$u(-1)=u(0)$ $u(M)=u(M-1)$	$\cos\frac{m(2i+1)\pi}{2M}$	$-4\sin^2\frac{m\pi}{2M}$
anti-zero flux anti-zero flux	$u(-1)=-u(0)$ $u(M)=-u(M-1)$	$\sin\frac{(m+1)(2i+1)\pi}{2M}$	$-4\sin^2\frac{(m+1)}{2M}$
zero zero	$u(-1)=0$ $u(M)=0$	$\sin\frac{(m+1)(i+1)\pi}{M+1}$	$-4\sin^2\frac{(m+1)}{2(M+}$
quasi zero flux quasi zero flux	$u(-1)=u(1)$ $u(M)=u(M-2)$	$\cos\frac{mi\pi}{M-1}$	$-4\sin^2\frac{m\pi}{2(M-1}$
zero zero flux	$u(-1)=0$ $u(M)=u(M-1)$	$\sin\frac{(2m+1)(i+1)\pi}{2M+1}$	$-4\sin^2\frac{(2m+1)}{2(2M+}$
anti-zero flux zero flux	$u(-1)=-u(0)$ $u(M)=u(M-1)$	$\sin\frac{(2m+1)(2i+1)\pi}{4M}$	$-4\sin^2\frac{(2m+1)\pi}{4M}$
zero quasi-zero flux	$u(-1)=0$ $u(M)=u(M-2)$	$\sin\frac{(2m+1)(i+1)\pi}{2M}$	$-4\sin^2\frac{(2m+1)}{4M}$
anti-zero flux quasi-zero flux	$u(-1)=-u(0)$ $u(M)=u(M-2)$	$\sin\frac{(2m+1)(2i+1)\pi}{2(2M-1)}$	$-4\sin\frac{(2m+1)\pi}{2(2M-1)}$
zero anti-zero flux	$u(-1)=0$ $u(M)=-u(M-1)$	$\sin\frac{2(m+1)(i+1)\pi}{2M+1}$	$-4\sin^2\frac{(m+1)\pi}{2M+1}$

3.4 Dispersion Curve

The dynamics of the CNN is significantly determined by the roots of the characteristic equations, even though the results are valid only for the linear central part. The crucial aspect regarding pattern formation is that, in certain conditions, at least one of the roots of the characteristic equation has positive real part. This will cause the corresponding spatial mode(s) to grow until some nonlinearity will limit the growth. The *dispersion curve* represents the real part of the temporal eigenvalues versus the spatial eigenvalues.

$$\text{Re }\lambda_{1,2}(k_m^2)=\text{Re }\{\gamma\frac{f_u+g_v}{2}-k_m^2\frac{D_u+D_v}{2}+\sqrt{\left[\gamma\frac{(g_v-f_u)}{2}+k_m^2\frac{D_u-D_v}{2}\right]^2+\gamma^2 f_v g_u}\} \quad (57)$$

A typical dispersion curve is presented below.

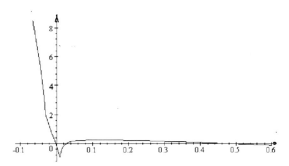

Fig. 25 Typical dispersion curve

The curve has been computed for the following parameters: $\gamma=5$, $f_u=0.1$, $f_v=-1$, $g_u=0.1$, $g_v=-2$, $D_u=1$, $D_v=150$, M=30. The curve satisfies Turing conditions that will be discussed next, i.e. it exhibits a "band" of unstable modes.

3.5 Turing Pattern Formation Mechanism

In principle, a pattern, i.e., a stable equilibrium points towards which the network emerges, can develop when the characteristic equation has at least one root with positive real part corresponding to a *nonzero* spatial frequency. Indeed, the instability of the zero spatial frequency spatial mode will determine that *all* cell voltages will go either to a positive or negative saturation value or simultaneously oscillate– situations which will not be called patterns.

An interesting situation is that when the origin is a **stable** equilibrium point for an isolated cell and an **unstable** equilibrium point for the whole array. The necessary conditions (Turing) that ensure the *instability* of an array built of *stable* cells linked together through resistive grids are [52]:

$$f_u + g_v < 0$$
$$f_u g_v - f_v g_u > 0$$
$$D_v f_u + D_u g_v > 0 \qquad (58)$$
$$(D_v f_u - D_u g_v)^2 + 4 D_u D_v f_v g_u > 0$$

The first two conditions ensure the stability of an isolated cell while the last two, the potential instability of the array. In fact, Turing patterns in CNN's are dependent on the following aspects:

a – fulfillment of Turing conditions, b – dispersion curve [52], c – initial conditions [56], d – boundary conditions [55], e – biasing sources signal, when they exist [57].

Beside, the shape of the nonlinear characteristic of the cell resistor influences the pattern as well but this is an aspect that cannot be easily handled. However, it has been observed by simulations that in the piecewise linear case the patterns based on mode competition predicted by the linear theory fit remarkably well with the simulations especially for 1D arrays, which means that the nonlinearity plays mainly the role of limiting the growing process of the unstable spatial modes. In the 2D case, however, the linear theory is often unable to predicting the final pattern.

3.6 Boundary Conditions in 2D CNN's

Extension of the results related to boundary conditions to the 2D case is straightforward. Denoting by $-k_m^2$, $-k_n^2$ and $-k_{mn}^2$ the spatial eigenvalues for the 1D and 2D case respectively and by $\phi_M(m,i)$, $\phi_N(n,i)$ and $\phi_{MN}(m,n;i,j)$ the corresponding eigenfunctions, the following relationships are satisfied [54]:

$$-k_{mn}^2 = -k_m^2 - k_n^2$$
$$\Phi_{MN}(m,n,i,j) = \Phi_M(m,i)\Phi_N(n,j)$$

(59)

where M and N are the dimensions of the array m=0,...,M-1 and n=0,...,N-1. From the above relations it is apparent that, for a given dispersion curve, corresponding to a particular choice of the cell and diffusion parameters, the number and the position of the unstable modes can be controlled using various combinations of boundary conditions. Compared to the one-dimensional case, the number of possibilities is obviously much greater. Again, the control is more efficient in the case of small dimensional arrays. The analytical results corresponding to various BC's can be obtained using relations (59) and the results in Table 1.

3.7 An Application

As already shown previously, if the CNN functions only in the central linear part of the cell piecewise linear resistor, it behaves like a time dependent frequency characteristic linear filter [41], [58].

   If ring boundary conditions are used then the eigenvectors are 2D Discrete Fourier Transform complex exponentials (Table 2). In the following $\hat{u}_{mn}(t)$ and $\hat{v}_{mn}(t)$ will represent the spectra of the signals $u_{ij}(t)$ and $v_{ij}(t)$ respectively, where the indexes (i, j); $i = 0..M - 1$ and $j = 0..M - 1$ stand for the cell number and (m,n); $m = 0..M - 1$, n $= 0..M - 1$ for the modes values.

Considering that the CNN works in the central linear part of the cell characteristics, and the second layer of the network *has zero initial conditions* ($v_{mn}(0) = 0$), it is possible to define a time dependent frequency characteristic:

$$H_{mn}(t_0) = \frac{\hat{u}_{mn}(t_0)}{\hat{u}_{mn}(0)} \tag{60}$$

where

$$\hat{u}_{mn}(t) = \frac{\hat{v}_{mn}(0) - q_{mn}\hat{u}_{mn}(0)}{p_m - q_m} e^{\lambda_{mn1}t} + \frac{\hat{v}_{mn}(0) - p_{mn}\hat{u}_{mn}(0)}{q_m - p_m} e^{\lambda_{mn2}t}$$

$$\hat{v}_{mn}(t) = p_{mn}\frac{\hat{v}_{mn}(0) - q_{mn}\hat{u}_{mn}(0)}{p_{mn} - q_{mn}} e^{\lambda_{mn1}t} + q_{mn}\frac{\hat{v}_{mn}(0) - p_{mn}\hat{u}_{mn}(0)}{q_{mn} - p_{mn}} e^{\lambda_{mn2}t} \tag{61}$$

and the constants satisfy similar constraints as is the 1D case.

Since $v_{mn}(0) = 0$ and thus $\hat{v}_{mn}(0) = 0$, the ratio between the amplitudes of the u_{mn} modes at a given moment t_0 and their initial amplitudes is:

$$H_{mn}(t_0) = \frac{(\lambda_{mn1} - \gamma f_u + D_u k_{mn}^2)e^{\lambda_{mn2}t_0} - (\lambda_{mn2} - \gamma f_u + D_u k_{mn}^2)e^{\lambda_{mn1}t_0}}{\lambda_{mn1} - \lambda_{mn2}} \tag{62}$$

The above relation reflects a time-dependent frequency characteristic corresponding to a dynamic that has been frozen at the moment t_0, before any nonlinearity has been reached. It is important to note that the above relation is conditioned by the existence of a band of unstable modes. If that condition is not fulfilled, there are no modes with positive real part and the cell voltages will tend toward zero.

Thus, using the freezing technique combined with different CNN parameters, a family of 2D circular filters with controlled selectivity can be obtained [59].

In Fig. 26 a) and Fig. 26 b) the frequency characteristics of several 2D and 1D linear filters and a 1D linear filter with the parameters $f_u = 0.1$, $f_v = -1$, $g_u = 0.1$, $g_v = -0.2$, $D_u = 1$, $D_v = 50$, $\gamma = 14$ at   the moments $t_0 = \{1, 2, 3, 4.5\}$ are presented. Their center frequency does not change and the band decreases as time increases. However, the freezing time cannot increase indefinitely since cell saturation will limit the linear behavior.

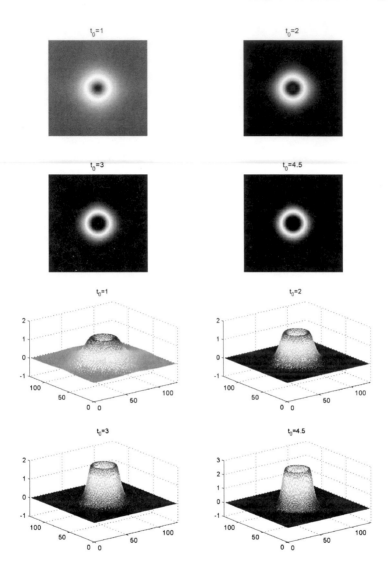

Fig. 26 a): Frequency characteristics of 2D circular band pass circular filters using the parameters $f_u = 0.1, f_v = -1, g_u = 0.1, g_v = -0.2, D_u = 1, D_v = 50, \gamma = 14$ for the moments $t_0 = \{1, 2, 3, 4.5\}$

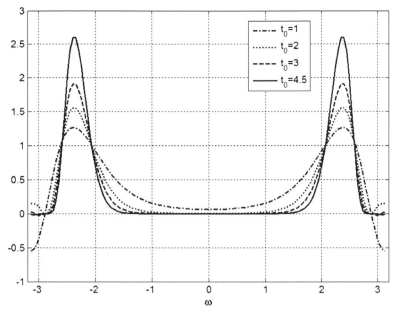

Fig. 26 b): Frequency characteristics of 1D band pass filters using the parameters $f_u = 0.1, f_v = -1, g_u = 0.1, g_v = -0.2, D_u = 1, D_v = 50, \gamma = 14$ for the moments $t_0 = \{1, 2, 3, 4.5\}$

In the following, several results obtained with a bank of circular filters in rotated texture recognition will be presented. The main advantage of such filters is the fact that the output energy of a filtered texture does not change with rotation.

The classification method is a classical one and is presented in Fig. 27:

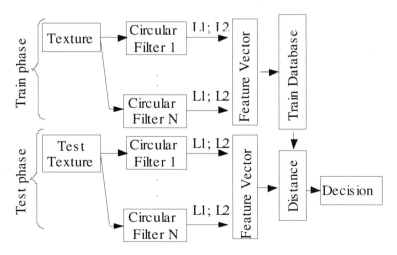

Fig. 27 Classification method

A texture is filtered with N circular filters. For each of the resulted N images the L1 or L2 norms are computed; they will form an N dimensional feature vector of the respective texture previously normalized.

In the following, results obtained with a bank of ideal circular filters and with a bank of CNN's are presented.

In both cases, for the "train" and test phases, 16 texture types from the Brodatz database [59] were used having the resolution 128x128. Each texture type was represented by 28 different images each being rotated with 10 angles $(0^0, 20^0, 30^0, 45^0, 60^0, 70^0, 90^0, 120^0, 135^0, 150^0$). The used textures and an example of a texture which is rotated under 10 angles are presented in Fig. 28.

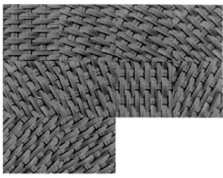

Fig. 28 Example from the Brodatz database

From the texture database, 15% of the images with 0^0 orientation were used in the "train" phase and the corresponding feature vectors form the "train" database. For each test image the normalized feature vector was determined and the L1 or L2 distances to the normalized feature vectors from the "train" database were calculated.

3.7.1 Results Obtained with Ideal Circular Filters

For the ideal circular filters the parameters were chosen so that the frequency characteristics approximate circular Gaussian filters with parameters π/σ and $\mu\sigma$

$$G(\omega_x, \omega_y) = \exp\left(-\frac{(\sqrt{\omega_x^2 + \omega_y^2} - \pi/\sigma)^2 \cdot (\mu\sigma)^2}{2}\right)$$ (63)

The central frequencies of the ideal circular filters were chosen so that the filter bank has a wavelet propriety (the product between the central frequency and filter selectivity is constant and $\sigma = a^i$ where $i = 1..N$). In Fig. 29 such a filter bank for $N = 5$ is presented.

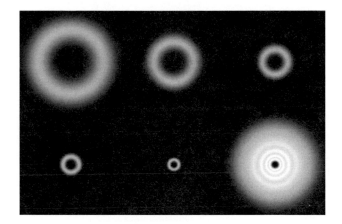

Fig. 29 Ideal filter bank ($a = 1.6$ si $\mu = 1.3$)

The textures recognition system performances depend on a , μ and N. For the considered Brodatz database, the best results were obtained for $\mu = 1.3$. In Table 2, the performances for N=4, 5 and 16 filters are presented.

Table 3 Performance classification for ideal circular filters

Filters number	N=4	N=5	N=16
Filters parameters	$\mu = 1.3$ $a = 2$	$\mu = 1.3$ $a = 1.6$	$\mu = 1.3$ $a = 1.3$
Classification performance	96.6%	97.36%	99.34%

From Table 3, it can be observed that increasing the filters number, of the recognition system performances increase. The poorest results were obtained for wood (N=4 and N=5) and for mat (N=16). Also there are small differences for the case when L1 or L2 norm were used in order to determine vector features.

In Table 4, the confusion matrix for N=5 is presented. The following texture numbers were used: 1-canvas, 2 - cloth, 3 - cotton, 4 - grass, 5 – leather, 6 - matting, 7 – paper, 8 - pigskin, 9 - raffia, 10 - rattan, 11 - reptile, 12 - sand, 13 - straw, 14 - weave, 15 - wood, 16 - wool.

Table 4 Confusion matrix obtained when N=5 ideal circular filters

	1	2	3	4	5	6	7	8	9	10	11	12	13	14	15	16
1	207															
2		207														
3			207													
4				183											2	
5				24	207										5	
6						199										
7						7	207				2				22	
8								207				1				
9									207							
10										199						
11											205					
12												206			15	
13					1								207			
14														207		
15															163	
16									8							207
(%)	100	100	100	88.4	100	96.1	100	100	100	96.1	99	99.5	100	100	78.7	100

3.7.2 Results with CNN Spatial Filters

In order to test the double layer CNN capacity to recognize rotated textures, 5 filters, specified in Table 5, were used. For changing the filters central frequencies, only the γ parameter of the cell shown in Fig. 23 [52] was modified while the bandwidth was obtained by means of appropriately choosing the moment t_0 when the pattern evolutions were stopped.

Table 5 CNN parameters values of the 5 circular filters

filter 1	$f_u = 0.1, f_v = -1, g_u = 0.1, g_v = -0.2, D_u = 1, D_v = 50, \gamma = 69, t_0 = 2.3$
filter 2	$f_u = 0.1, f_v = -1, g_u = 0.1, g_v = -0.2, D_u = 1, D_v = 50, \gamma = 33, t_0 = 2.5$
filter 3	$f_u = 0.1, f_v = -1, g_u = 0.1, g_v = -0.2, D_u = 1, D_v = 50, \gamma = 14, t_0 = 4.5$
filter 4	$f_u = 0.1, f_v = -1, g_u = 0.1, g_v = -0.2, D_u = 1, D_v = 50, \gamma = 5.5, t_0 = 11$
filter 5	$f_u = 0.1, f_v = -1, g_u = 0.1, g_v = -0.2, D_u = 1, D_v = 50, \gamma = 2, t_0 = 30$

In Fig. 30, the frequency filters characteristics, implemented with the two-grid CNN for N=5 that approximate those from Fig. 29 are presented.

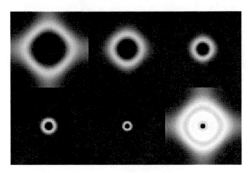

Fig. 30 The filters frequency characteristics implemented with double layer CNN

The classification results obtained when filters implemented with CNN are presented in Table 5. The classification performance is only 1.56% lower than that obtained with ideal circular filters. The smallest classification percentage was obtained like in the case of ideal filters for wood (73.9%).

Table 6 Confusion matrix obtained for N=5 circular filters implemented with double layer CNN

	1	2	3	4	5	6	7	8	9	10	11	12	13	14	15	16
1	159															
2		207														
3	48		207													
4				186											1	
5				20	207										14	
6						200										
7						7	207				2					
8			1					207								
9									207							
10										207						
11											205					2
12												204			30	
13													207			
14														207		
15												2			162	
16												1				205
%	76.8	100	100	89.8	100	96.6	100	100	100	100	99	98.5	100	100	78.2	99

The maximum fall in performances is for canvas (from 100% at 76.8%). This decrease in performances can be explained by the fact that this texture has the main part of the energy at high frequency where the CNN filter does not approximate well the ideal circular one.

References

[1] Chua, L.O., Yang, L.: Cellular Neural Networks: Theory. IEEE Trans. Circuits Syst. 35(10), 1257–1272 (1988)
[2] Chua, L.O., Yang, L.: Cellular Neural; Networks: Applications. IEEE Trans. Circuits Syst. 35(10), 1273–1290 (1988)
[3] Roska, T., Vanderwalle, J.: Cellular Neural Networks. John Wiley & Sons (1993)
[4] Huertas, J.L., Chen, W.-K., Madan, R.N. (eds.): Visions of the Nonlinear Science in the 21-st Century. World Scientific Publishing (1999)
[5] http://en.wikipedia.org/wiki/Cellular_neural_network
[6] Haykin, S.: Neural Networks: A Comprehensive Foundation. Prentice Hall Press (1998)
[7] Roska, T.: Cellular Wave Computers and CNN Technology – a SoC architecture with xK Processors and Sensor Arrays. In: Int'l Conference on Computer Aided Design, San Jose, CA, USA, November 6-10, pp. 557–564 (2005)
[8] Porod, W., Werblin, F., Chua, L., Roska, T., Rodriguez-Vázquez, A., Roska, B., Faya, R., Bernstein, G., Huang, Y., Csurgay, A.: Bio-Inspired Nano-Sensor-Enhanced CNN Visual Computer. Annals of the New York Academy of Sciences 1013, 92–109 (2004)
[9] Amenta, C., Arena, P., Baglio, S., Fortuna, L., Richiura, D., Xibilia, M., Vu, L.: SC-CNNs for Sensors Data Fusion and Control in Space Distributed Structures. In: Int'l Workshop on Cellular Neural Networks and Their Applications, Catania, Italy, pp. 147–152 (2000)
[10] Rekeczky, C., Timar, G.: Multiple Laser Dot Detection and Localization within an Attention Driven Sensor Fusion Framework. In: Int'l Workshop on Cellular Neural Networks and Their Applications, pp. 232–235 (2005)
[11] Haenggi, M.: Mobile Sensor-Actuator Networks: Opportunities and Challenges. In: Int'l Workshop on Cellular Neural Networks and Their Applications, pp. 283–290 (2002)
[12] Arena, P., Fortuna, L., Frasca, M., Patane, L.: CNN Based Central Pattern Generators with Sensory Feedback. In: Int'l Workshop on Cellular Neural Networks and Their Applications, p. 275 (2005)
[13] Shi, B., Roska, T., Chua, L.: Estimating Optical Flow with Cellular Neural Networks. Int'l Journal of Circuit Theory and Applications 26, 344–364 (1998)
[14] Ungureanu, P., David, E., Goras, L.: On Rotation Invariant Texture Classification Using Two-Grid Coupled CNNs. In: Neurel 2006, Belgrade, September 25-27, pp. 33–36 (2006) ISBN 1-4244-0432-0
[15] Roska, T., Chua, L.O.: The CNN Universal Machine: 10 Years Later. Journal of Circuits, Systems, and Computers. Int'l Journal of Bifurcation and Chaos 12(4), 377–388 (2003)
[16] Carmona, R., Jimenez-Garrido, F., Dominguez-Castro, R., Espejo, S., Rodriguez-Vazquez, A.: CMOS Realization of a 2-layer CNN Universal Machine. In: Int'l Workshop on Cellular Neural Networks and Their Applications (2002)
[17] Crounse, K., Wee, C., Chua, L.: Linear Spatial Filter Design for Implementation on the CNN Universal Machine. In: Int'l Workshop on Cellular Neural Networks and Their Applications, Italy, pp. 357–362 (2000)
[18] Szabot, T., Szolgay, P.: CNN-UM-Based Methods Using Deformable Contours on Smooth Boundaries. In: Int'l Workshop on Cellular Neural Networks and Their Applications (2006)

[19] Balya, D., Tímar, G., Cserey, G., Roska, T.: A New Computational Model for CNN-Ums and its Computational Complexity. In: Int'l Workshop on Cellular Neural Networks and Their Applications (2004)

[20] Pazienza, G., Gomez-Ramirezt, E., Vilasis-Cardona, X.: Genetic Programming for the CNN-UM. In: Int'l Workshop on Cellular Neural Networks and Their Applications (2006)

[21] Kincsest, Z., Nagyl, Z., Szolgay, P.: Implementation of Nonlinear Template Runner Emulated Digital CNN-UM on FPGA. In: Int'l Workshop on Cellular Neural Networks and Their Applications (2006)

[22] Voroshazit, Z., Nagyt, Z., Kiss, A., Szolgay, P.: An Embedded CNN-UM Global Analogic Programming Unit Implementation on FPGA. In: Int'l Workshop on Cellular Neural Networks and Their Applications (2006)

[23] Rodríguez-Vázquez, A., Liñán-Cembrano, G., Carranza, L., Roca-Moreno, E., Carmona-Galán, R., Jiménez-Garrido, F., Domínguez-Castro, R., Meana, S.: ACE16k: The Third Generation of Mixed-Signal SIMD-CNN ACE Chips Toward VSoCs. IEEE Trans. on Circuits and Systems - I 51(5), 851–863 (2004)

[24] Kek, L., Karacs, K., Roska, T. (eds.): Cellular Wave Computing Library (Templates, Algorithms, and Programs) Version 2.1 CSW-1-2007, Budapest, Hungary

[25] Boros, T., Lotz, K., Radványi, A., Roska, T.: Some Useful New Nonlinear and Delay-type Templates. Research report of the Analogical and Neural Computing Laboratory, Computer and Automation Research Institute, Hungarian Academy of Sciences (MTA SzTAKI), DNS-1-1991, Budapest (1991)

[26] Chua, L.O., Roska, T.: Stability of a class of nonreciprocal neural networks. IEEE TCAS 37, 1520–1527 (1990)

[27] Chua, L.O., Wu, C.W.: On the universe of stable cellular neural networks. Int. J. Circuit Theory Applicat. 20, 497–517 (1992)

[28] Thiran, P., Setti, G., Hasler, M.: An approach to information propagation in 1-D cellular neural networks-Part I: Local diffusion. IEEE TCAS-I 45, 777–789 (1998)

[29] Setti, G., Thiran, P., Serpico, C.: An approach to information propagation in 1-D cellular neural networks-Part II: Global propagation. IEEE TCAS-I 45(8), 790–811 (1998)

[30] Forti, M., Tesi, A.: A new method to analyze complete stability of PWL cellular neural networks. Int. J. Bifurcation and Chaos 11(3), 655–676 (2001)

[31] Shih, C.W.: Complete stability for a class of cellular neural networks. Int. J. Bifurcation and Chaos 11, 169–177 (2001)

[32] Shih, C.W., Weng, C.W.: On the templates corresponding to cycle-symmetric connectivity in cellular neural networks. Int. J. Bifurcation and Chaos 12, 2957–2966 (2002)

[33] Takahashi, N., Chua, L.O.: On the complete stability of nonsymmetric cellular neural networks. IEEE TCAS-I 45(7), 754–758 (1998)

[34] Gilli, M.: Stability of cellular neural networks and delayed cellular neural networks with nonpositive templates and nonmonotonic output functions. IEEE TCAS-I 41(8), 518–528 (1994)

[35] Civalleri, P.P., Gilli, M.: Practical stability criteria for cellular neural networks. Electronics Letters 33(11), 970–971 (1997)

[36] Gilli, M., Civalleri, P.P.: Template design methods for binary stable cellular neural networks. International Journal of Circuit Theory and Applications 30, 211–230 (2002)

[37] Zou, F., Nossek, J.A.: Stability of cellular neural networks with opposite-sign templates. IEEE TCAS 38(6), 675–677 (1991)
[38] Joy, M.P., Tavsanoglu, V.: A new parameter range for the stability of opposite-sign cellular neural networks. IEEE TCAS-I 40(3), 204–207 (1993)
[39] Di Marco, M., Forti, M., Grazzini, M., Nistri, P., Pancioni, L.: Global consistency of decisions and convergence of competitive cellular neural networks. Journal of Bifurcation and Chaos 17(9), 3127–3150 (2007)
[40] Balsi, M.: Stability of cellular neural networks with one-dimensional templates. International Journal of Circuit Theory and Applications 21, 293–297 (1993)
[41] Goraş, L.: On Pattern Formation in Cellular Neural Networks. In: Piuri, V., Gori, M., Ablameyko, S., Goras, L. (eds.) Limitations and Future Trends in Neural Computation. NATO Science Series: Computer & Systems Sciences, Siena, Italy, October 22-24, vol. 186 (2002)
[42] Goras, L., Alecsandrescu, I., Vornicu, I.: Spatial filtering using linear analog parallel architectures. In: International Symposium on Signals, Circuits and Systems ISSCS 2009, Iasi, Romania, vol. 2, pp. 409–412 (2009)
[43] Goras, L., Vornicu, I.: Spatial Filtering Using Analog Parallel Architectures and Their Log-Domain Implementation. Romanian Journal of Information Science and Technology 13(1), 73–83 (2010) ISSN 1453-8245
[44] Huijsing, J., van de Plassche, R., Sansen, W.: Analog circuit design – Volt electronics; Mixed-mode systems; Low-noise and RF power amplifiers for Telecommunication, Part I: 1-V Electronics, pp. 1–69. Kluwer Academic Publishers (1999)
[45] Frey, D.R.: Exponential state space filters: A Generic current mode design strategy. IEEE Transaction on Circuits and Systems – I: Fundamental Theory and Applications 43(1) (January 1996)
[46] Yu, P.C., Decker, S.J., Lee, H.-S., Sodini, C.G., Wyatt, J.L.: CMOS resistive fuse for image smoothing and segmentation. IEEE Journal of Solid-State Circuits 27(4) (April 1992)
[47] Naso, S., Storace, M., Pruzzo, G., Parodi, M.: CMOS implementation of a cellular nonlinear network for image segmentation. In: CNNA 2004 (2004)
[48] Ando, H., Morie, T., Miyake, M., Nagata, M., Iwata, A.: Image segmentation/ extraction using nonlinear cellular networks and their VLSI implementation using pulse-modulation techniques. IEICE Trans. Fundamentals E85-A(2) (February 2002)
[49] Schemmel, J., Meier, K., Loose, M.: A scalable switched capacitor realization of the resistive fuse network. Analog Integrated Circuits and Signal Processing 32, 135–148 (2002)
[50] Turing, A.M.: The Chemical Basis of Morphogenesis. Phil. Trans. Roy. Soc. Lond. B 237, 37–72 (1952)
[51] Goras, L., Chua, L.O., Leenearts, D.: Turing Patterns in CNNs – Part I: Once Over Lightly. IEEE Trans. on Circuits and Systems – I 42(10), 602–611 (1995)
[52] Goras, L., Chua, L.O., Leenearts, D.: Turing Patterns in CNNs – Part II: Equations and Behaviors. IEEE Trans. on Circuits and Systems – I 42(10), 612–626 (1995)
[53] Goras, L., Chua, L.O., Leenearts, D.: Turing Patterns in CNNs – Part III: Computer Simulation Results. IEEE Trans. on Circuits and Systems – I 42(10), 627–637 (1995)
[54] Goras, L., Chua, L.O.: On the Influence of CNN Boundary Conditions in Turing Pattern Formation. In: Proc. ECCTD 1997, Budapest, pp. 383–388 (1997)

[55] Goras, L., Teodorescu, T.D.: On CNN Boundary Conditions in Turing Pattern Formation. In: Proc. of the Fifth International Workshop on Cellular, Neural Networks and Their Applications, CNNA 1998 (1998)

[56] Goraş, L., Chua, L.O.: On the Role of CNN Initial Conditions in Turing Pattern Formation. In: Proc. SCS 1997, Iasi, Romania, pp. 105–108 (1997)

[57] Goras, L., Teodorescu, T., Ghinea, R.: On the Spatio-Temporal Dynamics of a Class of Cellular Neural Networks. Journal of Circuits, Systems and Computers Section I (Theory) (Special Issue on "CNN Technology and Visual Microprocessors") JCSC 12(4) (August 2003)

[58] Goras, L., Ungureanu, P.: On the Possibilities of Using Two-Grid Coupled CNN's for Face Features Extraction. In: Proceedings of the 8th IEEE International Workshop on Cellular Neural Networks and their Applications, CNNA 2004, Budapest, Hungary, July 22-24, pp. 381–386 (2004)

[59] Brodatz, P.: Textures: A photographic album for artists and designers. Dover, New York (1966)

Chapter 5
Approximating Multivariable Functions by Feedforward Neural Nets

Paul C. Kainen, Věra Kůrková, and Marcello Sanguineti

Abstract. Theoretical results on approximation of multivariable functions by feedforward neural networks are surveyed. Some proofs of universal approximation capabilities of networks with perceptrons and radial units are sketched. Major tools for estimation of rates of decrease of approximation errors with increasing model complexity are proven. Properties of best approximation are discussed. Recent results on dependence of model complexity on input dimension are presented and some cases when multivariable functions can be tractably approximated are described.

Keywords: multivariable approximation, feedforward neural networks, network complexity, approximation rates, variational norm, best approximation, tractability of approximation.

1 Introduction

Many classification, pattern recognition, and regression tasks can be formulated as mappings between subsets of multidimensional vector spaces, using

Paul C. Kainen
Department of Mathematics and Statistics, Georgetown University
Washington, D.C. 20057-1233, USA
e-mail: kainen@georgetown.edu

Věra Kůrková
Academy of Sciences of the Czech Republic
Pod Vodárenskou věží 2, 182 07, Prague 8, Czech Republic
e-mail: vera@cs.cas.cz

Marcello Sanguineti
DIBRIS, University of Genova,
via Opera Pia 13 - 16145 Genova, Italy
e-mail: marcello.sanguineti@unige.it

M. Bianchini et al. (Eds.): *Handbook on Neural Information Processing*, ISRL 49, pp. 143–181.
DOI: 10.1007/978-3-642-36657-4_5   © Springer-Verlag Berlin Heidelberg 2013

a suitable encoding of inputs and outputs. The goal of a computer scientist modeling such tasks is to find a mathematical structure which operates by adjusting parameters in order to classify or recognize different patterns and to approximate the regression function associated to data.

Mathematical formalizations have shown that many types of feedforward networks (including all standard types that are popular in applications as well as many others that may not have been considered by experimentalists) satisfy these requirements. Provided they contain sufficiently many basic computational units, it is possible to adjust their parameters so that they approximate to any accuracy desired a wide variety of mappings between subsets of multidimensional spaces. In neural network terminology, such classes of networks are called *universal approximators*. When network parameters are adjusted using suitable learning methods (e.g., gradient-descent, incremental or genetic algorithms), there is no need for explicit formulation of rules or feature extraction. However, we remark that *implicitly* some estimate of the number and nature of features may guide the practitioners choice of architectural parameters such as the number of hidden units and their type.

The universal approximation property has been proven for one-hidden layer networks with almost all types of reasonable computational units. Various interesting proof techniques have been used such as integral representations (Carrol and Dickenson [7], Ito [25], Park and Sandberg [60]), the Hahn-Banach Theorem (Cybenko [11]), the Stone-Weierstrass theorem (Hornik, Stinchcombe and White [24]), and orthogonal polynomials (Mhaskar [56], Leshno et al. [54]).

But universality cannot be proved within reasonable bounds on complexity. Each kind of network implementation determines a different measure of complexity. Currently, feedforward networks are mostly simulated on classical computers. For such simulations, the limiting factor is the number of hidden units and the size of their parameters. Thus, a suitable type of computational units and a structure for their connections has to be found that can solve a given task within the feasible limits of network complexity. Some guidelines for the choice of a type of neural network can be derived from mathematically grounded complexity theory of neural networks. Its recent achievements include estimates of rates of approximation by various types of feedforward networks and comparisons of complexity requirements of approximation by neural networks with linear approximation.

In this chapter, we first briefly sketch some ideas used in proofs of universal approximation capabilities of networks with perceptrons and radial units. Then the focus changes to network complexity. Major tools are described for estimating rates of decrease of approximation errors with increasing model complexity. We start with the Maurey-Jones-Baron Theorem holding in Hilbert spaces, present its extension to $\mathcal{L}^p$-spaces, and finally give an improvement to geometric rate due to Kůrková and Sanguineti [51]. We discuss other improvements and their limitations and show how estimates of rates of approximation of multivariable functions can be reformulated in terms of worst-case errors in sets of functions defined as balls in norms tailored to computational units.

Further, we sketch the results of Kainen, Kůrková, and Vogt [32, 33, 34] that similarly to linear approximation, approximation by a variable basis of half-space characteristic functions always has a best approximant, but unlike linear approximation, neural network approximation does not have *continuous* best approximation. This leads to the initially surprising result that lower bounds on the possible rate of continuous methods for approximation do not apply to neural networks (cf. [13]).

Finally, recent results on dependence of model complexity on input dimension (so-called "tractability") are considered; see [31]. We focus on those which were found by Kainen, Kůrková, and Vogt [35], dealing with representing the Gaussian via half-space characteristic functions (i.e., by a perceptron network using the Heaviside function as its sigmoidal activation function), and on those found by the current authors, utilizing Gaussian networks and the Bessel Potential function; see [30].

The chapter is organized as follows. In Section 2, we introduce a general model of approximation from a dictionary which includes one-hidden-layer networks. Section 3 sketches some proofs of universal approximation property of radial-basis function and perceptron networks, while the next section, Section 4, presents quadratic estimates of model complexity following from the Maurey-Jones-Barron Theorem and its extension by Darken et. al. In Section 5 we give geometric estimates of model complexity. Then in Section 6 these estimates are reformulated in terms of norms tailored to computational units. Section 7 shows that neural network approximation does not have a continuous best approximation and Section 8 is devoted to tractability of neural-network approximation. Section 9 is a brief discussion. A summary of the main notations used in the paper is given in Section 10.

2 Dictionaries and Variable-Basis Approximation

Feedforward neural networks compute parametrized sets of functions depending both on the type of computational units and on the type of their interconnections. *Computational units* compute functions depending on two vector variables: an *input vector* and a *parameter vector*. Generally, such units compute functions of the form $\phi : X \times Y \to \mathbb{R}$, where ϕ is a function of two variables, an input vector $x \in X \subseteq \mathbb{R}^d$ and a parameter $y \in Y \subseteq \mathbb{R}^s$, where $\mathbb{R}$ denotes the set of real numbers.

Sets of input-output functions of one-hidden-layer networks with one linear output unit can be formally described as

$$\text{span}_n \, G := \left\{ \sum_{i=1}^{n} w_i g_i \,\middle|\, w_i \in \mathbb{R}, g_i \in G \right\},$$

where the set G is called a *dictionary* [20], n is the *number of hidden units*, and w_i, $i = 1, \ldots, n$, are output weights. We write span G for the linear space consisting of all finite linear combinations of elements from G; that is,

$$\text{span } G = \bigcup_{n=1}^{\infty} \text{span}_n G.$$

Such a computational model $\text{span}_n G$ is sometimes called a *variable-basis* scheme [45, 46, 51], in contrast to *fixed-basis schemes* consisting of linear combinations of first n elements from a set G with a fixed linear ordering. Note that also networks with several hidden layers and one linear unit belong to this scheme (however in this case, the set G depends on the number of units in the previous hidden layers). Kernel models, splines with free nodes, and trigonometric polynomials with variable frequencies and phases are also special cases of variable schemes (see the references in [46].

The number n of computational units is often interpreted as *model complexity*. Note that the set of input-output functions of networks with an arbitrary (finite) number of hidden units is just span G.

Dictionaries are usually given as parameterized families of functions modelling computational units. They can be described as sets of the form

$$G_\phi = G_\phi(X, Y) := \{\phi(., y) : X \to \mathbb{R} \mid y \in Y\},$$

where $\phi : X \times Y \to \mathbb{R}$ is a function of two variables, an input vector $x \in X \subseteq \mathbb{R}^d$, where d is called *input dimension*, and a parameter $y \in Y \subseteq \mathbb{R}^s$. The most popular dictionaries used in neurocomputing include perceptrons, radial, and kernel units.

An element h of $\text{span}_n G_\phi(X, Y)$ then has the form

$$h = \sum_{i=1}^{n} w_i \phi(\cdot, y_i), \ w_i \in \mathbb{R}, \ y_i \in Y,$$

so h is determined by the function ϕ and $n(s + 1)$ real parameters.

In practical applications, inputs are bounded so one often studies networks with inputs in some compact (i.e., closed and bounded) subset X of $\mathbb{R}^d$. However, even if inputs are bounded, one may not know a priori what the bound is, so unbounded inputs are also considered, and the theoretical analysis is sometimes easier in this case.

For suitable choices of ϕ, sets $\text{span}_n G_\phi$ model families of input-output functions implemented by one-hidden-layer neural networks of various types. For example, the *perceptron* with *activation function* $\psi : \mathbb{R} \to \mathbb{R}$ takes

$$\phi(x, (v, b)) = \psi(v \cdot x + b).$$

Geometrically, perceptrons compute functions which are constant on all hyperplanes parallel to the hyperplane $\{x \in \mathbb{R}^d \,|\, v \cdot x = -b\}$. The scalar b is called *bias* while the components of v are called *inner weights*.

The terms "weights" or "parameters" are used to refer to both inner weights and biases, as well as the *outer* weights, which are the linear coefficients of combination of multiple units. The units might be perceptrons or other types. Thus, the distinction between inner and outer weights for these one-hidden-layer neural networks is just the distinction between the inputs and the outputs of the hidden units.

The most common activation functions are *sigmoidals*, i.e., bounded and measurable functions $\sigma : \mathbb{R} \to \mathbb{R}$ with limits 0 and 1 at $-\infty$ and ∞, resp. In some literature, sigmoidals are also required to be non-decreasing. Widely-used sigmoidals are the *logistic sigmoid* $\sigma(t) := 1/(1 + \exp(-t))$ and the *Heaviside function*, defined as $\vartheta(t) := 0$ for $t < 0$ and $\vartheta(t) := 1$ for $t \geq 0$. For a sigmoid σ, we let

$$P_d^\sigma := \{x \mapsto \sigma(v \cdot x + b) \,|\, v \in \mathbb{R}^d, b \in \mathbb{R}\}, \tag{1}$$

We write H_d instead of P_d^ϑ. Since for $t \neq 0$, $\vartheta(t) = \vartheta(t/|t|)$,

$$H_d := \{x \mapsto \vartheta(e \cdot x + b) \,|\, e \in \mathcal{S}^{d-1}, b \in \mathbb{R}\}, \tag{2}$$

where $\mathcal{S}^{d-1} := \{x \in \mathbb{R}^d \,|\, \|x\| = 1\}$ denotes the sphere in $\mathbb{R}^d$. Thus, H_d is the set of characteristic functions of closed half-spaces of $\mathbb{R}^d$ parameterized by the pair (e, b), where e is the direction of the orthogonal vector to the hyperplane and b is the distance from 0 to the hyperplane along a perpendicular.

Radial-basis functions (RBF) are given by

$$\phi(x, (v, b)) := \psi(b\|x - v\|),$$

where $\psi : \mathbb{R} \to \mathbb{R}$, $v \in \mathbb{R}^d$, and $b \in \mathbb{R}$. For a radial unit, the parameter b is called the *width*, and v the *center*. An RBF unit is constant on the set of all x at each, fixed distance from its center. The corresponding sets $\text{span}_n G_\phi$ are called *RBF networks*; a typical activation function for a radial unit is the Gaussian function $\psi(t) = \exp(-t^2) := e^{-t^2}$. This leads to the usual picture of "Gaussian hills" which makes plausible the density of their linear combination.

Note that sigmoidal perceptrons and RBF units are geometrically opposite: perceptrons apply a sigmoidal function to a weighted sum of inputs plus a bias, so they respond to *non-localized* regions of the input space by partitioning it with fuzzy hyperplanes (or sharp ones if the sigmoid is Heaviside's step function). The functions computed by perceptrons belong to the class of *plane waves*. In contrast, RBF units calculate the distance to a center, multiply it by a width factor and finally apply an activation function which is often an even function – hence they respond to *localized* regions. The functions computed by radial units belong to the class of *spherical waves*.

Although perceptrons were inspired neurobiologically (e.g., [64]), plane waves have long been studied by mathematicians, motivated by various problems from physics (e.g., [10]).

3 The Universal Approximation Property

The first theoretical question concerning a given type of a feedforward network architecture is whether a sufficiently elaborate network of this type can approximate all reasonable functions encountered in applications. In neural network terminology, this capability of a class of neural networks is called the *universal approximation property*, while mathematically it is defined as density of the set of input-output functions of the given class of networks. Recall that a subset F of a normed linear space is *dense* if $\operatorname{cl} F = \mathcal{X}$, where the closure cl is defined by the topology induced by the norm $\|.\|_{\mathcal{X}}$, i.e.,

$$\operatorname{cl} G := \{ f \in \mathcal{X} \mid (\forall \varepsilon > 0)(\exists g \in G)(\|f - g\|_{\mathcal{X}} < \varepsilon) \} \,.$$

Density of sets of input-output functions has been studied in both the sup-norm and $\mathcal{L}^p(X)$-cases. We write $(\mathcal{C}(X), \|\cdot\|_{\sup})$ for the space of all continuous bounded functions on a subset X of $\mathbb{R}^d$ with the supremum norm $\|\cdot\|_{\sup}$, defined for every continuous function on X as

$$\|f\|_{\sup} := \sup_{x \in X} |f(x)| \,.$$

For $p \in [1, \infty)$ let $(\mathcal{L}^p(X), \|\cdot\|_p)$ denote the set of all equivalence classes (w.r.t. equality up to sets of Lebesgue measure zero) of Lebesgue-measurable functions f on X such that the following $\mathcal{L}^p$-norm is finite:

$$\|f\|_p := \left(\int_X |f(x)|^p dx \right)^{1/p} < \infty \,.$$

Choice of norm is problem-dependent. Predicting the movement of a robotic welding tool might best utilize the supremum norm, while minimizing cost might be more likely over $\mathcal{L}^2$.

For RBF networks, the universal approximation property is intuitively quite clear - imagine the surface as a combination of Gaussian hills of various widths and heights. The classical method of approximation by convolutions with a suitable sequence of kernels enables one to prove this property for many types of radial functions. For $d \in \mathbb{N}_+$, where $\mathbb{N}_+$ denotes the set of positive integers, and ψ an even function, let $F_d^{\psi}(X)$ denote the dictionary

$$F_d^{\psi}(X) := \{ f : X \to \mathbb{R} \mid f(x) = \psi(b\|x - v\|), \, v \in \mathbb{R}^d, \, b \in \mathbb{R} \}$$

of functions on $X \subseteq \mathbb{R}^d$ computable by RBF networks with the radial function ψ and the distance from centers measured by a norm $\|\cdot\|$ on $\mathbb{R}^d$. In the following, we shall consider the Euclidean norm.

First, Hartman et al. [22] proved density of RBF networks with Gaussian radial function in $(\mathcal{C}(X), \|.\|_{\sup})$ for X compact convex. This proof used the special property of the Gaussian function that a d-dimensional Gaussian is the product of d one-dimensional Gaussians. Later, Park and Sandberg [61] extended the density property to RBFs with fairly general radial functions in $(\mathcal{L}^p(\mathbb{R}^d), \|.\|_p)$. Their proof exploits classical results on approximation of functions by convolutions with a sequence of kernel functions converging in the distributional sense to the Dirac delta function δ (see, e.g., [73]). The next theorem is from [61]; we sketch the idea of the proof.

Theorem 1 (Park and Sandberg). *For every positive integer d, every $p \in (1, \infty)$, every integrable bounded function $\psi : \mathbb{R} \to \mathbb{R}$ with finite non-zero integral and such that $\psi \in \mathcal{L}^p(\mathbb{R})$, span $F_d^\psi(\mathbb{R}^d)$ is dense in $(\mathcal{L}^p(\mathbb{R}^d), \|\cdot\|_p)$.*

Proof. When $\int_{\mathbb{R}} \psi(t)dt = c \neq 0$, by letting $\psi_0 = \frac{\psi}{c}$ we define a sequence $\{\psi_n(t) \mid n \in \mathbb{N}_+\}$ as $\psi_n(t) = n^d \psi_0(nt)$. By a classical result [4, p. 101] (see also [60, Lemma 1]), for every $f \in \mathcal{L}^p(\mathbb{R}^d)$ one has $f = \lim_{n \to \infty} f * \psi_n$ in $\|.\|_p$. Approximating the integrals

$$\int_{\mathbb{R}^d} f(x) \frac{n^d}{c} \psi(n(x - v))dv$$

by Riemann sums we get a sequence of functions of the form of RBF networks with ψ as a radial function.

Exploiting a similar classical result on approximation of functions in $(\mathcal{C}(X), \|.\|_{\sup})$ with X compact by convolutions with a sequence of bump functions, one gets an analogous proof of universal approximation property for RBF networks in $(\mathcal{C}(X), \|.\|_{\sup})$. Note that these arguments can be extended to other norms on $\mathbb{R}^d$ than the Euclidean one. Using a more sophisticated proof technique based on Hermite polynomials, Mhaskar [56] showed that for the Gaussian radial function, the universal approximation property (in sup norm) can even be achieved using networks with a given fixed width.

In one dimension, perceptron networks can also be localized as a pair of overlapping sigmoidal units with opposite-sign weights create a "bump" function. Hence, for $d = 1$, every "reasonable" function can be written as a limit of linear combinations of Heaviside perceptron units.

However, in contrast to localized Gaussian radial units and the one-dimensional case, for d greater than 1, the universal approximation property is far from obvious for perceptrons. But mathematics extends the range of visualization and offers tools that enable us to prove universal approximation for perceptrons with various types of activations.

One such tool is the Stone-Weierstrass theorem (Stone's extension of the classical result of Weierstrass regarding density of polynomials on a compact interval). A family of real-valued functions on a set X *separates points* if for any two distinct points in X there is a function in the family which takes on distinct values at the two points (i.e., for each pair of distinct points $x, y \in X$ there exists f in the family such that $f(x) \neq f(y)$). A family $\mathcal{A}$ of real-valued functions on X is an *algebra* if it is closed with respect to scalar multiplication and with respect to pointwise addition and multiplication of functions, e.g., for $f, g \in \mathcal{A}$, and $r \in \mathbb{R}$, the functions rf, $x \mapsto f(x) + g(x)$, and $x \mapsto f(x)g(x)$ belong to $\mathcal{A}$. The Stone-Weierstrass Theorem (e.g., [65, pp. 146-153]) states that an algebra $\mathcal{A}$ of real-valued functions on any compact set X is dense in $(\mathcal{C}(X), \|\cdot\|_{\sup})$ if and only if $\mathcal{A}$ separates points and is nontrivial in the sense that it contains a nonzero constant function.

For a function $\psi : \mathbb{R} \to \mathbb{R}$ we denote by $P_d^{\psi}(X)$ the dictionary

$$P_d^{\psi}(X) := \{f : X \to \mathbb{R} \mid f(x) = \psi(v \cdot x + b), \ v \in \mathbb{R}^d, \ b \in \mathbb{R}\}$$

of functions on $X \subseteq \mathbb{R}^d$ computable by perceptron networks with the activation function ψ. The linear span of this dictionary, span $P_d^{\psi}(X)$, is closed under addition and for reasonable ψ, it also separate points, but for many ψ it is not closed under multiplication. An exception is the function $\exp(t) = e^t$ which does produce a multiplicatively closed set $P_{\exp}^d(X)$ and so can serve as a tool to prove the universal approximation property for other activation functions.

The following "input-dimension-reduction theorem" by Stinchombe and White [69] exploits properties of $P_{\exp}^d(X)$ to show that for perceptron networks, it suffices to check the universal approximation property for networks with a single input.

Theorem 2 (Stinchombe and White). *Let* $\psi : \mathbb{R} \to \mathbb{R}$, d *be a positive integer, and* X *a compact subset of* $\mathbb{R}^d$. *Then* span $P_{\psi}^1(X)$ *is dense in* $(\mathcal{C}(X), \|\cdot\|_{\sup})$ *if and only if* span $P_{\psi}^d(X)$ *is dense in* $(\mathcal{C}(X), \|\cdot\|_{\sup})$.

Proof. By the Stone-Weierstrass Theorem, span $P_{\exp}^d(X)$ is dense in $(\mathcal{C}(X), \|\cdot\|_{\sup})$. Using composition of two approximations, the first one approximating $f \in \mathcal{C}(X)$ by an element of span $P_{\exp}^d(X)$, and the second one approximating $\exp$ on a suitable compact subset $Y \subset \mathbb{R}$ by an element of $P_{\psi}^1(Y)$, one gets density of span $P_{\psi}^d(X)$ in $(\mathcal{C}(X), \|\cdot\|_{\sup})$.

The Stone-Weierstrass Theorem was first used by Hornik, Stinchcombe, and White [24] to prove universal approximation property for one-hidden-layer sigmoidal perceptron networks. Later, Leshno et al. [54] characterized activation functions which determine perceptron networks with the universal approximation property. They showed that the universal approximation property is not restricted to (biologically motivated) sigmoidals but, with the exception of polynomials, it is satisfied by any reasonable activation function.

Theorem 3 (Leshno, Lin, Pinkus, and Schocken). *Let* $\psi : \mathbb{R} \to \mathbb{R}$ *be a locally bounded piecewise continuous function, d be a positive integer, and X a compact subset of* $\mathbb{R}^d$. *Then* span $P_{\psi}^d(X)$ *is dense in* $(\mathcal{C}(X), \|.\|_{\sup})$ *if and only if ψ is not an algebraic polynomial.*

Proof. We give only a sketch. The trick of Leshno et al.'s proof in [54] consists in expressing all powers as limits of higher-order partial derivatives with respect to the weight parameter v of the function $\psi(v \cdot x + b)$ (ψ being analytic guarantees that all the derivatives exist). It follows directly from the definition of iterated partial derivative that

$$\frac{\partial^k \psi(v \cdot x + b)}{\delta v^k}$$

can be expressed as a limit of functions computable by perceptron networks with activation ψ. More precisely,

$$\frac{\partial \psi(v \cdot x + b)}{\partial v} = \lim_{\eta \to 0} \frac{\psi((v + \eta)x + b) - \psi(v \cdot x + b)}{\eta},$$

and similarly for $k > 1$. Since $\frac{\partial^k \psi(v \cdot x + b)}{\partial v^k} = x^k \psi^{(k)}(v \cdot x + b)$, and for ψ non-polynomial, none of the $\psi^{(k)}$ is identically equal to zero for all k, setting $v = 0$ and choosing some b_k, for which $\psi^{(k)}(b_k) = c_k \neq 0$, one gets a sequence of functions from span $P_\psi^d(X)$ converging to $c_k x^k$. As all polynomials are linear combinations of powers, they can be obtained as limits of sequences of functions from span $P_\psi^d(X)$. So by Weierstrass' theorem and Theorem 2, span $P_d^\psi(X)$ is dense in $(\mathcal{C}(X), \| \cdot \|_{\sup})$ for any ψ which is analytic and non-polynomial. The statement can be extended to nonanalytic functions satisfying the assumptions of the theorem using suitable convolutions with analytic functions.

Inspection of the proof given by Leshno et al. [54] shows that the theorem is valid even when input weights are bounded by an arbitrarily small upper bound. However to achieve density, the set of hidden unit parameters must have either a finite or an infinite accumulation point.

Another standard method for treating approximation problems is based on the Hahn-Banach Theorem: to verify density of a set of functions, it is sufficient to show that every bounded linear functional that vanishes on this set must be equal to zero on the whole linear space. This method was used by Cybenko [11] to establish universal approximation for sigmoidal perceptron networks.

Other proofs of universal approximation property of perceptron networks took advantage of integral representations based on Radon transform (Carroll and Dickinson [7] and Ito [25]) and on Kolmogorov's representation of continuous functions of several variables by means of superpositions of continuous one-variable functions (Kůrková [37]).

In practical applications the domain of the function to be computed by a network is finite and so one can apply results from interpolation theory, which show that for finite domain functions, one can replace *arbitrarily close approximation* by *exact representation*. A major result of interpolation theory, Micchelli's theorem [58], proves that any function on a finite subset of $\mathbb{R}^d$ can be exactly represented as a network with Gaussian RBF units. An analogous result for sigmoidal perceptron networks has been proven by Ito [26].

However, development of interpolation theory has been motivated by the need to construct surfaces with certain characteristics fitted to a small number of points. Although its results are valid for any number of pairs of data, the application to neurocomputing leads to networks with the same number of hidden units as the number of input-output pairs. For large sets of data, this requirement may prevent implementation. In addition, it is well-known that fitting input-output function to all the training data may produce "overfitting", in which characteristics of noise and other random artifacts mask the intrinsic nature of the function. In such cases, estimates of accuracy achievable using networks with fewer hidden units are needed. These can be derived from estimates of rates of approximation by variable basis schemes that apply to both finite and infinite domains.

4 Quadratic Rates of Approximation

The cost of universality is arbitrarily large model complexity and hence the set of network parameters has to also be sufficiently large. Dependence of accuracy of approximation on the number of hidden units can be studied in the framework of approximation theory in terms of *rates of approximation*. In other words, rates of approximation characterize the trade-off between accuracy of approximation and model complexity.

Jones [27] introduced a recursive construction of approximants with rates of order $\mathcal{O}(1/\sqrt{n})$. Together with Barron [2] he proposed to apply it to sets of functions computable by one-hidden-layer sigmoidal perceptron networks. The following theorem is a version of Jones' result as improved by Barron [2]. It was actually discovered and proved first by Maurey (see [63]) using a probabilistic argument to guarantee existence rather than the incremental algorithm given here.

The following theorem restates the Maurey-Jones-Barron's estimate. Its proof is from [41], where the argument of Barron [2, p. 934, Lemma 1] is simplified. Let $\mathrm{conv}_n\, G$ denote the set of all convex combinations of n elements from the set G, i.e.,

$$\mathrm{conv}_n\, G := \left\{ \sum_{i=1}^{n} a_i g_i \;\middle|\; a_i \in [0,1],\; \sum_{i=1}^{n} a_i = 1,\; g_i \in G \right\},$$

while $\mathrm{conv}\, G$ denotes the *convex hull* of G

$$\mathrm{conv}\, G := \bigcup_{i=1}^{n} \mathrm{conv}_n\, G.$$

For $\mathcal{X}$ a normed linear space, in approximating an element $f \in \mathcal{X}$ by elements from a subset A of $\mathcal{X}$, the error is the *distance from f to A*,

$$\|f - A\|_{\mathcal{X}} := \inf_{g \in A} \|f - g\|_{\mathcal{X}}.$$

Theorem 4 (Maurey-Jones-Barron). *Let $(\mathcal{X}, \|.\|_{\mathcal{X}})$ be a Hilbert space, G be a non empty bounded subset of $\mathcal{X}$, and $s_G := \sup_{g \in G} \|g\|_{\mathcal{X}}$. Then for every $f \in \mathrm{cl\,conv}\, G$ and for every positive integer n,*

$$\|f - \mathrm{conv}_n G\|_{\mathcal{X}} \le \sqrt{\frac{s_G^2 - \|f\|_{\mathcal{X}}^2}{n}}.$$

Proof. Since distance from $\mathrm{conv}_n G$ is continuous on $(\mathcal{X}, \|.\|_{\mathcal{X}})$, i.e., $f \mapsto \|f - \mathrm{conv}_n G\|_{\mathcal{X}}$ is continuous for $f \in \mathcal{X}$ [67, p. 391], it suffices to verify the statement for $f \in \mathrm{conv}\, G$. Let

$$f = \sum_{j=1}^{m} a_j h_j$$

be a representation of f as a convex combination of elements of G. Set

$$c := s_G^2 - \|f\|^2.$$

By induction we construct $\{g_n\}_{n \ge 1} \subseteq G$ such that for all n

$$e_n^2 = \|f - f_n\|^2 \le \frac{c}{n}, \text{ where } f_n = \sum_{i=1}^{n} \frac{g_i}{n}.$$

Indeed, for the basis case of the induction, note that

$$\sum_{j=1}^{m} a_j \|f - h_j\|^2 = \|f\|^2 - 2\langle f, \sum_{j=1}^{m} a_j h_j \rangle + \sum_{j=1}^{m} a_j \|h_j\|^2 \le s_G^2 - \|f\|^2 - c,$$

so there exists $j \in \{1, \ldots, m\}$ for which $\|f - h_j\|^2 \le c$. Take $f_1 = g_1 := h_j$. Suppose $g_1, \ldots, g_n$ satisfy the error-bound. Then

$$e_{n+1}^2 = \|f - f_{n+1}\|^2 = \|\frac{n}{n+1}(f - f_n) + \frac{1}{n+1}(f - g_{n+1})\|^2 =$$

$$= \frac{n^2}{(n+1)^2} e_n^2 + \frac{2n}{(n+1)^2} \langle f - f_n, f - g_{n+1} \rangle + \frac{1}{(n+1)^2} \|f - g_{n+1}\|^2.$$

As in the basis case,

$$\sum_{j=1}^{m} a_j \left(\frac{2n}{(n+1)^2} \langle f - f_n, f - h_j \rangle + \frac{1}{(n+1)^2} \|f - h_j\|^2 \right) =$$

$$= \frac{1}{(n+1)^2}\left(\sum_{j=1}^{m} a_j g_j - \|f\|^2\right) \leq \frac{1}{(n+1)^2}(s_G^2 - \|f\|^2) = \frac{c}{(n+1)^2}$$

So there must exist $j \in \{1, \ldots, m\}$ such that

$$\frac{2n}{(n+1)^2}\langle f - f_n, f - h_j\rangle + \frac{1}{(n+1)^2}\|f - h_j\|^2 \leq \frac{c}{(n+1)^2}.$$

Setting $g_{n+1} := h_j$, we get $e_{n+1}^2 \leq \frac{n^2}{(n+1)^2}e_n^2 + \frac{c}{(n+1)^2} \leq c/(n+1)^2$.

Inspection of the proof shows that approximants in the convex hull of G can always be taken to be barycenters of simplices - i.e., with all parameters of the convex combination equal.

Theorem 4 actually gives an upper bound on *incremental* learning algorithms (i.e., algorithms which at each step add a new hidden unit but do not change the previously-determined inner parameters for the already chosen units, see e.g., Kůrková [40]). In principle, non-incremental algorithms might do better. Darken at al. [12] extended Maurey-Jones-Barron's estimate to $\mathcal{L}^p$-spaces, $p \in (1, \infty)$. They used a more sophisticated argument replacing inner products with peak functionals and taking advantage of Clarkson's inequalities (see [23, pp. 225, 227]). Recall that for a Banach space $(\mathcal{X}, \|\cdot\|_{\mathcal{X}})$ and $f \in \mathcal{X}$, we denote by Π_f a *peak functional for* f, i.e., a continuous linear functional such that $\|\Pi_f\|_{\mathcal{X}} = 1$ and $\Pi_f(f) = \|f\|_{\mathcal{X}}$ [5, p.1].

The next theorem is a slight reformulation of [12, Theorem 5] with a simplified proof.

Theorem 5 (Darken-Donahue-Gurvits-Sontag). *Let $\Omega \subseteq \mathbb{R}^d$ be open, G a subset of $(\mathcal{L}_p(\Omega), \|\cdot\|_p)$, $p \in (1, \infty)$, $f \in \mathrm{cl\,conv}\,G$, and $r > 0$ such that $G \subseteq B_r(f, \|\cdot\|)$. Then for every positive integer n*

$$\|f - \mathrm{span}_n G\|_p \leq \frac{2^{1/a} r}{n^{1/b}},$$

where $q := p/(p-1)$, $a := \min(p, q)$, and $b := \max(p, q)$.

Proof. As in the proof of Theorem 4, it is sufficient to verify the statement for $f \in \mathrm{conv}\,G$. Let $f = \sum_{j=1}^{m} w_j h_j$ be a representation of f as a convex combination of elements of G. We show by induction that there exist a sequence $\{g_i\}$ of elements of G such that the barycenters $f_n = \sum_{i=1}^{n} \frac{g_i}{n}$ satisfy $e_n := \|f - f_n\| \leq \frac{2^{1/a} r}{n^{1/b}}$.

First we check that there exists $g_1 \in G$ such that $f_1 = g_1$ satisfies $e_1 = \|f - f_1\|_p \leq 2^{1/a} r$. This holds trivially as $G \subseteq B_r(f, \|\cdot\|)$, so for any $g \in G$ we have $\|f - g\| \leq r < 2^{1/a} r$. Hence we can set $f_1 := g_1$ for any $g_1 \in G$.

Assume that we already have $g_1, \ldots, g_n$. Then

$$f_{n+1} = \frac{n}{n+1} f_n + \frac{1}{n+1} g_{n+1} = \frac{1}{n+1}\sum_{i=1}^{n+1} g_i.$$

We shall express e_{n+1}^a in terms of e_n^a.

Let Π_n be a peak functional for $f - f_n$. Since $\sum_{j=1}^{m} w_j \left(f - h_j \right) = 0$, by linearity of Π_n we have $0 = \Pi_n \left(\sum_{j=1}^{m} w_j \left(f - h_j \right) \right) = \sum_{j=1}^{m} w_j \Pi_n (f - h_j)$. Thus, there must exist $j \in \{1, \dots, m\}$ such that $\Pi_n(f - h_j) \leq 0$. Set $g_{n+1} = h_j$, so $\Pi_n(f - g_{n+1}) \leq 0$. For every $p \in (1, \infty)$, $q =: p/(p-1)$, $a := \min(p, q)$, and $f, g \in \mathcal{L}^p(\Omega)$, Clarkson's inequalities imply

$$\|f + g\|_p^a + \|f - g\|_p^a \leq 2 \left(\|f\|_p^a + \|g\|_p^a \right)$$

(see, e.g., [47, Proposition A3 (iii)]. Hence we get

$$
e_{n+1}^a = \|f - f_{n+1}\|_p^a = \left\| \frac{n}{n+1}(f - f_n) + \frac{1}{n+1}(f - g_{n+1}) \right\|_p^a
$$

$$
\leq 2 \left(\left\| \frac{n}{n+1}(f - f_n) \right\|_p^{\bar{p}} + \left\| \frac{1}{n+1}(f - g_{n+1}) \right\|_p^a \right)
$$

$$
\times \left\| \frac{n}{n+1}(f - f_n) - \frac{1}{n+1}(f - g_{n+1}) \right\|_p^a. \tag{3}
$$

As $\|\Pi_n\| = 1$ and $\Pi_n(f - g_{n+1}) \leq 0$, we have $\left\| \frac{n}{n+1}(f - f_n) - \frac{1}{n+1}(f - g_{n+1}) \right\|_p \geq \left\| \Pi_n \left(\frac{n}{n+1}(f - f_n) - \frac{1}{n+1}(f - g_{n+1}) \right) \right\|_p > \left\| \Pi_n \left(\frac{n}{n+1}(f - f_n) \right) \right\|_p = \frac{n}{n+1} \|\Pi_n(f - f_n)\|_p = \frac{n}{n+1} \|f - f_n\|_p$. Hence

$$
- \left\| \frac{n}{n+1}(f - f_n) - \frac{1}{n+1}(f - g_{n+1}) \right\|_p^a \leq - \left(\frac{n}{n+1} \|f - f_n\|_p \right)^a. \tag{4}
$$

By (3) and (4),

$$
e_{n+1}^a = \|f - f_{n+1}\|_p^a \leq
$$

$$
2 \left(\left\| \frac{n}{n+1}(f - f_n) \right\|_p^a + \left\| \frac{1}{n+1}(f - g_{n+1}) \right\|_p^a \right) - \left(\frac{n}{n+1} \|f - f_n\|_p \right)^a =
$$

$$
\frac{2}{(n+1)^a} \|f - g_{n+1}\|_p^a + \left(\frac{2}{n+1} \right)^a \|f - f_n\|_p^a =
$$

$$
\frac{2}{(n+1)^a} \|f - g_{n+1}\|_p^a + \left(\frac{2}{n+1} \right)^a e_n^a.
$$

As $e_n = \|f - f_n\| \leq \frac{2^{1/a} r}{n^{1/b}}$, we get $e_{n+1}^a \leq \frac{2 r^a}{(n+1)^a} + \left(\frac{n}{n+1} \right)^a \left(\frac{2^{1/a} r}{n^{1/b}} \right)^a = \frac{2 r^a}{(n+1)^a} \left(1 + \frac{n^a}{n^{a/b}} \right) = \frac{2 r^a}{(n+1)^a} \left(1 + n^{a - a/b} \right)$. It can easily be verified that $a - \frac{a}{b} = 1$ in both cases, $a = p$ (and so $b = q = \frac{p}{p-1}$) and $a = q$ (and so $b = p$). Thus $e_{n+1}^a \leq \frac{2 r^a}{(n+1)^a} (n + 1)$. Hence, $e_{n+1}^a \leq \frac{2 r^a}{(n+1)^{a-1}} = \frac{2 r^a}{(n+1)^{a/b}}$, i.e., $e_{n+1} \leq \frac{2^{1/a} r}{(n+1)^{1/b}}$.

The Maurey-Jones-Barron Theorem and its extension to $\mathcal{L}_p$-spaces received a lot of attention because they imply an estimate of model complexity of the order $\mathcal{O}(\varepsilon^{-2})$. Several authors derived improvements and investigated tightness of the improved bounds for suitable sets G such as G orthogonal [45, 52], G formed by functions computable by sigmoidal perceptrons, and precompact G with certain covering number properties [49, 55]. In particular, for a dictionary G of functions computable by perceptrons with certain sigmoidal activations, impossibility of improving the exponent $-\frac{1}{2}$ in the upper bound from Theorem (4) over $-(\frac{1}{2} + \frac{1}{d})$ (where d is the number of variables) was proved in [1, 55, 49].

To illustrate an improvement of the Maurey-Jones-Barron Theorem, we state an upper bound by Makovoz [55, Theorem 1], which brings in the entropy numbers of G. Recall that the n-th entropy number $e_n(G)$ of a subset G of a normed linear space is defined as

$$e_n(G) := \inf\{\varepsilon > 0 \,|\, (G \subseteq \cup_{i=1}^n U_i) \,\&\, (\forall i = 1, \ldots, n)\,(\operatorname{diam}(U_i) \leq \varepsilon)\},$$

where $\operatorname{diam}(U) = \sup_{x,y \in U} \|x - y\|$.

Theorem 6 (Makovoz). *Let G be a bounded subset of a Hilbert space $(\mathcal{X}, \|\cdot\|_{\mathcal{X}})$. Then for every $f \in \operatorname{span} G$ of the form $f = \sum_{i=1}^\infty c_i g_i$ such that $\sum_{i=1}^\infty |c_i| < \infty$ and every positive integer n there exists $g = \sum_{i=1}^n a_i\, g_i \in \operatorname{span}_n G$ such that*

$$\|f - g\|_{\mathcal{X}} \leq \frac{2\, e_n(G) \sum_{i=1}^\infty |c_i|}{\sqrt{n}},$$

where $\sum_{i=1}^n |a_i| \leq \sum_{i=1}^\infty |c_i|$.

5   Geometric Rates of Approximation

Throughout this section let $(\mathcal{X}, \|.\|_{\mathcal{X}})$ be a Hilbert space with G a non empty bounded subset of $\mathcal{X}$. The Maurey-Jones-Barron Theorem and its improvements presented in the previous section are worst-case estimates (i.e., they give upper bounds holding for all functions from the closure of the symmetric convex hull of G). Thus, one can expect that for suitable subsets of this hull, better rates may hold.

Lavretsky [53] noticed that a certain geometric condition would allow substantial improvement in the Jones-Barron's iterative construction [2, 27]. More precisely, for $\delta > 0$, he defined the set

$$F_\delta(G) := \big\{ f \in \operatorname{cl} \operatorname{conv} G \,\big|\, \forall h \in \operatorname{conv} G,\ f \neq h\ \exists g \in G :$$
$$(f - g) \cdot (f - h) \leq -\delta \|f - g\|_{\mathcal{X}} \|f - h\|_c X \big\} \tag{5}$$

and proved the following result.

Theorem 7 (Lavretsky). *Let* $(\mathcal{X}, \|.\|_{\mathcal{X}})$ *be a Hilbert space with* G *any bounded symmetric subset containing* 0, $s_G := \sup_{g \in G} \|g\|$, *and* $\delta > 0$. *Then for every* $f \in F_\delta(G)$ *and every positive integer* n,

$$\|f - \text{conv}_n G\|_{\mathcal{X}} \leq \sqrt{(1 - \delta^2)^{n-1}(s_G^2 - \|f\|_{\mathcal{X}}^2)}\,.$$

In [51], Kůrková and Sanguineti improved the idea of Lavretsky by showing that for every function f in the convex hull of a bounded subset G of a Hilbert space there exists $\tau_f \in [0, 1)$ such that the rate of approximation of f by convex combinations of n functions from G is bounded from above by $\sqrt{\tau_f^{n-1}(s_G^2 - \|f\|_{\mathcal{X}}^2)}$.

Theorem 8 (Kůrková-Sanguineti). *Let* $(\mathcal{X}, \| \cdot \|_{\mathcal{X}})$ *be a Hilbert space,* G *its bounded nonempty subset, and* $s_G := \sup_{g \in G} \|g\|_{\mathcal{X}}$. *For every* $f \in \text{conv}\, G$ *there exists* $\tau_f \in [0, 1)$ *such that for every positive integer* n

$$\|f - \text{conv}_n G\|_{\mathcal{X}} \leq \sqrt{\tau_f^{n-1}(s_G^2 - \|f\|_{\mathcal{X}}^2)}\,.$$

Proof. Let $f = \sum_{j=1}^m a_j\, g_j$ be a representation of f as a convex combination of elements of G with all $a_j > 0$ (and $\sum_j a_j = 1$). Let G' be the set of elements combined; i.e.,

$$G' := \{g_1, \ldots, g_m\}\,.$$

For each $n = 1, \ldots, m$, we find $f_n \in \text{conv}_n G$, and $\rho_n > 0$ such that

$$\|f - f_n\|_{\mathcal{X}}^2 \leq (1 - \rho_n^2)^{n-1}\left(s_G^2 - \|f\|_{\mathcal{X}}^2\right)\,. \tag{6}$$

Let $g_{j_1} \in G'$ be nearest to f, i.e.,

$$\|f - g_{j_1}\|_{\mathcal{X}} - \min_{g \in G'} \|f - g\|_{\mathcal{X}},$$

and set $f_1 := g_{j_1}$. As

$$\sum_{j=1}^m a_j \|f - g_j\|_{\mathcal{X}}^2 = \|f\|_{\mathcal{X}}^2 - 2f \cdot \sum_{i=1}^m a_j g_j + \sum_{j=1}^m a_j \|g_j\|_{\mathcal{X}}^2$$
$$\leq s_G^2 - \|f\|_{\mathcal{X}}^2\,,$$

we get $\|f - f_1\|_{\mathcal{X}}^2 \leq s_G^2 - \|f\|_{\mathcal{X}}^2$ and so (6) holds for $n = 1$ with any $\rho_1 \in (0, 1)$.

Assuming that we have f_{n-1}, we define f_n. When $f_{n-1} = f$, we set $f_n := f_{n-1}$ and the estimate holds trivially.

When $f_{n-1} \neq f$, we define f_n as the convex combination

$$f_n := \alpha_n f_{n-1} + (1 - \alpha_n) g_{j_n}, \tag{7}$$

with $g_{j_n} \in G'$ and $\alpha_n \in [0, 1]$ chosen in such a way that for some $\rho_n > 0$

$$\|f - f_n\|_{\mathcal{X}}^2 \leq (1 - \rho_n^2)^{n-1}\|f - f_{n-1}\|_{\mathcal{X}}^2.$$

First, we choose a suitable g_{j_n} and then we find α_n depending on our choice of g_{j_n}. Denoting $e_n := \|f - f_n\|_{\mathcal{X}}$, by (7) we get

$$e_n^2 = \alpha_n^2 e_{n-1}^2 + 2\alpha_n(1 - \alpha_n)(f - f_{n-1}) \cdot (f - g_{j_n}) + (1 - \alpha_n)^2\|f - g_{j_n}\|_{\mathcal{X}}^2. \quad (8)$$

For all $j \in \{1, \dots, m\}$, set

$$\eta_j := -\frac{(f - f_{n-1}) \cdot (f - g_j)}{\|f - f_{n-1}\|_{\mathcal{X}} \|f - g_j\|_{\mathcal{X}}}$$

(both terms in the denominator are nonzero: the first one because we are considering the case when $f \neq f_{n-1}$ and the second one because we assume that for all j, $a_j > 0$ and thus $f \neq g_j$). Note that for all j, $\eta_j \in [-1, 1]$ as it is the cosine of the angle between the vectors $f - f_{n-1}$ and $f - g_j$.
As $f = \sum_{j=1}^m a_j g_j$, we have

$$\sum_{j=1}^m a_j(f - f_{n-1}) \cdot (f - g_j) = (f - f_{n-1}) \cdot (f - \sum_{j=1}^m a_j g_j) = 0.$$

Thus
(i) either there exists $g \in G'$, for which $(f - f_{n-1}) \cdot (f - g) < 0$
(ii) or for all $g \in G'$, $(f - f_{n-1}) \cdot (f - g) = 0$.
   We show that case (ii) can't happen since it would imply that $f = f_{n-1}$. Indeed, $f_{n-1} \in \mathrm{conv}_{n-1}G'$ and thus can be expressed as

$$f_{n-1} = \sum_{k=1}^{n-1} b_k g_{j_k}$$

with all $b_k \in [0, 1]$ and $\sum_{k=1}^{n-1} b_k = 1$. If for all $g \in G'$, $(f - f_{n-1}) \cdot (f - g) = 0$, then $\|f - f_{n-1}\|_{\mathcal{X}}^2$ is equal to

$$(f - f_{n-1}) \cdot (f - \sum_{k=1}^{n-1} b_k g_{j_k}) = \sum_{k=1}^{n-1} b_k(f - f_{n-1}) \cdot (f - g_{j_k}) = 0.$$

By assumption, $f \neq f_{n-1}$, so case (i) must hold. Therefore, the subset

$$G'' := \{g \in G' \mid (f - f_{n-1}) \cdot (f - g) < 0\}$$

is nonempty. Let $g_{j_n} \in G''$ be chosen so that

$$\eta_{j_n} = \max_{j=1,\dots,m} \eta_j$$

and set $\rho_n := \eta_{j_n}$. As $G'' \neq \emptyset$, we have $\rho_n > 0$. Let $r_n := \|f - g_{j_n}\|_{\mathcal{X}}$. By (8) we get

$$e_n^2 = \alpha_n^2 e_{n-1}^2 - 2\alpha_n(1 - \alpha_n)\rho_n e_{n-1} r_n + (1 - \alpha_n)^2 r_n^2. \tag{9}$$

To define f_n as a convex combination of f_{n-1} and g_{j_n}, it remains to find $\alpha_n \in [0, 1]$ for which e_n^2 is minimal as a function of α_n. By (9) we have

$$e_n^2 = \alpha_n^2 \left(e_{n-1}^2 + 2\rho_n e_{n-1} r_n + r_n^2\right) - 2\alpha_n \left(\rho_n e_{n-1} r_n + r_n^2\right) + r_n^2. \tag{10}$$

Thus

$$\frac{\partial e_n^2}{\partial \alpha_n} = 2\alpha_n \left(e_{n-1}^2 + 2\rho_n e_{n-1} r_n + r_n^2\right) - 2 \left(\rho_n e_{n-1} r_n + r_n^2\right)$$

and

$$\frac{\partial^2 e_n^2}{\partial^2 \alpha_n} = 2 \left(e_{n-1}^2 + 2\rho_n e_{n-1} r_n + r_n^2\right).$$

As now we are considering the case when $f \neq f_{n-1}$, we have $e_{n-1} > 0$ and hence $\frac{\partial e_n^2}{\partial^2 \alpha_n} > 0$. So the minimum is achieved at

$$\alpha_n = \frac{\rho_n e_{n-1} r_n + r_n^2}{e_{n-1}^2 + 2\rho_n e_{n-1} r_n + r_n^2}. \tag{11}$$

Plugging (11) into (10) we get

$$e_n^2 = \frac{(1 - \rho_n^2)e_{n-1}^2 r_n^2}{e_{n-1}^2 + 2\rho_n e_{n-1} r_n + r_n^2} < \frac{(1 - \rho_n^2)e_{n-1}^2 r_n^2}{r_n^2} = (1 - \rho_n^2)e_{n-1}^2. \tag{12}$$

Let

$$k := \max\{n \in \{1, \ldots, m\} \mid f_n \neq f_{n-1}\}.$$

Setting

$$\rho_f := \min\{\rho_n \mid n = 1, \ldots, k\},$$

by induction we get the upper bound

$$\|f - \mathrm{conv}_n G\|_{\mathcal{X}}^2 \leq (1 - \rho_f^2)^{n-1} \left(s_G^2 - \|f\|_{\mathcal{X}}^2\right)$$

holding for all n (for $n > m$ it holds trivially with $f_n = f$). We conclude by setting $\tau_f := 1 - \rho_f^2$.

We illustrated Theorem 8 in [51] by estimating values of parameters of geometric rates when G is an orthonormal basis. We derived also insights into the structure of sets of functions with fixed values of parameters of such rates. As for Theorem 4, the proof of Theorem 8 is based on a constructive incremental procedure, described in [51]. For every function $f \in \operatorname{conv} G$ and its any representation $f = \sum_{j=1}^{m} a_j g_j$ as a convex combination of elements of G, the proof constructs a linear ordering

$$\{g_{j_1}, \ldots, g_{j_m}\}$$

of the subset

$$G' := \{g_1, \ldots, g_m\} \, .$$

Then it shows that for every positive integer $n \leq m$ and for some $\tau_f \in [0, 1)$ one has

$$\| f - \operatorname{span}\{g_{j_1}, \ldots, g_{j_m}\} \|_{\mathcal{X}}^2 \leq \tau_f^{n-1} \left(s_G^2 - \|f\|_{\mathcal{X}}^2 \right) \, .$$

Table 1 describes this procedure.

The geometric bound would be more useful if one could calculate τ_f as a functional of f. Nevertheless, this bound shows the possibility of substantial improvement for suitably nice functions f.

The speed of decrease of the estimate depends on $\tau_f \in [0, 1)$ which is obtained as the smallest cosine of the angles between functions used in the construction of approximants. Inspection of the proof shows that the parameter τ_f is not defined uniquely. It depends on the choice of a representation of $f = \sum_{j=1}^{m} a_j g_j$ as a convex combination of elements of G and on the choice of g_{j_n} for those positive integers n, for which there exist more than one g_j with the same cosine ρ_n. However, the *minimal parameter*, for which the geometric upper bound from Theorem 8 holds, is unique.

Let

$$\tau(f) := \min \left\{ \tau > 0 \mid \| f - \operatorname{conv}_n G \|_{\mathcal{X}}^2 \leq \tau^{n-1}(s_G^2 - \|f\|^2) \right\}. \qquad (13)$$

By Theorem 8, for every $f \in \operatorname{conv} G$ the set over which the minimum in (13) is taken is nonempty and bounded. It follows from the definition of this set that its infimum is achieved, i.e., it is a minimum. Therefore,

$$\| f - \operatorname{conv}_n G \|_{\mathcal{X}} \leq \sqrt{\tau(f)^{n-1}(s_G^2 - \|f\|_{\mathcal{X}}^2)} \, .$$

6 Approximation of Balls in Variational Norms

The Maurey-Jones-Barron Theorem is a useful tool for estimation of rates of variable-basis approximation. Since $\operatorname{conv}_n G \subseteq \operatorname{span}_n G$, the upper bound from Theorem 4 on approximation by $\operatorname{conv}_n G$ also applies to approximation by $\operatorname{span}_n G$. When G is bounded, $\operatorname{conv} G$ is a proper subset of $\operatorname{span} G$ and so $\operatorname{cl} \operatorname{conv} G$ is a proper subset of $\operatorname{cl} \operatorname{span} G$; thus, Theorem 4 cannot be applied to all elements of $\mathcal{X}$. However, its corollary on approximation by $\operatorname{span}_n G$ applies to all functions in $\mathcal{X}$. Indeed, replacing the set G by sets of the form

Table 1 The incremental construction used in the proof of Theorem 8

1 CHOOSE $g_{j_1} \in \{g_j \,|\, j = 1, \ldots, m\}$ SUCH THAT
$$\|f - g_{j_1}\| = \min_{j=1,\ldots,m} \|f - g_j\|;$$

2 $f_1 := g_{j_1};$

FOR $n = 2, \ldots, m - 1$:

BEGIN

FOR $j = 1, \ldots, m,$

3   COMPUTE $\eta_j := -\dfrac{(f - f_{n-1}) \cdot (f - g_j)}{\|f - f_{n-1}\| \, \|f - g_j\|}$

IF FOR $j = 1, \ldots, m$ ONE HAS $\eta_j = 0,$ THEN

END

ELSE

BEGIN

4      $\rho_n := \max\{\eta_j > 0 \,|\, j = 1, \ldots, m\};$

5   CHOOSE g_{j_n} SUCH THAT $\rho_n = \eta_{j_n};$

6   COMPUTE $e_{n-1} := \|f - f_{n-1}\|;$

7   COMPUTE $r_n := \|f - g_{j_n}\|;$

8   COMPUTE $\alpha_n := \dfrac{\rho_n e_{n-1} r_n + r_n^2}{e_{n-1}^2 + 2\rho_n e_{n-1} r_n + r_n^2};$

9      $f_n := \alpha_n f_{n-1} + (1 - \alpha_n) g_n;$

$n := n + 1.$

END

END

LET

$$k := \max\{n \in \{1, \ldots, m\} \,|\, f_n \neq f_{n-1}\}$$

AND

$$\rho_f := \min\{\rho_n \,|\, n = 1, \ldots, k\}$$

$$\tau_f := 1 - \rho_f^2$$

$$G(c) := \{wg \mid w \in \mathbb{R}, |w| \leq c, \ g \in G\}$$

with $c > 0$, we get $\mathrm{conv}_n G(c) \subset \mathrm{span}_n G(c) = \mathrm{span}_n G$ for any $c \in \mathbb{R}$. This approach can be mathematically formulated in terms of a norm tailored to a set G (in particular to sets G_ϕ corresponding to various computational units ϕ).

Let $(\mathcal{X}, \|\cdot\|_\mathcal{X})$ be a normed linear space and G be its bounded non empty subset, then *G-variation* (variation with respect to G) is defined as the Minkowski functional of the set $\mathrm{cl}\, \mathrm{conv}(G \cup -G)$, where $-G := \{f \in \mathcal{X} \mid f = -g, \ g \in G\}$, i.e.,

$$\|f\|_G := \inf\{c > 0 \mid f/c \in \mathrm{cl}\, \mathrm{conv}(G \cup -G)\}.$$

Note that G-variation can be infinite and that it is a norm on the subspace of $\mathcal{X}$ formed by those $f \in \mathcal{X}$, for which $\|f\|_G$ is finite. The closure in its definition depends on the topology induced on $\mathcal{X}$ by the norm $\|\cdot\|_\mathcal{X}$. However, when $\mathcal{X}$ is finite dimensional, G-variation does not depend on the choice of a norm on $\mathcal{X}$, since all norms on a finite-dimensional space induce the same topology.

Intuitively, G-variation of f measures how much the set G needs to be inflated to contain f in the closure of its symmetric convex hull. It is easy to check that

$$\|\cdot\|_\mathcal{X} \leq s_G \|\cdot\|_G, \tag{14}$$

where $s_G := \sup_{g \in G} \|g\|_\mathcal{X}$. Indeed, if for $b > 0$, $f/b \in \mathrm{cl}\, \mathrm{conv}(G \cup -G)$, then $f/b = \lim_{\varepsilon \to 0} h_\varepsilon$, where $h_\varepsilon \in \mathrm{conv}(G \cup -G)$ and so $\|h_\varepsilon\| \leq s_G$. Thus, $\|f\|_\mathcal{X} \leq s_G\, b$. Hence, by the definition of $\|f\|_G$ we have $\|f\|_\mathcal{X} \leq s_G \|f\|_G$.

Variation with respect to G was introduced by Kůrková [39] as an extension of Barron's [1] concept of *variation with respect to half-spaces* corresponding to G formed by functions computable by Heaviside perceptrons, defined as

$$H_d := \{x \mapsto \vartheta(e \cdot x + b) \mid e \in \mathcal{S}^{d-1}, b \in \mathbb{R}\},$$

where $\mathcal{S}^{d-1} := \{x \in \mathbb{R}^d \mid \|x\| = 1\}$ denotes the sphere of radius 1 in $\mathbb{R}^d$ (recall that ϑ is the Heaviside function, defined as $\vartheta(t) := 0$ for $t < 0$ and $\vartheta(t) := 1$ for $t \geq 0$). In particular, if f is an input-output function of a one-hidden-layer network with Heaviside perceptrons, then variation with respect to half-spaces of f is equal to the sum of absolute values of output weights. For $d = 1$, variation with respect to half-spaces is up to a constant equal to total variation which plays an important role in integration theory. G-variation is also an extension of the notion of ℓ_1-norm. When G is an orthonormal basis of a separable Hilbert space, G-variation is equal to the ℓ_1-*norm with respect to* G, which is defined for every $f \in X$ as $\|f\|_{1,G} := \sum_{g \in G} |f \cdot g|$ [18, 19, 45, 52].

Approximation capabilities of sets of functions can be studied in terms of *worst-case errors*, formalized by the concept of *deviation*. For two subsets A and M of $\mathcal{X}$, the deviation of M from A is defined as

$$\delta(M, A) = \delta(M, A; \mathcal{X}) = \delta(M, A; (\mathcal{X}, \|\cdot\|_{\mathcal{X}})) := \sup_{f \in M} \|f - A\|_{\mathcal{X}}. \tag{15}$$

We use the abbreviated notations when the ambient space and the norm are clear from the context. When the supremum in (15) is achieved, deviation is the *worst-case error* in approximation of functions from M by functions from A. In this section, we consider the case in which the set M of functions to be approximated is a ball in G-variation. By the definition, the unit ball in G-variation is the closure in the norm $\|\cdot\|_{\mathcal{X}}$ of the symmetric convex hull of G, i.e.,

$$B_1(\|\cdot\|_G) := \mathrm{cl}\left(\mathrm{conv}\left(G \cup -G\right)\right). \tag{16}$$

The estimates given in the next corollary follow by the Maurey-Jones-Barron Theorem (here Theorem 4) and its extension by Darken et al. (Theorem 5). These estimates give upper bounds on rates of approximation by $\mathrm{span}_n G$ for all functions in a Hilbert space and in $\mathcal{L}^p(X)$ spaces with $p \in (1, \infty)$.

Corollary 1. *Let* $(\mathcal{X}, \|\cdot\|_{\mathcal{X}})$ *be a Banach space,* G *its bounded nonempty subset,* $s_G := \sup_{g \in G} \|g\|_{\mathcal{X}}$. *Then for every* $f \in \mathcal{X}$ *and every positive integer* n,
(i) for $(\mathcal{X}, \|\cdot\|_{\mathcal{X}})$ *a Hilbert space,*

$$\|f - \mathrm{span}_n G\|_{\mathcal{X}} \le \sqrt{\frac{s_G^2 \|f\|_G^2 - \|f\|_{\mathcal{X}}^2}{n}},$$

so

$$\delta(B_r(\|\cdot\|_G), \mathrm{span}_n G) \le \frac{r\, s(G)}{n^{1/2}};$$

(ii) for $(\mathcal{X}, \|\cdot\|_{\mathcal{X}}) = (\mathcal{L}^p(X), \|\cdot\|_{\mathcal{L}^p(X)})$ *with* $p \in (1, \infty)$,

$$\|f - \mathrm{span}_n G\|_{\mathcal{X}} \le \frac{2^{1+1/a}\, s_G \|f\|_G}{n^{1/b}},$$

where $a := \min(p, \frac{p}{p-1})$ *and* $b := \max(p, \frac{p}{p-1})$, *so*

$$\delta(B_r(\|\cdot\|_{G, \mathcal{L}^p(X)}), \mathrm{span}_n G_d)_{\mathcal{L}^p(X)} \le 2^{1+1/a} \frac{r\, s(G)}{n^{1/b}}.$$

By the upper bounds from Corollary 1, all functions from the unit ball in G-variation can be approximated within $s_G/\sqrt{n}$ or $2^{1+1/a} s_G/n^{1/b}$ by networks with n hidden units from the dictionary G, independently on the number d of variables. For this reason, such estimates are sometimes called "dimension-independent", which is misleading since with increasing number of variables, the condition of being in the unit ball in G-variation becomes more and more constraining.

Note that since $0 \in \mathrm{span}_n G$, we have for all $f \in \mathcal{X}$, $\|f - \mathrm{span}_n G\|_{\mathcal{X}} \le \|f\|$, thus the bound from Corollary 1 is nontrivial only when $\|f\|_{\mathcal{X}}^2 \ge \frac{(s_G \|f\|_G)^2 - \|f\|_{\mathcal{X}}^2}{n}$ or equivalently $\frac{\|f\|_{\mathcal{X}}}{s_G \|f\|_G} \ge \frac{1}{\sqrt{n+1}}$. For example, for $s_G = 1$ and $\|f\|_G = 1$, this implies that $\|f\|_{\mathcal{X}} \ge \frac{1}{\sqrt{n+1}}$.

Properties of balls in variation corresponding to standard hidden units are not yet well understood. Such balls can be described as subsets of images of balls in $\mathcal{L}_1$-norm under certain integral transformations (see, e.g., [2, 15, 43]).

In [52], examples of functions with variation with respect to half-spaces (i.e., with respect to Heaviside perceptrons) growing exponentially with the number of variables d were given. However, such exponentially-growing lower bounds on variation with respect to half-spaces are merely lower bounds on an upper bound on rates of approximation. They do not prove that such functions cannot be approximated by networks with n perceptrons with faster rates than $\frac{s_G \|f\|_G}{\sqrt{n}}$.

Combining Theorem 6 with estimates of entropy numbers, Makovoz [55] disproved the possibility of a substantial improvement of the upper bound from Corollary 1 (i) for the set G corresponding to perceptrons with certain activations. We say that a sigmoidal is *polynomially quickly approximating the Heaviside* if there exist η, $C > 0$ such that for all $t \in \mathbb{R}$ one has $|\sigma(t) - \vartheta(t)| \leq C |t|^{\eta}$. The following theorem states Makovoz' result in terms of variation norm.

Theorem 9 (Makovoz). *Let d be a positive integer, σ either the Heaviside function or a Lipschitz continuous sigmoidal polynomially quickly approximating the Heaviside, and $X \subset \mathbb{R}^d$ compact. If $\tau > 0$ is such that for some $c > 0$ and all positive integers n one has*

$$\delta\big(B_1(\|.\|_{P_\sigma^d(X)})), \operatorname{conv}_n P_\sigma^d(X)\big) \leq \frac{c}{n^\tau},$$

then $\tau \leq \frac{1}{2} + \frac{1}{d}$.

Hence, for a wide family of sigmoidal perceptron networks the term $n^{-1/2}$ cannot be improved beyond $n^{-1/2-1/d}$, so in high dimension, $n^{-1/2}$ is essentially best possible.

In [49], Kůrková and Sanguineti extended this tightness result to more general approximating sets. Recall that the *ε-covering number* of a subset G of $(\mathcal{X}, \|\cdot\|_{\mathcal{X}})$ is the cardinality of a minimal ε-net in G, i.e.,

$$\mathcal{N}(G, \varepsilon) := \min\Big\{ m \in \mathbb{N}_+ \,|\, \exists f_1, \ldots, f_m \in G \text{ such that } G \subseteq \bigcup_{i=1}^{m} B_\varepsilon(f_i, \|\cdot\|_{\mathcal{X}}) \Big\}.$$

If the set over which the minimum is taken is empty, then $\mathcal{N}(G, \varepsilon) = +\infty$. When there exists $\beta > 0$ such that $\mathcal{N}(G, \varepsilon) \leq \left(\frac{1}{\varepsilon}\right)^\beta$ for $\varepsilon \downarrow 0$, G is said to have *power-type covering numbers*.

For a subset A of a normed linear space $(\mathcal{X}, \|.\|_{\mathcal{X}})$ and a positive integer r, we denote

$$A_r := \left\{ f \in A \,\Big|\, \|f\|_{\mathcal{X}} \geq \frac{1}{r} \right\}.$$

The larger the sets A_r, the slower the decrease of the norms of the elements of A. When A_r is finite for all positive integers r, we call the function $\alpha_A : \mathbb{N}_+ \to \mathbb{N}_+$ defined as

$$\alpha_A(r) := \operatorname{card} A_r$$

the *decay function of A*, where $\operatorname{card} A_r$ denotes the number of elements of A_r. A set A such that A_r is finite for all positive integers r is *slowly decaying with respect to γ* if there exists $\gamma > 0$ such that $\alpha_A(r) = r^\gamma$. Note that if A is a precompact subset of a Hilbert space and A_r is orthogonal, then A_r must be finite. Thus decay functions are defined for all precompact orthogonal subsets of Hilbert spaces and also for subsets $A = \bigcup_{r=1}^\infty A_r$ with all A_r orthogonal but A not necessarily orthogonal. Finally, we call *slowly decaying* a set A formed by d-variable functions with the decay function $\alpha_A(r) = r^d$.

Theorem 10 (Kůrková-Sanguineti). *Let $(\mathcal{X}, \|.\|_{\mathcal{X}})$ be a Hilbert space, G its bounded precompact subset with $s_G = \sup_{g \in G} \|g\|$ and power-type covering numbers; let $t > 0$ and $\gamma > 0$, and $B_1(\|.\|_G) \supseteq t\,A$, where A is slowly decaying with respect to γ. If $\tau > 0$ is such that for some $c > 0$ and all positive integers n one has*

$$\delta\big(B_1(\|.\|_G), \operatorname{conv}_n(G \cup -G)\big) \leq \frac{c}{n^\tau}\,,$$

then $\tau \leq \frac{1}{2} + \frac{1}{\gamma}$.

The proof of Theorem 10 exploits characteristics of generalized Hadamard matrices. A Hadamard matrix is a $d \times d$ matrix of ± 1 entries such that the rows are pairwise-orthogonal; i.e., they have dot product of zero. An $r \times d$ matrix of ± 1s is called quasi-orthogonal if the dot product of any two distinct rows is small compared to d. When the dot product is bounded in absolute value by some constant t, then Kainen and Kůrková [29] showed that as d goes to infinity, the maximum number of rows in a quasi orthogonal matrix grows exponentially.

It was proven in [49] that Theorem 6 follows by Theorem 10 applied to the set $P_d^\sigma(X)$ of functions computable by perceptrons, where σ is either the Heaviside function or a Lipschitz continuous sigmoidal polynomially quickly approximating the Heaviside.

7   Best Approximation and Non-continuity of Approximation

To estimate rates of variable-basis approximation, it is helpful to study properties like existence, uniqueness, and continuity of corresponding approximation operators.

Existence of a best approximation has been formalized in approximation theory by the concept of proximinal set (sometimes also called "existence" set). A subset M of a normed linear space $(\mathcal{X}, \|.\|_{\mathcal{X}})$ is called *proximinal* if for every $f \in X$ the distance $\|f - M\|_{\mathcal{X}} = \inf_{g \in M} \|f - g\|_{\mathcal{X}}$ is achieved for some element of M, i.e., $\|f - M\|_{\mathcal{X}} = \min_{g \in M} \|f - g\|_{\mathcal{X}}$ (Singer [67]). Clearly a proximinal subset must be closed. On the other hand, for every f in $\mathcal{X}$, the distance-from-f function is continuous on $\mathcal{X}$ [67, p. 391] and hence on any subset M. When M is compact, therefore, it is necessarily proximinal.

Two generalizations of compactness also imply proximinality. A set M is called *boundedly compact* if the closure of its intersection with any bounded set is compact. A set M is called *approximatively compact* if for each $f \in X$ and any sequence $\{g_i\}$ in M such that $\lim_{i \to \infty} \|f - g_i\|_{\mathcal{X}} = \|f - M\|_{\mathcal{X}}$, there exists $g \in M$ such that $\{g_i\}$ converges subsequentially to g [67, p. 368]. Any closed, boundedly compact set is approximatively compact, and any approximatively compact set is proximinal [67, p. 374].

We investigate the existence property for one-hidden-layer Heaviside perceptron networks. Gurvits and Koiran [21] have shown that for all positive integers d the set H_d of characteristic functions of closed half-spaces in $\mathbb{R}^d$ intersected with the unit cube $[0,1]^d$ is compact in $(\mathcal{L}^p([0,1]^d), \|.\|_p)$ with $p \in [1, \infty)$. This can be easily verified once the set H_d is reparametrized by elements of the unit sphere S^d in $\mathbb{R}^{d+1}$. Indeed, a function $\vartheta(v \cdot x + b)$, with the vector $(v_1, \ldots, v_d, b) \in \mathbb{R}^{d+1}$ nonzero, is equal to $\vartheta(\hat{v} \cdot x + \hat{b})$, where $(\hat{v}_1, \ldots, \hat{v}_d, \hat{b}) \in S^d$ is obtained from $(v_1, \ldots, v_d, b) \in \mathbb{R}^{d+1}$ by normalization. Since S^d is compact, so is H_d. However, $\mathrm{span}_n H_d$ is neither compact nor boundedly compact for any positive integers n, d.

The following theorem from [34] shows that $\mathrm{span}_n H_d$ is approximatively compact in $\mathcal{L}^p$-spaces. It extends a result of Kůrková [38], who showed that $\mathrm{span}_n H_d$ is closed in $\mathcal{L}^p$-spaces with $p \in (1, \infty)$.

Theorem 11 (Kainen-Kůrková-Vogt). *Let d be any positive integer. Then* $\mathrm{span}_n H_d$ *is an approximatively compact subset of* $(\mathcal{L}^p([0,1]^d, \|.\|_p)$ *for $n \geq 1$ and $p \in [1, \infty)$.*

Theorem 11 shows that for all positive integers n, d a function in $\mathcal{L}^p([0,1]^d)$ with $p \in [1, \infty)$ has a best approximation among functions computable by one-hidden-layer networks with a single linear output unit and n Heaviside perceptrons in the hidden layer. Thus for any p-integrable function on $[0,1]^d$ there exists a linear combination of n characteristic functions of closed half-spaces that is nearest in the $\mathcal{L}^p$-norm. In other words, in the space of parameters of networks of this type, there exists a global minimum of the error functional defined as $\mathcal{L}^p$-distance from the function to be approximated. A related proposition is proved by Chui, Li, and Mhaskar in [9], where certain sequences are shown to have subsequences that converge almost everywhere (a. e.). These authors work in $\mathbb{R}^d$ rather than $[0,1]^d$ and show a. e. convergence rather than $\mathcal{L}^p$-convergence.

Theorem 11 cannot be extended to perceptron networks with differentiable activation functions, e.g., the logistic sigmoid or hyperbolic tangent. For such functions, the sets

$$\mathrm{span}_n P_d(\psi),$$

where

$$P_d(\psi) := \{f : [0,1]^d \to \mathbb{R} \mid f(x) = \psi(v \cdot x + b), v \in \mathbb{R}^d, b \in \mathbb{R}\},$$

are not closed and hence cannot be proximinal. This was first observed by Girosi and Poggio [16] and later exploited by Leshno et al. [54] for a proof of the universal approximation property.

Cheang and Barron [8] showed that linear combinations of characteristic functions of closed half-spaces with relatively few terms can yield good approximations of such functions as the characteristic function χ_B of a ball. However, χ_B is not approximated by the linear combination itself but rather by the characteristic function of the set where the linear combination exceeds a certain threshold. This amounts to replacing a linear output in the corresponding neural network by a threshold unit.

Note that Theorem 11 does not offer any information on the error of the best approximation. Estimates on such errors available in the literature (e.g., DeVore, Howard, and Micchelli [13], Pinkus [62]) are based on the assumption that the best approximation operators are continuous. However, it turns out that continuity of such operators may not hold [33], [32] as we now explain.

Recall [67] that for a subset M of a normed linear space $(\mathcal{X}, \|\cdot\|_{\mathcal{X}})$ and $f \in \mathcal{X}$, the *(metric) projection* $\mathrm{Pr}_M(f)$ of f to M is the set of elements in M at minimum distance from f; i.e.,

$$\mathrm{Pr}_M(f) := \{g \in M \mid \|f - g\|_{\mathcal{X}} = \|f - M\|_{\mathcal{X}} = \inf_{h \in M} \|f - h\|_{\mathcal{X}}\}.$$

When f is in M, it is its own metric projection. An element of $\mathrm{Pr}_M(f)$ is called a *best approximation* to f *from* M. A mapping $\Psi : \mathcal{X} \to M$ is called a *best approximation mapping* (to elements of $\mathcal{X}$ from M) with respect to $\|\cdot\|_{\mathcal{X}}$ if it maps every element of $\mathcal{X}$ into its projection in M, i.e., for every $f \in \mathcal{X}$ one has $\Psi(f) \in \mathrm{Pr}_M(f)$, that is, $\|f - \Psi(f)\|_{\mathcal{X}} = \|f - M\|_{\mathcal{X}}$.

A classical result from approximation theory [67] states that when $\mathcal{X}$ is a uniformly convex Banach space (for example an $\mathcal{L}^2$-space), the best approximation mapping to a closed convex subset is unique and continuous. This has a basic consequence in linear approximation: it means that for every element f of such a space, there exists a unique linear combination of fixed basis functions (i.e., a unique element of a linear approximating subspace) that minimizes the distance from f and that such a best approximation varies continuously as f is varied.

The situation is different when one considers approximation by neural networks. This is mainly due to the fact that, instead of a finite-dimensional subspace, the approximating functions belong to the union $\mathrm{span}_n G$ of finite-dimensional subspaces spanned by all n-tuples of elements of G. The following result from [33, Theorem 2.2] (see also [32]) states the non-existence of continuous best approximation by $\mathrm{span}_n G$ in $\mathcal{L}^p$-spaces, $p \in (1, \infty)$. By card G we denote the number of elements of G.

Theorem 12 (Kainen-Kůrková-Vogt). *Let X be a measurable subset of $\mathbb{R}^d$, n a positive integer, and G a linearly independent subset of $\mathcal{L}^p(X)$, $p \in (1, \infty)$, with $\operatorname{card} G > n$. Then there exists no continuous best approximation of $\mathcal{L}^p(X)$ by $\operatorname{span}_n G$.*

According to Theorem 12, in $\mathcal{L}^p$-spaces with $p \in (1, \infty)$, for every positive integer n and every linearly independent subset G with $\operatorname{card} G > n$ there is no continuous best approximation mapping to $\operatorname{span}_n G$. As regards the requirement of linear independence of sets of functions representing neural networks, it was proved for the hyperbolic tangent as hidden unit, for certain Heaviside networks, and for Gaussian radial-basis functions. A characterization of linearly independent families for different types of activation functions was given in [44].

Combining Theorem 11 with Theorem 12, we conclude that while best approximation operators exist from $\mathcal{L}^p([0,1]^d)$ to $\operatorname{span}_n H_d$, they cannot be continuous for $p \in (1, \infty)$. This loss of continuity has important consequences on estimates of approximation rates by neural networks. In particular, the lack of continuous dependence in approximation by neural networks does not allow one to apply *a priori* the lower bounds available for linear approximators. In contrast to deviation from a single subspace, deviation from $\operatorname{span}_n G$ which is a union of many such subspaces is much more difficult to estimate since, as we have seen, with the exception of some marginal cases, best approximation mappings to such unions do not posess the good properties of best approximation mapping to a single linear subspace.

8 Tractability of Approximation

8.1 A Shift in Point-of-View: Complexity and Dimension

Only recently has the influence of input dimension on approximation accuracy and rate been studied. Input dimension d is the number of distinct one-dimensional input channels to the computational units. So if a chip-bearing structure like an airplane's wing is providing $400,000$ distinct channels of information, then $d = 400,000$. Some experimental results have shown that optimization over connectionistic models built from relatively few computational units with a simple structure can obtain surprisingly good performances in selected optimization tasks (seemingly high-dimensional); see, e.g., [17, 28, 48, 50, 51, 59, 68, 74, 75] and the references therein. Due to the fragility and lack of theoretical understanding even for these examples, together with the ever-growing amount of data provided by new technology, we believe it is important to explicitly consider the role of d in the theory. Algorithms might require an exponential growth in time and resources as d increases [3] and so even powerful computers would be unable to handle them - hence, they would not be feasible.

On the other hand, in applications, functions of hundreds of variables have been approximated quite well by networks with only a moderate number of hidden units (see, e.g., NETtalk in [66]). Estimates of rates of approximation by neural networks derived from constructive proofs of the universal approximation property have not been able to explain such successes since the arguments in these papers lead to networks with complexity growing exponentially with the number of input units, e.g., $\mathcal{O}(1/\sqrt[d]{n})$. Current theory only predicts that to achieve an accuracy within ε, approximating functions of complexity of order $(1/\varepsilon)^d$ are required.

Some insights into properties of sets of multivariable functions that can be approximated by neural networks with good rates can be derived from the results of the previous sections. The approximation rates that we presented typically include several factors, one of which involves the number n of terms in the linear combinations, while another involves the number d of inputs to computational units. Dependence on dimension d can be implicit; i.e., estimates involve parameters that are constant with respect to n but *do* depend on d and the manner of dependence is not specified; see, e.g., [1, 2, 6, 12, 14, 15, 21, 27, 55]. Terms depending on d are referred to as "constants" since these papers focus on the number n of computational units and assume a fixed value for the dimension d of the input space. Such estimates are often formulated as $O(\kappa(n))$, where dependence on d is hidden in the "big O" notation [36]. However, in some cases, such "constants" actually grow at an exponential rate in d [52, 42]. Moreover, the families of functions for which the estimates are valid may become negligibly small for large d [46].

In general dependence of approximation errors on d may be harder to estimate than dependence on n [71] and few such estimates are available. Deriving them can help to determine when machine-learning tasks are feasible. The role of d is considered explicitly in information-based complexity (see [70, 71, 72]) and more recently this situation has been studied in the context of functional approximation and neural networks [30, 57].

8.2 *Measuring Worst-Case Error in Approximation*

We focus on upper bounds on worst-case errors in approximation from dictionaries, formalized by the concept of deviation defined in equation (15). An important case is when the deviation of a set A_d of functions of d-variables from the set $\mathrm{span}_n\, G_d$ takes on the factorized form

$$\delta(A_d, \mathrm{span}_n\, G_d)_{\mathcal{X}_d} \le \xi(d)\, \kappa(n)\,. \tag{17}$$

In the bound (17), dependence on the number d of variables and on model complexity n are separated and expressed by the functions $\xi : \mathbb{N}' \to \mathbb{R}_+$ and $\kappa : \mathbb{N}_+ \to \mathbb{R}$, respectively, with $\mathbb{N}'$ an infinite subset of the set $\mathbb{N}_+$ of positive integers and κ nonincreasing nonnegative. Such estimates have been derived, e.g., in [1, 2, 6, 12, 27, 35].

Definition 1. *The problem of approximating A_d by elements of* $\text{span}_n\, G_d$ *is called* **tractable** *with respect to d in the worst case or simply tractable if in the upper bound (17) for every $d \in \mathbb{N}'$ one has $\xi(d) \le d^\nu$ for some $\nu > 0$ and κ is a nonincreasing function. We call the problem* **hyper-tractable** *if the upper bound (17) holds with $\lim_{d\to\infty} \xi(d) = 0$ and κ is nonincreasing.*

Thus, if approximation of A_d by $\text{span}_n\, G_d$ is hyper-tractable, then the scaled problem of approximating $r_d A_d$ by $\text{span}_n\, G_d$ is tractable, unless r_d grows faster than $\xi(d)^{-1}$. If $\xi(d)$ goes to zero at an exponential rate, then the scaled problem is hyper-tractable if r_d grows polynomially.

When $\kappa(n) = n^{-1/s}$, the input-output functions of networks with

$$n \ge \left(\frac{\xi(d)}{\varepsilon}\right)^s$$

computational units can approximate given class of functions within ε. If $\xi(d)$ is polynomial, model complexity is a polynomial in d.

In [31], we derived conditions on sets of functions which can be tractably approximated by various dictionaries, including cases where such sets are large enough to include many smooth functions on $\mathbb{R}^d$ (for example, d-variable Gaussian functions on $\mathbb{R}^d$ in approximation by perceptron networks). Even if $\xi(d)$ is a polynomial in d, this may not provide sufficient control of model complexity unless the degree is quite small. For large dimension d of the input space, even quadratic approximation may not be sufficient. But there are situations where dependence on d is *linear* or better [31], and cases are highlighted in which the function $\xi(d)$ decreases exponentially fast with dimension.

As the arguments and proof techniques exploited to derive the results in [31] are quite technical, here we report only some results on tractability of approximation by Gaussian RBF networks and certain perceptron networks, presented in two tables. In the following, for a norm $\| \cdot \|$ in a space of d-variable functions and $r_d > 0$, we denote by $B_{r_d}(\| \cdot \|)$ the ball of radius r_d in such a norm.

8.3   Gaussian RBF Network Tractability

We first consider tractability of approximation by Gaussian RBF networks. The results are summarized in Table 2. To explain and frame the results, we need to discuss two functions - the Gaussian and the Bessel Potential. The former is involved since we choose it as activation function for the RBF network, the latter because it leads to a norm which is equivalent to the Sobolev norm (that is, both Sobolev and Bessel Potential norms are bounded by a multiple of the other as non-negative functionals).

Let $\gamma_{d,b} : \mathbb{R}^d \to \mathbb{R}$ denote the d-dimensional Gaussian function of *width* $b > 0$ and *center* $0 = (0, \ldots, 0)$ in $\mathbb{R}^d$:

$$\gamma_{d,b}(x) := e^{-b\|x\|^2}.$$

We write γ_d for $\gamma_{d,1}$. Width is parameterized inversely: larger values of b correspond to sharper peaks. For $b > 0$, let

$$G_d^\gamma(b) := \left\{ \tau_y(\gamma_{d,b}) \mid y \in \mathbb{R}^d \right\},$$

denote the set of d-variable Gaussian RBFs with width $b > 0$ and all possible centers, where, for each vector y in $\mathbb{R}^d$, τ_y is the translation operator defined for any real-valued function $g : \mathbb{R}^d \to \mathbb{R}$ by

$$\tau_y(g)(x) := g(x - y) . \forall x \in \mathbb{R}^d$$

The set of Gaussians with varying widths is denoted

$$G_d^\gamma := \bigcup_{b>0} G_d^\gamma(b).$$

For $m > 0$, the *Bessel potential of order* m on $\mathbb{R}^d$ is the function $\beta_{d,m}$ with the Fourier transform

$$\hat{\beta}_{d,m}(\omega) - (1 + \|\omega\|^2)^{-m/2},$$

where we consider the *Fourier transform* $\mathcal{F}(f) := \hat{f}$ as

$$\hat{f}(\omega) := (2\pi)^{-d/2} \int_{\mathbb{R}^d} f(x) e^{ix \cdot \omega} dx.$$

For $m > 0$ and $q \in [1, \infty)$, let

$$L^{q,m}(\mathbb{R}^d) := \{ f \mid f = w * \beta_d, m, \ w \in \mathcal{L}^q(\mathbb{R}^d) \}$$

be the *Bessel potential space* which is formed by convolutions of functions from $\mathcal{L}^q(\mathbb{R}^d)$ with $\beta_{d,m}$. The Bessel norm is defined as

$$\|f\|_{L^{q,m}(\mathbb{R}^d)} := \|w_f\|_{\mathcal{L}^q(\mathbb{R}^d)} \quad \text{for} \quad f = w_f * \beta_{d,m}.$$

In row 1 of Table 2, $\xi(d) = (\pi/2b)^{d/4} r_d$. Thus for $b = \pi/2$, the estimate implies tractability for r_d growing with d polynomially, while for $b > \pi/2$, it implies tractability even when r_d increases exponentially fast. Hence, the width b of Gaussians has a strong impact on the size of radii r_d of balls in $G_d^\gamma(b)$-variation for which $\xi(d)$ is a polynomial. The narrower the Gaussians, the larger the balls for which the estimate in row 1 of Table 2 implies tractability.

For every $m > d/2$, the upper bound from row 2 of Table 2 on the worst-case error in approximation by Gaussian-basis-function networks is of the factorized form $\xi(d)\kappa(n)$, where $\kappa(n) = n^{-1/2}$ and

Table 2 Factorized approximation rates for Gaussian RBF networks. In row 1, $b > 0$; in row 2, $m > d/2$.

ambient space	dictionary	approximated functions	$\xi(d)$	$\kappa(n)$
$(\mathcal{L}^2(\mathbb{R}^d), \|\cdot\|_{\mathcal{L}^2(\mathbb{R}^d)})$	$G_d^\gamma(b)$	$B_{r_d}(\|\cdot\|_{G_d^\gamma(b)})$	$r_d \left(\frac{\pi}{2b}\right)^{d/4}$	$n^{-1/2}$
$(\mathcal{L}^2(\mathbb{R}^d), \|\cdot\|_{\mathcal{L}^2(\mathbb{R}^d)})$	G_d^γ	$B_{r_d}(\|\cdot\|_{L^{1,m}}) \cap L^{2,m}$	$\left(\frac{\pi}{2}\right)^{d/4} \frac{\Gamma(m/2-d/4)}{\Gamma(m/2)} r_d$	$n^{-1/2}$

$$\xi(d) = r_d \left(\frac{\pi}{2}\right)^{d/4} \frac{\Gamma(m/2 - d/4)}{\Gamma(m/2)}.$$

Let $h > 0$ and put $m_d = d/2 + h$. Then $\xi(d)/r_d = \left(\frac{\pi}{2}\right)^{d/4} \frac{\Gamma(h/2)}{\Gamma(h/2+d/4)}$, which goes to zero exponentially fast with increasing d. So for $h > 0$ and $m_d \geq d/2 + h$, the approximation of functions in $B_{r_d}(\|\cdot\|_{L^{1,m_d}(\mathbb{R}^d)}) \cap L^{2,m_d}(\mathbb{R}^d)$ by $\mathrm{span}_n G_d^\gamma$ is hyper-tractable in $\mathcal{L}^2(\mathbb{R}^d)$.

8.4 Perceptron Network Tractability

Let us now consider tractability of approximation by perceptron networks. The results are summarized in Table 3. This subsection also requires some technical machinery. We describe a class of real-valued functions on $\mathbb{R}^d$, the functions of weakly-controlled decay, defined by Kainen, Kůrková, and Vogt in [35], which have exactly the weakest possible constraints on their behavior at infinity to guarantee finiteness of a certain semi-norm and we show that functions in this class have a nice integral formula, leading to our results. This subsection provides an instance in which $\mathbb{N}'$, the domain of the dimensional complexity function ξ, is the odd positive integers.

The dictionary of functions on $X \subseteq \mathbb{R}^d$ computable by perceptron networks with the activation function ψ is denoted by

$$P_d^\psi(X) := \{f : X \to \mathbb{R} \mid f(x) = \psi(v \cdot x + b), \, v \in \mathbb{R}^d, \, b \in \mathbb{R}\},$$

so $P_d^\psi(\mathbb{R}^d) = P_d^\psi$ as defined in (1). For ϑ the Heaviside function, as in (2),

$$P_d^\vartheta(X) = H_d(X) := \{f : X \to \mathbb{R} \mid f(x) = \vartheta(e \cdot x + b), \, e \in S^{d-1}, \, b \in \mathbb{R}\},$$

where S^{d-1} is the sphere constituted by the unit-euclidean-norm vectors in $\mathbb{R}^d$ and $H_d(X)$ is the set of characteristic functions of closed half-spaces of $\mathbb{R}^d$ restricted to X. Of course, all these functions, and their finite linear combinations, have finite sup-norms. The integral formula we develop below shows that the nice functions (of weakly controlled decay) are tractably approximated by the linear span of $H_d(X)$.

For $\mathcal{F}$ any family of functions on $\mathbb{R}^d$ and $\Omega \subseteq \mathbb{R}^d$, let

$$\mathcal{F}|_\Omega := \{f|_\Omega \mid f \in \mathcal{F}\},$$

where $f|_\Omega$ is the restriction of f to Ω. We also use the phrase "variation with respect to half-spaces" for the restrictions of H_d. For simplicity, we may write H_d instead of $H_d|_\Omega$. When $\Omega_d \subset \mathbb{R}^d$ has finite Lebesgue measure, for each continuous nondecreasing sigmoid σ, variation with respect to half-spaces is equal to $P_d^\sigma|_{\Omega_d}$-variation in $\mathcal{L}^2(\Omega_d)$ [43]. Hence, investigating tractability of balls in variation with respect to half-spaces has implications for approximation by perceptron networks with arbitrary continuous nondecreasing sigmoids.

A real-valued function f on $\mathbb{R}^d$, d odd, is of *weakly-controlled decay* [35] if f is d-times continuously differentiable and for all multi-indices $\alpha \in \mathbb{N}^d$ with $|\alpha| = \sum_{i=1}^{d} \alpha_i$ and $D^\alpha = \partial^{\alpha_1} \cdot \ldots \cdot \partial^{\alpha_d}$, such that

$$\lim_{\|x\|\to\infty} D^\alpha f(x) - 0, \ \forall \alpha, \ |\alpha| < d \quad (i)$$

$$\exists \varepsilon > 0 \setminus \lim_{\|x\|\to\infty} D^\alpha f(x)\|x\|^{d+1+\varepsilon} = 0, \ \forall \alpha, \ |\alpha| = d. \quad (ii)$$

Let $\mathcal{V}(\mathbb{R}^d)$ denote the set of functions of weakly controlled decay on $\mathbb{R}^d$. This set includes the Schwartz class of smooth functions rapidly decreasing at infinity as well as the class of d-times continuously differentiable functions with compact supports. In particular, it includes the Gaussian function. Also, if $f \in \mathcal{V}(\mathbb{R}^d)$, then $\|D^\alpha f\|_{\mathcal{L}^1(\mathbb{R}^d)} < \infty$ if $|\alpha| = d$. The maximum over all α with $|\alpha| = d$ is called the *Sobolev seminorm* of f and is denoted $\|f\|_{d,1,\infty}$. We denote by A_{r_d} the intersection of $\mathcal{V}(\mathbb{R}^d)$ with the ball $B_{r_d}(\|\cdot\|_{d,1,\infty})$ of radius r_d in the Sobolev seminorm $\|.\|_{d,1,\infty}$. Then

$$A_{r_d} := \mathcal{V}(\mathbb{R}^d) \cap B_{r_d}(\|\cdot\|_{d,1,\infty}) = r_d A_1.$$

The estimates in rows 1 and 2 of Table 3 imply that approximation of functions from balls of radii r_d in variation with respect to half-spaces is tractable in the space $\mathcal{M}(\mathbb{R}^d)$ of bounded measurable functions on $\mathbb{R}^d$ with respect to supremum norm, when the radius r_d grows polynomially. In $(\mathcal{L}^2(\Omega_d), \|.\|_{\mathcal{L}^2(\Omega_d)})$, this approximation is tractable when r_d times $\lambda(\Omega_d)$ grows polynomially with d. If for all $d \in \mathbb{N}'$, Ω_d is the unit ball in $\mathbb{R}^d$, then this approximation is hyper-tractable unless r_d is exponentially growing.

In row 5, we denote by $G_d^{\gamma,1} := \{\tau_y(\gamma_d) \mid y \in \mathbb{R}^d\}$ the set of d-variable Gaussians with widths equal to 1 and varying centers. Using a result of Kainen, Kůrková, and Vogt [35], we have $\xi(d) = (2\pi d^{3/4}) \lambda(\Omega_d)^{1/2}$. This implies that approximation of d-variable Gaussians on a domain Ω_d by perceptron

Table 3 Factorized approximation rates for perceptron networks. In rows 1 and 2, $B_{r_d}(\|.\|_{H_d,\mathcal{M}(\mathbb{R}^d)})$ and $B_{r_d}(\|.\|_{H_d|_{\Omega_d},\mathcal{L}^2(\Omega_d)})$ denote the balls of radius r_d in H_d-variations with respect to the ambient $\|.\|_{\mathcal{M}(\mathbb{R}^d)}$- and $\|.\|_{\mathcal{L}^2(\Omega_d)}$-norms, respectively. In rows 3 and 4, $k_d = 2^{1-d}\pi^{1-d/2}d^{d/2}/\Gamma(d/2) \sim (\pi d)^{1/2}(e/2\pi)^{d/2}$. We assume that $\lambda(\Omega_d) < \infty$ and $\Omega_d \neq \emptyset$, where λ denotes the Lebesgue measure.

ambient space	dictionary G_d	target set $\mathcal{F}$ to be approx.	$\xi(d)$	$\kappa(n)$
$(\mathcal{M}(\mathbb{R}^d),\|.\|_{\mathcal{M}(\mathbb{R}^d)})$	$H_d(\mathbb{R}^d)$	$B_{r_d}(\|.\|_{H_d(\mathbb{R}^d),\mathcal{M}(\mathbb{R}^d)})$	$6\sqrt{3}\,r_d\,d^{1/2}$	$(\log n)^{1/2}n^{-1/2}$
$(\mathcal{L}^2(\Omega_d),\|.\|_{\mathcal{L}^2(\Omega_d)})$	$H_d|_{\Omega_d}$	$B_{r_d}(\|.\|_{H_d|_{\Omega_d},\mathcal{L}^2(\Omega_d)})$	$\lambda(\Omega_d)\,r_d$	$n^{-1/2}$
$(\mathcal{L}^2(\Omega_d),\|.\|_{\mathcal{L}^2(\Omega_d)})$ $\Omega_d \subset \mathbb{R}^d$, d odd	$H_d|_{\Omega_d}$	A_{r_d}	$r_d\,k_d\lambda(\Omega_d)^{1/2}$	$n^{-1/2}$
$(\mathcal{L}^2(\Omega_d),\|.\|_{\mathcal{L}^2(\Omega_d)})$ $\Omega_d \subset \mathbb{R}^d$, d odd	$P_d^\sigma(\Omega_d)$ σ continuous nondecr. sigmoid	A_{r_d}	$r_d\,k_d\lambda(\Omega_d)^{1/2}$	$n^{-1/2}$
$(\mathcal{L}^2(\Omega_d),\|.\|_{\mathcal{L}^2(\Omega_d)})$ $\Omega_d \subset \mathbb{R}^d$, d odd	$H_d|_{\Omega_d}$	$G_d^{\gamma,1}|_{\Omega_d}$	$(2\pi d)^{3/4}\lambda(\Omega_d)^{1/2}$	$n^{-1/2}$

networks is tractable when the Lebesgue measure $\lambda(\Omega_d)$ grows polynomially with d, while if the domains Ω_d are unit balls in $\mathbb{R}^d$, then the approximation is hyper-tractable.

9 Discussion

A number of years ago, Lotfi Zadeh remarked to two of the current authors that the developed countries must control their "under-coordinated technology" and this 21st century need is a driving force behind neural nets and other new computational approaches. It is hoped that modern methods will permit greater control and coordination of technology, and the techniques described in this article represent a key part of current understanding.

As we have shown, neural network theory unifies and embodies a substantial portion of the real and functional analysis which has been developed during the 19th and 20th centuries. This analysis is based on deep and hard-won knowledge and so presents the possibility of intellectual leverage in tackling the problems which arise in the large-scale practical application of computation. Thus, the abundance of powerful mathematical tools which are utilized gives modern approaches a possibility of overcoming previous obstacles.

Another good reason to consider the methodologies discussed in this article is their promise to enable much larger dimensionality to be considered in applications through a more sophisticated model capable of being put into special-purpose hardware. One of the obstructions to rapid and accurate neural computations may be that most work has dealt with static, rather than dynamic, problem instances. It is clear that human pattern recognition is strongly facilitated by dynamics. Through functional analysis, which has begun to be more strongly utilized in neural network theory, one can properly analyze high-dimensional phenomena in time as well as space. There will always be limitations due to hardware itself, but neural network approaches via integral formulae of the sort encountered in mathematical physics hold out the possibility of direct instantiation of neural networks by analog computations, optical or implemented in silicon.

Finally, the notion of hyper-tractable computations appears to show that for some problems, increase of dimension can improve computational performance. As lower bounds to neural network accuracy are not currently known, it may be that the heuristic successes discovered in a few well-chosen examples can be extended to a much broader domain of problem types.

10   Summary of Main Notations

$\mathbb{R}$	Set of real numbers.
d	Input dimension.
$\mathcal{S}^{d-1}$	Sphere of radius 1 in $\mathbb{R}^d$.
$\mathbb{N}_+$	Set of positive integers.
$(\mathcal{X}, \|\cdot\|_{\mathcal{X}})$	Normed linear space.
$B_r(\|\cdot\|_{\mathcal{X}})$	Ball of radius r in the norm $\|\cdot\|_{\mathcal{X}}$.
G	Subset of $(\mathcal{X}, \|\cdot\|_{\mathcal{X}})$, representing a generic dictionary.
$G(c)$	$\{wg \mid g \in G,\ w \in \mathbb{R},\ |w| \le c\}$.
$\mathrm{card}(G)$	Cardinality of the set G.
$e_n(G)$	n-th entropy number of the set G.
$\mathrm{diam}(G)$	Diameter of the set G.
$\mathrm{span}\,G$	Linear span of G.

$\text{span}_n G$	Set of all linear combinations of at most n elements of G.
$\text{conv } G$	Convex hull of G.
$\text{conv}_n G$	Set of all convex combinations of at most n elements of G.
$\text{cl } G$	Closure of G in the norm $\|\cdot\|_{\mathcal{X}}$.
s_G	$\sup_{g \in G} \|g\|_{\mathcal{X}}$.
G_ϕ	Dictionary of functions computable by a unit of type ϕ.
$\psi(v \cdot x + b)$	Perceptron with an activation ψ, bias b, and weight vector v.
σ	Sigmoidal function.
ϑ	Heaviside function.
H_d	Set of characteristic functions of closed half-spaces of $\mathbb{R}^d$ (Heaviside perceptrons).
$P_d^\psi(X)$	Dictionary of functions on $X \subseteq \mathbb{R}^d$ computable by perceptron networks with activation ψ.
$\psi(b\|x - v\|)$	Radial-basis function with activation ψ, width b, and center v.
$F_d^\psi(X)$	Dictionary of functions on $X \subseteq \mathbb{R}^d$ computable by RBF networks with activation ψ.
$\gamma_{d,b}$	d-dimensional Gaussian of width b and center 0.
$G_d^\gamma(b)$	Set of d-variable Gaussian RBFs with width b and all possible centers.
G_d^γ	Set of Gaussians with varying widths.
$\|f - A\|_{\mathcal{X}}$	Distance from f to the set A in the norm $\|\cdot\|_{\mathcal{X}}$.
Π_f	Peak functional for f.
$\delta(M, A)$	Deviation of M from A in the norm $\|\cdot\|_{\mathcal{X}}$.
$\mathcal{N}(G, \varepsilon)$	ε-covering number of G in the norm $\|\cdot\|_{\mathcal{X}}$.

$\alpha_A(r)$	Decay function of the set A.
$\mathrm{Pr}_M(f)$	Projection of f to the set M.
$f\|_\Omega$	Restriction of f to the set Ω.
$\hat{f}$	Fourier transform of f.
$\beta_{d,m}$	Bessel potential of order m on $\mathbb{R}^d$.
$L^{q,m}(\mathbb{R}^d)$	Bessel potential space of order m in $\mathcal{L}^q(\mathbb{R}^d)$.
$\|\cdot\|_{1,A}$	ℓ_1-norm with respect to A.
$\|\cdot\|_G$	G-variation with respect to $\|\cdot\|_\mathcal{X}$.
$\|\cdot\|_{H_d}$	Variation with respect to half-spaces (Heaviside perceptrons).
$(\mathcal{C}(X), \|\cdot\|_{\mathrm{sup}})$	Space of all continuous functions on a subset $X \subseteq \mathbb{R}^d$, with the supremum norm.
λ	Lebesgue measure.
$(\mathcal{L}^p(X), \|\cdot\|_p)$, $p \in [1,\infty]$	Space of all Lebesgue-measurable and p-integrable functions f on X, with the $\mathcal{L}^p(X)$-norm.
$\mathcal{V}(\mathbb{R}^d)$	Set of functions of weakly controlled decay on $\mathbb{R}^d$.
$\|\cdot\|_{d,1,\infty}$	Sobolev seminorm.

Acknowledgement. V. Kůrková was partially supported by GA ČR grant P202/11/1368 and institutional support 67985807. M. Sanguineti was partially supported by a PRIN grant from the Italian Ministry for University and Research, project "Adaptive State Estimation and Optimal Control". Collaboration of V. Kůrková and P. C. Kainen was partially supported by MŠMT project KONTAKT ALNN ME10023 and collaboration of V. Kůrková and M. Sanguineti was partially supported by CNR - AVČR 2010-2012 project "Complexity of Neural-Network and Kernel Computational Models".

References

1. Barron, A.R.: Neural net approximation. In: Narendra, K. (ed.) Proc. 7th Yale Workshop on Adaptive and Learning Systems, pp. 69–72. Yale University Press (1992)

2. Barron, A.R.: Universal approximation bounds for superpositions of a sigmoidal function. IEEE Trans. on Information Theory 39, 930–945 (1993)
3. Bellman, R.: Dynamic Programming. Princeton University Press, Princeton (1957)
4. Bochner, S., Chandrasekharan, K.: Fourier Transform. Princeton University Press, Princeton (1949)
5. Braess, D.: Nonlinear Approximation Theory. Springer (1986)
6. Breiman, L.: Hinging hyperplanes for regression, classification and function approximation. IEEE Trans. Inform. Theory 39(3), 999–1013 (1993)
7. Carrol, S.M., Dickinson, B.W.: Construction of neural nets using the Radon transform. In: Proc. the Int. Joint Conf. on Neural Networks, vol. 1, pp. 607–611 (1989)
8. Cheang, G.H.L., Barron, A.R.: A better approximation for balls. J. of Approximation Theory 104, 183–203 (2000)
9. Chui, C.K., Li, X., Mhaskar, H.N.: Neural networks for localized approximation. Math. of Computation 63, 607–623 (1994)
10. Courant, R., Hilbert, D.: Methods of Mathematical Physics, vol. II. Interscience, New York (1962)
11. Cybenko, G.: Approximation by superpositions of a sigmoidal function. Mathematics of Control, Signals, and Systems 2, 303–314 (1989)
12. Darken, C., Donahue, M., Gurvits, L., Sontag, E.: Rate of approximation results motivated by robust neural network learning. In: Proc. Sixth Annual ACM Conf. on Computational Learning Theory, pp. 303–309. ACM, New York (1993)
13. DeVore, R.A., Howard, R., Micchelli, C.: Optimal nonlinear approximation. Manuscripta Mathematica 63, 469–478 (1989)
14. Donahue, M., Gurvits, L., Darken, C., Sontag, E.: Rates of convex approximation in non-Hilbert spaces. Constructive Approximation 13, 187–220 (1997)
15. Girosi, F., Anzellotti, G.: Rates of convergence for Radial Basis Functions and neural networks. In: Mammone, R.J. (ed.) Artificial Neural Networks for Speech and Vision, pp. 97–113. Chapman & Hall (1993)
16. Girosi, F., Poggio, T.: Networks and the best approximation property. Biological Cybernetics 63, 169–176 (1990)
17. Giulini, S., Sanguineti, M.: Approximation schemes for functional optimization problems. J. of Optimization Theory and Applications 140, 33–54 (2009)
18. Gnecco, G., Sanguineti, M.: Estimates of variation with respect to a set and applications to optimization problems. J. of Optimization Theory and Applications 145, 53–75 (2010)
19. Gnecco, G., Sanguineti, M.: On a variational norm tailored to variable-basis approximation schemes. IEEE Trans. on Information Theory 57, 549–558 (2011)
20. Gribonval, R., Vandergheynst, P.: On the exponential convergence of matching pursuits in quasi-incoherent dictionaries. IEEE Trans. on Information Theory 52, 255–261 (2006)
21. Gurvits, L., Koiran, P.: Approximation and learning of convex superpositions. J. of Computer and System Sciences 55, 161–170 (1997)
22. Hartman, E.J., Keeler, J.D., Kowalski, J.M.: Layered neural networks with Gaussian hidden units as universal approximations. Neural Computation 2, 210–215 (1990)
23. Hewit, E., Stromberg, K.: Abstract Analysis. Springer, Berlin (1965)
24. Hornik, K., Stinchcombe, M., White, H.: Universal approximation of an unknown mapping and its derivatives using multilayer feedforward networks. Neural Networks 3, 551–560 (1990)

25. Ito, Y.: Representation of functions by superpositions of a step or sigmoidal function and their applications to neural network theory. Neural Networks 4, 385–394 (1991)
26. Ito, Y.: Finite mapping by neural networks and truth functions. Mathematical Scientist 17, 69–77 (1992)
27. Jones, L.K.: A simple lemma on greedy approximation in Hilbert space and convergence rates for projection pursuit regression and neural network training. Annals of Statistics 20, 608–613 (1992)
28. Juditsky, A., Hjalmarsson, H., Benveniste, A., Delyon, B., Ljung, L., Sjöberg, J., Zhang, Q.: Nonlinear black-box models in system identification: Mathematical foundations. Automatica 31, 1725–1750 (1995)
29. Kainen, P.C., Kůrková, V.: Quasiorthogonal dimension of Euclidean spaces. Applied Mathematics Letters 6, 7–10 (1993)
30. Kainen, P.C., Kůrková, V., Sanguineti, M.: Complexity of Gaussian radial basis networks approximating smooth functions. J. of Complexity 25, 63–74 (2009)
31. Kainen, P.C., Kurkova, V., Sanguineti, M.: Dependence of Computational Models on Input Dimension: Tractability of Approximation and Optimization Tasks. IEEE Transactions on Information Theory 58, 1203–1214 (2012)
32. Kainen, P.C., Kůrková, V., Vogt, A.: Geometry and topology of continuous best and near best approximations. J. of Approximation Theory 105, 252–262 (2000)
33. Kainen, P.C., Kůrková, V., Vogt, A.: Continuity of approximation by neural networks in L_p-spaces. Annals of Operational Research 101, 143–147 (2001)
34. Kainen, P.C., Kůrková, V., Vogt, A.: Best approximation by linear combinations of characteristic functions of half-spaces. J. of Approximation Theory 122, 151–159 (2003)
35. Kainen, P.C., Kůrková, V., Vogt, A.: A Sobolev-type upper bound for rates of approximation by linear combinations of Heaviside plane waves. J. of Approximation Theory 147, 1–10 (2007)
36. Knuth, D.E.: Big omicron and big omega and big theta. SIGACT News 8, 18–24 (1976)
37. Kůrková, V.: Kolmogorov's theorem and multilayer neural networks. Neural Networks 5, 501–506 (1992)
38. Kůrková, V.: Approximation of functions by perceptron networks with bounded number of hidden units. Neural Networks 8, 745–750 (1995)
39. Kůrková, V.: Dimension-independent rates of approximation by neural networks. In: Warwick, K., Kárný, M. (eds.) Computer-Intensive Methods in Control and Signal Processing. The Curse of Dimensionality, Birkhäuser, Boston, MA, pp. 261–270 (1997)
40. Kůrková, V.: Incremental approximation by neural networks. In: Warwick, K., Kárný, M., Kůrková, V. (eds.) Complexity: Neural Network Approach, pp. 177–188. Springer, London (1998)
41. Kůrková, V.: High-dimensional approximation and optimization by neural networks. In: Suykens, J., et al. (eds.) Advances in Learning Theory: Methods, Models and Applications, ch. 4, pp. 69–88. IOS Press, Amsterdam (2003)
42. Kůrková, V.: Minimization of error functionals over perceptron networks. Neural Computation 20, 252–270 (2008)
43. Kůrková, V., Kainen, P.C., Kreinovich, V.: Estimates of the number of hidden units and variation with respect to half-spaces. Neural Networks 10, 1061–1068 (1997)

44. Kůrková, V., Neruda, R.: Uniqueness of functional representations by Gaussian basis function networks. In: Proceedings of ICANN 1994, pp. 471–474. Springer, London (1994)
45. Kůrková, V., Sanguineti, M.: Bounds on rates of variable-basis and neural-network approximation. IEEE Trans. on Information Theory 47, 2659–2665 (2001)
46. Kůrková, V., Sanguineti, M.: Comparison of worst case errors in linear and neural network approximation. IEEE Trans. on Information Theory 48, 264–275 (2002)
47. Kůrková, V., Sanguineti, M.: Error estimates for approximate optimization by the extended Ritz method. SIAM J. on Optimization 15, 261–287 (2005)
48. Kůrková, V., Sanguineti, M.: Learning with generalization capability by kernel methods of bounded complexity. J. of Complexity 21, 350–367 (2005)
49. Kůrková, V., Sanguineti, M.: Estimates of covering numbers of convex sets with slowly decaying orthogonal subsets. Discrete Applied Mathematics 155, 1930–1942 (2007)
50. Kůrková, V., Sanguineti, M.: Approximate minimization of the regularized expected error over kernel models. Mathematics of Operations Research 33, 747–756 (2008)
51. Kůrková, V., Sanguineti, M.: Geometric upper bounds on rates of variable-basis approximation. IEEE Trans. on Information Theory 54, 5681–5688 (2008)
52. Kůrková, V., Savický, P., Hlaváčková, K.: Representations and rates of approximation of real–valued Boolean functions by neural networks. Neural Networks 11, 651–659 (1998)
53. Lavretsky, E.: On the geometric convergence of neural approximations. IEEE Trans. on Neural Networks 13, 274–282 (2002)
54. Leshno, M., Lin, V.Y., Pinkus, A., Schocken, S.: Multilayer feedforward networks with a nonpolynomial activation function can approximate any function. Neural Networks 6, 861–867 (1993)
55. Makovoz, Y.: Random approximants and neural networks. J. of Approximation Theory 85, 98–109 (1996)
56. Mhaskar, H.N.: Versatile Gaussian networks. In: Proc. of IEEE Workshop of Nonlinear Image Processing, pp. 70–73 (1995)
57. Mhaskar, H.N.: On the tractability of multivariate integration and approximation by neural networks. J. of Complexity 20, 561–590 (2004)
58. Micchelli, C.A.: Interpolation of scattered data: Distance matrices and conditionally positive definite functions. Constructive Approximation 2, 11–22 (1986)
59. Narendra, K.S., Mukhopadhyay, S.: Adaptive control using neural networks and approximate models. IEEE Trans. on Neural Networks 8, 475–485 (1997)
60. Park, J., Sandberg, I.W.: Universal approximation using radial–basis–function networks. Neural Computation 3, 246–257 (1991)
61. Park, J., Sandberg, I.W.: Approximation and radial-basis-function networks. Neural Computation 5, 305–316 (1993)
62. Pinkus, A.: Approximation theory of the MLP model in neural networks. Acta Numerica 8, 143–195 (1999)
63. Pisier, G.: Remarques sur un résultat non publié de B. Maurey. In: Séminaire d'Analyse Fonctionnelle 1980-1981, Palaiseau, France. École Polytechnique, Centre de Mathématiques, vol. I(12) (1981)
64. Rosenblatt, F.: The perceptron: A probabilistic model for information storage and organization of the brain. Psychological Review 65, 386–408 (1958)

65. Rudin, W.: Principles of Mathematical Analysis. McGraw-Hill (1964)
66. Sejnowski, T.J., Rosenberg, C.: Parallel networks that learn to pronounce English text. Complex Systems 1, 145–168 (1987)
67. Singer, I.: Best Approximation in Normed Linear Spaces by Elements of Linear Subspaces. Springer, Heidelberg (1970)
68. Smith, K.A.: Neural networks for combinatorial optimization: A review of more than a decade of research. INFORMS J. on Computing 11, 15–34 (1999)
69. Stinchcombe, M., White, H.: Approximation and learning unknown mappings using multilayer feedforward networks with bounded weights. In: Proc. Int. Joint Conf. on Neural Networks IJCNN 1990, pp. III7–III16 (1990)
70. Traub, J.F., Werschulz, A.G.: Complexity and Information. Cambridge University Press (1999)
71. Wasilkowski, G.W., Woźniakowski, H.: Complexity of weighted approximation over $\mathbb{R}^d$. J. of Complexity 17, 722–740 (2001)
72. Woźniakowski, H.: Tractability and strong tractability of linear multivariate problems. J. of Complexity 10, 96–128 (1994)
73. Zemanian, A.H.: Distribution Theory and Transform Analysis. Dover, New York (1987)
74. Zoppoli, R., Parisini, T., Sanguineti, M., Baglietto, M.: Neural Approximations for Optimal Control and Decision. Springer, London (in preparation)
75. Zoppoli, R., Sanguineti, M., Parisini, T.: Approximating networks and extended Ritz method for the solution of functional optimization problems. J. of Optimization Theory and Applications 112, 403–439 (2002)

Chapter 6
Bochner Integrals and Neural Networks

Paul C. Kainen and Andrew Vogt

Abstract. A Bochner integral formula $f = \mathcal{B} - \int_Y w(y)\Phi(y)\,d\mu(y)$ is derived that presents a function f in terms of weights w and a parametrized family of functions $\Phi(y)$, y in Y. Comparison is made to pointwise formulations, norm inequalities relating pointwise and Bochner integrals are established, G-variation and tensor products are studied, and examples are presented.

Keywords: Variational norm, essentially bounded, strongly measurable, Bochner integration, tensor product, L^p spaces, integral formula.

1 Introduction

A neural network utilizes data to find a function consistent with the data and with further "conceptual" data such as desired smoothness, boundedness, or integrability. The weights for a neural net and the functions embodied in the hidden units can be thought of as determining a finite sum that approximates some function. This finite sum is a kind of quadrature for an integral formula that would represent the function exactly.

This chapter uses abstract analysis to investigate neural networks. Our approach is one of *enrichment*: not only is summation replaced by integration, but also numbers are replaced by real-valued functions on an input set Ω, the functions lying in a function space $\mathcal{X}$. The functions, in turn, are replaced by $\mathcal{X}$-valued measurable functions Φ on a measure space Y of parameters. The goal is to understand approximation of functions by neural networks so that one can make effective choices of the parameters to produce a good approximation.

Paul C. Kainen · Andrew Vogt
Department of Mathematics and Statistics, Georgetown University
Washington, D.C. 20057-1233, USA
e-mail: {kainen,vogta}@georgetown.edu

M. Bianchini et al. (Eds.): *Handbook on Neural Information Processing*, ISRL 49, pp. 183–214.
DOI: 10.1007/978-3-642-36657-4_6   © Springer-Verlag Berlin Heidelberg 2013

To achieve this, we utilize Bochner integration. The idea of applying this tool to neural nets is in Girosi and Anzellotti [14] and we developed it further in Kainen and Kůrková [23]. Bochner integrals are now being used in the theory of support vector machines and reproducing kernel Hilbert spaces; see the recent book by Steinwart and Christmann [42], which has an appendix of more than 80 pages of material on operator theory and Banach-space-valued integrals. Bochner integrals are also widely used in probability theory in connection with stochastic processes of martingale-type; see, e.g., [8, 39]. The corresponding functional analytic theory may help to bridge the gap between probabilistic questions and deterministic ones, and may be well-suited for issues that arise in approximation via neural nets.

Training to replicate given numerical data does not give a useful neural network for the same reason that parrots make poor conversationalists. The phenomenon of overfitting shows that achieving fidelity to data at all costs is not desirable; see, e.g., the discussion on interpolation in our other chapter in this book (Kainen, Kůrková, and Sanguineti). In approximation, we try to find a function close to the data that achieves desired criteria such as sufficient smoothness, decay at infinity, etc. Thus, a method of integration which produces functions *in toto* rather than numbers could be quite useful.

Enrichment has lately been utilized by applied mathematicians to perform image analysis and even to deduce global properties of sensor networks from local information. For instance, the Euler characteristic, ordinarily thought of as a discrete invariant, can be made into a variable of integration [7]. In the case of sensor networks, such an analysis can lead to effective computations in which theory determines a minimal set of sensors [40].

By modifying the traditional neural net focus on training sets of data so that we get to families of functions in a natural way, we aim to achieve methodological insight. Such a framework may lead to artificial neural networks capable of performing more sophisticated tasks.

The main result of this chapter is Theorem 5 which characterizes functions to be approximated in terms of pointwise integrals and Bochner integrals, and provides inequalities that relate corresponding norms. The relationship between integral formulas and neural networks has long been noted; e.g., [20, 6, 37, 13, 34, 29] We examine integral formulas in depth and extend their significance to a broader context.

An earlier version of the Main Theorem, including the bounds on variational norm by the L^1-norm of the weight function in a corresponding integral formula, was given in [23] and it also utilized *functional* (i.e., Bochner) integration. However, the version here is more general and further shows that if ϕ is a real-valued function on $\Omega \times Y$ (the cartesian product of input and parameter spaces), then the associated map Φ which maps the measure space to the Banach space defined by $\Phi(y)(x) = \phi(x, y)$ is measurable; cf. [42, Lemma 4.25, p. 125] where Φ is the "feature map."

Other proof techniques are available for parts of the Main Theorem. In particular, Kůrková [28] gave a different argument for part (iv) of the theorem, using a characterization of variation via peak functionals [31] as well as the theorem of Mazur (Theorem 8) used in the proof of Lemma 1. But the Bochner integral approach reveals some unexpected aspects of functional approximation which may be relevant for neural network applications.

Furthermore, the treatment of analysis and topology utilizes a number of basic theorems from the literature and provides an introduction to functional analysis motivated by its applicability. This is a case where neural nets provide a fresh perspective on classical mathematics. Indeed, theoretical results proved here were obtained in an attempt to better understand neural networks.

An outline of the paper is as follows: In section 2 we discuss variational norms; sections 3 and 4 present needed material on Bochner integrals. The Main Theorem (Theorem 5) on integral formulas is given in Section 5. In section 6 we show how to apply the Main Theorem to an integral formula for the Bessel potential function in terms of Gaussians. In section 7 we show how this leads to an inequality involving Gamma functions and provide an alternative proof by classical means. Section 8 interprets and extends the Main Theorem in the language of tensor products. Using tensor products, we replace individual $\mathcal{X}$-valued Φ's by families $\{\Phi_j : j \in J\}$ of such functions. This allows more nuanced representation of the function to be approximated. In section 9 we give a detailed example of concepts related to G-variation, while section 10 considers the relationship between pointwise integrals and evaluation of the corresponding Bochner integrals. Remarks on future directions are in section 11, and the chapter concludes with two appendices and references.

2   Variational Norms and Completeness

We assume that the reader has a reasonable acquaintance with functional analysis but have attempted to keep this chapter self-contained. Notations and basic definitions are given in Appendix I, while Appendix II has the precise statement of several important theorems from the literature which will be needed in our development.

Throughout this chapter, all linear spaces are over the reals $\mathbb{R}$. For A any subset of a linear space X, $b \in X$, and $r \in \mathbb{R}$,

$$b + rA := \{b + ra \mid a \in A\} = \{y \in X : y = b + ra, a \in A\}.$$

Also, we sometimes use the abbreviated notation

$$\| \cdot \|_1 = \| \cdot \|_{L^1(Y,\mu)} \quad \text{and} \quad \| \cdot \|_\infty = \| \cdot \|_{L^\infty(Y,\mu;\mathcal{X})}; \tag{1}$$

the standard notations on the right are explained in sections 12 and 4, resp. The symbol "∋" stands for "such that."

A set G in a normed linear space $\mathcal{X}$ is *fundamental* (with respect to $\mathcal{X}$) if $\mathrm{cl}_{\mathcal{X}}$ (span G) $= \mathcal{X}$, where closure depends only on the topology induced by the norm. We call G *bounded* with respect to $\mathcal{X}$ if $s_{G,\mathcal{X}} := \sup_{g \in G} \|g\|_{\mathcal{X}} < \infty$.

We now review G-*variation* norms. These norms, which arise in connection with approximation of functions, were first considered by Barron [5], [6]. He treated a case where G is a family of characteristic functions of sets satisfying a special condition. The general concept, formulated by Kůrková [24], has been developed in such papers as [30, 26, 27, 15, 16, 17].

Consider the set

$$B_{G,\mathcal{X}} := \mathrm{cl}_{\mathcal{X}} \left(\mathrm{conv}(\pm G) \right), \text{where} \pm G := G \cup -G. \tag{2}$$

This is a symmetric, closed, convex subset of $\mathcal{X}$, with Minkowski functional

$$\|f\|_{G,\mathcal{X}} := \inf\{\lambda > 0 : f/\lambda \in B_{G,\mathcal{X}}\}.$$

The subset $\mathcal{X}_G$ of $\mathcal{X}$ on which this functional is finite is given by

$$\mathcal{X}_G := \{f \in \mathcal{X} : \exists \lambda > 0 \ni f/\lambda \in B_{G,\mathcal{X}}\}.$$

If G is bounded, then $\| \cdot \|_{G,\mathcal{X}}$ is a norm on $\mathcal{X}_G$ and $B_{G,\mathcal{X}}$ is the closed unit ball centered at the origin, in this norm. In general $\mathcal{X}_G$ may be a proper subset of $\mathcal{X}$ even if G is bounded and fundamental w.r.t. $\mathcal{X}$. See the example at the end of this section. The inclusion $\iota : \mathcal{X}_G \subseteq \mathcal{X}$ is linear and for every $f \in \mathcal{X}_G$

$$\|f\|_{\mathcal{X}} \le \|f\|_{G,\mathcal{X}} \, s_{G,\mathcal{X}} \tag{3}$$

Indeed, if $f/\lambda \in B_{G,\mathcal{X}}$, then f/λ is a convex combination of elements of $\mathcal{X}$-norm at most $s_{G,\mathcal{X}}$, so $\|f\|_{\mathcal{X}} \le \lambda s_{G,\mathcal{X}}$ establishing (3) by definition of variational norm. Hence, if G is bounded in $\mathcal{X}$, the operator ι is bounded with operator norm not exceeding $s_{G,\mathcal{X}}$.

Proposition 1. *Let nonempty $G \subseteq \mathcal{X}$ a normed linear space. Then*

(i) $\mathrm{span}\, G \subseteq \mathcal{X}_G \subseteq \mathrm{cl}_{\mathcal{X}} \,\mathrm{span}\, G$;

(ii) G is fundamental if and only if $\mathcal{X}_G$ is dense in $\mathcal{X}$;

(iii) For G bounded and $\mathcal{X}$ complete, $(\mathcal{X}_G, \| \cdot \|_{G,\mathcal{X}})$ is a Banach space.

Proof. (i) Let $f \in \mathrm{span}\, G$, then $f = \sum_{i=1}^{n} a_i g_i$, for real numbers a_i and $g_i \in G$. We assume the a_i are not all zero since 0 is in $\mathcal{X}_G$. Then $f = \lambda \sum_{i=1}^{n} \frac{|a_i|}{\lambda}(\pm g_i)$, where $\lambda = \sum_{i=1}^{n} |a_i|$. Thus, f is in $\lambda \,\mathrm{conv}(\pm G) \subseteq \lambda B_{G,\mathcal{X}}$. So $\|f\|_{G,\mathcal{X}} \le \lambda$ and f is in $\mathcal{X}_G$.

Likewise if f is in $\mathcal{X}_G$, then for some $\lambda > 0$, f/λ is in

$$B_{G,\mathcal{X}} = \mathrm{cl}_{\mathcal{X}}(\mathrm{conv}(\pm G)) \subseteq \mathrm{cl}_X(\mathrm{span}(G)),$$

so f is in $cl_{\mathcal{X}}(\text{span}(G))$.

(ii) Suppose G is fundamental. Then $\mathcal{X} = cl_{\mathcal{X}}(\text{span } G) = cl_{\mathcal{X}}(\mathcal{X}_G)$ by part (i). Conversely, if $\mathcal{X}_G$ is dense in $\mathcal{X}$, then $\mathcal{X} = cl_{\mathcal{X}}(\mathcal{X}_G) \subseteq cl_{\mathcal{X}}(\text{span } G) \subseteq \mathcal{X}$, and G is fundamental.

(iii) Let $\{f_n\}$ be a Cauchy sequence in $\mathcal{X}_G$. By (3) $\{f_n\}$ is a Cauchy sequence in $\mathcal{X}$ and has a limit f in $\mathcal{X}$. The sequence $\|f_n\|_{G,\mathcal{X}}$ is bounded in $\mathcal{X}_G$, that is, there is a positive number M such that for all n $f_n/M \in B_{G,\mathcal{X}}$. Since $B_{G,\mathcal{X}}$ is closed in $\mathcal{X}$, f/M is also in $B_{G,\mathcal{X}}$. Hence $\|f\|_{G,\mathcal{X}} \leq M$ and f is in $\mathcal{X}_G$. Now given $\epsilon > 0$ choose a positive integer N such that $\|f_n - f_k\|_{G,\mathcal{X}} < \epsilon$ for $n, k \geq N$. In particular fix $n \geq N$, and consider a variable integer $k \geq N$. Then $\|f_k - f_n\|_{G,\mathcal{X}} < \epsilon$. So $(f_k - f_n)/\epsilon \in B_{G,\mathcal{X}}$, and $f_k \in f_n + \epsilon B_{G,\mathcal{X}}$ for all $k \geq N$. But $f_n + \epsilon B_{G,\mathcal{X}}$ is closed in $\mathcal{X}$. Hence $f \in f_n + \epsilon B_{G,\mathcal{X}}$, and $\|f - f_n\|_{G,\mathcal{X}} \leq \epsilon$. So the sequence converges to f in $\mathcal{X}_G$. □

The following example illustrates several of the above concepts. Take $\mathcal{X}$ to be a real separable Hilbert space with orthonormal basis $\{e_n : n = 0, 1, ...\}$. Let $G = \{e_n : n = 0, 1, ...\}$. Then

$$B_{G,\mathcal{X}} = \left\{ \sum_{n \geq 1} c_n e_n - \sum_{n \geq 1} d_n e_n : \forall n, \ c_n \geq 0, d_n \geq 0, \sum_{n \geq 1} (c_n + d_n) = 1 \right\}.$$

Now $f \in \mathcal{X}$ is of the form $\sum_{n \geq 1} a_n e_n$ where $\|f\|_{\mathcal{X}} = \sqrt{\sum_{n \geq 1} a_n^2}$, and if $f \in \mathcal{X}_G$, then $a_n = \lambda(c_n - d_n)$ for all n and suitable c_n, d_n. The minimal λ can be obtained by taking $a_n = \lambda c_n$ when $a_n \geq 0$, and $a_n = -\lambda d_n$ when $a_n < 0$. It then follows that $\|f\|_{G,\mathcal{X}} = \sum_{n \geq 1} |a_n|$. Hence when $\mathcal{X}$ is isomorphic to ℓ_2, $\mathcal{X}_G$ is isomorphic to ℓ_1. As G is fundamental, by part(ii) above, the closure of ℓ_1 in ℓ_2 is ℓ_2. This provides an example where $\mathcal{X}_G$ is not a closed subspace of $\mathcal{X}$ and so, while it is a Banach space w.r.t. the variational norm, it is not complete in the ambient-space norm.

3   Bochner Integrals

The Bochner integral replaces numbers with functions and represents a broad-ranging extension, generalizing the Lebesgue integral from real-valued functions to functions with values in an arbitrary Banach space. Key definitions and theorems are summarized here for convenience, following the treatment in [44] (cf. [33]). Bochner integrals are used here (as in [23]) in order to prove a bound on variational norm.

Let (Y, μ) be a measure space. Let $\mathcal{X}$ be a Banach space with norm $\|\cdot\|_{\mathcal{X}}$. A function $s : Y \to \mathcal{X}$ is *simple* if it has a finite set of nonzero values $f_j \in \mathcal{X}$, each on a measurable subset P_j of Y with $\mu(P_j) < \infty$, $1 \leq j \leq m$, and the P_j are pairwise-disjoint. Equivalently, a function s is simple if it can be written in the following form:

$$s = \sum_{j=1}^{m} \kappa(f_j)\chi_{P_j}, \qquad (4)$$

where $\kappa(f_j) : Y \to \mathcal{X}$ denotes the constant function with value f_j and χ_P denotes the characteristic function of a subset P of Y. This decomposition is nonunique and we identify two functions if they agree μ-almost everywhere - i.e., the subset of Y on which they disagree has μ-measure zero.

Define an $\mathcal{X}$-valued function I on the simple functions by setting for s of form (4)

$$I(s,\mu) := \sum_{j=1}^{m} \mu(P_j)f_j \in \mathcal{X}.$$

This is independent of the decomposition of s [44, pp.130–132]. A function $h : Y \to \mathcal{X}$ is *strongly measurable* (w.r.t. μ) if there exists a sequence $\{s_k\}$ of simple functions such that for μ-a.e. $y \in Y$

$$\lim_{k \to \infty} \|s_k(y) - h(y)\|_{\mathcal{X}} = 0.$$

A function $h : Y \to \mathcal{X}$ is *Bochner integrable* (with respect to μ) if it is strongly measurable and there exists a sequence $\{s_k\}$ of simple functions $s_k : Y \to \mathcal{X}$ such that

$$\lim_{k \to \infty} \int_Y \|s_k(y) - h(y)\|_{\mathcal{X}} d\mu(y) = 0. \qquad (5)$$

If h is strongly measurable and (5) holds, then the sequence $\{I(s_k,\mu)\}$ is Cauchy and by completeness converges to an element in $\mathcal{X}$. This element, which is independent of the sequence of simple functions satisfying (5), is called the *Bochner integral of h* (w.r.t. μ) and denoted

$$I(h,\mu) \quad \text{or} \quad \mathcal{B} - \int_Y h(y)d\mu(y).$$

The Bochner integral coincides with the Lebesgue integral when $\mathcal{X} = \mathbb{R}$.

Let $\mathcal{L}^1(Y,\mu;\mathcal{X})$ denote the linear space of all strongly measurable functions from Y to $\mathcal{X}$ which are Bochner integrable w.r.t. μ; let $L^1(Y,\mu;\mathcal{X})$ be the corresponding set of equivalence classes (modulo μ-a.e. equality). It is easily shown that equivalent functions have the same Bochner integral. Then the following elegant characterization holds.

Theorem 1 (Bochner). *Let $(\mathcal{X}, \|\cdot\|_{\mathcal{X}})$ be a Banach space and (Y,μ) a measure space. Let $h : Y \to \mathcal{X}$ be strongly measurable. Then*

$$h \in \mathcal{L}^1(Y,\mu;\mathcal{X}) \text{ if and only if } \int_Y \|h(y)\|_{\mathcal{X}} d\mu(y) < \infty.$$

A consequence of this theorem is that $I : L^1(Y,\mu;\mathcal{X}) \to \mathcal{X}$ is a continuous linear operator and

$$\|I(h,\mu)\|_{\mathcal{X}} = \left\|\mathcal{B} - \int_Y h(y)\, d\mu(y)\right\|_{\mathcal{X}} \le \|h\|_{L^1(Y,\mu;\mathcal{X})} := \int_Y \|h(y)\|_{\mathcal{X}}\, d\mu(y). \quad (6)$$

In particular, the Bochner norm of s, $\|s\|_{L^1(Y,\mu;\mathcal{X})}$, is $\sum_i \mu(P_i)\|f_i\|_{\mathcal{X}}$, where s is a simple function satisfying (4) and $L^1(Y,\mu;\mathcal{X})$ is a Banach space under the norm.

For Y a measure space and $\mathcal{X}$ a Banach space, $h : Y \to \mathcal{X}$ is *measurable* if the inverse image of each Borel set in $\mathcal{X}$ is measurable in Y; h is *weakly measurable* if for every continuous linear functional F on X the composite real-valued function $F \circ h$ is measurable [43, pp. 130–134]. Strong measurability implies measurability provided that the measure μ on Y is complete [33, p. 114]. If h is measurable, then it is weakly measurable since measurable followed by continuous is measurable: for U open in $\mathbb{R}$, $(F \circ h)^{-1}(U) = h^{-1}(F^{-1}(U))$.

Recall that a topological space is *separable* if it has a countable dense subset. Let λ denote Lebesgue measure on $\mathbb{R}^d$ and let $\Omega \subseteq \mathbb{R}^d$ be λ-measurable, $d \ge 1$. Then $L^q(\Omega, \lambda)$ is separable when $1 \le q < \infty$; e.g., [36, pp. 208]. A function $h : Y \to \mathcal{X}$ is *μ-almost separably valued* (μ-a.s.v.) if there exists a μ-measurable subset $Y_0 \subset Y$ with $\mu(Y_0) = 0$ and $h(Y \setminus Y_0)$ is a separable subset of $\mathcal{X}$. The function h is *strongly measurable* if it is measurable and is μ-a.s.v. [33, p. 114]. For the following theorem, see [19, p. 72].

Theorem 2 (Pettis). *Let $(\mathcal{X}, \|\cdot\|_{\mathcal{X}})$ be a Banach space and (Y, μ) a measure space. Suppose $h : Y \to \mathcal{X}$. Then h is strongly measurable if and only if h is weakly measurable and μ-a.s.v.*

The following basic result (see, e.g., [9]) says that Bochner integration commutes with continuous linear operators. It was later extended by Hille to all closed operators.

Theorem 3. *Let (Y, ν) be a measure space, let $\mathcal{X}$, $\mathcal{X}'$ be Banach spaces, and let $h \in \mathcal{L}^1(Y, \nu; \mathcal{X})$. If $T : \mathcal{X} \to \mathcal{X}'$ is a bounded linear operator, then $T \circ h \in \mathcal{L}^1(Y, \nu; \mathcal{X}')$ and*

$$T\left(\mathcal{B} - \int_Y h(y)\, d\nu(y)\right) = \mathcal{B} - \int_Y (T \circ h)(y)\, d\nu(y).$$

We only need the above result for bounded linear functionals, in which case the Bochner integral on the right coincides with ordinary integration.

There is a mean-value theorem for Bochner integrals (Diestel and Uhl [12, Lemma 8, p. 48]). We give their argument with a slightly clarified reference to the Hahn-Banach theorem.

Lemma 1. *Let (Y, ν) be a finite measure space, let $\mathcal{X}$ be a Banach space, and let $h : Y \to \mathcal{X}$ be Bochner integrable w.r.t. ν. Then*

$$\mathcal{B} - \int_Y h(y)\, d\nu(y) \in \nu(Y)\, \mathrm{cl}_{\mathcal{X}}(\mathrm{conv}(\{\pm h(y) : y \in Y\})).$$

Proof. Without loss of generality, $\nu(Y) = 1$. Suppose $f := I(h, \nu) \notin \mathrm{cl}_X(\mathrm{conv}(\{\pm h(y) : y \in Y\}))$. By a consequence of the Hahn-Banach theorem given as Theorem 8 in Appendix II below), there is a continuous linear functional F on X such that $F(f) > \sup_{y \in Y} F(h(y))$. Hence, by Theorem 3,

$$\sup_{y \in Y} F(h(y)) \geq \int_Y F(h(y))d\nu(y) = F(f) > \sup_{y \in Y} F(h(y)).$$

which is absurd.   □

4   Spaces of Bochner Integrable Functions

In this section, we derive a few consequences of the results from the previous section which we shall need below.

A measurable function h from a measure space (Y, ν) to a normed linear space X is called *essentially bounded* (w.r.t. ν) if there exists a ν-null set N for which

$$\sup_{y \in Y \setminus N} \|h(y)\|_X < \infty.$$

Let $\mathcal{L}^\infty(Y, \nu; X)$ denote the linear space of all measurable, essentially bounded functions from (Y, ν) to X. Let $L^\infty(Y, \nu; X)$ be its quotient space mod the relation of equality ν-a.e. This is a Banach space with norm

$$\|h\|_{L^\infty(Y,\nu;X)} := \inf\{B \geq 0 : \exists \nu\text{-null } N \subset Y \ni \|h(y)\|_X \leq B, \ \forall y \in Y \setminus N\}.$$

To simplify notation, we sometimes write $\|h\|_\infty$ for $\|h\|_{L^\infty(Y,\nu;X)}$ Note that if $\|h\|_\infty = c$, then $\|h(y)\|_X \leq c$ for ν-a.e. y. Indeed, for positive integers k, $\|h(y)\|_X \leq c + (1/k)$ for y not in a set of measure zero N_k so $\|h(y\|_X \leq c$ for y not in the union $\bigcup_{k \geq 1} N_k$ also a set of measure zero.

We also have a useful fact whose proof is immediate.

Lemma 2. *For every measure space (Y, μ) and Banach space X, the natural map*

$$\kappa_X : X \to L^\infty(Y, \mu; X)$$

associating to each element $g \in X$ the constant function from Y to X given by $(\kappa_X(g))(y) \equiv g$ for all y in Y is an isometric linear embedding.

Lemma 3. *Let X be a separable Banach space, let (Y, μ) be a measure space, and let $w : Y \to \mathbb{R}$ and $\Psi : Y \to X$ be μ-measurable functions. Then $w\Psi$ is strongly measurable as well as measurable.*

Proof. By definition, $w\Psi$ is the function from Y to X defined by

$$w\Psi : \ y \mapsto w(y)\Psi(y),$$

where the multiplication is that of a Banach space element by a real number. Then $w\Psi$ is measurable because it is obtained from a pair of measurable functions by applying scalar multiplication which is continuous. Hence, by separability, Pettis' Theorem 2, and the fact that measurable implies weakly measurable, or by application of [33, p. 114], we have strong measurability for $w\Psi$.   □

If (Y, ν) is a finite measure space, $\mathcal{X}$ is a Banach space, and $h : Y \to \mathcal{X}$ is strongly measurable and essentially bounded, then h is Bochner integrable by Theorem 1. The following lemma, which follows from Lemma 3, allows us to weaken the hypothesis on the function by further constraining the space $\mathcal{X}$.

Lemma 4. *Let (Y, ν) be a finite measure space, $\mathcal{X}$ a separable Banach space, and $h : Y \to \mathcal{X}$ be ν-measurable and essentially bounded w.r.t. ν. Then $h \in \mathcal{L}^1(Y, \nu; \mathcal{X})$ and*

$$\int_Y \|h(y)\|_{\mathcal{X}} d\nu(y) \le \nu(Y) \|h\|_{L^\infty(Y,\nu;\mathcal{X})}.$$

Let $w \in \mathcal{L}^1(Y, \mu)$, and let μ_w be defined for μ-measurable $S \subseteq Y$ by $\mu_w(S) := \int_S |w(y)| d\mu(y)$. For $t \ne 0$, $\operatorname{sgn}(t) := t/|t|$.

Theorem 4. *Let (Y, μ) be a measure space, $\mathcal{X}$ a separable Banach space; let $w \in \mathcal{L}^1(Y, \mu)$ be nonzero μ-a.e., let μ_w be the measure defined above, and let $\Phi : Y \to \mathcal{X}$ be μ-measurable. If one of the Bochner integrals*

$$\mathcal{B} - \int_Y w(y) \Phi(y) d\mu(y), \quad \mathcal{B} - \int_Y \operatorname{sgn}(w(y)) \Phi(y) d\mu_w(y)$$

exists, then both exist and are equal.

Proof. By Lemma 3, both $w\Phi$ and $(\operatorname{sgn} \circ w)\Phi$ are strongly measurable. Hence, by Theorem 1, the respective Bochner integrals exist if and only if the $\mathcal{X}$-norms of the respective integrands have finite ordinary integral. But

$$\int_Y \|[(\operatorname{sgn} \circ w)\Phi](y)\|_{\mathcal{X}} d\mu_w(y) = \int_Y \|w(y)\Phi(y)\|_{\mathcal{X}} d\mu(y), \qquad (7)$$

so the Bochner integral $I((\operatorname{sgn} \circ w)\Phi, \mu_w)$ exists exactly when $I(w\Phi, \mu)$ does. Further, the respective Bochner integrals are equal since for any continuous linear functional F in $\mathcal{X}^*$, by Theorem 3 (used twice)

$$F\left(\mathcal{B} - \int_Y w(y) \Phi(y) d\mu(y)\right) = \int_Y F(w(y)\Phi(y)) d\mu(y)$$

$$= \int_Y w(y) F(\Phi(y)) d\mu(y) = \int_Y \operatorname{sgn}(w(y)) |w(y)| F(\Phi(y)) d\mu(y)$$

$$= \int_Y sgn(w(y))F(\varPhi(y))d\mu_w(y) = \int_Y F(sgn(w(y))\varPhi(y))d\mu_w(y)$$

$$= F\left(\mathcal{B} - \int_Y sgn(w(y))\varPhi(y)d\mu_w(y)\right).$$

$\square$

Corollary 1. *Let (Y,μ) be a σ-finite measure space, $\mathcal{X}$ a separable Banach space, $w : Y \to \mathbb{R}$ be in $\mathcal{L}^1(Y,\mu)$ and $\varPhi : Y \to \mathcal{X}$ be in $\mathcal{L}^\infty(Y,\mu;\mathcal{X})$. Then $w\varPhi$ is Bochner integrable w.r.t. μ.*

Proof. By Lemma 3, $(sgn \circ w)\varPhi$ is strongly measurable, and Lemma 4 then implies that the Bochner integral $I((sgn \circ w)\varPhi, \mu_w)$ exists since $\mu_w(Y) = \|w\|_{L^1(Y,\mu)} < \infty$. So $w\varPhi$ is Bochner integrable by Theorem 4. $\square$

5   Main Theorem

In the next result, we show that certain types of integrands yield integral formulas for functions f in a Banach space of L^p-type both pointwise and at the level of Bochner integrals. Furthermore, the variational norm of f is shown to be bounded by the L^1-norm of the weight function from the integral formula. Equations (9) and (10) and part (iv) of this theorem were derived in a similar fashion by one of us with Kůrková in [23] under more stringent hypotheses; see also [13, eq. (12)].

Theorem 5. *Let $(\varOmega, \rho)$, (Y,μ) be σ-finite measure spaces, let w be in $\mathcal{L}^1(Y,\mu)$, let $\mathcal{X} = L^q(\varOmega, \rho)$, $q \in [1,\infty)$, be separable, let $\phi : \varOmega \times Y \to \mathbb{R}$ be $\rho \times \mu$-measurable, let $\varPhi : Y \to \mathcal{X}$ be defined for each y in Y by $\varPhi(y)(x) := \phi(x,y)$ for ρ-a.e. $x \in \varOmega$ and suppose that for some $M < \infty$, $\|\varPhi(y)\|_{\mathcal{X}} \le M$ for μ-a.e. y. Then the following hold:*

(i) *For ρ-a.e. $x \in \varOmega$, the integral $\int_Y w(y)\phi(x,y)d\mu(y)$ exists and is finite.*

(ii) *The function f defined by*

$$f(x) = \int_Y w(y)\phi(x,y)d\mu(y) \tag{8}$$

is in $\mathcal{L}^q(\varOmega, \rho)$ and its equivalence class, also denoted by f, is in $L^q(\varOmega, \rho) = \mathcal{X}$ and satisfies

$$\|f\|_{\mathcal{X}} \le \|w\|_{L^1(Y,\mu)} M. \tag{9}$$

(iii) *The function $\varPhi$ is measurable and hence in $\mathcal{L}^\infty(Y,\mu;\mathcal{X})$, and f is the Bochner integral of $w\varPhi$ w.r.t. μ, i.e.,*

$$f = \mathcal{B} - \int_Y (w\varPhi)(y)d\mu(y). \tag{10}$$

(iv) For $G = \{\Phi(y) : \|\Phi(y)\|_{\mathcal{X}} \leq \|\Phi\|_{L^\infty(Y,\mu;\mathcal{X})}\}$, f is in $\mathcal{X}_G$, with

$$\|f\|_{G,\mathcal{X}} \leq \|w\|_{L^1(Y,\mu)} = \|w_1\| \tag{11}$$

and as in (1)

$$\|f\|_{\mathcal{X}} \leq \|f\|_{G,\mathcal{X}} s_{G,\mathcal{X}} \leq \|w\|_1 \|\Phi\|_\infty. \tag{12}$$

Proof. (i) Consider the function $(x,y) \longmapsto |w(y)| |\phi(x,y)|^q$. This is a well-defined $\rho \times \mu$-measurable function on $\Omega \times Y$. Furthermore its repeated integral

$$\int_Y \int_\Omega |w(y)| |\phi(x,y)|^q d\rho(x) d\mu(y)$$

exists and is bounded by $\|w\|_1 M^q$ since $\Phi(y) \in L^q(\Omega, \rho)$ and $\|\Phi(y)\|_q^q \leq M^q$ for a. e. y. and $w \in L^1(Y, \mu)$. By Fubini's Theorem 9 the function $y \longmapsto |w(y)| |\phi(x,y)|^q$ is in $L^1(Y, \mu)$ for a.e. x. The inequality

$$|w(y)| |\phi(x,y)| \leq \max\{|w(y)| |\phi(x,y)|^q, |w(y)|\} \leq (|w(y)| |\phi(x,y)|^q + |w(y)|),$$

for all x and y, shows that for each x the function $y \longmapsto |w(y)| |\phi(x,y)|$ is dominated by the sum of two integrable functions. Hence the integrand in the definition of $f(x)$ is integrable for a. e. x, and f is well-defined almost everywhere.

(ii) For each $q \in [1, \infty)$, the function $G(u) = u^q$ is a convex function for $u \geq 0$. Accordingly by Jensen's inequality (Theorem 10 below),

$$G\left(\int_Y |\phi(x,y)| d\sigma(y)\right) \leq \int_Y G(|\phi(x,y)|) d\sigma(y)$$

provided both integrals exist and σ is a probability measure on the measurable space Y. We take σ to be defined by the familiar formula:

$$\sigma(A) = \frac{\int_A |w(y)| d\mu(y)}{\int_Y |w(y)| d\mu(y)}$$

for μ-measurable sets A in Y, so that integration with respect to σ reduces to a scale factor times integration of $|w(y)| d\mu(y)$. Since we have established that both $|w(y)| |\phi(x,y)|$ and $|w(y)| |\phi(x,y)|^q$ are integrable with respect to μ for a.e. x, we obtain:

$$|f(x)|^q \leq \|w\|_1^q G\left(\int_Y |\phi(x,y)| d\sigma(y)\right) \leq \|w\|_1^q \int_Y G(|\phi(x,y)|) d\sigma(y)$$

$$= \|w\|_1^{q-1} \int_Y |w(y)| |\phi(x,y)|^q d\mu(y)$$

for a.e. x. But we can now integrate both side with respect to $d\rho(x)$ over Ω because of the integrability noted above in connection with Fubini's Theorem. Thus $f \in \mathcal{X} = L^q(\Omega, \rho)$ and $\|f\|_{\mathcal{X}}^q \leq \|w\|_1^q M^q$, again interchanging order.

(iii) We show that the inverse image under Φ of any open ball in $\mathcal{X}$ is measurable; that is, $\{y : \|\Phi(y) - g\|_{\mathcal{X}} < \varepsilon\}$ is a μ-measurable subset of Y for each g in $\mathcal{X}$ and $\varepsilon > 0$. Note that

$$\|\Phi(y) - g\|_{\mathcal{X}}^q = \int_\Omega |\phi(x, y) - g(x)|^q d\rho(x)$$

for all y in Y where $x \mapsto \phi(x, y)$ and $x \mapsto g(x)$ are ρ-measurable functions representing the elements $\Phi(y)$ and g belonging to $\mathcal{X} = L^q(Y, \mu)$. Since (Y, μ) is σ-finite, we can find a strictly positive function w_0 in $\mathcal{L}^1(Y, \mu)$. (For example, let $w_0 = \sum_{n \geq 1} (1/n^2) \frac{\chi_{Y_n}}{\mu(Y_n) + 1}$, where $\{Y_n : n \geq 1\}$ is a countable disjoint partition of Y into μ-measurable sets of finite measure.) Then $w_0(y)|\phi(x, y) - g(x)|^q$ is a $\rho \times \mu$-measurable function on $\Omega \times Y$, and

$$\int_Y \int_\Omega w_0(y)|\phi(x, y) - g(x)|^q d\rho(x) d\mu(y) \leq \|w_0\|_{L^1(Y, \mu)} \varepsilon^q.$$

By Fubini's Theorem 9, $y \mapsto w_0(y)\|\Phi(y) - g\|_{\mathcal{X}}^q$ is μ-measurable. Since w_0 is μ-measurable and strictly positive, $y \mapsto \|\Phi(y) - g\|_{\mathcal{X}}^q$ is also μ-measurable.

Since $\{y : \|\Phi(y) - g\|_{\mathcal{X}}^q < \varepsilon^q\}$ is measurable and coincides with the original set $(q = 1)$, it follows that $\Phi : Y \to \mathcal{X}$ is measurable.

Thus, Φ is essentially bounded, with essential sup $\|\Phi\|_{L^\infty(Y, \mu; \mathcal{X})} \leq M$. (In (9), M can be replaced by this essential sup.)

By Corollary 1, $w\Phi$ is Bochner integrable. To prove that f is the Bochner integral, using Theorem 4, we show that for each bounded linear functional $F \in \mathcal{X}^*$, $F(I(\text{sgn} \circ w\Phi, \mu_w)) = F(f)$. By the Riesz representation theorem [35, p. 316], for any such F there exists a (unique) $g_F \in \mathcal{L}^p(\Omega, \rho)$, $p = 1/(1 - q^{-1})$, such that for all $g \in \mathcal{L}^q(\Omega, \rho)$,

$$F(g) = \int_\Omega g_F(x) g(x) d\rho(x).$$

By Theorem 3,

$$F(I((\text{sgn} \circ w)\Phi, \mu_w)) = \int_Y F(\text{sgn}(w(y))\Phi(y)) d\mu_w(y).$$

But for $y \in Y$, $F(\text{sgn}(w(y))\Phi(y)) = \text{sgn}(w(y))F(\Phi(y))$, so

$$F(I((\text{sgn} \circ w)\Phi, \mu_w)) = \int_Y \int_\Omega w(y) g_F(x) \phi(x, y) d\rho(x) d\mu(y).$$

Also, using (8), we have:

$$F(f) = \int_\Omega g_F(x)f(x)d\rho(x) = \int_\Omega \int_Y w(y)g_F(x)\phi(x,y)d\mu(y)d\rho(x).$$

The integrand of the iterated integrals is measurable with respect to the product measure $\rho \times \mu$, so by Fubini's Theorem the iterated integrals are equal provided that one of the corresponding absolute integrals is finite. Indeed,

$$\int_Y \int_\Omega |w(y)g_F(x)\phi(x,y)|d\rho(x)d\mu(y) = \int_Y \|g_F\Phi(y)\|_{L^1(\Omega,\rho)}d\mu_w(y). \quad (13)$$

By Hölder's inequality, for every y,

$$\|g_F\Phi(y)\|_{L^1(\Omega,\rho)} \leq \|g_F\|_{L^p(\Omega,\rho)}\|\Phi(y)\|_{L^q(\Omega,\rho)},$$

using the fact that $\mathcal{X} = L^q(\Omega,\rho)$. Therefore, by the essential boundedness of Φ w.r.t. μ, the integrals in (13) are at most

$$\|g_F\|_{L^p(\Omega,\rho)}\|\Phi\|_{\mathcal{L}^\infty(Y,\mu;\mathcal{X})}\|w\|_{L^1(Y,\mu)} < \infty.$$

Hence, f is the Bochner integral of $w\Phi$ w.r.t. μ.

(iv) We use Lemma 1. Let Y_0 be a measurable subset of Y with $\mu(Y_0) = 0$ and for $Y' = Y \setminus Y_0$, $\Phi(Y') = G$; see the remark following the definition of essential supremum. But restricting $\mathrm{sgn} \circ w$ and Φ to Y', one has

$$f = \mathcal{B} - \int_{Y'} \mathrm{sgn}(w(y))\Phi(y)d\mu_w(y);$$

hence, $f \subset \mu_w(Y)\mathrm{cl}_\mathcal{X}\mathrm{conv}(\pm G)$. Thus, $\|w\|_{L^1(Y,\mu)} = \mu_w(Y) \geq \|f\|_{G,\mathcal{X}}$. □

6 An Example Involving the Bessel Potential

Here we review an example related to the Bessel functions which was considered in [21] for $q = 2$. In the following section, this Bessel-potential example is used to find an inequality related to the Gamma function.

Let $\mathcal{F}$ denote the Fourier transform, given for $f \in L^1(\mathbb{R}^d,\lambda)$ and $s \in \mathbb{R}^d$ by

$$\hat{f}(s) = \mathcal{F}(f)(s) = (2\pi)^{-d/2}\int_{\mathbb{R}^d} f(x)\exp(-is \cdot x)\,dx,$$

where λ is Lebesgue measure and dx means $d\lambda(x)$. For $r > 0$, let

$$\hat{\beta}_r(s) = (1 + \|s\|^2)^{-r/2}.$$

Since the Fourier transform is an isometry of $\mathcal{L}^2$ onto itself (Parseval's identity), and $\hat{\beta}_r$ is in $\mathcal{L}^2(\mathbb{R}^d)$ for $r > d/2$ (which we now assume), there is a unique function β_r, called the *Bessel potential* of order r, having $\hat{\beta}_r$ as its

Fourier transform. See, e.g., [2, p. 252]. If $1 \leq q < \infty$ and $r > d/q$, then $\hat{\beta}_r \in \mathcal{L}^q(\mathbb{R}^d)$ and

$$\|\hat{\beta}_r\|_{\mathcal{L}^q} = \pi^{d/2q} \left(\frac{\Gamma(qr/2 - d/2)}{\Gamma(qr/2)} \right)^{1/q}. \qquad (14)$$

To see this, observe that by radial symmetry, $(\|\hat{\beta}_r\|_{\mathcal{L}^q})^q = \int_{\mathbb{R}^d} (1 + \|x\|^2)^{-qr/2} dx = \omega_d I$, where $I = \int_0^\infty (1 + \rho^2)^{-qr/2} \rho^{d-1} d\rho$ and $\omega_d = 2\pi^{d/2}/\Gamma(d/2)$ is the area of the unit sphere in $\mathbb{R}^d$ [11, p. 303]. Substituting $\sigma = \rho^2$ and $d\rho = (1/2)\sigma^{-1/2} d\sigma$, and using [10, p. 60], we find that

$$I = (1/2) \int_0^\infty \frac{\sigma^{d/2-1}}{(1+\sigma)^{qr/2}} d\sigma = \frac{\Gamma(d/2)\Gamma(qr/2 - d/2)}{2\Gamma(qr/2)},$$

establishing (14).

For $b > 0$, let $\gamma_b : \mathbb{R}^d \to \mathbb{R}$ denote the scaled Gaussian $\gamma_b(x) = e^{-b\|x\|^2}$. A simple calculation shows that the $\mathcal{L}^q$-norm of γ_b is given by:

$$\|\gamma_b\|_{\mathcal{L}^q} = (\pi/qb)^{d/2q}. \qquad (15)$$

Indeed, using $\int_{-\infty}^\infty \exp(-t^2) dt = \pi^{1/2}$, we obtain:

$$\|\gamma_b\|_{\mathcal{L}^q}^q = \int_{\mathbb{R}^d} \exp(-b\|x\|^2)^q dx = \left(\int_{\mathbb{R}} \exp(-qb\,t^2) dt \right)^d = (\pi/qb)^{d/2}.$$

We now express the Bessel potential as an integral combination of Gaussians. The Gaussians are normalized in $\mathcal{L}^q$ and the corresponding weight function w is explicitly given. The integral formula is similar to one in Stein [41]. By our main theorem, this is an example of (8) and can be interpreted either as a pointwise integral or as a Bochner integral.

Proposition 2. *For d a positive integer, $q \in [1, \infty)$, $r > d/q$, and $s \in \mathbb{R}^d$*

$$\hat{\beta}_r(s) = \int_0^\infty w_r(t) \gamma_t^o(s)\, dt\,,$$

where

$$\gamma_t^o(s) = \gamma_t(s)/\|\gamma_t\|_{\mathcal{L}^q}$$

and

$$w_r(t) = (\pi/qt)^{d/2q}\, t^{r/2-1}\, e^{-t}/\Gamma(r/2).$$

Proof. Let

$$I = \int_0^\infty t^{r/2-1}\, e^{-t}\, e^{-t\|s\|^2}\, dt.$$

Putting $u = t(1 + \|s\|^2)$ and $dt = du(1 + \|s\|^2)^{-1}$, we obtain

$$I = (1 + \|s\|^2)^{-r/2} \int_0^\infty u^{r/2-1} e^{-u} \, du = \hat{\beta}_r(s) \Gamma(r/2).$$

Using the norm of the Gaussian (15), we arrive at

$$\hat{\beta}_r(s) = I/\Gamma(r/2) = \left(\int_0^\infty (\pi/qt)^{d/2q} t^{r/2-1} e^{-t} \gamma_t^o(s) dt \right) / \Gamma(r/2),$$

which is the result desired. □

Now we apply Theorem 5 with $Y = (0, \infty)$ and $\phi(s, t) = \gamma_t^o(s) = \gamma_t(s)/\|\gamma_t\|_{L^q(\mathbb{R}^d)}$ to bound the variational norm of $\hat{\beta}_r$ by the L^1-norm of the weight function.

Proposition 3. *For d a positive integer, $q \in [1, \infty)$, and $r > d/q$,*

$$\|\hat{\beta}_r\|_{G, \mathcal{X}} \leq (\pi/q)^{d/2q} \frac{\Gamma(r/2 - d/2q)}{\Gamma(r/2)},$$

where $G = \{\gamma_t^o : 0 < t < \infty\}$ and $\mathcal{X} = L^q(\mathbb{R}^d)$.

Proof. By (11) and Proposition 2, we have

$$\|\hat{\beta}_r\|_{G, \mathcal{X}} \leq \|w_r\|_{L^1(Y)} = k \int_0^\infty e^{-t} t^{r/2 + d/2q - 1} dt,$$

where $k = (\pi/q)^{d/2q}/\Gamma(r/2)$, and by definition, the integral is $\Gamma(r/2 - d/2q)$. □

7 Application: A Gamma Function Inequality

The inequalities among the variational norm $\|\cdot\|_{G, \mathcal{X}}$, the Banach space norm $\|\cdot\|_{\mathcal{X}}$, and the L^1-norm of the weight function, established in the Main Theorem, allow us to derive other inequalities. The Bessel potential β_r of order r considered above provides an example.

Let d be a positive integer, $q \in [1, \infty)$, and $r > d/q$. By Proposition 3 and (14) of the last section, and by (12) of the Main Theorem, we have

$$\pi^{d/2q} \left(\frac{\Gamma(qr/2 - d/2)}{\Gamma(qr/2)} \right)^{1/q} \leq (\pi/q)^{d/2q} \frac{\Gamma(r/2 - d/2q)}{\Gamma(r/2)}. \tag{16}$$

Hence, with $a = r/2 - d/2q$ and $s = r/2$, this becomes

$$q^{d/2q} \left(\frac{\Gamma(qa)}{\Gamma(qs)} \right)^{1/q} \le \frac{\Gamma(a)}{\Gamma(s)}. \tag{17}$$

In fact, (17) holds if s, a, d, q satisfy (i) $s > a > 0$ and (ii) $s - a = d/2q$ for some $d \in Z^+$ and $q \in [1, \infty)$. As $a > 0$, $r > d/q$. If $T = \{t > 0 : t = d/2q$ for some $d \in Z^+, q \in [1, \infty)\}$, then $T = (0, \frac{1}{2}] \cup (0, 1] \cup (0, \frac{3}{2}] \cup \ldots = (0, \infty)$, so there always exist d, q satisfying (ii); the smallest such d is $\lceil 2(s - a) \rceil$.

The inequality (17) suggests that the Main Theorem can be used to establish other inequalities of interest among classical functions. We now give a direct argument for the inequality. Its independent proof confirms our function-theoretic methods and provides additional generalization.

We begin by noting that in (17) it suffices to take $d = 2q(s - a)$. If the inequality is true in that case, it is true for all real numbers $d \le 2q(s - a)$. Thus, we wish to establish that

$$s \longmapsto \frac{\Gamma(qs)}{\Gamma(s)^q q^{sq}}$$

is a strictly increasing function of s for $q > 1$ and $s > 0$. (For $q = 1$ this function is constant.)

Equivalently, we show that

$$H_q(s) := \log \Gamma(qs) - q \log \Gamma(s) - sq \log q$$

is a strictly increasing function of s for $q > 1$ and $s > 0$.

Differentiating with respect to s, we obtain:

$$\begin{aligned} \frac{dH_q(s)}{ds} &= q \frac{\Gamma'(qs)}{\Gamma(qs)} - q \frac{\Gamma'(s)}{\Gamma(s)} - q \log q \\ &= q(\psi(qs) - \psi(s) - \log q) \\ &=: q A_s(q) \end{aligned}$$

where ψ is the digamma function. It suffices to establish that $A_s(q) > 0$ for $q > 1$, $s > 0$. Note that $A_s(1) = 0$. Now consider

$$\frac{dA_s(q)}{dq} = s\psi'(qs) - \frac{1}{q}.$$

This derivative is positive if and only if $\psi'(qs) > \frac{1}{qs}$ for $q > 1$, $s > 0$.

It remains to show that $\psi'(x) > \frac{1}{x}$ for $x > 0$. Using the power series for ψ' [1, 6.4.10], we have for $x > 0$,

$$\psi'(x) = \sum_{n=0}^{\infty} \frac{1}{(x+n)^2} = \frac{1}{x^2} + \frac{1}{(x+1)^2} + \frac{1}{(x+2)^2} + \ldots$$

$$> \frac{1}{x(x+1)} + \frac{1}{(x+1)(x+2)} + \frac{1}{(x+2)(x+3)} + \cdots$$

$$= \frac{1}{x} - \frac{1}{x+1} + \frac{1}{x+1} - \frac{1}{x+2} + \cdots = \frac{1}{x}.$$

8   Tensor-Product Interpretation

The basic paradigm of feedforward neural nets is to select a single type of computational unit and then build a network based on this single type through a choice of controlling internal and external parameters so that the resulting network function approximates the target function; see our companion chapter in this book. However, a single type of hidden unit may not be as effective as one based on a *plurality* of hidden-unit types. Here we explore a tensor-product interpretation which may facilitate such a change in perspective.

Long ago Hille and Phillips [19, p. 86] observed that the Banach space of Bochner integrable functions from a measure space (Y, μ) into a Banach space $\mathcal{X}$ has a fundamental set consisting of two-valued functions, achieving a single non-zero value on a measurable set of finite measure. Indeed, every Bochner integrable function is a limit of simple functions, and each simple function (with a finite set of values achieved on disjoint parts P_j of the partition) can be written as a sum of characteristic functions, weighted by members of the Banach space. If s is such a simple function, then

$$s = \sum_{i=1}^{n} \chi_j g_j,$$

where the χ_j are the characteristic functions of the P_j and the g_j are in $\mathcal{X}$. (If, for example, Y is embedded in a finite-dimensional Euclidean space, the partition could consist of generalized rectangles.)

Hence, if $f = \mathcal{B} - \int_Y h(y)d\mu(y)$ is the Bochner integral of h with respect to some measure μ, then f can be approximated as closely as desired by elements in $\mathcal{X}$ of the form

$$\sum_{i=1}^{n} \mu(P_i)g_i,$$

where $Y = \bigcup_{i=1}^{n} P_i$ is a μ-measurable partition of Y.

Note that given a σ-finite measure space (Y, μ) and a separable Banach space $\mathcal{X}$, every element f in $\mathcal{X}$ is (trivially) the Bochner integral of any integrand $w \cdot \kappa(f)$, where w is a nonnegative function on Y with $\|w\|_{L^1(Y,\mu)} = 1$ (see part (iii) of Theorem 5) and $\kappa(f)$ denotes the constant function on Y with value f. In effect, f is in $\mathcal{X}_G$ when $G = \{f\}$. When Φ is chosen first (or more precisely ϕ as in our Main Theorem), then f may or may not be in $\mathcal{X}_G$. According to the Main Theorem, f is in $\mathcal{X}_G$ when it is given by an integral formula involving Φ and some L^1 weight function. In this case, $G = \Phi(Y) \cap B$ where B is the ball in $\mathcal{X}$ of radius $\|\Phi\|_{L^\infty(Y,\mu;\mathcal{X})}$.

In general, the elements $\Phi(y)$, $y \in Y$ of the Banach space involved in some particular approximation for f will be distinct functions of some general type obtained by varying the parameter y. For instance, kernels, radial basis functions, perceptrons, or various other classes of computational units can be used, and when these computational-unit-classes determine fundamental sets, by Proposition 1, it is possible to obtain arbitrarily good approximations. However, Theorem 6 below suggests that having a finite set of distinct types $\Phi_i : Y \to \mathcal{X}$ may allow a smaller "cost" for approximation, if we regard

$$\sum_{i=1}^{n} \|w_i\|_1 \|\Phi_i\|_\infty$$

as the cost of the approximation

$$f = \mathcal{B} - \int_Y \left(\sum_{i=1}^{n} w_i \Phi_i \right) (y) d\mu(y).$$

We give a brief sketch of the ideas, following Light and Cheney [33].

Let $\mathcal{X}$ and $\mathcal{Z}$ be Banach spaces. Let $\mathcal{X} \otimes \mathcal{Z}$ denote the linear space of equivalence classes of formal expressions

$$\sum_{i=1}^{n} f_i \otimes h_i, \ f_i \in \mathcal{X}, \ h_i \in \mathcal{Z}, \ n \in \mathbb{N}, \quad \sum_{i=1}^{m} f_i' \otimes h_i', \ f_i' \in \mathcal{X}, \ h_i' \in \mathcal{Z}, \ m \in \mathbb{N},$$

where these expressions are equivalent if for every $F \in \mathcal{X}^*$

$$\sum_{i=1}^{n} F(f_i) h_i = \sum_{i=1}^{m} F(f_i') h_i',$$

that is, if the associated operators from $\mathcal{X}^* \to \mathcal{Z}$ are identical, where $\mathcal{X}^*$ is the algebraic dual of $\mathcal{X}$. The resulting linear space $\mathcal{X} \otimes \mathcal{Z}$ is called the *algebraic tensor product* of $\mathcal{X}$ and $\mathcal{Z}$. We can extend $\mathcal{X} \otimes \mathcal{Z}$ to a Banach space by completing it with respect to a suitable norm. Consider the norm defined for $t \in \mathcal{X} \otimes \mathcal{Z}$,

$$\gamma(t) = \inf \left\{ \sum_{i=1}^{n} \|f_i\|_\mathcal{X} \|h_i\|_\mathcal{Z} \ : \ t = \sum_{i=1}^{n} f_i \otimes h_i \right\}. \tag{18}$$

and complete the algebraic tensor product with respect to this norm; the result is denoted $\mathcal{X} \otimes_\gamma \mathcal{Z}$.

In [33, Thm. 1.15, p. 11], Light and Cheney noted that for any measure space (Y, μ) and any Banach space $\mathcal{X}$ the linear map

$$\Lambda_\mathcal{X} : L^1(Y, \mu) \otimes \mathcal{X} \to \mathcal{L}^1(Y, \mu; \mathcal{X})$$

given by

$$\sum_{i=1}^{r} w_i \otimes g_i \mapsto \sum_{i=1}^{r} w_i g_i.$$

is well-defined on equivalence classes. They showed that $\Lambda_{\mathcal{X}}$ extends to a map

$$\Lambda_{\mathcal{X}}^{\gamma} : L^1(Y, \mu) \otimes_{\gamma} \mathcal{X} \to L^1(Y, \mu; \mathcal{X}),$$

which is an isometric isomorphism of the completed tensor product onto the space $L^1(Y, \mu; X)$ of Bochner-integrable functions.

The following theorem extends the function $\Lambda_{\mathcal{X}}$ via the natural embedding $\kappa_{\mathcal{X}}$ of $\mathcal{X}$ into the space of essentially bounded $\mathcal{X}$-valued functions defined in Lemma 2 of Section 4.

Theorem 6. *Let $\mathcal{X}$ be a separable Banach space and let (Y, μ) be a σ-finite measure space. Then there exists a continuous linear surjection*

$$\Lambda_{\mathcal{X}}^{\infty,\gamma} : L^1(Y, \mu) \otimes_{\gamma} L^{\infty}(Y, \mu; \mathcal{X}) \to L^1(Y, \mu; \mathcal{X}),$$

denoted below by e, which makes the following diagram commute:

$$
\begin{array}{ccc}
L^1(Y, \mu) \otimes_{\gamma} \mathcal{X} & \xrightarrow{\;a\;} & L^1(Y, \mu; \mathcal{X}) \\
\downarrow{\scriptstyle b} & \nearrow{\scriptstyle e} & \\
L^1(Y, \mu) \otimes_{\gamma} L^{\infty}(Y, \mu; \mathcal{X}) & &
\end{array}
\qquad (19)
$$

where the horizontal arrow a is the isometric isomorphism $\Lambda_{\mathcal{X}}^{\gamma}$, and the vertical arrow b is induced by $1 \otimes \kappa_{\mathcal{X}}$.

Proof. The map

$$e' : \sum_{i-1}^{n} w_i \otimes \Phi_i \mapsto \sum_{i=1}^{n} w_i \Phi_i$$

defines a linear function $L^1(Y, \mu) \otimes L^{\infty}(Y, \mu; \mathcal{X}) \to L^1(Y, \mu; \mathcal{X})$; indeed, by Cor. 1 of Section 4 (or by our Main Theorem under suitable hypotheses), each summand is Bochner integrable.

To define the continuous linear surjection e given in (19), let $t = \sum_{i=1}^{n} w_i \Phi_i$ be an element of $L^1(Y, \mu)$. Then

$$\|e'(t)\|_{L^1(Y,\mu;\mathcal{X})} = \int_Y \|e'(t)\|_{\mathcal{X}} d\mu(y) =$$

$$\int_Y \|\sum_i w_i(y)\Phi_i(y)\|_{\mathcal{X}} d\mu(y) \le \int_Y \sum_i |w_i(y)| \|\Phi_i(y)\|_{\mathcal{X}} d\mu(y)$$

$$\le \sum_i \|w_i\|_1 \|\Phi_i\|_{\infty}$$

Hence, $\|e'(t)\|_{L^1(Y,\mu;\mathcal{X})} \leq \gamma(t)$. So the map e' is continuous on the algebraic tensor product with respect to the completion norm (18) and has a unique continuous extension to the completed tensor product. The easily verified equation $\Lambda_{\mathcal{X}} = e' \circ (1 \otimes \kappa_{\mathcal{X}})$ now implies that $a = e \circ b$. $\qquad\square$

9   An Example Involving Bounded Variation on an Interval

The following example, more elaborate than the one following Proposition 1, is treated in part by Barron [6] and Kůrková [25].

Let $\mathcal{X}$ be the set of equivalence classes of (essentially) bounded Lebesgue-measurable functions on $[a, b]$, $-\infty < a < b < \infty$, i.e., $\mathcal{X} = L^\infty([a, b])$, with norm $\|f\|_{\mathcal{X}} := \inf\{M : |f(x)| \leq M \text{ for almost every } x \in [a, b]\}$. Let G be the set of equivalence classes of all characteristic functions of closed intervals of the forms $[a, b]$, or $[a, c]$ or $[c, b]$ with $a < c < b$. These functions are the restrictions of characteristic functions of closed half-lines to $[a, b]$. The equivalence relation is $f \sim g$ if and only if $f(x) = g(x)$ for almost every x in $[a, b]$ (with respect to Lebesgue measure).

Let $BV([a, b])$ be the set of all equivalence classes of functions on $[a, b]$ with bounded variation; that is, each equivalence class contains a function f such that the *total variation* $V(f, [a, b])$ is finite, where total variation is the largest possible total movement of a discrete point which makes a finite number of stops as x varies from a to b, maximized over all possible ways to choose a finite list of intermediate points, that is,

$$V(f, [a, b]) := \sup\{\sum_{i=1}^{n-1} |f(x_{i+1}) - f(x_i)| : n \geq 1, a \leq x_1 < x_2 < \cdots < x_n \leq b\}.$$

In fact, each equivalence class $[f]$ contains exactly one function f^* of bounded variation that satisfies the *continuity conditions*:

(i) f^* is right-continuous at c for $c \in [a, b)$, and

(ii) f^* is left-continuous at b.

Moreover, $V(f^*, [a, b]) \leq V(f, [a, b])$ for all $f \sim f^*$.

To see this, recall that every function f of bounded variation is the difference of two nondecreasing functions $f = f_1 - f_2$, and f_1, f_2 are necessarily right-continuous except at a countable set. We can take $f_1(x) := V(f, [a, x]) + K$, where K is an arbitrary constant, and $f_2(x) := V(f, [a, x]) + K - f(x)$ for $x \in [a, b]$. Now redefine both f_1 and f_2 at countable sets to form f_1^* and f_2^* which satisfy the continuity conditions and are still nondecreasing on $[a, b]$. Then $f^* := f_1^* - f_2^*$ also satisfies the continuity conditions. It is easily shown that $V(f^*, [a, b]) \leq V(f, [a, b])$. Since any equivalence class in $\mathcal{X}$ can contain at most one function satisfying (i) and (ii) above, it follows that f^* is unique and that $V(f^*, [a, b])$ minimizes the total variation for all functions in the equivalence class.

Proposition 4. *Let $\mathcal{X} = L^\infty([a, b])$ and let $G \subset \mathcal{X}$ be the set of equivalence classes of the family of characteristic functions*

$$\{\chi_{[a,c]} : a \leq c < b\} \cup \{\chi_{(c,b]} : a < c \leq b\}.$$

Then $\mathcal{X}_G = BV([a, b])$, and

$$\|[f]\|_{\mathcal{X}} \leq \|[f]\|_{G,\mathcal{X}} \leq 2V(f^*, [a, b]) + |f^*(a)|,$$

where f^* is the member of $[f]$ satisfying the continuity conditions (i) and (ii).

Proof. Let $C_{G,\mathcal{X}}$ be the set of equivalence classes of functions of the form

$$(q - r)\chi_{[a,b]} + \sum_{n=1}^{k}(s_n - t_n)\chi_{[a,c_n]} + \sum_{n=1}^{k}(u_n - v_n)\chi_{[c_n,b]}, \qquad (20)$$

where k is a positive integer, $q, r \geq 0$, for $1 \leq n \leq k$, $s_n, t_n, u_n, v_n \geq 0$, and

$$(q + r) + \sum_{n=1}^{k}(s_n + t_n + u_n + v_n) = 1.$$

All of the functions so exhibited have bounded variation ≤ 1 and hence $C_{G,\mathcal{X}} \subseteq BV([a, b])$.

We will prove that a sequence in $C_{G,\mathcal{X}}$, convergent in the $\mathcal{X}$-norm, converges to a member of $BV([a, b])$ and this will establish that $B_{G,\mathcal{X}}$ is a subset of $BV([a, b])$ and hence that $\mathcal{X}_G$ is a subset of $BV([a, b])$.

Let $\{[f_k]\}$ be a sequence in $C_{G,\mathcal{X}}$ that is Cauchy in the $\mathcal{X}$-norm. Without loss of generality, we pass to the sequence $\{f_k^*\}$, which is Cauchy in the sup-norm since $x \mapsto |f_k^*(x) - f_j^*(x)|$ satisfies the continuity conditions (i) and (ii). Thus, $\{f_k^*\}$ converges pointwise-uniformly and in the sup-norm to a function f on $[a, b]$ also satisfying (i) and (ii) with finite sup-norm and whose equivalence class has finite $\mathcal{X}$-norm. This implies $\{[f_k]\}$ converges to $[f]$ in the $\mathcal{X}$-norm.

Let $\{x_1, \ldots, x_n\}$ satisfy $a \leq x_1 < x_2 < \cdots < x_n \leq b$. Then

$$\sum_{i=1}^{n-1} |f_k^*(x_{i+1}) - f_k^*(x_i)| \leq V(f_k^*, [a, b]) \leq V(f_k, [a, b]) \leq 1$$

for every k, where *par abus de notation* f_k denotes the member of $[f_k]$ satisfying (20). Letting k tend to infinity and then varying n and $x_1, \ldots, x_n$, we obtain $V(f, [a, b]) \leq 1$ and so $[f] \in BV([a, b])$

It remains to show that everything in $BV([a,b])$ is actually in $\mathcal{X}_G$. Let g be a nonnegative nondecreasing function on $[a,b]$ satisfying the continuity conditions (i) and (ii) above. Given a positive integer n, there exists a positive integer $m \geq 2$ and $a = a_1 < a_2 < \cdots < a_m = b$ such that $g(a_{i+1}^-) - g(a_i) \leq 1/n$ for $i = 1, \ldots, m-1$. Indeed, for $2 \leq i \leq m-1$, let $a_i := \min\{x | g(a) + \frac{(i-1)}{n} \leq g(x)\}$. (Moreover, it follows that the set of a_i's include all points of left-discontinuity of g such that the jump $g(a_i) - g(a_i^-)$ is greater than $1/n$.) Let $g_n : [a,b] \to \mathbb{R}$ be defined as follows:

$$g_n := g(a_1)\chi_{[a_1,a_2)} + g(a_2)\chi_{[a_2,a_3)} + \cdots + g(a_{m-1})\chi_{[a_{m-1},a_m]}$$

$$= g(a_1)(\chi_{[a_1,b]} - \chi_{[a_2,b]}) + g(a_2)(\chi_{[a_2,b]} - \chi_{[a_3,b]}) + \cdots + g(a_{m-1})(\chi_{[a_{m-1},b]})$$

$$= g(a_1)\chi_{[a_1,b]} + (g(a_2) - g(a_1))\chi_{[a_2,b]} + \cdots + (g(a_{m-1}) - g(a_{m-2}))\chi_{[a_{m-1},b]}.$$

Then $[g_n]$ belongs to $g(a_{m-1})C_{G,\mathcal{X}}$, and *a fortiori* to $g(b)C_{G,\mathcal{X}}$ as well as of $\mathcal{X}_G$, and $\|[g_n]\|_{G,\mathcal{X}} \leq g(b)$. Moreover, $\|[g_n] - [g]\|_{\mathcal{X}} \leq 1/n$. Therefore, since $B_{G,\mathcal{X}} = \mathrm{cl}_{\mathcal{X}}(C_{G,\mathcal{X}})$, $[g]$ is in $g(b)B_{G,\mathcal{X}}$ and accordingly $[g]$ is in $\mathcal{X}_G$ and

$$\| [g] \|_{G,\mathcal{X}} \leq g(b).$$

Let $[f]$ be in $BV([a,b])$ and let $f^* = f_1^* - f_2^*$, as defined above, for this purpose we take $K = |f^*(a)|$. This guarantees that both f_1^* and f_2^* are nonnegative. Accordingly, $[f] = [f_1^*] - [f_2^*]$, and is in $\mathcal{X}_G$. Furthermore, $\|[f]\|_{G,\mathcal{X}} \leq \|[f_1^*]\|_{G,\mathcal{X}} + \|[f_2^*]\|_{G,\mathcal{X}} \leq f_1^*(b) + f_2^*(b) = V(f^*,[a,b]) + |f^*(a)| + V(f^*,[a,b]) + |f^*(a)| - f^*(b) \leq 2V(f^*,[a,b]) + |f^*(a)|$. The last inequality follows from the fact that $V(f^*,[a,b]) + |f^*(a)| - f^*(b) \geq |f^*(b) - f^*(a)| + |f^*(a)| - f^*(b) \geq 0$. □

An argument similar to the above shows that $BV([a,b])$ is a Banach space under the norm $2V(f^*,[a,b]) + |f^*(a)|$ (with or without the 2). The identity map from $BV([a,b])$ (with this norm) to $(\mathcal{X}_G, \|\cdot\|_{G,\mathcal{X}}$, is continuous (by Proposition 4) and it is also onto. Accordingly, by the Open Mapping Theorem (e.g., Yosida [43, p. 75]) the map is open, hence a homeomorphism, so the norms are equivalent. Thus, in this example, $\mathcal{X}_G$ is a Banach space under these two equivalent norms.

Note however that the $\mathcal{X}$-norm restricted to $\mathcal{X}_G$ does not give a Banach space structure; i.e., $\mathcal{X}_G$ is not complete in the $\mathcal{X}$-norm. Indeed, with $\mathcal{X} = L^\infty([0,1])$, let f_n be $1/n$ times the characteristic function of the disjoint union of n^2 closed intervals contained within the unit interval. Then each f_n is in $\mathcal{X}_G$ and $\|[f_n]\|_{\mathcal{X}} = 1/n$ but $\|[f_n]\|_{G,\mathcal{X}} \geq Cn$, some $C > 0$, since the $\|\cdot\|_{G,\mathcal{X}}$ is equivalent to the total-variation norm. While $\{f_n\}$ converges to zero in one norm, in the other it blows up. If $\mathcal{X}_G$ were a Banach space under $\|\cdot\|_{\mathcal{X}}$, it would be another Cauchy sequence, a contradiction.

10   Pointwise-Integrals vs. Bochner Integrals

10.1   *Evaluation of Bochner Integrals*

A natural conjecture is that the Bochner integral, evaluated pointwise, is the pointwise integral; that is, if $h \in \mathcal{L}^1(Y, \mu, \mathcal{X})$, where $\mathcal{X}$ is any Banach space of functions defined on a measure space Ω, then

$$\left(\mathcal{B} - \int_Y h(y) \, d\mu(y) \right)(x) = \int_Y h(y)(x) \, d\mu(y) \qquad (21)$$

for all $x \in \Omega$. Usually, however, one is dealing with equivalence classes of functions and thus can expect the equation (21) to hold only for almost every x in Ω. Furthermore, to specify $h(y)(x)$, it is necessary to take a particular function representing $h(y) \in \mathcal{X}$

The Main Theorem implies that (21) holds for ρ-a.e. $x \in \Omega$ when $\mathcal{X} = L^q(\Omega, \rho)$, for $1 \le q < \infty$, is separable and $h = w\Phi$. Here $w : Y \to \mathbb{R}$ is a weight function with finite L^1-norm, and $\Phi : Y \to \mathcal{X}$ is essentially bounded, and defined by requiring that for each $y \subset Y$, $\Phi(y)(x) - \phi(x, y)$ for ρ-a.e. $x \subset \Omega$ and $\phi : \Omega \times Y \to \mathbb{R}$ is $\rho \times \mu$-measurable. More generally, we can show the following.

Theorem 7. *Let (Ω, ρ), (Y, μ) be σ-finite measure spaces, let $\mathcal{X} = L^q(\Omega, \rho)$, $q \in [1, \infty]$, and let $h \in \mathcal{L}^1(Y, \mu; \mathcal{X})$ so that for each y in Y, $h(y)(x) = H(x, y)$ for ρ-a.e. x, where H is a $\rho \times \mu$-measurable real-valued function on $\Omega \times Y$. Then*
(i) $y \mapsto H(x, y)$ is integrable for ρ-a.e. $x \in \Omega$,
(ii) the equivalence class of $x \mapsto \int_Y H(x, y) \, d\mu(y)$ is in $\mathcal{X}$, and
(iii) for ρ-a.e. $x \in \Omega$

$$\left(\mathcal{B} - \int_Y h(y) \, d\mu(y) \right)(x) = \int_Y H(x, y) \, d\mu(y).$$

Proof. We first consider the case $1 \le q < \infty$. Let g be in $\mathcal{L}^p(\Omega, \rho)$, where $1/p + 1/q = 1$. Then

$$\int_Y \int_\Omega |g(x) H(x, y)| \, d\rho(x) d\mu(y) \le \int_Y \|g\|_p \|h(y)\|_q \, d\mu(y) \qquad (22)$$

$$= \|g\|_p \int_Y \|h(y)\|_{\mathcal{X}} \, d\mu(y) < \infty. \qquad (23)$$

Here we have used Young's inequality and Bochner's theorem. By Fubini's theorem, (i) follows. In addition, the map $g \mapsto \int_\Omega g(x) \left(\int_Y H(x, y) \, d\mu(y) \right) d\rho(x)$ is a continuous linear functional F on L^p with $\|F\|_{\mathcal{X}^*} \le \int_Y \|h(y)\|_{\mathcal{X}} \, d\mu(y)$. Since $(L^p)^*$ is L^q for $1 < q < \infty$, then the function $x \mapsto \int_Y H(x, y) \, d\mu(y)$ is in $\mathcal{X} = L^q$ and has norm $\le \int_Y \|h(y)\|_{\mathcal{X}} \, d\mu(y)$. The case $q = 1$ is covered by taking $g \equiv 1$, a member of

L^∞, and noting that $\| \int_Y H(x,y) \, d\mu(y) \|_{L^1} = \int_\Omega | \int_Y H(x,y) \, d\mu(y) | d\rho(x) \le \int_\Omega \int_Y |H(x,y)| \, d\mu(y) d\rho(x) = \int_Y \|h(y)\|_{\mathcal{X}} \, d\mu(y)$. Thus, (ii) holds for $1 \le q < \infty$.

Also by Fubini's theorem and Theorem 3, for all $g \in \mathcal{X}^*$,

$$\int_\Omega g(x) \left(\mathcal{B} - \int_Y h(y) \, d\mu(y) \right)(x) d\rho(x) = \int_Y \left(\int_\Omega g(x) H(x,y) d\rho(x) \right) d\mu(y)$$

$$= \int_\Omega g(x) \left(\int_Y H(x,y) d\mu(y) \right) d\rho(x).$$

Hence (iii) holds for all $q < \infty$, including $q = 1$.

Now consider the case $q = \infty$. For $g \in L^1(\Omega, \rho) \subseteq (L^\infty(\Omega, \rho))^*$, the inequality (23) holds, and by [18, pp. 348–9], (i) and (ii) hold and $\| \int_Y H(x,y) \, d\mu(y) \|_\infty \le \int_Y \|h(y)\|_\infty \, d\mu(y) < \infty$. For $g \in \mathcal{L}^1(\Omega, \rho)$,

$$\int_\Omega g(x) \left(\mathcal{B} - \int_Y h(y) \, d\mu(y) \right)(x) d\rho(x) = \int_Y \left(\int_\Omega g(x) h(y)(x) d\rho(x) \right) d\mu(y)$$

$$= \int_Y \left(\int_\Omega g(x) H(x,y) d\rho(x) \right) d\mu(y) = \int_\Omega g(x) \left(\int_Y H(x,y) \, d\mu(y) \right) d\rho(x).$$

The two functions integrated against g are in $L^\infty(\Omega, \rho)$ and agree, and since $\left(L^1(\Omega, \rho) \right)^* = L^\infty(\Omega, \rho)$, these functions must be the same ρ-a.e. □

There are cases where $\mathcal{X}$ consists of pointwise-defined functions and (21) can be taken literally.

If $\mathcal{X}$ is a separable Banach space of pointwise-defined functions from Ω to $\mathbb{R}$ in which the evaluation functionals are bounded (and so in particular if $\mathcal{X}$ is a reproducing kernel Hilbert space [4]), then (21) holds for all x (not just ρ-a.e.). Indeed, for each $x \in \Omega$, the evaluation functional $E_x : f \mapsto f(x)$ is bounded and linear, so by Theorem 3, E_x commutes with the Bochner integral operator. As non-separable reproducing kernel Hilbert spaces exist [3, p.26], one still needs the hypothesis of separability.

In a special case involving Bochner integrals with values in Marcinkiewicz spaces, Nelson [38] showed that (21) holds. His result involves going from equivalence classes to functions, and uses a "measurable selection." Reproducing kernel Hilbert spaces were studied by Le Page in [32] who showed that (21) holds when μ is a probability measure on Y under a Gaussian distribution assumption on variables in the dual space. Another special case of (21) is derived in Hille and Phillips [19, Theorem 3.3.4, p. 66], where the parameter space is an interval of the real line and the Banach space is a space of bounded linear transformations (i.e., the Bochner integrals are operator-valued).

10.2   *Essential Boundedness Is Needed for the Main Theorem*

The following is an example of a function $h : Y \to \mathcal{X}$ which is not Bochner integrable. Let $Y = (0,1) = \Omega$ with $\rho = \mu =$ Lebesgue measure and $q = 1 = d$ so $\mathcal{X} = L^1((0,1))$. Put $h(y)(x) = y^{-x}$. Then for all $y \in (0,1)$

$$\|h(y)\|_X = \int_0^1 y^{-x}dx = \frac{1 - \frac{1}{y}}{\log y}.$$

By l'Hospital's rule

$$\lim_{y \to 0^+} \|h(y)\|_X = +\infty.$$

Thus, the function $y \mapsto \|h(y)\|_X$ is not essentially bounded on $(0,1)$ and Theorem 5 does not apply. Furthermore, for $y \leq 1/2$,

$$\|h(y)\|_{\mathcal{X}} \geq \frac{1}{-2y \log y}$$

and

$$\int_0^1 \|h(y)\|_{\mathcal{X}} \, dy \geq \int_0^{1/2} \frac{1}{-2y \log y} \, dy = -(1/2) \log (\log y)|_0^{1/2} = \infty.$$

Hence, by Theorem 1, h is not Bochner integrable. Note however that

$$f(x) = \int_Y h(y)(x)d\mu(y) = \int_0^1 y^{-x}dy = \frac{1}{1-x}$$

for every $x \in \Omega$. Thus $h(y)(x)$ has a pointwise integral $f(x)$ for all $x \in (0,1)$, but f is not in $\mathcal{X} = L^1((0,1))$.

We have shown that $y \mapsto \|h(y)\|_{\mathcal{X}}$ is continuous. In fact, h is continuous at each y in $(0,1)$ since, using $\frac{d}{dy}y(y^{-x}) = (-x)y^{-x-1}$ and the ordinary mean value theorem, we get for $t \in (0,y)$, $\int_0^1 (|(y+t)^{-x} - y^{-x}|dx \leq \frac{1}{y} \int_0^1 \frac{t}{y^x} dx$, and also $\int_0^1 |(y-t)^{-x} - y^{-x}|dx \leq \frac{t}{y-t} \int_0^1 \frac{dx}{(y-t)^x}$. Thus, h is measurable. As $\mathcal{L}^1((0,1))$ is separable, h is strongly measurable.

10.3   *Connection with Sup Norm*

In [22], we take $\mathcal{X}$ to be the space of bounded measurable functions on $\mathbb{R}^d$, Y equal to the product $S^{d-1} \times \mathbb{R}$ with measure ν which is the (completion of the) product measure determined by the standard (unnormalized) measure $d(e)$ on the sphere and ordinary Lebesgue measure on $\mathbb{R}$. Recalling that $\vartheta : \mathbb{R} \to \mathbb{R}$ is the characteristic function of $[0, \infty)$, we take $\phi(x,y) := \phi(x,e,b) = \vartheta(e \cdot x + b)$, and thus $x \mapsto \vartheta(e \cdot x + b)$ is the characteristic function of the closed half-space $\{x : e \cdot x + b \geq 0\}$.

We showed that if a function f on $\mathbb{R}^d$ decays, along with its partial derivatives of order $\leq d$, at a sufficient rate (i.e., f is of weakly controlled decay - see [22] or the chapter of Kainen, Kůrková, and Sanguineti in this book), then there is an integral formula expressing $f(x)$ as an integral combination of the characteristic functions of closed half-spaces weighted by iterated Laplacians integrated over half-spaces. The characteristic functions all have sup-norm of 1 and the weight-function is in $L^1(S^{d-1} \times \mathbb{R}, \nu)$, as above. For example, when d is odd,

$$f(x) = \int_{S^{d-1} \times \mathbb{R}} w_f(e, b) \vartheta(e \cdot x + b) d\nu(e, b),$$

where

$$w_f(e, b) := a_d \int_{H_{e,b}} D_e^{(d)} f(y) d_H(y),$$

with a_d a scalar exponentially decreasing with d. The integral is of the iterated directional derivative over the hyperplane with normal vector e and offset b,

$$H_{e,b} := \{y \in \mathbb{R}^d : e \cdot y + b = 0\}.$$

For $\mathcal{X} = \mathcal{M}(\mathbb{R}^d)$, the space of bounded Lebesgue-measurable functions on $\mathbb{R}^d$, which is a Banach space w.r.t. sup-norm, and G the family H_d consisting of the set of all characteristic functions for closed half-spaces in $\mathbb{R}^d$, it follows from (iv) of the Main Theorem (Theorem 5) that $f \in \mathcal{X}_G$ and by (iii) that f is a Bochner integral

$$f = \mathcal{B} - \int_{S^{d-1} \times \mathbb{R}} w_f(e, b) \Theta(e, b) \, d\nu(e, b),$$

where $\Theta(e, b) \in \mathcal{M}(\mathbb{R}^d)$ is given by

$$\Theta(e, b)(x) := \vartheta(e \cdot x + b).$$

Application of the Main Theorem requires only that w_f be in L^1, but [22] gives explicit formulas for w_f (in both even and odd dimensions) provided that f satisfies the decay conditions described above and in our paper; see also the other chapter in this book referenced earlier.

11   Some Concluding Remarks

Neural networks express a function f in terms of a combination of members of a given family G of functions. It is reasonable to expect that a function f can be so represented if f is in $\mathcal{X}_G$. The choice of G thus dictates the f's that can be represented (if we leave aside what combinations are permissible). Here we have focused on the case $G = \{\Phi(y) : y \in Y\}$. The form $\Phi(y)$ is usually associated with a specific family such as Gaussians or Heavisides. The tensor-product interpretation suggests the possibility of using multiple families $\{\Phi_j : j \in J\}$ or multiple G's to represent a larger class of f's. Alternatively, one may replace Y by $Y \times J$ with a suitable extension of the measure.

The Bochner integral approach also permits $\mathcal{X}$ to be an arbitrary Banach space (not necessarily an L^p-space). For example, if $\mathcal{X}$ is a space of bounded linear transformations and $\Phi(Y)$ is a family of such transformations, we can approximate other members f of this Banach space $\mathcal{X}$ in a neural-network-like manner. Even more abstractly, we can approximate an *evolving* function f_t, where t is time, using weights that evolve over time and/or a family $\Phi_t(y)$ whose members evolve in a prescribed fashion. Such an approach would require some axiomatics about permissible evolutions of f_t, perhaps similar to methods used in time-series analysis and stochastic calculus. See, e.g., [8].

Many of the restrictions we have imposed in earlier sections are not truly essential. For example, the separability constraints can be weakened. Moreover, σ-finiteness of Y need not be required since an integrable function w on Y must vanish outside a σ-finite subset. More drastically, the integrable function w can be replaced by a distribution or a measure. Indeed, we believe that both finite combinations and integrals can be subsumed in generalized combinations derived from Choquet's theorem. The abstract transformations of the concept of neural network discussed here provide an "enrichment" that may have practical consequences.

12 Appendix I: Some Banach Space Background

The following is a brief account of the machinery of functional analysis used in this chapter. See, e.g., [43]. For $G \subseteq \mathcal{X}$, with $\mathcal{X}$ any linear space, let

$$\mathrm{span}_n(G) := \left\{ x \in \mathcal{X} : \exists w_i \in \mathbb{R}, g_i \in G,\ 1 \leq i \leq n,\ \ni\ x = \sum_{i-1}^{n} w_i g_i \right\}$$

denote the set of all n-fold linear combinations from G. If the w_i are nonnegative with sum 1, then the combination is called a *convex* combination; $\mathrm{conv}_n(G)$ denotes the set of all n-fold convex combinations from G. Let

$$\mathrm{span}(G) := \bigcup_{n=1}^{\infty} \mathrm{span}_n(G)\ \text{ and }\ \mathrm{conv}(G) := \bigcup_{n=1}^{\infty} \mathrm{conv}_n(G).$$

A *norm* on a linear space $\mathcal{X}$ is a function which associates to each element f of $\mathcal{X}$ a real number $\|f\| \geq 0$ such that

(1) $\|f\| = 0 \iff f = 0$;

(2) $\|rf\| = |r| \|f\|$ for all $r \in \mathbb{R}$; and

(3) the triangle inequality holds: $\|f + g\| \leq \|f\| + \|g\|$, $\forall f, g \in \mathcal{X}$.

A metric $d(x, y) := \|x - y\|$ is defined by the norm, and both addition and scalar multiplication become continuous functions with respect to the topology induced by the norm-metric. A metric space is *complete* if every sequence in the space that satisfies the Cauchy criterion is convergent. In particular, if a normed linear space is complete in the metric induced by its norm, then it is called a *Banach* space.

Let (Y, μ) be a measure space; it is called σ-finite provided that there exists a countable family $Y_1, Y_2, \ldots$ of subsets of Y pairwise-disjoint and measurable with finite μ-measure such that $Y = \bigcup_i Y_i$. The condition of σ-finiteness is required for Fubini's theorem. A set N is called a μ-null set if it is measurable with $\mu(N) = 0$. A function from a measure space to another measure space is called *measurable* if the pre-image of each measurable subset is measurable. When the range space is merely a topological space, then functions are measurable if the pre-image of each open set is measurable.

Let (Ω, ρ) be a measure space. If $q \in [1, \infty)$, we write $L^q(\Omega, \rho)$ for the Banach space consisting of all equivalence classes of the set $\mathcal{L}^q(\Omega, \rho)$ of all ρ-measurable functions from Ω to $\mathbb{R}$ with absolutely integrable q-th powers, where f and g are equivalent if they agree ρ-almost everywhere (ρ-a.e.) - that is, if the set of points where f and g disagree has ρ-measure zero, and $\|f\|_{L^q(\Omega,\rho)} := (\int_\Omega |f(x)|^q d\rho(x))^{1/q}$, or $\|f\|_q$ for short.

13   Appendix II: Some Key Theorems

We include, for the reader's convenience, the statements of some crucial theorems cited in the text.

The following consequence of the Hahn-Banach Theorem, due to Mazur, is given by Yosida [43, Theorem 3', p. 109]. The hypotheses on $\mathcal{X}$ are satisfied by any Banach space, but the theorem holds much more generally. See [43] for examples where $\mathcal{X}$ is not a Banach space.

Theorem 8. *Let X be a real locally convex linear topological space, M a closed convex subset, and $x_0 \in X \setminus M$. Then $\exists$ continuous linear functional*

$$F : X \to \mathbb{R} \ni F(x_0) > 1, \ F(x) \leq 1 \ \forall x \in M.$$

Fubini's Theorem relates iterated integrals to product integrals. Let Y, Z be sets and $\mathcal{M}$ be a σ-algebra of subsets of Y and $\mathcal{N}$ a σ-algebra of subsets of Z. If $M \in \mathcal{M}$ and $N \in \mathcal{N}$, then $M \times N \subseteq Y \times Z$ is called a *measurable rectangle*. We denote the smallest σ-algebra on $Y \times Z$ which contains all the measurable rectangles by $\mathcal{M} \times \mathcal{N}$. Now let $(Y, \mathcal{M}, \mu)$ and $(Z, \mathcal{N}, \nu)$ be σ-finite measure spaces, and for $E \in \mathcal{M} \times \mathcal{N}$, define

$$(\mu \times \nu)(E) := \int_Y \nu(E_y) d\mu(y) = \int_Z \mu(E^z) d\nu(z),$$

where $E_y := \{z \in Z : (y, z) \in E\}$ and $E^z := \{y \in Y : (y, z) \in E\}$. Also, $\mu \times \nu$ is a σ-finite measure on $Y \times Z$ with $\mathcal{M} \times \mathcal{N}$ as the family of measurable sets. For the following, see Hewitt and Stromberg [18, p. 386].

Theorem 9. Let $(Y, \mathcal{M}, \mu)$ and $(Z, \mathcal{N}, \nu)$ be σ-finite measure spaces. Let f be a complex-valued $\mathcal{M} \times \mathcal{N}$-measurable function on $Y \times Z$, and suppose that at least one of the following three absolute integrals is finite: $\int_{Y \times Z} |f(y, z)| d(\mu \times \nu)(y, z)$, $\int_Z \int_Y |f(y, z)| d\mu(y) d\nu(z)$, $\int_Y \int_Z |f(y, z)| d\nu(z) d\mu(y)$. Then the following statements hold:
(i) $y \mapsto f(y, z)$ is in $\mathcal{L}^1(Y, \mathcal{M}, \mu)$ for ν-a.e. $z \in Z$;
(ii) $z \mapsto f(y, z)$ is in $\mathcal{L}^1(Z, \mathcal{N}, \nu)$ for μ-a.e. $y \in Y$;
(iii) $z \mapsto \int_Y f(y, z) d\mu(y)$ is in $\mathcal{L}^1(Z, \mathcal{N}, \nu)$;
(iv) $y \mapsto \int_Z f(y, z) d\nu(z)$ is in $\mathcal{L}^1(Y, \mathcal{M}, \mu)$;
(v) all three of the following integrals are equal:

$$\int_{Y \times Z} f(y, z) d(\mu \times \nu)(y, z) =$$

$$\int_Z \int_Y f(y, z) d\mu(y) d\nu(z) =$$

$$\int_Y \int_Z f(y, z) d\nu(z) d\mu(y).$$

A function $G : I \to \mathbb{R}$, I any subinterval of $\mathbb{R}$, is called *convex* if

$$\forall x_1, x_2 \in I, 0 \le t \le 1, \quad G(tx_1 + (1 - t)x_2) \le tG(x_1) + (1 - t)G(x_2).$$

The following formulation is from Hewitt and Stromberg [18, p. 202].

Theorem 10 (Jensen's Inequality). Let (Y, σ) be a probability measure space. Let G be a convex function from an interval I into $\mathbb{R}$ and let f be in $\mathcal{L}^1(Y, \sigma)$ with $f(Y) \subseteq I$ such that $G \circ f$ is also in $\mathcal{L}^1(Y, \sigma)$. Then $\int_Y f(y) d\sigma(y)$ is in I and

$$G \left(\int_Y f(y) d\sigma(y) \right) \le \int_Y (G \circ f)(y) d\sigma(y).$$

Acknowledgements. We thank Victor Bogdan for helpful comments on earlier versions.

References

1. Abramowitz, M., Stegun, I.A.: Handbook of Mathematical Functions. National Bureau of Standards, Washington, DC (1972)
2. Adams, R.A., Fournier, J.J.F.: Sobolev Spaces. Academic Press, Amsterdam (2003)
3. Alpay, D.: The Schur Algorithm, Reproducing Kernel Spaces, and System Theory. American Mathematical Society, Providence (2001)
4. Aronszajn, N.: Theory of reproducing kernels. Trans. of AMS 68, 337–404 (1950)
5. Barron, A.R.: Neural net approximation. In: Narendra, K. (ed.) Proc. 7th Yale Workshop on Adaptive and Learning Systems, pp. 69–72. Yale University Press (1992)
6. Barron, A.R.: Universal approximation bounds for superpositions of a sigmoidal function. IEEE Trans. on Information Theory 39, 930–945 (1993)
7. Baryshnikov, Y., Ghrist, R.: Target enumeration via euler characteristic integrals. SIAM J. Appl. Math. 70, 825–844 (2009)
8. Bensoussan, A.: Stochastic control by functional analysis methods. N. Holland, Amsterdam (1982)
9. Bochner, S.: Integration von funktionen, deren werte die elemente eines vectorraumes sind. Fundamenta Math. 20, 262–276 (1933)
10. Carlson, B.C.: Special Functions of Applied Mathematics. Academic Press, New York (1977)
11. Courant, R.: Differential and Integral Calculus, vol. II. Wiley, New York (1960)
12. Diestel, J., Uhl Jr., J.J.: Vector Measures. American Mathematical Society, Providence (1977)
13. Girosi, F.: Approximation error bounds that use VC-bounds. In: Fogelman-Soulied, F., Gallinari, P. (eds.) Proceedings of the International Conference on Artificial Neural Networks, Paris, pp. 295–302. EC & Cie (October 1995)
14. Girosi, F., Anzellotti, G.: Rates of convergence for Radial Basis Functions and neural networks. In: Mammone, R.J. (ed.) Artificial Neural Networks for Speech and Vision, pp. 97–113. Chapman and Hall (1993)
15. Giulini, S., Sanguineti, M.: Approximation schemes for functional optimization problems. J. of Optimization Theory and Applications 140, 33–54 (2009)
16. Gnecco, G., Sanguineti, M.: Estimates of variation with respect to a set and applications to optimization problems. J. of Optimization Theory and Applications 145, 53–75 (2010)
17. Gnecco, G., Sanguineti, M.: On a variational norm tailored to variable-basis approximation schemes. IEEE Trans. on Information Theory 57(1), 549–558 (2011)
18. Hewitt, E., Stromberg, K.: Real and abstract analysis. Springer, New York (1965)
19. Hille, E., Phillips, R.S.: Functional analysis and semi-groups, vol. XXVI. AMS, AMS Colloq. Publ., Providence (1957)
20. Ito, Y.: Representation of functions by superpositions of a step or sigmoid function and their applications to neural network theory. Neural Networks 4, 85–394 (1991)

21. Kainen, P.C., Kůrková, V., Sanguineti, M.: Complexity of gaussian radial basis networks approximating smooth functions. J. of Complexity 25, 63–74 (2009)
22. Kainen, P.C., Kůrková, V., Vogt, A.: A Sobolev-type upper bound for rates of approximation by linear combinations of Heaviside plane waves. J. of Approximation Theory 147, 1–10 (2007)
23. Kainen, P.C., Kůrková, V.: An integral upper bound for neural network approximation. Neural Computation 21(10), 2970–2989 (2009)
24. Kůrková, V.: Dimension-independent rates of approximation by neural networks. In: Warwick, K.M.,, K. (eds.) Computer-Intensive Methods in Control and Signal Procession: Curse of Dimensionality, pp. 261–270. Birkhauser, Boston (1997)
25. Kůrková, V.: Neural networks as universal approximators. In: Arbib, M. (ed.) The Handbook of Brain Theory and Neural Networks, pp. 1180–1183. MIT Press, Cambridge (2002)
26. Kůrková, V.: High-dimensional approximation and optimization by neural networks. In: Suykens, J., Horváth, G., Basu, S., Micchelli, C., Vandewalle, J. (eds.) Advances in Learning Theory: Methods, Models and Applications, ch. 4, pp. 69–88. IOS Press, Amsterdam (2003)
27. Kůrková, V.: Minimization of error functionals over perceptron networks. Neural Computation 20, 252–270 (2008)
28. Kůrková, V.: Model Complexity of Neural Networks and Integral Transforms. In: Alippi, C., Polycarpou, M., Panayiotou, C., Ellinas, G. (eds.) ICANN 2009, Part I. LNCS, vol. 5768, pp. 708–717. Springer, Heidelberg (2009)
29. Kůrková, V., Kainen, P.C., Kreinovich, V.: Estimates of the number of hidden units and variation with respect to half-spaces. Neural Networks 10, 1061–1068 (1997)
30. Kůrková, V., Sanguineti, M.: Comparison of worst case errors in linear and neural network approximation. IEEE Trans. on Information Theory 48, 264–275 (2002)
31. Kůrková, V., Savický, P., Hlaváčková, K.: Representations and rates of approximation of real-valued boolean functions by neural networks. Neural Networks 11, 651–659 (1998)
32. Le Page, R.D.: Note relating bochner integrals and reproducing kernels to series expansions on a gaussian banach space. Proc. American Math. Soc. 32, 285–288 (1972)
33. Light, W.A., Cheney, E.W.: Approximation theory in tensor product spaces. Lecture Notes in Math., vol. 1169. Springer, Berlin (1985)
34. Makovoz, Y.: Random approximants and neural networks. J. of Approximation Theory 85, 98–109 (1996)
35. Martínez, C., Sanz, M.: The Theory of Fractional Powers of Operators. Elsevier, Amsterdam (2001)
36. McShane, E.J., Botts, T.A.: Real Analysis. Van Nostrand, Princeton (1959)
37. Mhaskar, H.N., Micchelli, C.A.: Dimension-independent bounds on the degree of approximation by neural networks. IBM J. of Research and Development 38, 277–284 (1994)

38. Nelson, R.R.: Pointwise evaluation of bochner integrals in marcinkiewicz spaces. Nederl. Akad. Vetensch. Indag. Math. 44, 365–379 (1982)
39. Prévôt, C., Röckner, M.: A concise course on stochastic partial differential equations. Springer, Berlin (2007)
40. de Silva, V., Ghrist, R.: Homological sensor networks. Notices of the A.M.S. 54, 10–17 (2007)
41. Stein, E.M.: Singular Integrals and Differentiability Properties of Functions. Princeton University Press, Princeton (1970)
42. Steinwart, I., Christmann, A.: Support Vector Machines. Springer, New York (2008)
43. Yosida, K.: Functional Analysis. Academic Press, New York (1965)
44. Zaanen, A.C.: An Introduction to the Theory of Integration. N. Holland, Amsterdam (1961)

Chapter 7
Semi-supervised Learning

Mohamed Farouk Abdel Hady and Friedhelm Schwenker

Abstract. In traditional supervised learning, one uses "labeled" data to build a model. However, labeling the training data for real-world applications is difficult, expensive, or time consuming, as it requires the effort of human annotators sometimes with specific domain experience and training. There are implicit costs associated with obtaining these labels from domain experts, such as limited time and financial resources. This is especially true for applications that involve learning with large number of class labels and sometimes with similarities among them. Semi-supervised learning (SSL) addresses this inherent bottleneck by allowing the model to integrate part or all of the available unlabeled data in its supervised learning. The goal is to maximize the learning performance of the model through such newly-labeled examples while minimizing the work required of human annotators. Exploiting unlabeled data to help improve the learning performance has become a hot topic during the last decade and it is divided into four main directions: SSL with graphs, SSL with generative models, semi-supervised support vector machines and SSL by disagreement (SSL with committees). This survey article provides an overview to research advances in this branch of machine learning.

1 Introduction

Supervised learning algorithms require a large amount of labeled training data in order to construct models with high prediction performance, see Figure 1. In many practical data mining applications such as computer-aided medical diagnosis [38], remote sensing image classification [49], speech recognition [32], email classification [33], or automated classification of text documents [44, 45], there is often an extremely inexpensive large pool of unlabeled data available. However, the data

Mohamed Farouk Abdel Hady · Friedhelm Schwenker
Institute of Neural Information Processing
University of Ulm
D-89069 Ulm, Germany
e-mail: {mohamed.abdel-hady,friedhelm.schwenker}@uni-ulm.de

M. Bianchini et al. (Eds.): *Handbook on Neural Information Processing*, ISRL 49, pp. 215–239.
DOI: 10.1007/978-3-642-36657-4_7   © Springer-Verlag Berlin Heidelberg 2013

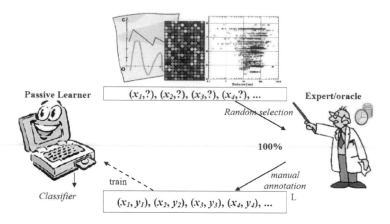

Fig. 1 Graphical illustration of traditional supervised learning

labeling process is often difficult, tedious, expensive, or time consuming, as it requires the efforts of human experts or special devices. Due to the difficulties in incorporating unlabeled data directly into traditional supervised learning algorithms such as support vector machines and RBF neural networks and the lack of a clear understanding of the value of unlabeled data in the learning process, the study of semi-supervised learning attracted attention only after the middle of 1990s. As the demand for automatic exploitation of unlabeled data increases, semi-supervised learning has become a hot topic.

In computer-aided diagnosis (CAD), mammography is a specific type of imaging that uses a low-dose x-ray system to examine breasts and it is used to aid in the early detection and diagnosis of breast diseases in women. There is a large number of mammographic images that can be obtained from routine examination but it is difficult to ask a physician or radiologist to search all images and highlight the abnormal areas of calcification that may indicate the presence of cancer. If we use supervised learning techniques to build a computer software to highlight these areas on the images, based on limited amount of diagnosed training images, it may be difficult to get an accurate diagnosis software. Then a question arises: can we exploit the abundant undiagnosed images [38] with the few diagnosed images to construct a more accurate software.

For remote sensing applications, the remote sensing sensors can produce data in large number of spectral bands. The objective of using such high resolution sensors is to discriminate among more ground cover classes and hence obtain a better understanding about the nature of the materials that cover the surface of the Earth. This large number of classes and large number of spectral bands require a large number of labeled training examples (pixels) from all the classes of interest. The class labels of such training examples are usually very expensive and time consuming to acquire [49]. The reason is that identifying the ground truth of the data must be gathered by visual inspection of the scene near the same time that the data is being taken, by using an experienced analyst based on their spectral responses, or by other means. In any case, usually only a limited number of training examples can be obtained. The purpose of SSL is to study how to reduce the small sample size problem by using unlabeled data that may be available in large number and with no extra cost. In the machine learning literature, there are mainly three paradigms for addressing the problem of combining labeled and unlabeled data to boost the performance: *semi-supervised learning*, *transductive learning* and *active learning*. *Semi-supervised learning* (*SSL*) refers to methods that attempt to take advantage of unlabeled data for supervised learning (*semi-supervised classification*), see Figure 2, or to incorporate prior information such as class labels, pairwise constraints or cluster membership in the context of unsupervised learning (*semi-supervised clustering*). *Transductive learning* [52] refers to methods which also attempt to exploit unlabeled examples but assuming that the unlabeled examples are exactly the test examples. That is, the test data set is known in advance and the goal of learning is to optimize the classification performance on the given test set. *Active learning* [48], sometimes called *selective sampling* refers to methods which assume that the given learning algorithm has control on the selection of the input training data such that it can select the most important examples from a pool of unlabeled examples, then an oracle such as a human expert is asked for labeling these examples, where the aim is to minimize data utilization. The most popular algorithms are *uncertainty sampling* (*US*) and *query by committee* (*QBC*) sampling. The former trains a single classifier and then query the unlabeled example on which the classifier is least confident [37]; the latter constructs multiple classifiers and then query the unlabeled example on which the classifiers disagree to the most [24]. In the remainder of this article is organized as follows: brief introduction to semi-supervised learning is given in the next section, and then the different semi-supervised learning techniques are introduced in the following sections. Section 8 presents the combination of semi-supervised learning with active learning. Finally, we conclude in section 9.

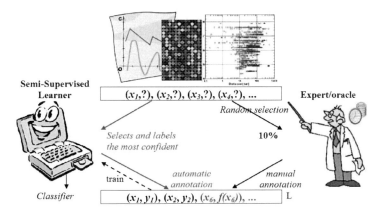

Fig. 2 Graphical illustration of semi-supervised learning

2 Semi-supervised Learning

Let $L = \{(x_i, y_i)|x_i \in \mathbb{R}^d, y_i \in \Omega, i = 1, \ldots, l\}$ be the set of labeled training examples where each example is described by a d-dimensional feature vector $x_i \in \mathbb{R}^d$, y_i denotes the class label of x_i and $\Omega = \{\omega_1, \ldots, \omega_K\}$ is the set of target classes (ground truth). Also let $U = \{x_j^*|j = 1, \ldots, u\}$ be the set of unlabeled data; usually $l << u$. The recent research on semi-supervised learning (*SSL*) concentrates into four directions: *semi-supervised classification* [13, 45, 33, 57, 61, 38], *semi-supervised regression* [60], *semi-supervised clustering* such as constrained and seeded k-means clustering [53, 51, 6] and *semi-supervised dimensionality reduction* [7, 62]. In this survey semi-supervised learning refers to semi-supervised classification, where one has additional unlabeled data and the goal is classification. Many semi-supervised classification algorithms have been developed. They can be divided into five categories according to [63]: (1) *Self-Training* [44], (2) semi-supervised learning with generative models [40, 45, 49], (3) S3VMs (Semi-Supervised Support Vector Machines) [31, 18, 27, 35], (4) semi-supervised learning with graphs [8, 56, 64], and (5) semi-supervised learning with committees (semi-supervised by disagreement) [13, 45, 33, 57, 61, 38, 59].

3 Self-Training

Self-Training [44] is an incremental algorithm that initially builds a single classifier using a small amount of labeled data. Then it iteratively predicts the labels of the unlabeled examples, rank the examples by confidence in their prediction and permenantly adds the *most confident examples* into the labeled training set. It retrains the underlying classifier with the augmented training set and the process is repeated for a given number of iterations or until some heuristic convergence criterion is satisfied. The classification accuracy can be improved over iterations only if the initial and subsequent classifiers correctly label most of the unlabeled examples. Unfortunately, adding mislabeling noise is not avoidable. In practical applications, more accurate confidence measures and predefined confidence thresholds are used in order to limit the number of mislabeled examples.

Self-Training is a wrapper algorithm that can be applied to many learning algorithms. It has been appeared in the literature with several names: self-learning [47, 42], self-corrective recognition [42], naive labelling [30], and decision-directed [55]. One drawback when *Self-Training* is applied on linear classifiers such as *support vector machines* is that the most confident examples often lie away from the target decision boundary (non informative examples). Therefore, in many cases this process does not create representative training sets as it selects non informative examples. Note that an example is called informative, if it lies close to the separating hyperplane and therefore it can influence its position. Another drawback is that *Self-Training* is sensitive to outliers.

4 SSL with Generative Models

In generative approaches, it is assumed that both labeled and unlabeled examples come from the same parametric model where the number of components, prior $p(y)$, and conditional $p(x|y)$ are all known and correct. Once the model parameters are learned, unlabeled examples are classified using the mixture components associated to each class. Methods in this category such as in [45, 43] usually treat the class labels of the unlabeled data $\{x_j\}_{j=1}^u$ as missing values and employ the *EM* (Expectation-Maximization) algorithm [21] to conduct maximum likelihood estimation (MLE) of the model parameters θ. It begins with an initial model trained on the labeled examples $\{(x_i, y_i)\}_{i=1}^l$. It then iteratively uses the current model to temporarily estimate the class probabilities of all the unlabeled examples and then maximizes the likelihood of the parameters (trains a new model) on all labeled examples (the original and the newly labeled) until it converges.

$$\log p(X_l, Y_l, X_u | \theta) = \sum_{i=1}^{l} \log p(y_i|\theta)p(x_i|y_i, \theta) + \lambda \sum_{j=1}^{u} \log(\sum_{y=1}^{2} p(y|\theta)p(x_j|y, \theta)) \quad (1)$$

The methods differ from each other by the generative models used to fit the data, for example, mixture of Gaussian distributions (GMM) is used for image classification [49], mixture of multinomial distributions (Naive Bayes) [45, 43] is used for text categorization and Hidden Markov Models (HMM) [30] is used for speech recognition. Although the generative models are simple and easy to implement and may be more accurate than discriminative models when the amount of labeled examples is very small, the methods in this category suffer from a serious problem. That is, when the model assumption is incorrect, fitting the model using a large amount of unlabeled data will result in performance degradation [19]. Thus, in order to reduce the danger [63], one needs to carefully construct the generative model, for instance to construct more than one Gaussian component per class. Also, one can down weight the unlabeled examples in the maximum likelihood estimation (set $\lambda < 1$).

5 Semi-supervised SVMs (*S3VMs*)

Considering that the training set is divided into two disjoint subsets L for labeled data and U for unlabeled data. The aim of *S3VM* learning algorithm is to exploit the abundant unlabeled data $U = \{x_j\}_{j=1}^u$ to adjust the decision boundary initially constructed from a small amount of labeled data $L = \{(x_i, y_i)\}_{i=1}^l, y_i = \pm 1$, such that it goes through the low density regions while keeping the labeled examples correctly classified [31, 18], see Figure 3. The following optimization problem is solved over both the decision boundary parameters (w, b) and a vector of binary labels assigned to unlabeled examples $\hat{y}_U = (\hat{y}_1, \ldots, \hat{y}_u)^T \in \{-1, 1\}^u$:

$$\min_{w, b, \hat{y}_U} \Psi(w, b, \hat{y}_U) = \frac{1}{2}\|w\|^2 + C \sum_{i=1}^{l} V(y_i, f(x_i)) + C^* \sum_{j=1}^{u} V(\hat{y}_j, f(x_j)), \quad (2)$$

where $f(x_i) = \langle w, \phi(x_i) \rangle - b$ is the decision function of *SVM*, ϕ is a nonlinear function that maps an input vector x_i into a high-dimensional dot-product feature space where it is possible to construct an optimal separating hyperplane with better generalization ability and V is a margin loss function. The Hinge loss is a popular loss function that is defined as,

$$V(y_i, f(x_i)) = \max(0, 1 - y_i f(x_i))^p \text{ and } V(\hat{y}_j, f(x_j)) = \max(0, 1 - \hat{y}_j f(x_j))^p \quad (3)$$

where $p=1$ or 2. It is an extension of the standard support vector machines. In the standard *SVM*, only the labeled data is used while in *S3VM* the unlabeled data is also used. The first two terms in the objective function in Eq. (2) define a standard *SVM*. The third term incorporates unlabeled data. The loss over labeled and unlabeled examples is weighted by two parameters, C and C^*, which reflect confidence in class labels and in the cluster assumption respectively. The minimization problem in Eq. (2) is solved under the following class balancing constraint

$$\frac{1}{u} \sum_{j=1}^{u} \max(0, \hat{y}_j) = r \quad (4)$$

This constraint helps to avoid unbalanced solutions by enforcing that a certain user-specified fraction, r, of the unlabeled data should be assigned to the positive class. The minimization techniques of Ψ can be divided into two broad strategies:

1. Combinatorial Optimization: For a given fixed $\hat{y}_U$, find the optimal solution for (w, b) which is the standard SVM training. The goal now is to minimize Γ over a set of binary variables.

$$\Gamma(\hat{y}_U) = \min_{w,b} \Psi(w, b, \hat{y}_U) \quad (5)$$

2. Continuous Optimization: For a given fixed (w, b), find the optimal $\hat{y}_U$. The unknown variables $\hat{y}_U$ are excluded from optimization, which leads to the following continuous objective function over (w, b):

$$\frac{1}{2} \|w\|^2 + C \sum_{i=1}^{l} \max(0, 1 - y_i f(x_i))^2 + C^* \sum_{j=1}^{u} \max(0, 1 - |f(x_j)|)^2 \quad (6)$$

That can be solved by continuous optimization techniques. The balancing constraint becomes as follows

$$\frac{1}{u} \sum_{j=1}^{u} f(x_j) = r \quad (7)$$

After optimization, $\hat{y}_U$ is simply found by applying the decision function on the unlabeled example, $\hat{y}_j = arg\min_{y \in \{-1,1\}} V(y, f(x_j)) = sign(f(x_j))$.

Since its first implementation by Joachims [31], the non-convexity of the problem associated with *S3VM* motivates the development of a number of optimization techniques, for instance, local combinatorial search [31], gradient descent [18],

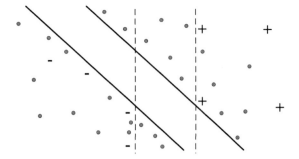

Fig. 3 Graphical illustration of S3VMs: The unlabeled examples help to put the decision boundary in low density regions. Using labeled data only, the maximum margin separating hyperplane is plotted with the versicle dashed lines. Using both labeled and unlabeled data (dots), the maximum margin separating hyperplane is plotted with the oblique solid lines.

continuation techniques [16], convex-concave procedures [25], semi-definite programming [10], deterministic annealing [50], genetic algorithm optimization [4] and branch-and-bound algorithms [17]. In [31], an initial SVM classifier is firstly constructed using the available labeled examples and then the labels of the unlabeled examples y_U^* are iteratively predicted. Then it maximizes the margin over both labeled and the (newly labeled) unlabeled examples $\{(x_j^*, y_j^*)\}_{j=1}^{u}$. The optimal decision boundary is the one that has the minimum training error on both labeled and unlabeled data. *S3VM*, sometimes called *Transductive SVM*, assumes that unlabeled data from different classes are separated with large margin. In addition, it assumes there is a low density region through which the separating hyperplane passes. Thus, it does not work for domains in which this assumption is not fulfilled.

6 Semi-supervised Learning with Graphs

Blum and Chawla [11] proposed the first graph-based semi-supervised learning method. They constructed a graph whose nodes represent both labeled and unlabeled training examples and the edges between nodes are weighted according to the similarity between the corresponding examples. Based on the graph, the aim is to find the minimum cut of the graph such that nodes in each connected component have the same label. Later, Blum et al. [12] added random noise to the edge weights and the labels of the unlabeled examples are predicted using majority voting. The procedure is similar to bagging and produces a soft minimum cut. Note that in both [11] and [12] a discrete predictive function is used that assigns one of the possible labels to each unlabeled example. Zhu et al. [64] introduced a continuous prediction function. They modeled the distribution of the prediction function over the graph with Gaussian random fields and analytically proved that the prediction function with the lowest energy should have the harmonic property. They designed a label propagation strategy over the graph using such a harmonic property where the labels propagate from the labeled nodes to the unlabeled ones, see Figure 4. It is worth noting that all graph-based methods assume that examples connected by strong edges tend to have the same class label and vice versa [63]. It is noteworthy that most of the graph-based semi-supervised learning usually focus on how to conduct semi-supervised learning over a given graph. A key that will seriously influence the learning performance is how to construct a graph which reflects the essential similarities among examples.

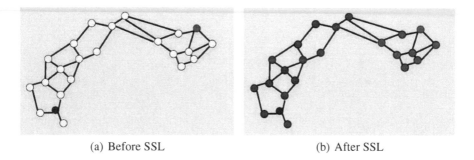

(a) Before SSL   (b) After SSL

Fig. 4 Graphical illustration of label propagation

7 Semi-supervised Learning with Committees (*SSLC*)

The main factor for the success of any committee-based semi-supervised learning, sometimes called semi-supervised learning by disagreement [59], is to construct an ensemble of diverse and accurate classifiers, let them collaborate to exploit unlabeled examples, and maintain a large disagreement (diversity) between these classifiers. In this section, existing committee-based semi-supervised learning techniques are divided into three categories, that is, learning with multiple views and learning with single view multiple classifiers.

7.1 SSLC with Multiple Views

Multi-view learning is based on the assumption that the instance input space $X = X_1 \times X_2$, where $X_1 \subset \mathbb{R}^{D_1}$ and $X_2 \subset \mathbb{R}^{D_2}$ represent two different descriptions of an instance, called views. These views are obtained through different physical sources/sensors or are derived by different feature extraction procedures and are giving different types of discriminating information about the instance. For instance, in visual objective recognition tasks, an image can be described by color, shape or texture. In emotion recognition tasks, an emotion can be recognized from either speech and facial expressions.

Multi-view learning was first introduced for semi-supervised learning by Blum and Mitchell in the context of *Co-Training* [13]. They state two strong requirements for successful *Co-Training*: the two sets of features should be conditionally independent given the class and either of them sufficient to learn the classification task.. The pseudo-code is shown in Algorithm 1 (see Figure 5). At the initial iteration, two classifiers are trained using a small amount of labeled training data. Then at each further iteration, each classifier predicts the class label of the unlabeled examples, estimates the confidence in its prediction, ranks the examples by confidence, adds the examples about which it is *most confident* into the labeled training set. The aim is that the *most confident* examples with respect to one classifier can be *informative* with respect to the other. An example is informative with respect to a classifier if it carries a new discriminating information. That is, it lies close to the decision boundary and thus adding it to the training set can improve the classification performance of this classifier. Nigam and Ghani [44] showed that *Co-Training* is sensitive to the view independence requirement.

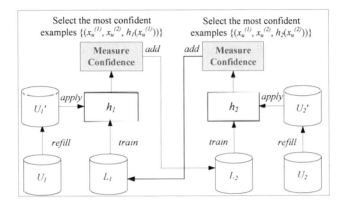

Fig. 5 Graphical illustration of *Co-Training*

Nigam and Ghani [44] proposed another multi-view semi-supervised algorithm, called *Co-EM*. It uses the model learned in one view to probabilistically label the unlabeled examples in the other model. Intuitively, *Co-EM* runs *EM* (Section 4) in each view and before each new *EM* iteration, inter-changes the probabilistic labels predicted in each view. *Co-EM* is considered as a probabilistic variant of *Co-Training*. Both algorithms are based on the same idea: they use the knowledge acquired in one view, in the form of soft class labels for the unlabeled examples, to train the other view. The major difference between the two algorithms is that *Co-EM* does not commit to the labels predicted in the previous iteration because it uses probabilistic labels that may change from one iteration to the other. On the other hand, Co-Training commits to the most confident predictions that are once added into the training set are never revisited. Thus, it may add to the training set a large number of mislabeled examples.

The standard *Co-Training* was applied in domains with truly independent feature splits satisfying its conditions. In [33], Kiritchenko et al. applied *Co-Training* for email classification where the bags of words that represent email messages were split into two sets: the words from headers (V_1) and the words from bodies (V_2).

Algorithm 1. Pseudo code of *Standard Co-Training*

Require: set of labeled training examples (L), set of unlabeled training examples (U), maximum number of iterations (T),base learning algorithm (*BaseLearn*), two feature sets (views) representing an example (V_1, V_2), sample size (n), number of unlabeled examples in the pool (u) and number of classes (C)

Training Phase

1: Get the class prior probabilities, $\{Pr_c\}_{c=1}^C$

2: Set the class growth rate, $n_c = n \times Pr_c$ where $c = 1, \ldots, C$

3: Train initial classifiers $h_1^{(0)}$ and $h_2^{(0)}$ on the initial L

$\quad h_1^{(0)} = BaseLearn(V_1(L)) \quad$ and $\quad h_2^{(0)} = BaseLearn(V_2(L))$

4: **for** $t \in \{1, \ldots, T\}$ **do**

5:     **if** U is empty **then**

6:         $T \leftarrow t\text{-}1$ and abort loop

7:     **end if**

8:     **for** $v \in \{1, 2\}$ **do**

9:         Apply $h_v^{(t-1)}$ on U.

10:         Select a subset S_v as follows: for each class ω_c, select the n_c most confident examples assigned to class ω_c

11:         Move S_v from U to L

12:     **end for**

13:     Re-train classifiers $h_1^{(t)}$ and $h_2^{(t)}$ on the new L

$\quad h_1^{(t)} = BaseLearn(V_1(L)) \quad$ and $\quad h_2^{(t)} = BaseLearn(V_2(L))$

14: **end for**

Prediction Phase

15: **return** combination of the predictions of $h_1^{(T)}$ and $h_2^{(T)}$

Levin et al. [36] have used *Co-Training* to improve visual detectors for cars in traffic surveillance video where one classifier detects cars in the original gray level images (V_1). The second one uses images where the background has been removed (V_2).

Abdel Hady et al. [3] have combined *Co-Training* with tree-structured approach for multi-class decomposition through two different architectures. In the first architecture, *cotrain-of-trees*, a tree-structured ensemble of binary RBF networks is trained on each given view. Then, using Co-Training the most confident unlabeled examples labeled by each tree ensemble classifier are added to the training set of the other tree classifier. A combination method based on Dempster-Schafer evidence theory provides class probability estimates that were used to measure confidence on prediction. In the second architecture, *tree-of-cotrains*, first the given K-class problem is decomposed into K-1 simpler binary problems using the tree-structured approach. Then using Co-Training a binary RBF network is trained on each given view to solve each binary problem. In order to combine the intermediate results of the internal nodes within each tree, the above mentioned evidence-theoretic combination method is used. Then cotrain-of-trees and tree-of-cotrains were evaluated on three real-world 2D and 3D visual object recognition tasks where one classifier is based on color histograms (V_1) while the second uses orientation histograms (V_2).

Although there are some cases in which there are two or more independent and redundant views, there exist many real-world applications in which multiple views are not available or it is computationally inefficient to extract more than one feature set for each example.

Co-Training was applied in domains without natural feature splits through splitting the available feature set into two views V_1 and V_2. Nigam and Ghani [44] investigated the influence of the views independence. They found that *Co-Training* works better on truly independent views than on random views. Also, *Co-Training* was found to outperform EM when the views are truly independent. It was also shown that if there is sufficient redundancy in data, the performance of *Co-Training* with random splits is comparable to *Co-Training* with a natural split. Of course there is no guarantee that random splitting will produce independent views.

Feger and Koprinska [22] introduced a method, called *maxInd*, for splitting the feature set into two views. The aim is to minimize the dependence between the two feature subsets (*inter-dependence*), measured by conditional mutual information *CondMI*. The result is represented as an undirected graph, with features as nodes and the *CondMI* between each pair of features as weight on the edge between them. In the second step the graph is cut into two disjoint parts of the same size. This split is performed in such a way that minimizes the sum of the cut edges in order to minimize the dependence between the two parts of the graph. They had found that *maxInd* does not outperform the random splits. A possible explanation from their perspective is that *Co-Training* is sensitive to the dependence of the features within each view (*intra-dependence*). The random split leads to *intra-dependence* lower than that of *maxInd* and the truly independent split. Their study states that there is a trade-off between the *intra-dependence* of each view, and the *inter-dependence* between the views. That is minimizing the *inter-dependence* leads to maximizing the *intra-dependence* of each view. In addition, the measurement of *CondMI* is not accurate enough because it is based on only a small number of labeled examples.

Salaheldin and El Gayar [46] introduced three new criteria for splitting features in *Co-Training* and compare them to existing artificial splits and natural split. The first feature split criterion is based on maximizing the confidence of the views. The second criterion maximizes both confidence and independence of the views. The independence of a view is measured by conditional mutual information as in [22]. For each view, a classifier is trained using the labeled data; it is then used to predict the class of the unlabeled data. The entropy of the classifier output for each input example is calculated and the average of entropies indicates the confidence of the view. They showed that splitting the features with a mixed criterion is better than using each criterion alone. Finally, they proposed a third criterion based on maximizing the views diversity. A genetic algorithm is used to optimize the fitness functions based on the three proposed criteria. The experimental results on two data sets show that the proposed splits are promising alternatives to random splitting.

7.2 SSLC with Single View

7.2.1 For Classification

In a number of recent studies [26, 57, 61, 38], the applicability of *Co-Training* using a single view without feature splitting has been investigated. Goldman and Zhou [26] first presented a single-view *SSL* method, called *Statistical Co-learning*. It used two different supervised learning algorithms with the assumption that each of them produce a hypothesis that partition the input space into a set of equivalence classes. For example, a decision tree partitions the input space with one equivalence class per leaf. They used 10-fold cross validation:(1) to select the most confident examples to label at each iteration and (2) to combine the two hypotheses producing the final decision. Its drawbacks are: first the assumptions concering the used algorithms limits its applicability. Second the amount of available labeled data was insufficient for applying cross validation which is time-consuming. Zhou and Goldman [57] then presented another single view method, called *Democratic Co-learning* which is applied to three or more supervised learning algorithms and reduce the need for statistical tests. Therefore, it resolves the drawbacks of *Statistical Co-learning* but it still uses the time-consuming cross-validation technique to measure confidence intervals. These confidence intervals are used to select the most confident unlabeled examples and to combine the hypotheses decisions.

Zhou and Li [61] present a new *Co-Training* style *SSL* method, called *Tri-Training*, where three classifiers are initially trained on bootstrap subsamples generated from the original labeled training set. These classifiers are then refined during the *Tri-Training* process, and the final hypothesis is produced via majority voting. The construction of the initial classifiers looks like training an ensemble from the labeled data with *Bagging* [14]. At each *Tri-Training* iteration, an unlabeled example is added to the training set of a classifier if the other two classifiers agree on their prediction under certain conditions. *Tri-Training* is more applicable than previous *Co-Training*-Style algorithms because it neither requires multiple views as in [13, 44] nor does it depend on different supervised learning algorithms as in [26, 57]. There are two limitations for *Tri-Training*: the ensemble size is limited to three classifiers and it hurts the diversity as *Bagging* is used as ensemble learner. Therefore, the classifiers become identical through the iterations because their training sets become similar. The reason is that the unlabeled examples added to one classifier are not removed from the unlabeled data set therefore the same examples can be selected and added to another classifier at the same iteration or in further iterations.

Li and Zhou [38] proposed an extension to *Tri-Training*, called *Co-Forest*. The aim is to maintain the diversity during the SSL process through using *Random Forest* instead of *Bagging*. That is an initial ensemble of random trees is trained on bootstrap subsamples generated from the given labeled data set L. To select new training examples from a given unlabeled data set U for each ensemble member h_i ($i = 1, \ldots, N$), a new ensemble H_i, called the concomitant ensemble of h_i, is defined that contains all the classifiers except h_i. At each iteration t and for each ensemble member h_i, first the error rate of H_i, $\hat{\varepsilon}_{i,t}$, is estimated. If $\hat{\varepsilon}_{i,t}$ is less than $\hat{\varepsilon}_{i,t-1}$

(1th condition), H_i predicts the class label of the unlabeled examples in $U'_{i,t}$ (random subsample of U of size $\frac{\hat{\varepsilon}_{i,t-1}W_{i,t-1}}{\hat{\varepsilon}_{i,t}}$). A set $L'_{i,t}$ is defined that contains the unlabeled examples in $U'_{i,t}$ where the confidence of H_i about their prediction exceeds a prede-fined threshold (θ) and $W_{i,t}$ is the sum of the confidences of the examples in $L'_{i,t}$. If $W_{i,t}$ is greater than $W_{i,t-1}$ (2nd condition) and $\hat{\varepsilon}_{i,t}W_{i,t}$ is less than $\hat{\varepsilon}_{i,t-1}W_{i,t-1}$ (3rd condition), the ith random tree will be re-trained using the original labeled data set L and $L'_{i,t}$. Note that the bootstrap sample used to train the ith random tree at iteration 0 is discarded and $L'_{i,t}$ is not added *permenantly* into L. The algorithm will stop if there is no classifier h_i satisfying the three conditions.

d'Alché et al. [20] generalized *MarginBoost* to semi-supervised classification. *MarginBoost* is a variant of AdaBoost [23] based on the minimization of an explicit cost function. Such function is defined for any scalar decreasing function of the mar-gin. As the usual definition of margin cannot be used for unlabeled data, the authors extend the margin notion to unlabeled data. In practice, the margin is estimated us-ing the *MarginBoost* classification output. Then, they reformulate the cost function of *MarginBoost* to include both the labeled and unlabeled data. A generative model is used as a base classifier and the unlabeled data is used by EM algorithms. The results have shown that *SSMBoost* outperforms the classical AdaBoost when a few amount of labeled data is available (only 5% of the training data is labeled).

Bennet et al. [9] proposed another committee-based SSL method, called *ASSEM-BLE*, which iteratively constructs ensemble classifiers using both labeled and unla-beled data. The aim of *ASSEMBLE* is to overcome some limitations of *SSMBoost*. For example, while *SSMBoost* requires the base classifier to be a generative mix-ture model in order to apply EM for semi-supervision, *ASSEMBLE* is more general that can be used with any cost-sensitive base learning algorithm. At each iteration of *ASSEMBLE*, the unlabeled examples are assigning pseudo-classes using the cur-rent ensemble before constructing the next base classifier using both the labeled and newly-labeled examples. The experiments show that *ASSEMBLE* works well and it won the NIPS 2001 unlabeled data competition using decision trees as base classifiers.

Abdel Hady and Schwenker [1] introduced a new *committee-based single-view Co-Training* style algorithm, *CoBC*, for application domains in which the available data is not described by multiple redundant and independent views. *CoBC* works as follows: firstly the class prior probabilities are determined then an initial committee of N diverse accurate classifiers $H^{(0)}$ is trained on L using the given ensemble learn-ing algorithm *EnsembleLearn* and base learning algorithm *BaseLearn*. Then the fol-lowing steps are repeated until the maximum number of iterations T is reached or U becomes empty. For each iteration t and for each classifier i, a set $U'_{i,t}$ of u examples drawn randomly from U without replacement. It is computationally more efficient to use $U'_{i,t}$ instead of using the whole set U. The method *SelectCompetentExamples* (see Algorithm 3) is applied to estimate the competence of each unlabeled example in $U'_{i,t}$ given the companion committee $H_i^{(t-1)}$. Note that $H_i^{(t-1)}$ is the ensemble of all base classifiers trained in the previous iteration except $h_i^{(t-1)}$. A set $\pi_{i,t}$ is created that contains the n_c most competent examples assigned to each class ω_c.

Algorithm 2. Pseudo code of *CoBC* for classification

Require: set of labeled training examples (L), set of unlabeled training examples (U), maximum number of iterations (T), ensemble learning algorithm (*EnsembleLearn*), base
learning algorithm (*BaseLearn*), ensemble size (N), number of unlabeled examples in
the pool (u), number of nearest neighbors (k), sample size (n), number of classes (C) and
an initial committee ($H^{(0)}$)

   Training Phase

1: Get the class prior probabilities, $\{Pr_c\}_{c=1}^C$
2: Set the class growth rate, $n_c = n \times Pr_c$ where $c = 1,\dots,C$
3: **if** $H^{(0)}$ is not given **then**
4:    Construct an initial committee of N classifiers,
      $H^{(0)} = EnsembleLearn(L, BaseLearn, N)$
5: **end if**
6: **for** $t \in \{1,\dots,T\}$ **do**
7:    $L'_t \leftarrow \emptyset$
8:    **if** U is empty **then** $T = t$-1 and abort loop   **end if**
      {Get most confident examples ($\pi_{i,t}$) using companion committee $H_i^{(t-1)}$ }
9:    **for** $i \in \{1,\dots,N\}$ **do**
10:      $U'_{i,t} \leftarrow RandomSubsample(U, u)$
11:      $\pi_{i,t} \leftarrow SelectCompetentExamples(i, U'_{i,t}, H_i^{(t-1)}, k, \{n_c\}_{c=1}^C, C)$
12:      $L'_t \leftarrow L'_t \cup \pi_{i,t}$, $U'_{i,t} \leftarrow U'_{i,t} \setminus \pi_{i,t}$ and $U \leftarrow U \cup U'_{i,t}$
13:   **end for**
14:   **if** L'_t is empty **then** $T = t$-1 and abort loop   **end if**
      {Re-train the N classifiers using their augmented training sets }
15:   **for** $i \in \{1,\dots,N\}$ **do**
16:      $L_i = L_i \cup L'_t$
17:      $h_i^{(t)} = BaseLearn(L_i)$ (for incremental learning, $h_i^{(t)} = BaseLearn(h_i^{(t-1)}, L'_t)$)
18:   **end for**
19: **end for**
   Prediction Phase
20: **return** $H^{(T)}(x) = \frac{1}{N}\sum_{i=1}^N h_i^{(T)}(x)$ for a given example x

Then $\pi_{i,t}$ is removed from $U'_{i,t}$ and inserted into the set L'_t that contains all the examples labeled at iteration t. The remaining examples in $U'_{i,t}$ are returned to U.
There are two options: (1) if the underlying ensemble learner depends on training
set perturbation to promote diversity, then insert $\pi_{i,t}$ only into L_i. Otherwise, $h_i^{(t)}$ and
$h_j^{(t)}$ ($i \neq j$) will be identical because they are refined with the same newly labeled
examples. This will degrade the ensemble diversity and therefore degrades the relative improvement expected due to exploiting the unlabeled data. One can observe
that if the ensemble members are identical, *CoBC* will degenerate to *Self-Training*.
(2) If ensemble learner employs another source of diversity, then it is not a problem to insert $\pi_{i,t}$ into the training sets of all classifiers as shown in step 16. Then,
CoBC does not recall *EnsembleLearn* but only the N committee members are re-
trained using their updated training sets L_i. It is worth noting that: (1) *CoBC* can
improve the recognition rate only if the most confident examples with respect to the

companion committee H_i are informative examples with respect to h_i. (2) Although *CoBC* selects the most confident examples, adding mislabeled examples to the training set (*noise*) is unavoidable but the negative impact of this *noise* could be compensated by augmenting the training set with sufficient amount of newly labeled examples.

A new confidence measure is proposed (Algorithm 3) in order to compensate the inaccurate probability-based ranking provided by traditional decision trees. That is, all unlabeled examples x_u which lie into a particular leaf node (region), will have the same class probability estimates (*CPE*)s because the *CPE* depends on class frequencies and not the distance between x_u and the decision boundaries. The new measure depends on estimating the companion committee accuracy on labeling the neighborhood of an unlabeled example x_u. This local accuracy represents the probability that the companion committee correctly predicts the class label of x_u. The local competence of an unlabeled example x_u given a companion committee $H_i^{(t-1)}$ can be defined as follows:

$$Comp(x_u, H_i^{(t-1)}) = \sum_{\substack{(x_n, y_n) \in N_k(x_u) \\ y_n = \hat{y}_u}} W_n . H_i^{(t-1)}(x_n, \hat{y}_u) \tag{8}$$

where

$$W_n = \frac{1}{||x_n - x_u||_2 + \varepsilon}, \tag{9}$$

$$\hat{y}_u = arg \max_{1 \leq c \leq C} H_i^{(t-1)}(x_u, \omega_c), \tag{10}$$

$H_i^{(t-1)}(x_n, \hat{y}_u)$ is the probability given by $H_i^{(t-1)}$ that neighbor x_n belongs to the same class assigned to x_u ($\hat{y}_u$), W_n is the reciprocal of the Euclidean distance between x_u and its neighbor x_n and ε is a constant added to avoid zero denominator. The neighborhood could also be determined using a separate validation set (a set of labeled examples that is not used for training the classifiers), but it may be impractical to spend a part from the small-sized labeled data for validation. To avoid the inaccurate estimation of local accuracy that may result due to overfitting, the newly-labeled training examples $\pi_{i,t}$ will not be involved in the estimation. That is, only the initially (manually) labeled training examples are taken into account. Then, the set $N_k(x_u)$ is defined as the set of k nearest labeled examples to x_u.

The local competence assumes that the actual data distribution satisfies the well-known cluster assumption: examples with similar inputs should belong to the same class. Therefore, the local competence of x_u is zero if there is not any neighbor belongs to the predicted class label $\hat{y}_u$ which contradicts the cluster assumption. Therefore, one can observe that $\hat{y}_u$ is an incorrect class label of x_u ($\hat{y}_u \neq y_u$). In addition, the local competence increases as the number of neighbors that belong to $\hat{y}_u$ increases and as the distances between these neighbors and x_u decreases.

Experiments were conducted on ten image recognition tasks in which the random subspace method [28] is used to construct ensembles of diverse 1-nearest neighbor classifiers and C4.5 decision trees. The results verify the effectiveness of *CoBC* to exploit the unlabeled data given a small amount of labeled examples.

Algorithm 3. Pseudo Code of the *SelectCompetentExamples* method

Require: pool of unlabeled examples $(U'_{i,t})$, the companion committee of classifier $h_i^{(t-1)}$
   $(H_i^{(t-1)})$, number of nearest neighbors k, growth rate $(\{n_c\}_{c=1}^C)$ and number of classes
   (C)
1: $\pi_{i,t} \leftarrow \emptyset$
2: **for** each class $\omega_c \in \{\omega_1, \ldots, \omega_C\}$ **do**
3:    $count_c \leftarrow 0$
4: **end for**
5: **for** each $x_u \in U'_{i,t}$ **do**
6:    $H_i^{(t-1)}(x_u) = \frac{1}{N-1} \sum_{j=1,\ldots,N, j \neq i} h_j^{(t-1)}(x_u)$
7:    Apply the companion committee $H_i^{(t-1)}$ to x_u,
       $\hat{y}_u \leftarrow \arg\max_{1 \leq c \leq C} H_i^{(t-1)}(x_u, \omega_c)$
8:    Find the k nearest neighbors of x_u,
       $N_k(x_u) = \{(x_n, y_n) | (x_n, y_n) \in Neighbors(x_u, k, L)\}$
9:    Calculate $Comp(x_u, H_i^{(t-1)})$ as defined in Eq. (8) and Eq. (9)
10: **end for**
11: Rank the examples in $U'_{i,t}$ based on competence (in descending order)
     {Select the n_c examples with the maximum competence for class ω_c}
12: **for** each $x_u \in U'_{i,t}$ **do**
13:    **if** $Comp(x_u, H_i^{(t-1)}) > 0$ and $count_{\hat{y}_u} < n_{\hat{y}_u}$ **then**
14:       $\pi_{i,t} = \pi_{i,t} \cup \{(x_u, \hat{y}_u)\}$ and $count_{\hat{y}_u} = count_{\hat{y}_u} + 1$
15:    **end if**
16: **end for**
17: **return** $\pi_{i,t}$

7.2.2   For Regression

Previous studies on semi-supervised learning mainly focus on classification tasks. Although regression is almost as important as classification, semi-supervised regression has rarely been studied. One reason is that for real-valued outputs the cluster assumption is not applicable. Although methods based on manifold assumption can be extended to regression, as pointed out by [63], these methods are essentially transductive instead of really semi-supervised since they assume that the unlabeled examples are exactly test examples.

Abdel Hady et al. [2] introduced an extension of *CoBC* for regression, *CoBCReg*. There are two potential problems that can prevent any *Co-Training* style algorithm from exploiting the unlabeled data to improve the performance and these problems are the motivation for *CoBCReg*. Firstly the outputs of unlabeled examples may be incorrectly estimated by a regressor. This leads to adding noisy examples to the training set of the other regressor and therefore *SSL* will degrade the performance. Secondly there is no guarantee that the newly-predicted examples selected by a regressor as *most confident examples* will be *informative examples* for the other regressor. In order to mitigate the former problem, a committee of predictors is used in *CoBCReg* to predict the unlabeled examples instead of a single predictor. For the latter problem, each regressor selects the most informative examples for itself.

Let L and U represent the labeled and unlabeled training set respectively, which are drawn randomly from the same distribution where for each instance x_μ in L is associated with the target real-valued output while the real-valued outputs of examples in U are unknown. The pseudo-code of *CoBCReg* is shown in Algorithm 4. *CoBCReg* works as follow: initially an ensemble consists of N regressors, which is denoted by H, is constructed from L using *Bagging*. Then the following steps will be repeated until the maximum number of iterations T is reached or U becomes empty. For each iteration t and for each ensemble member h_i, a set U' of u examples is drawn randomly from U without replacement. It is computationally more efficient to use a pool U' instead of using the whole set U. The *SelectRelevantExamples* method (Algorithm 5) is applied to estimate the relevance of each unlabeled example in U' given the *companion committee* H_i. H_i is the ensemble consisting of all member regressors except h_i. A set π_j is created that contains the gr most relevant examples. Then π_j is removed from U' and inserted into the training set of h_i (L_i) such that h_i is refined using the augmented training set L_i. In the prediction phase, the regression estimate for a given example is the weighted average of the outputs of the N regressors created at the final *CoBCReg* iteration. The combination of an ensemble of regressors is only effective if they are diverse. Clearly, if they are identical, then for each regressor, the outputs estimated by the other regressors will be the same as these estimated by the regressor for itself. That is, there is no more knowledge to be transfered among regressors. In *CoBCReg*, there are three sources for diversity creation, the *RBF* network regressors are trained using: (1) different bootstrap samples, (2) different random initialization of *RBF* centers and (3) different distance measures. The Minkowski distance between two D-dimensional feature vectors x_1 and x_2, as defined in Eq. (11), is used with different distance order p to train different *RBF* network regressors. In general, the smaller the order, the more robust the resulting distance metric to data variations. Another benefit of this setting, is that, since it is difficult to find in advance the best p value for a given task, then regressors based on different p values might show complementary behavior.

$$\|x_1 - x_2\|_p = \left(\sum_{i=1}^{D} |x_{1i} - x_{2i}|^p \right)^{1/p} \tag{11}$$

Unlike *Co-Forest* [38], *CoBCReg* does not hurt the diversity among regressors because the examples selected by a regressor are removed from U. Thus, they can not be selected further by other regressors which keeps the training sets of regressors not similar. Even if the training sets become similar, the regressors could still be diverse because they are instantiated with different distance measures, for some data sets this acts like using different feature spaces.

The main challenge for *CoBCReg* is the mechanism for estimating the confidence because the number of possible predictions in regression is unknown. For regression, in [34], variance is used as an effective selection criterion for active learning because a high variance between the estimates of the ensemble members leads to a

---

Algorithm 4. Pseudo Code of CoBC for Regression

Require: set of l labeled training examples (L), set of u unlabeled examples (U), maximum
    number of Co-Training iterations (T), *ensemble size* (N), pool size (u), growth rate (gr),
    number of RBF hidden nodes (k), RBF width parameter (α), distance order of the ith
    regressor (p_i)

    Training Phase
1: **for** $i = 1$ to N **do**
2:     $\{L_i, V_i\} \leftarrow BootstrapSample(L)$ $\{L_i$ is bag and V_i is out-of-bag$\}$
3:     $h_i = RBFNN(L_i, k, \alpha, p_i)$
4: **end for**
5: **for** $t \in \{1 \dots T\}$ **do**
6:     **if** U is empty **then** $T = t\text{-}1$ and abort loop   **end if**
7:     **for** $i \in \{1 \dots N\}$ **do**
8:        Create a pool U' of u examples by random sampling from U
9:        $\pi_i = SelectRelevantExamples(i, U', V_i, gr)$
10:      $U' = U' \setminus \pi_i$ and $U = U \cup U'$
11:     **end for**
12:     **for** $i \in \{1 \dots N\}$ **do**
13:        **if** π_i is not empty **then**
14:           $L_i = L_i \cup \pi_i$
15:           $h_i = RBFNN(L_i, k, \alpha, p_i)$
16:        **end if**
17:     **end for**
18: **end for**
    Prediction Phase
19: **return** $H(x) = \sum_{i=1}^{N} w_i h_i(x)$ for a given sample x

---

high average error. Unfortunately, a low variance does not necessarily imply a low average error. That is, it can not be used as a selection criterion for *SSL* because agreement of committee members does not imply that the estimated output is close to the target output. In fact, we will not measure the *labeling confidence* but we will provide another confidence measure called *selection confidence* (See Algorithm 5).

The most relevantly selected example should be the one which minimizes the regressor error on the validation set. Thus, for each regressor h_j, create a pool U' of u unlabeled examples. Then, the root mean squared error (*RMSE*) of h_j is evaluated first (ε_j). Then for each example x_u in U', h_j is refined with $(x_u, H_j(x_u))$ creating new regressor h'_j. So the *RMSE* of h'_j can be evaluated (ε'_j), where $H_j(x_u)$ is the real-valued output estimated by the *companion committee* of h_j (H_j denotes all other ensemble members in H except h_j). Finally, the unlabeled example $\tilde{x}_j$ which maximizes the relative improvement of the *RMSE* (Δ_{x_u}) is selected as the most relevant example labeled by *companion committee* H_j.

It is worth mentioning that the *RMSEs* ε_j and ε'_j should be estimated accurately. If the training data of h_j is used, this will under-estimate the *RMSE*. Fortunately, since the bootstrap sampling [14] is used to construct the committee, the *out-of-bootstrap* examples are considered for a more accurate estimate of ε'_j.

Algorithm 5. Pseudo Code of of the *SelectRelevantExamples* method

---

Require: the index of the regressor excluded from the committee (j), pool of u unlabeled
    examples (U'), validation set (V_j), growth rate (gr)

1: Calculate validation error of h_j using V_j, ε_j
2: **for** each $x_u \in U'$ **do**
3:    $H_j(x_u) = \frac{1}{N-1} \sum_{i=1, i \neq j}^{N} h_i(x_u)$
4:    $h'_j = RBFNN(L_j \cup \{(x_u, H_j(x_u))\}, k, \alpha, p_j)$
5:    Calculate validation error ε'_j of h'_j using V_j, then $\Delta_{x_u} = (\varepsilon_j - \varepsilon'_j)/\varepsilon_j$
6: **end for**
7: $\pi_j \leftarrow \emptyset$
8: **for** gr times **do**
9:    **if** there exists $x_u \in U' \setminus \pi_j$ with $\Delta_{x_u} > 0$ **then**
10:      $\tilde{x}_j = arg\max_{x_u \in U' \setminus \pi_j} \Delta_{x_u}$
11:      $\pi_j = \pi_j \cup \{(\tilde{x}_j, H_j(\tilde{x}_j))\}$
12:    **end if**
13: **end for**
14: **return** π_j

---

8 Combination with Active Learning

Both semi-supervised learning and active learning tackle the same problem but from
different directions. That is, they aim to improve the generalization error and at the
same time minimize the cost of data annotation through exploiting the abundant
unlabeled data.

8.1 SSL with Graphs

Zhu et al. [65] combine *semi-supervised learning* and *active learning* under a Gaussian random field model. Labeled and unlabeled data are represented as nodes in a
weighted graph, with edge weights encoding the similarity between examples. Then
the semi-supervised learning problem is formulated, in another work by the same
authors [64], in terms of a Gaussian random field on this graph, the mean of which
is characterized in terms of harmonic functions. *Active learning* was performed on
top of the *semi-supervised learning* scheme by greedily selecting queries from the
unlabeled data to minimize the estimated expected classification error (risk); in the
case of Gaussian fields the risk is efficiently computed using matrix methods. They
present experimental results on synthetic data, handwritten digit recognition, and
text classification tasks. The active learning scheme requires a much smaller number of queries to achieve high accuracy compared with random query selection. Hoi
et al. [29] proposed a novel framework that combine support vector machines and
semi-supervised active learning for image retrieval. It is based on the Gaussian fields
and harmonic functions semi-supervised approach proposed by Zhu et al. [64].

8.2   SSL with Generative Models

McCallum and Nigam [39] present a Bayesian probabilistic framework for text classification that reduces the need for labeled training documents by taking advantage of a large pool of unlabeled documents. First they modified the *Query-by-Committee* method of active learning (*QBC*) to use the unlabeled pool for explicitly estimating document density when selecting examples for labeling. Then the modified *QBC* is combined with *Expectation-Maximization* (*EM*) in order to predict the class labels of those documents that remain unlabeled. They proposed two approaches to combine *QBC* and *EM*, called *QBC-then-EM* and *QBC-with-EM*. *QBC-then-EM* runs *EM* to convergence after actively selecting all the training examples that will be labeled. This means to use *QBC* to select a better starting point for *EM* hill climbing, instead of randomly selecting documents to label for the starting point. *QBC-with-EM* is a more interesting approach to interleave *EM* with *QBC* so that *EM* not only builds on the results of *QBC*, but *EM* also informs *QBC*. To do this, *EM* runs to convergence on each committee member before performing the disagreement calculations. The aim is (1) to avoid requesting labels for examples whose label can be reliably predicted by EM, and (2) to encourage the selection of examples that will help *EM* find a local maximum likelihood with higher classification accuracy. This directs *QBC* to pick more informative documents to label because it has more accurate committee members. Experimental results show that using the combination of *QBC* and *EM* performs better than using either individually and requires only slightly half the number of labeled training examples required by either *QBC* or EM alone to achieve the same accuracy.

8.3   SSL with Committees

Muslea et al. [41] combined *Co-Testing* and *Co-EM* in order to produce an active multi-view semi-supervised algorithm, called *Co-EMT*. The experimental results on web page classification show that *Co-EMT* outperforms other non-active multi-view algorithms (*Co-Training* and *Co-EM*) without using more labeled data and it is more robust to the violation of the requirements of independent and redundant views.

Zhou et al. [58] proposed an approach, called *SSAIR* (Semi-Supervised Active Image Retrieval), that attempts to exploit unlabeled data to improve the performance of content-based image retrieval (*CBIR*). In detail, in each iteration of relevance feedback, two simple classifiers are trained from the labeled data, i.e. images result from user query and user feedback. Each classifier then predicts the class labels of the unlabeled images in the database and passes the most relevant/irrelevant images to the other classifier. After re-training with the additional labeled data, the classifiers classify the images in the database again and then their classifications are combined. Images judged to be relevant with high confidence are returned as the retrieval result, while these judged with low confidence are put into the pool which is used in the next iteration of relevance feedback. Experiments show that semi-supervised learning and active learning mechanisms are both beneficial to *CBIR*. It is worth mentioning that *SSAIR* depends on single-view versions of *Co-Testing* and

Co-Training that require neither two independent and redundant views nor two different supervised learning algorithm. In order to create the diversity, the two classifiers used for *Co-Testing* and *Co-Training* are trained using the Minkowsky distance metric with different distance order.

Abdel Hady and Schwenker [1] introduced two new approaches, *QBC-then-CoBC* and *QBC-with-CoBC*, that combine the merits of *committee-based active learning* and *committee-based semi-supervised learning*. The first approach is the most straightforward way of combining *CoBC* and active learning where *CoBC* is run after active learning completes (denoted by *QBC-then-CoBC*). The objective is that active learning can help *CoBC* through providing it with a better starting point instead of randomly selecting examples to label for the starting point. A more interesting approach, denoted *QBC-with-CoBC*, is to interleave *CoBC* with *QBC*, so that *CoBC* not only runs on the results of active learning, but *CoBC* also helps *QBC* in the sample selection process as it augments the labeled training set with the most competent examples selected by *CoBC*. Thus, mutual benefit can be achieved. Experiments were conducted on the ten image recognition tasks. The results have shown that both *QBC-then-CoBC* and *QBC-with-CoBC* can enhance the performance of *standalone QBC* and *standalone CoBC*. Also they outperform other non committee-based combinations of semi-supervised and active learning algorithms such that *US-then-ST*, *US-then-CoBC* and *QBC-then-ST*.

9 Conclusion

During the past decade, many semi-supervised learning approaches have been introduced, many theoretical supports have been discovered, and many successful real-world applications have been reported. The work in [13, 5] has theoretically studied *Co-Training* with two views, but could not explain why the single-view variants can work. Wang and Zhou [54] provided a theoretical analysis that emphasizes that the important factor for the success of single-view committee-based *Co-Training style* algorithms is the creation of a large diversity (disagreement) among the co-trained classifiers, regardless of the method used to create diversity, for instance through: sufficiently redundant and independent views as in standard *Co-Training* [13, 44], artificial feature splits in [22, 46], different supervised learning algorithms as in [26, 57], training set manipulation as in [9, 61], different parameters of the same supervised learning algorithms [60] or feature set manipulation as in [38, 1].

Brown et al. presented in [15] an extensive survey of the various techniques used for creating diverse ensembles, and categorized them, forming a preliminary taxonomy of diversity creation methods. One can see that multi-view *Co-Training* is a special case of semi-supervised learning with committees. Therefore, the data mining community is interested in a more general *Co-Training style* framework that can exploit the diversity among the members of an ensemble for correctly predicting the unlabeled data in order to boost the generalization ability of the ensemble.

There is no *SSL* algorithm that is the best for all real-world data sets. Each *SSL* algorithm has its strong assumptions because labeled data is scarce and there is no guarantee that unlabeled data will always help. One should use the method whose assumptions match the given problem. Inspired by [63], we have the following checklist: If the classes produce well clustered data, then EM with generative mixture models may be a good choice; If the features are naturally divided into two or more redundant and independent sets of features, then standard *Co-Training* may be appropriate; If *SVM* is already used, then *Transductive SVM* is a natural extension; In all cases, *Self-Training* and *CoBC* are practical wrapper methods.

References

1. Abdel Hady, M.F., Schwenker, F.: Combining committee-based semi-supervised learning and active learning. Journal of Computer Science and Technology (JCST): Special Issue on Advances in Machine Learning and Applications 25(4), 681–698 (2010)
2. Abdel Hady, M.F., Schwenker, F., Palm, G.: Semi-supervised Learning for Regression with Co-training by Committee. In: Alippi, C., Polycarpou, M., Panayiotou, C., Ellinas, G. (eds.) ICANN 2009, Part I. LNCS, vol. 5768, pp. 121–130. Springer, Heidelberg (2009)
3. Abdel Hady, M.F., Schwenker, F., Palm, G.: Semi-supervised learning for tree-structured ensembles of RBF networks with co-training. Neural Networks 23(4), 497–509 (2010)
4. Adankon, M., Cheriet, M.: Genetic algorithm–based training for semi-supervised svm. Neural Computing and Applications 19, 1197–1206 (2010)
5. Balcan, M.-F., Blum, A., Yang, K.: Co-Training and expansion: Towards bridging theory and practice. In: Advances in Neural Information Processing Systems 17, pp. 89–96 (2005)
6. Basu, S., Banerjee, A., Mooney, R.: Semi-supervised clustering by seeding. In: Proc. of the 19th International Conference on Machine Learning (ICML 2002), pp. 19–26 (2002)
7. Basu, S., Bilenko, M., Mooney, R.: A probabilistic framework for semi-supervised clustering. In: Proc. of the 10th ACM SIGKDD Conference on Knowledge Discovery and Data Mining (KDD 2004), pp. 59–68 (2004)
8. Belkin, M., Niyogi, P., Sindhwani, V.: Manifold regularization: A geometric framework for learning from labeled and unlabeled examples. Journal of Machine Learning Research 7, 2399–2434 (2006)
9. Bennet, K., Demiriz, A., Maclin, R.: Exploiting unlabeled data in ensemble methods. In: Proc. of the 8th ACM SIGKDD International Conference on Knowledge Discovery and Data Mining, pp. 289–296 (2002)
10. De Bie, T., Cristianini, N.: Semi-supervised learning using semi-definite programming. In: Semi-supervised Learning. MIT Press (2006)
11. Blum, A., Chawla, S.: Learning from labeled and unlabeled data using graph mincuts. In: Proc. of the 18th International Conference on Machine Learning (ICML 2001), pp. 19–26 (2001)
12. Blum, A., Lafferty, J., Rwebangira, M., Reddy, R.: Semi-supervised learning using randomized mincuts. In: Proc. of the 21st International Conference on Machine Learning (ICML 2004), pp. 13–20 (2004)
13. Blum, A., Mitchell, T.: Combining labeled and unlabeled data with co-training. In: Proc. of the 11th Annual Conference on Computational Learning Theory (COLT 1998), pp. 92–100. Morgan Kaufmann (1998)
14. Breiman, L.: Bagging predictors. Machine Learning 24(2), 123–140 (1996)

15. Brown, G., Wyatt, J., Harris, R., Yao, X.: Diversity creation methods: a survey and categorisation. Information Fusion 6(1), 5–20 (2005)
16. Chapelle, O., Chi, M., Zien, A.: A continuation method for semi-supervised svms. In: International Conference on Machine Learning (2006)
17. Chapelle, O., Sindhwani, V., Keerthi, S.: Branch and bound for semi-supervised support vector machines. In: Advances in Neural Information Processing Systems (2006)
18. Chapelle, O., Zien, A.: Semi-supervised learning by low density separation. In: Proc. of the 10th International Workshop on Artificial Intelligence and Statistics, pp. 57–64 (2005)
19. Cozman, F.G., Cohen, I.: Unlabeled data can degrade classification performance of generative classifiers. In: Proc. of the 15th International Conference of the Florida Artificial Intelligence Research Society (FLAIRS), pp. 327–331 (2002)
20. d'Alché-Buc, F., Grandvalet, Y., Ambroise, C.: Semi-supervised MarginBoost. In: Neural Information Processing Systems Foundation, NIPS 2002 (2002)
21. Dempster, A.P., Laird, N.M., Rubin, D.B.: Maximum likelihood from incomplete data via the em algorithm. Journal of the Royal Statistical Society. Series B (Methodological) 39(1), 1–38 (1977)
22. Feger, F., Koprinska, I.: Co-training using RBF nets and different feature splits. In: Proc. of the International Joint Conference on Neural Networks (IJCNN 2006), pp. 1878–1885 (2006)
23. Freund, Y., Schapire, R.E.: A decision-theoretic generalization of on-line learning and an application to boosting. Journal of Computer and System Sciences 55(1), 119–139 (1997)
24. Freund, Y., Seung, H.S., Shamir, E., Tishby, N.: Selective sampling using the query by committee algorithm. Machine Learning 28, 133–168 (1997)
25. Fung, G., Mangasarian, O.: Semi-supervised support vector machines for unlabeled data classification. Optimization Methods and Software 15, 29–44 (2001)
26. Goldman, S., Zhou, Y.: Enhancing supervised learning with unlabeled data. In: Proc. of the 17th International Conference on Machine Learning (ICML 2000), pp. 327–334 (2000)
27. Grandvalet, Y., Bengio, Y.: Semi-supervised learning by entropy minimization. Advances in Neural Information Processing Systems 17, 529–536 (2005)
28. Ho, T.K.: The random subspace method for constructing decision forests. IEEE Transactions Pattern Analysis and Machine Intelligence 20(8), 832–844 (1998)
29. Hoi, S.C.H., Lyu, M.R.: A semi-supervised active learning framework for image retrieval. In: Proc. of IEEE Computer Society Conference on Computer Vision and Pattern Recognition (CVPR), pp. 302–309 (2005)
30. Inoue, M., Ueda, N.: Exploitation of unlabeled sequences in hidden markov models. IEEE Transactions On Pattern Analysis and Machine Intelligence 25(12), 1570–1581 (2003)
31. Joachims, T.: Transductive inference for text classification using support vector machines. In: Proc. of the 16th International Conference on Machine Learning, pp. 200–209 (1999)
32. Kemp, T., Waibel, A.: Unsupervised training of a speech recognizer: Recent experiments. In: Proc. EUROSPEECH, pp. 2725–2728 (1999)
33. Kiritchenko, S., Matwin, S.: Email classification with co-training. In: Proc. of the 2001 Conference of the Centre for Advanced Studies on Collaborative research (CASCON 2001), pp. 8–19. IBM Press (2001)
34. Krogh, A., Vedelsby, J.: Neural network ensembles, cross validation, and active learning. Advances in Neural Information Processing Systems 7, 231–238 (1995)

35. Lawrence, N.D., Jordan, M.I.: Semi-supervised learning via gaussian processes. Advances in Neural Information Processing Systems 17, 753–760 (2005)
36. Levin, A., Viola, P., Freund, Y.: Unsupervised improvement of visual detectors using co-training. In: Proc. of the International Conference on Computer Vision, pp. 626–633 (2003)
37. Lewis, D., Catlett, J.: Heterogeneous uncertainty sampling for supervised learning. In: Proc. of the 11th International Conference on Machine Learning (ICML 1994), pp. 148–156 (1994)
38. Li, M., Zhou, Z.-H.: Improve computer-aided diagnosis with machine learning techniques using undiagnosed samples. IEEE Transactions on Systems, Man and Cybernetics- Part A: Systems and Humans 37(6), 1088–1098 (2007)
39. McCallum, A.K., Nigam, K.: Employing EM and pool-based active learning for text classification. In: Proc. of the 15th International Conference on Machine Learning (ICML 1998), pp. 350–358. Morgan Kaufmann (1998)
40. Miller, D.J., Uyar, H.S.: A mixture of experts classifier with learning based on both labelled and unlabelled data. Advances in Neural Information Processing Systems 9, 571–577 (1997)
41. Muslea, I., Minton, S., Knoblock, C.A.: Active + semi-supervised learning = robust multi-view learning. In: Proc. of the 19th International Conference on Machine Learning (ICML 2002), pp. 435–442 (2002)
42. Nagy, G., Shelton, G.L.: Self-corrective character recognition systems. IEEE Transactions on Information Theory, 215–222 (1966)
43. Nigam, K.: Using Unlabeled Data to Improve Text Classification. PhD thesis, School of Computer Science, Carnegie Mellon University, Pittsburgh, USA (2001)
44. Nigam, K., Ghani, R.: Analyzing the effectiveness and applicability of co-training. In: Proc. of the 9th International Conference on Information and Knowledge Management, New York, NY, USA, pp. 86–93 (2000)
45. Nigam, K., McCallum, A.K., Thrun, S., Mitchell, T.: Text classification from labeled and unlabeled documents using EM. Machine Learning 39(2-3), 103–134 (2000)
46. Salaheldin, A., El Gayar, N.: New Feature Splitting Criteria for Co-training Using Genetic Algorithm Optimization. In: El Gayar, N., Kittler, J., Roli, F. (eds.) MCS 2010. LNCS, vol. 5997, pp. 22–32. Springer, Heidelberg (2010)
47. Seeger, M.: Learning with labeled and unlabeled data. Technical report, University of Edinburgh, Institute for Adaptive and Neural Computation (2002)
48. Settles, B.: Active learning literature survey. Technical report, Department of Computer Sciences, University of Wisconsin-Madison, Madison, WI (2009)
49. Shahshahani, B., Landgrebe, D.: The effect of unlabeled samples in reducing the small sample size problem and mitigating the hughes phenomenon. IEEE Transactions on Geoscience and Remote Sensing 32(5), 1087–1095 (1994)
50. Sindhwani, V., Keerthi, S., Chapelle, O.: Deterministic annealing for semi-supervised kernel machines. In: International Conference on Machine Learning (2006)
51. Tang, W., Zhong, S.: Pairwise constraints-guided dimensinality reduction. In: Proc. of the SDM 2006 Workshop on Feature Selection for Data Mining (2006)
52. Vapnik, V.: The Nature of Statistical Learning Theory. Springer (1995)
53. Wagstaff, K., Cardie, C., Schroedl, S.: Constrained k-means clustering with background knowledge. In: Proc. of the 18th International Conference on Machine Learning (ICML 2001), pp. 577–584 (2001)
54. Wang, W., Zhou, Z.-H.: Analyzing Co-training Style Algorithms. In: Kok, J.N., Koronacki, J., Lopez de Mantaras, R., Matwin, S., Mladenič, D., Skowron, A. (eds.) ECML 2007. LNCS (LNAI), vol. 4701, pp. 454–465. Springer, Heidelberg (2007)

55. Young, T.Y., Farjo, A.: On decision directed estimation and stochastic approximation. IEEE Transactions on Information Theory, 671–673 (1972)
56. Zhou, D., Bousquet, O., Lal, T.N., Weston, J., Schölkopf, B.: Learning with local and global consistency. Advances in Neural Information Processing Systems 16, 753–760 (2004)
57. Zhou, Y., Goldman, S.: Democratic co-learning. In: Proc. of the 16th IEEE International Conference on Tools with Artificial Intelligence (ICTAI 2004), pp. 202–594. IEEE Computer Society, Washington, DC (2004)
58. Zhou, Z.-H., Chen, K.-J., Jiang, Y.: Exploiting Unlabeled Data in Content-Based Image Retrieval. In: Boulicaut, J.-F., Esposito, F., Giannotti, F., Pedreschi, D. (eds.) ECML 2004. LNCS (LNAI), vol. 3201, pp. 525–536. Springer, Heidelberg (2004)
59. Zhou, Z.-H., Li, M.: Semi-supervised learning by disagreement. Knowledge and Information Systems (in press)
60. Zhou, Z.-H., Li, M.: Semi-supervised regression with co-training. In: Proc. of the 19th International Joint Conference on Artificial Intelligence (IJCAI 2005), pp. 908–913 (2005)
61. Zhou, Z.-H., Li, M.: Tri-training: Exploiting unlabeled data using three classifiers. IEEE Transactions on Knowledge and Data Engineering 17(11), 1529–1541 (2005)
62. Zhou, Z.-H., Zhang, D., Chen, S.: Semi-supervised dimensionality reduction. In: Proc. of the 7th SIAM International Conference on Data Mining (SDM 2007), pp. 629–634 (2007)
63. Zhu, X.: Semi-supervised learning literature survey. Technical Report 1530 (2008)
64. Zhu, X., Ghahramani, Z., Lafferty, J.: Semi-supervised learning using gaussian fields and harmonic functions. In: Proc. of the 20th International Conference on Machine Learning (ICML 2003), pp. 912–919 (2003)
65. Zhu, X., Lafferty, J., Ghahramani, Z.: Combining active learning and semi-supervised learning using gaussian fields and harmonic functions. In: Proc. of the ICML 2003 Workshop on The Continuum from Labeled to Unlabeled Data (2003)

Chapter 8
Statistical Relational Learning

Hendrik Blockeel

Abstract. Relational learning refers to learning from data that have a complex structure. This structure may be either internal (a data instance may itself have a complex structure) or external (relationships between this instance and other data elements). Statistical relational learning refers to the use of statistical learning methods in a relational learning context, and the challenges involved in that. In this chapter we give an overview of statistical relational learning. We start with some motivating problems, and continue with a general description of the task of (statistical) relational learning and some of its more concrete forms (learning from graphs, learning from logical interpretations, learning from relational databases). Next, we discuss a number of approaches to relational learning, starting with symbolic (non-probabilistic) approaches, and moving on to numerical and probabilistic methods. Methods discussed include inductive logic programming, relational neural networks, and probabilistic logical or relational models.

1 Introduction

Machine learning approaches can be distinguished along a large number of dimensions. One can consider different *tasks*: classification, regression, and clustering are among the better known ones. One can also distinguish approaches according to what kind of *inputs* they can handle; the format of the *output* they produce; the algorithmic or mathematical description of the actual learning *method*; the *assumptions* made by that learning method (sometimes called its inductive bias); etc.

In this chapter, we will first have a closer look at the input format: what kind of input data can a learning system handle? This is closely related to

Hendrik Blockeel
Department of Computer Science, KU Leuven, Leuven, Belgium
Leiden Institute of Advanced Computer Science, Leiden, The Netherlands
e-mail: `hendrik.blockeel@cs.kuleuven.be`

M. Bianchini et al. (Eds.): *Handbook on Neural Information Processing*, ISRL 49, pp. 241–281.
DOI: 10.1007/978-3-642-36657-4_8   © Springer-Verlag Berlin Heidelberg 2013

properties of the learned model: this model is often a (predictive) function, which takes arguments of a particular type, and this type should be compatible with the input data. In this context, we will consider the *attribute-value learning* setting, which most machine learning systems use, and the *relational learning* setting, which is the setting in which statistical relational learning takes place. The distinction between attribute-value and relational learning is quite fundamental, and forms an important motivation for considering relational learning as a separate field. After discussing how relational learning is set apart from attribute-value learning (in Section 2), we will have a closer look at relational learning methods in general (Sections 3–4), and then zoom in on statistical relational learning (Section 5).

2   Relational Learning versus Attribute-Value Learning

2.1   Attribute-Value Learning

The term attribute-value learning (AVL) refers to a setting where the input data consists of a set of data elements, each of which is described by a fixed set of attributes, to which values are assigned. That is, the input data set D consists of elements $\mathbf{x}_i$, $i = 1, \ldots N$, with N denoting the total number of elements in D. These elements are also called *examples*, *instances* or *individuals*. Each element is described by a number of *attributes*, which we usually denote A_i, $i = 1, \ldots, n$. Each attribute A has a set of possible values, called its *domain*, and denoted $Dom(A)$. The domains of the attributes may vary: an attribute A may be boolean, in which case its domain is $Dom(A) = \{true, false\} = \mathbb{B}$;[1] it may be nominal, in which case its domain is a finite set of symbolic values $\{v_1, v_2, \ldots, v_{|Dom(A)|}\}$; it may be numerical, for instance $Dom(A) = \mathbb{N}$ (discrete) or $Dom(A) = \mathbb{R}$ (continuous); the domain may be some (totally or partially) ordered set; etc.

Thus, mathematically, the instances are points in an n-dimensional instance space $\mathcal{X}$, which is of the following form:

$$\mathcal{X} = Dom(A_1) \times \cdots \times Dom(A_n)$$

We also call these points *n-tuples*.

Many learning systems, such as decision tree learners, rule learners, or instance-based learners, can handle this data format directly. Other learning approaches, including artificial neural networks and support vector machines, treat the instance space as a vector space, and assume $Dom(A_i) = \mathbb{R}$ for all A_i. When the original data are not numerical, using the latter type of approaches requires encoding the data numerically. This can often be done

---

[1] We use $\mathbb{B}$ for the set of booleans, analogously to the use of $\mathbb{N}$ and $\mathbb{R}$ for natural numbers and reals, respectively.

in a relatively simple way (for instance, the boolean values true and false can be encoded as 1 and 0), and we will not discuss this in detail here.

For ease of discussion, let us now focus on the task of learning a classifier. Here, the instance space is typically of the form $\mathcal{X} \times \mathcal{Y}$ with $\mathcal{X} = Dom(A_1) \times \cdots \times Dom(A_{n-1})$ and $\mathcal{Y} = Dom(A_n)$; the instances are of the form $(\mathbf{x}, y)$, and the task is to learn a function $f : \mathcal{X} \to \mathcal{Y}$ that, given some $\mathbf{x}$, predicts the corresponding y. If $Dom(A_i) = \mathbb{R}$ for all A_i, $\mathbf{x}$ is an $n-1$-dimensional vector; this means that we learn a function with $n-1$ input arguments, all of which are reals. This type of functions is well-understood, it is studied in great detail in calculus. Also when $Dom(A_i) \neq \mathbb{R}$, such functions are easy to define by referring to the different A_i. But the mathematical concept of a function is more general: we can also consider functions that take variable-length tuples, sets, sequences, or graphs as arguments. Such functions are generally much more difficult to specify. It is exactly that type of functions that we encounter in relational learning.

2.2   Relational Learning

Relational learning refers to learning from data that have a complex structure, either internally or externally. A complex *internal structure*, in this case, implies that an instance cannot be described as a single point in a predefined n-dimensional space; rather, the instance consists of a variable number of components and relationships between them (e.g., a graph). *External structure* refers to the fact that relationships exist between the different data elements; the instances cannot be considered independent and identically distributed (i.i.d.), and properties of one instance may depend not just on other properties of itself, but also on properties of other instances somehow related to it.

We first give some motivating examples for both settings (internal or external structure); next, we discuss the connection between them, and how they compare to attribute-value learning. Much of the discussion in this subsection is based on Struyf and Blockeel [56].

Learning from Examples with External Relationships. This setting considers learning from a set of examples where each example itself has a relatively simple description, for instance in the attribute-value format, and relationships may be present among these examples.

Example 1. Consider the task of web-page classification. Each web-page is described by a fixed set of attributes, such as a bag of words representation of the page. Web-pages may be related through hyper-links, and the class label of a given page typically depends on the labels of the pages to which it links.

Example 2. Consider the Internet Movie Database (www.imdb.com). Each movie is described by a fixed set of attributes, such as its title and genre.

Movies are related to entities of other types, such as *Studio, Director, Producer*, and *Actor*, each of which is in turn described by a different set of attributes. Movies can also be related to each other via entities of other types. For example, they can be made by the same studio or star the same well known actor. The learning task in this domain could be, for instance, predicting the opening weekend box office receipts of the movies.

If relationships are present among examples, then the examples may not be independent and identically distributed (i.i.d.). Many learning algorithm assume they are, and when this assumption is violated, this can be detrimental to learning performance, as Jensen and Neville [31] show. On the other hand, knowledge about the relationships among examples can be beneficially exploited by the learning algorithm. Collective classification techniques [32], for example, take the class labels of related examples into account when classifying a new instance, which can lead to better predictive results.

Thus, to an attribute-value learner, relations between instances may hamper accurate learning, but to a relational learner, this relations are a source of information that can be put to good use.

Learning from Examples with a Complex Internal Structure. In this setting, each example may have a complex internal structure, but no relationships exist that relate different examples to one another. Learning algorithms typically use individual-centered representations in this setting, such as logical interpretations [9] or strongly typed terms [42], which store all the data available about a given instance. Special cases of this setting include applications where the examples can be represented as graphs, trees, or sequences.

Example 3. Consider a database of candidate chemical compounds to be used in drugs. The molecular structure of each compound can be represented as a graph where the vertices are atoms and the edges are bonds. Each atom is labeled with its element type and the bonds can be single, double, triple, or aromatic bonds. Compounds are classified as active or inactive with regard to a given disease and the goal is to build models that are able to distinguish active from inactive compounds based on their molecular structure. Such models can, for instance, be used to gain insight in the common substructures, such as binding sites, that determine a compound's activity.

External or Internal: It Does Not Matter. In many applications, relationships are naturally viewed either as internal to instances, or external to them. If we need to classify nodes in a graph, for instance, since the instances to classify are individual nodes, the link structure is considered external. If we want to classify whole graphs, the link structure is internal. From a representation point of view, however, this difference does not matter. When we describe a single node to be classified, the node's (external) context is part of its description. So regardless of whether the relational information is internal or external to the object being described, it is always internal to the representation of the object.

Thus, from the point of view of representation of the input data, there is really only one relational learning setting (even if many different tasks can be defined in this setting). This setting is not equivalent to the attribute-value learning setting, however. There is no general way in which all the information available in a relational dataset can be mapped to a finite set of attributes (thus making attribute-value learning possible) without losing information. Relational learning cannot be reduced to attribute-value learning; it is inherently more difficult than the latter. In the next section we argue in more detail why this is the case.

2.3   Mapping Relational Data to Attribute-Value Data

As a concrete example of relational learning, consider the case where the function to be learned takes graphs as input, and produces a boolean output. An example of this setting is the pharmacophore problem discussed later on. Assume that the target function is of the form: "if the graph contains a subgraph isomorphic to G, then it is positive, otherwise negative", with G a particular graph. For brevity, we refer to this target function as CONTAINS-G, and we call the class of all such functions CONTAINS-$\mathcal{G}$. We use $\mathcal{G}$ to denote the "set of all finite graphs" (more formally, the set of all graphs whose node set is a finite subset of $\mathbb{N}$; each imaginable finite graph is then isomorphic to a graph in $\mathcal{G}$).

In the attribute-value framework, we assume that objects $\mathbf{x}$ are described by listing the values of a fixed set of attributes. If the object descriptions are not given in this format, the question arises whether it is possible to represent them with a fixed set of attributes such that any function definable on the original representation can be defined on the attribute-value representation. That is, we need to find a mapping p from the original description space $\mathcal{X}$ to a product space $\mathcal{X}'$ such that, given a class of functions $\mathcal{F}$ from $\mathcal{X}$ to $\mathbb{B}$, for all $f \in \mathcal{F}$ there is a corresponding $f' : \mathcal{X}' \to \mathbb{B}$ such that $f(\mathbf{x}) = f'(p(\mathbf{x}))$. In other words, applying f' to the attribute value representation of the objects gives the same result as applying f to the original representation.

Clearly, such a mapping should be injective; that is, it should not be the case that two different objects $(\mathbf{x}_1 \neq \mathbf{x}_2)$ are mapped to the same representation $\mathbf{x}'$. If that were the case, then any function f for which $f(\mathbf{x}_1) \neq f(\mathbf{x}_2)$ cannot possibly have an equivalent function f' in the attribute-value space.

Concretely, for the class of functions CONTAINS-$\mathcal{G}$, we would need an attribute-value representation where for each $G \in \mathcal{G}$, the condition "contains G as a subgraph" can be expressed in terms of the attributes. This is trivially possible if we make sure that for each G, a boolean attribute is defined that is true for a graph that contains G, and false otherwise. The problem is that there is an infinite number of such graphs G, hence, we would need an infinite number of such attributes. If the size of G is limited to some maximum number of nodes or edges, the number of different options for G is finite, but it can still be a huge number, to the extent that it may be practically impossible to represent graphs in this way.

Besides the trivial choice of attributes mentioned above, one can think of other attributes describing graphs, such as the number of nodes, the number of edges, the maximal degree of any node, and so on. The question is, is there a finite set of attributes such that two different graphs are never mapped onto the same tuple? When all these attributes have finite domains, this is clearly not the case: the number of different tuples is then finite, while the number of graphs is infinite, so the mapping can never be injective. If some attributes can have infinite domains, however, an injective mapping can be conceived. The set of all finite graphs is enumerable, which means a one-to-one mapping from graphs to the natural numbers exists. Hence, a single attribute with domain $\mathbb{N}$ suffices to encode any set of graphs in a single table without loss of information. Thus, strictly speaking, a one-to-one encoding from the set of graphs to attribute-value format exists. But now, the problem is that a condition of the form "has G as a subgraph", for a particular G, can not necessarily be expressed in terms of this one attribute in a way that might be learnable by an attribute-value learner. Attribute-value learners typically learn models that can be expressed as a function of the inputs using a limited number of mathematical operations. Neural networks, for instance, learn functions that can be expressed in terms of the input attributes using summation, multiplication, and application of a non-linear squashing function. Suppose that, for instance, for a particular encoding, the set of graphs containing graph 32 as a subgraph is the infinite set $S_{32} = \{282, 5929, 11292, \ldots\}$. It is not obvious that a formula exists that uses only the number 32, the operators $+$ and $\times$, a sigmoid function and a threshold function, and that results in true for exactly the numbers in the set (and false otherwise), let alone that such an expression could be found that works for each graph number n and the corresponding set S_n.

Informally, we call a learning problem *reducible to attribute-value learning* if an encoding is known (a transformation from the original space $\mathcal{X}$ to a product space $\mathcal{X}'$) such that an existing attribute-value learner exists that can express for each function $f : \mathcal{X} \to \mathbb{B}$ the corresponding $f' : \mathcal{X}' \to \mathbb{B}$.

Reducibility to AVL implies that a (relational) learning problem can be transformed into an attribute value learning problem, after which a standard attribute-value learner can be used to solve it. Many problems, including that of learning a target function in CONTAINS-$\mathcal{G}$, are not reducible to AVL. Generally, problems involving learning from data where data elements contain sets (this includes graphs, as these are defined using sets of nodes and edges) are not reducible to AVL.

Propositionalization versus Relational Learning. Among relational learners, we can distinguish systems that use so-called *propositionalization* as a preprocessing step, from learners that do not. The latter could be considered "truly" relational learners.

Propositionalization refers to the mapping of relational information onto an attribute-value (also known as "propositional") representation. This is often done in a separate phase that precedes the actual learning process. Propositionalization amounts to explicitly defining a set of features, and representing the data using that set. Formally, given a data set $D \subseteq \mathcal{X}$, the process of propositionalization consists of defining a set F of features $\phi_i : \mathcal{X} \to R_i$, where each R_i is nominal or numerical, and representing D using these features as attributes, i.e., representing D by $D' = \{(\phi_1(\mathbf{x}), \phi_2(\mathbf{x}), \ldots, \phi_n(\mathbf{x})) | \mathbf{x} \in D\}$. We call D' a propositional representation of D.

The feature set may be fixed for a particular algorithm, it may be variable and specified by the user, or it may be the result of some feature selection process that, from an initial set of features, retains only those that are likely to be relevant for the learning task.

When a learning problem is not reducible to AVL, then, no matter how the features are defined, propositionalization causes loss of information: some classifiers expressible in the original space may no longer be expressible in the feature space.

Truly relational learners do not suffer from this problem: they learn a function in the original space. For instance, consider again our example of learning functions in CONTAINS-$\mathcal{G}$. A truly relational learner uses the original graph representations and can express any function in CONTAINS-$\mathcal{G}$.

As argued earlier, in principle, we could define one feature for each $G \in \mathcal{G}$, which expresses whether the graph being described contains G as a subgraph (but we would need an infinite number of features for this). More generally, whatever the class of functions is that a relational learner uses, we could define for each function in that class a feature that shows the result of that function, when applied to a particular instance. Thus, each relational learner could be said to implicitly use a (possibly infinite) set of features. From this point of view, the main difference between a truly relational learner and a propositional learner (or a learner that uses propositionalization) is that a truly relational learner typically proceeds by gradually selecting more and more actual features from some initial set of potential features. The result of the relational learning is a function that can be described in terms of a small number of relevant features, and which is constant in all other. While the set of relevant features is finite, the number of features that can in principle be considered for inclusion may be infinite.

Note that propositionalization-based systems may also perform feature selection, just like truly relational learners. The difference is that a propositionalization-based system first constructs a representation of the data set in terms of a finite set of features, then reduces that set. A truly relational learner does not construct the propositionalized data set D' at any point.

Relational learners are more powerful in the sense that they can use a much larger feature set: even if not infinite, the cardinality of this set can easily be so high that even storing the values of all these features is not feasible, making propositionalization intractable. On the other hand, they necessarily search only a very small (and always finite) part of this space, which means that good heuristics are needed to steer the search, in order to ensure that all relevant features are indeed encountered.

Propositionalization is a step that is very often used in practice, when using machine learning on relational data. In fact, practitioners often assume implicitly that such a propositionalization step is necessary before one can learn. That is correct if one considers attribute-value learners only, but false when relational learners are also an option.

2.4 Summary of This Section

Most off-the-shelf learning systems assume the input data to be in the attribute-value format (sometimes also called the "standard" format). Relational data cannot be represented in the attribute-value format without loss of information. There are then two options: either the user converts the data into attribute-value format and accepts information loss, or a learning algorithm must be used that handles such relational data directly. Such a relational learner could be said to ultimately construct an attribute-value representation just as well, either explicitly (through propositionalization) or implicitly (in which case it may be searching a huge space of potential features for the most informative ones). The last type of learner is the most powerful, but also faces the most challenging task.

3 Relational Learning: Tasks and Formalisms

Many different kinds of learning tasks have been defined in relational learning, and an even larger number of approaches have been proposed for tackling these tasks. We give an overview of different learning settings and tasks that can be considered instances of relational learning. Where mentioning methods, we focus on symbolic and non-probabilistic methods; methods based on neural processing and probabilistic inference are treated in more detail in the next two sections.

3.1 Inductive Logic Programming

In inductive logic programming (ILP), the input and output knowledge of a learner are described in (variants of) first-order predicate logic. Languages based on first-order logic are highly expressive from the point of view of knowledge representation, and indeed, a language such as Prolog [2] can be used directly to represent objects and the relationships between them, as well as background knowledge that one may have about the domain.

Example 4. This example is based on the work by Finn et al. [21]. Consider a data set that describes chemical compounds. The active compounds in the set are ACE inhibitors, which are used in treatments for hypertension. The molecular structure of the compounds is represented as a set of Prolog facts, such as:

```
atom(m1, a1, o).
atom(m1, a2, c).
...
bond(m1, a1, a2, 1).
...
coord(m1, a1, 5.91, -2.44, 1.79).
coord(m1, a2, 0.57, -2.77, 0.33).
...
```

which states that molecule m1 includes an oxygen atom a1 and a carbon atom a2 that are single bonded. The coord/5 predicate lists the 3D coordinates of the atoms in the given conformer. Background knowledge, such as the concepts zinc site, hydrogen donor, and the distance between atoms, are defined by means of Prolog clauses. Fig. 1 shows a clause learned by the inductive logic programming system PROGOL ([18], Ch. 7) that makes use of these background knowledge predicates. This clause is the description of a pharmacophore, that is, a submolecular structure that causes a certain observable property of a molecule.

Note that, in Prolog, variables start with capitals, and constants with lower-case characters. We will use this convention also when writing logical clauses outside the Prolog context.

Research in inductive logic programming originally focused on concept learning. Concept learning, which is often considered a central task in artificial intelligence and was for a long time the main focus of machine learning research, concerns learning a definition of a concept from example instances. In the ILP setting, the concept to be learned is an n-ary relation or predicate, defined intentionally by a set of rules, and the task is to discover this set of rules by analyzing positive and examples, which are tuples said (not) to belong to the relation. The example in Fig. 1 is a concept learning task: an operational definition of the concept of an ACE-inhibitor is learned.

Pioneering work on concept learning in the first order logic context resulted in well-known ILP systems such as FOIL[51] and PROGOL [45]. Later, the focus has widened to include many other tasks such as clausal discovery, where the goal is to discover logical clauses that hold in a dataset, without focusing on clauses that define a particular concept[11]; regression, where

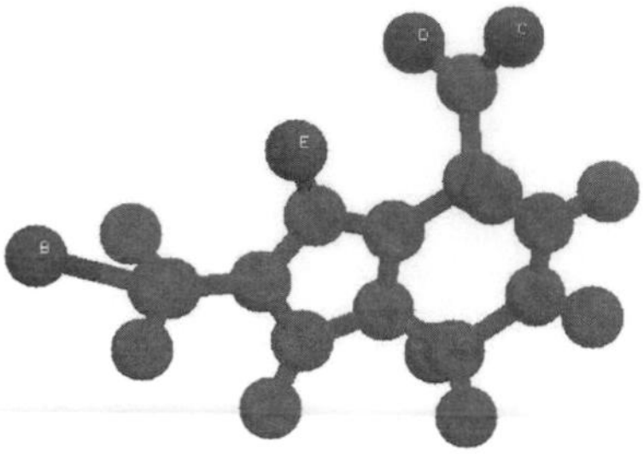

(a) ```
    ACE_inhibitor(A) :-
        zincsite(A, B),
        hacc(A, C),
        dist(A, B, C, 7.9, 1.0),
        hacc(A, D),
        dist(A, B, D, 8.5, 1.0),
        dist(A, C, D, 2.1, 1.0),
        hacc(A, E),
        dist(A, B, E, 4.9, 1.0),
        dist(A, C, E, 3.1, 1.0),
        dist(A, D, E, 3.8, 1.0).
    ```

(c)

(b)  Molecule $A$ is an ACE inhibitor if:
     molecule $A$ can bind to zinc at site $B$, and
     molecule $A$ contains a hydrogen acceptor $C$, and
     the distance between $B$ and $C$ is $7.9 \pm 1.0$Å, and
     molecule $A$ contains a hydrogen acceptor $D$, and
     the distance between $B$ and $D$ is $8.5 \pm 1.0$Å, and
     the distance between $C$ and $D$ is $2.1 \pm 1.0$Å, and
     molecule $A$ contains a hydrogen acceptor $E$, and
     the distance between $B$ and $E$ is $4.9 \pm 1.0$Å, and
     the distance between $C$ and $E$ is $3.1 \pm 1.0$Å, and
     the distance between $D$ and $E$ is $3.8 \pm 1.0$Å.

**Fig. 1** (a) Prolog clause modeling the concept of an ACE inhibitor in terms of the background knowledge predicates `zincsite/2`, `hacc/2`, and `dist/5`. The inductive logic programming system PROGOL automatically translates (a) into the "Sternberg English" rule (b), which can be easily read by human experts. (c) A molecule with the active site indicated by the dark colored atoms. (Example based on Finn et al. [21].)

clauses are learned that compute a numerical prediction [33]; frequent pattern discovery [16]; and reinforcement learning [17, 57]. Also additional learning paradigms, besides rule learning, have been explored, including decision trees, instance-based learners, etc. Extensive overviews of the theory of inductive logic programming, including descriptions of tasks, methods, and algorithms, are available in the literature [10, 18].

## 3.2 Learning from Graphs

A graph is a mathematical structure consisting of a set of nodes $V$ and a set of edges $E \subseteq V \times V$ between those nodes. The set of edges is by definition a binary relation defined over the nodes. Hence, for any learning problem where the relationships between examples can be described using a single binary relation, the training set can be represented straightforwardly as a graph. This setting covers a wide range of relational learning tasks, for
    ```

example, web mining (the set of links between pages is a binary relation), social network analysis (binary "friend" relation), etc. Non-binary relationships can be represented as hypergraphs; in a hypergraph, edges are defined as subsets of V of arbitrary size, rather than elements of $V \times V$.

In graph-based learning systems, there is a clear distinction between approaches that learn from examples with external relationships, where the whole data set is represented as a single graph and each node is an example, and individual-centered approaches, where each example by itself is a graph. In the first kind of approaches, the goal is often to predict properties of existing nodes or edges, to predict the existence or non-existence of edges ("link discovery"), to predict whether two nodes actually refer to the same object ("node identification"), detection of subgraphs that frequently occur in the graph, etc. When learning from multiple graphs, a typical goal is to learn a model for classifying the graphs, to find frequent substructures (where frequency is defined as the number of graphs a subgraphs occurs in), etc.

Compared to other methods for relational learning, graph-based methods typically focus more on the structure of the graph, and less on properties of single nodes. They may take node and edge labels into account, but often do not allow for more elaborate information to be associated with each node.

Graph mining methods are often more efficient than other relational mining methods because they avoid certain kinds of overhead, but are typically still NP-complete, as they generally rely on subgraph isomorphism testing. Nevertheless, researchers have been able to significantly improve efficiency or even avoid NP-completeness by looking only for linear or tree-shaped patterns, or by restricting the graphs analyzed to a relatively broad subclass. As an example, Horvath et al. [29] show that a large majority of molecules belong to the class of outerplanar graphs, and propose an efficient algorithm for subgraph isomorphism testing in this class.

Well-known systems for graph mining include gSpan [71], Gaston [47], and Subdue [6]. Excellent overviews of the field are provided by Cook and Holder [7] and Washio and Motoda [70].

3.3 Multi-relational Data Mining

Multi-relational data mining approaches relational learning from the relational database point of view. The term "multi-relational" refers to the fact that from the database perspective, one learns from information spread over multiple tables or relations, as opposed to attribute-value learning, where one learns from a single table.

Multi-relational data mining systems tightly integrate with relational databases. Mainly rule and decision tree learners have been developed in this setting. Because practical relational databases may be huge, most of these systems pay much attention to efficiency and scalability, and use techniques such as sampling and pre-computation (e.g., materializing views). An example of a scalable and efficient multi-relational rule learning system is CrossMine [72].

In the context of multi-relational data mining, propositionalization boils down to summarizing all the data relevant to a single instance, which may be spread over multiple tuples in multiple relations, into a single tuple. As explained before, this is in general not possible without loss of generalization. Krogel et al. [39] compare a number of methods for propositionalization in the context of multi-relational learning.

Most inductive logic programming systems are directly applicable to multi-relational data mining by representing each relational table as a predicate. This is possible because the relational representation is essentially a subset of first-order logic (known as datalog). Much research on multi-relational data mining originates within the ILP community [18]. Nevertheless, there is a clear difference between ILP and multi-relational learning in terms of typical biases of these methods, as will become clear in the next section.

4   Neural Network Based Approaches to Relational Learning

Among the many approaches to relational learning, a few neural network approaches have been proposed. We briefly summarize two of them here, and discuss a third one in more detail.

4.1 CIL^2P

The KBANN system (Knowledge-Based Artificial Neural Networks) [60] was one of the first to integrate logical and neural representations and inference. It used propositional logic only, however, and hence cannot be considered a truly relational learning system. Perhaps the earliest system that did combine first-order logic with neural networks, is CIL^2P, which stands for Connectionist Inductive Learning and Logic Programming [8]. This system is set in the first order logic context, and aims at integrating neural network inference and logic programming inference, both deductive (using the model to draw conclusions) and inductive (learning the model). This integration makes it possible to learn logic programs using neural network learning methods.

A limitation of this approach is that the neural network represents a ground version of the logic program.[2] As the grounding of a first order logic program may be large, the neural network is correspondingly large and may be inefficient because of this. Note, in particular, that a major advantage of a logic programming language such as Prolog is that it can reason on the level

---

[2] Given a logic program, grounding it means replacing every rule that contains variables with all the possible instantiations (i.e., applications to constants) of that rule; the resulting program is equivalent to the original one.

of variables, without instantiating those variables (a type of inference that in statistical relational learning is called *lifted inference*). For instance, if a rule $p(X) \leftarrow q(X, Y)$ is known, as well as a rule $q(X, a)$, Prolog can deduce that $p(X)$ is true whatever p is, without needing to prove this separately for each concrete case. A neural network that essentially operates on the level of ground instances must make this inference separately for each case, which makes it complex and inefficient.

The approach that will be discussed next, avoids the construction of a large ground network (i.e., a network the size of which depends on the size of the logic program's grounding), and at the same time aims at constructing a broader type of models than is typically considered when using logic-based representations.

4.2   Relational Neural Networks

Relational Neural Networks (RNNs) [62] are a neural network based approach to relational learning that is set in the context of multi-relational data mining. They were originally proposed to reconcile two rather different biases of relational learners; these biases have been called selection bias and aggregation bias [1], and they are related to how sets are handled by the system. We will first describe the motivation for the development of RNNs in more detail.

Motivation: Aggregation-Oriented versus Selection-Oriented Methods. The main problem with relational data is the existence of one-to-many relationships. We may need to classify an object $\mathbf{x}$, taking into account properties of other objects that it is related to via a particular relation, and there may be multiple such objects. If $S(\mathbf{x})$ is the set of these objects, then the question is what features we define on this set. As argued before, defining features for sets is a non-trivial task (learning from sets is not reducible to AVL).

As it turns out, within relational learning, we can consider two quite different approaches. Both of these learn features of the type $\mathcal{F}(\sigma_C(S(\mathbf{x})))$, where σ_C is the selection operator from relational algebra, and $\mathcal{F}$ is an aggregation function, which summarizes a set into a single value (for instance, $\mathcal{F}$ could be the average, maximum, minimum or variance of a set of reals; it could be the mode of a set of nominal values; it could be the cardinality of any set; etc.; it could also be any combination of these).

In the **first type of learner**, $S(\mathbf{x})$ is defined in a straightforward way; it could be, for instance, all objects $\mathbf{y}$ related to $\mathbf{x}$ through the relation $R(\mathbf{x}, \mathbf{y})$. This S is then summarized using a small set of features, each of which is one aggregation function (often just the standard functions are used, such as count, max, min, average). Sometimes the set is viewed as a sample from a distribution, and this distribution's parameters or other characteristic numbers are used to describe S. (For instance, besides the mean and variance, one could also estimate higher order moments of the distribution; the k-th moment of a distribution p is defined as $E(x^k) = \int_x x^k p(x) dx$.) This type of features has been considered by several researchers in relational learning; for instance, standard aggregation functions are used to represent sets of related objects in PRMs (probabilistic relational models, see next section) or used as features in the propositionalization method used by Krogel and Wrobel [40], whereas Perlich and Provost [48] consider aggregation functions based on a broader set of distributional characteristics.

The **second type of learner** we consider here is typical for inductive logic programming. When a clause is learned, such as $p(X) \leftarrow r(X, Y), s(Y, a)$, the set of objects **y** to which **x** is related is defined in a relatively complex way; it is a selection of tuples from the cartesian product of all relations mentioned in the body of the clause. The set of instantiations of Y that are obtained for a particular value of X can be written in relational algebra as $\sigma_{S.A2='a' \wedge R.A2=S.A1}(R \times S))$.[3] The clause body evaluates to true or false depending on whether this set of instantiations is empty or not. Thus, the features expressed by such a clause can be written as $\mathcal{F}_\exists(\sigma_C(R_1 \times \cdots \times R_k))$, where $\mathcal{F}_\exists$ represents the existential aggregate function (which returns true if its argument is non-empty, and false otherwise), the R_i are all the relations mentioned in the clause body, and C expresses the conditions explicitly or implicitly imposed on the variables in the clause.

We now see that ILP systems typically construct relatively complex selection conditions, but in the end simply test for emptiness of the resulting set. We call them **selection-oriented systems**: they invest effort in constructing a good selection condition, but ignore the fact that multiple aggregation functions might be useful for characterizing the resulting set. The first type of approaches we just mentioned, which we call **aggregation-oriented systems**, do the opposite: they consider the possible usefulness of multiple aggregation functions, but do not invest effort in building a more complex set $S(\mathbf{x})$ than what can be defined using a single relation, without further selection of the elements in the relation.

To illustrate this situation in somewhat more concrete terms: suppose a person is to be classified based on information about their children; one approach could consider the number of children, or their average age; another approach could consider whether the person has any daughters (assuming daughters to be those element in the children relation for which the attribute Sex has the value 'female'); but none of the mentioned approaches can consider, as a feature, the age of the oldest daughter.

Progress towards resolving this issue was presented by Vens et al. [69], who show how the systematic search performed by an ILP system can be extended towards clauses that can contain aggregation functions on conjunctions of literals. This essentially solves the problem of learning features of the

---

[3] As arguments of logical predicates have no fixed name, but attributes in relational algebra do, we use A_i to refer to the i'th argument of any predicate.

form $\mathcal{F}(\sigma_C(S(\mathbf{x})))$, where both C and $\mathcal{F}$ are non-trivial, for multiple standard aggregation functions (max, min, average, count). But the problem remains that $\mathcal{F}$ is restricted to a fairly small set of aggregation functions. Combinations of the results of these functions can be constructed afterwards by means of standard machine learning techniques, but any aggregation function that cannot be described as a combination of these basic building blocks, remains unlearnable.

Relational Neural Networks. The main motivation for learning relational neural networks was the desire to have a learner that learns features of the form $\mathcal{F}(\sigma_C(S(\mathbf{x})))$ with possibly complex $\mathcal{F}$ and C, though not necessarily in an explicit form. In the same way that a neural network can approximate many functions without building the actual expression that defines these functions, such relational neural networks should be able to approximate any feature of the form $\mathcal{F}(\sigma_C(S(\mathbf{x})))$, without being restricted to combinations of predefined aggregation functions, and without any restrictions regarding the form of C.

Since $S(\mathbf{x})$ is an unbounded set of tuples, such a neural network should be able to handle an unbounded set as a single input unit, and provide one or more numerical outputs for that set. To achieve this aim, Uwents and Blockeel [62] use recurrent neural networks, to which the elements of the set are presented one by one. As a recurrent neural network processes a sequence of elements, rather than an unordered set, its output may depend on the order in which the elements of the set are presented. To counter this effect, reshuffling is used; that is, each time a set is presented to the network during training, its elements are presented to it in a random sequence. This implies that the network is forced to become as order-independent as possible; however, it remains in essence a sequence processing system, and complete order-independence is not necessarily achievable. Uwents and Blockeel experimented with multiple architectures for recurrent neural networks, including standard architectures such as the Jordan architecture [62], but also cascade-correlation networks [64], and found that some of these could learn relatively complex features quite well.

In a more extensive comparison [63, 61], a toy dataset about classification of trains into eastbound and westbound is used, in which artificial target concepts of varying complexity are incorporated. One of the more complicated concepts is: "Trains having more than 45 wheels in total and at least 10 rectangle loads and maximum 27 cars are eastbound, the others are westbound". Note that trains are represented as sequences of cars, where each car has a number of properties listed; properties such as the total number of cars or wheels in the train, or the number of "cars with rectangle loads" are not represented explicitly, but need to be constructed through aggregation and selection. The comparison showed the cascade-correlation approach to work better than other neural network based approaches; it achieves near-perfect performance on simple concepts, and better performance on more complex concepts than state-of-the-art learners that are propositionalization-based [40] or try to learn a symbolic representation of the concept [66].

4.3 Graph Neural Networks

Graph neural networks (GNNs) are discussed elsewhere in this volume, and we refer to that chapter for more details on the formalism. We limit ourselves here to pointing out some of the main differences between relational neural networks and graph neural networks. First, the setting is different: consistent with the naming, RNNs are set in the context of relational databases, whereas GNNs are set in the context of graphs. While there is a connection between these two formalisms, there are also obvious differences. Consider a tuple in a relational database as a node in a graph, with foreign key relationships defining the edges in the graph. Since tuples can come from different relations, the nodes in this graph are typed. RNNs naturally define a separate model per type of node, whereas GNNs define the same model for all nodes. (RNN behavior can of course be simulated by introducing an attribute for each node that represents its type; the value of that attribute would then be used in the GNN model.) Further, RNNs aim more explicitly at approximating a particular type of features, and have been evaluated mostly in that context. GNNs work with undirected graphs, and as such create a more symmetric model. Uwents et al. [65] provide an extensive discussion of the relationship between GNNs and RNNs, as well as an experimental comparison.

5 Statistical Relational Learning

The above approaches to relational learning do not rely strongly on probability theory. Although many types of models, and their predictions, can be interpreted in a probabilistic manner (for instance, when a rule has a coverage of 15 positives and 5 negatives, it might be said to predict positive with 75% certainty), they do not necessarily define a unique probability distribution, and inference in these models is not based on probability theory. In statistical relational learning, the focus will be on models that by definition define a probability distribution, such as probabilistic graphical models (which includes Bayesian networks and Markov networks).

In the following we discuss a number of approaches to statistical relational learning. We start with approaches set in the context of graphical models; next, we discuss approaches set in the context of relational databases, and finally, of first order logic.

There is a plethora of alternative approaches to statistical relational learning, among which the relationships are not always clear. Many approaches seem to be differing mostly in syntax, yet there are often subtle differences in how easily certain knowledge can be expressed. We do not aim at giving an exhaustive overview here, or at indicating in exactly what way these methods differ from each other. Rather, we will discuss a few representative methods in detail.

5.1   *Structuring Graphical Models*

Graphical Models. Probabilistic graphical models define a joint distribution over multiple random variables in a compact way. They consist of a graph that implies certain independence relations over the variables. These independence relations imply that the joint distribution can be written as a product of a number of lower-dimensional factors. The graphical model consists of a graph together with these factors; the graph imposes a certain structure of the joint distribution, while the factors determine it uniquely.

The best-known examples of graphical models are Bayesian networks and Markov networks. In both cases, there is one node for each random variable. Bayesian networks use directed acyclic graphs (DAGs); with each node one factor is associated, and that factor is equal to the conditional probability distribution of the node given its parents. Markov networks use undirected graphs; here, one factor is associated with each maximal clique in the graph.

When learning graphical models, a distinction can be made between *structure learning*, which involves learning the graph structure of a graphical model, and *parameter learning*, which involves learning the factors associated with the graph structure.

In both cases, a distinction can be made between generative and discriminative learning. This difference is relevant when it is known in advance what variables will need to be predicted (the so-called target variables). Let $\mathbf{Y}$ denote the set of target variables, and $\mathbf{X}$ the set of all other variables that are not target variables. The task is to learn a predictive model that predicts $\mathbf{Y}$ from $\mathbf{X}$. In the probabilistic setting, one predicts not necessarily specific values for the variables in $\mathbf{Y}$, but a probability distribution for $\mathbf{Y}$. Given an observation $\mathbf{x}$ of $\mathbf{X}$, the probability that $\mathbf{Y} = \mathbf{y}$ (or, more generally, the probability density for $\mathbf{y}$) is then $p_{\mathbf{Y}|\mathbf{X}}(\mathbf{x}, \mathbf{y})$. The direct approach to predictive learning consists of learning a model of $p_{\mathbf{Y}|\mathbf{X}}$; this is called discriminative learning. An indirect approach consists of learning the joint distribution $p_{\mathbf{X},\mathbf{Y}}$. All other distributions can be derived from this joint distribution, including the conditional distribution of $\mathbf{Y}$ given $\mathbf{X}$: $p_{\mathbf{Y}|\mathbf{X}}(\mathbf{x}, \mathbf{y}) = p_{\mathbf{X},\mathbf{Y}}(\mathbf{x}, \mathbf{y})/p_{\mathbf{X}}(\mathbf{x})$ with $p_{\mathbf{X}}(\mathbf{x}) = \int_{\mathbf{y}} p_{\mathbf{X},\mathbf{Y}}(\mathbf{x}, \mathbf{y})d\mathbf{y}$. This indirect approach is called generative learning.

Generative learning is the most general approach; it does not require the target variables $\mathbf{Y}$ to be specified in advance, in contrast to discriminative learning. On the other hand, discriminative learning can be more efficient and more accurate because it focuses directly on the task at hand.

We refer to [37] for a more detailed introduction to graphical models.

Dynamic Bayesian Networks. Until now, we have simply assumed that a set of variables is given, without any structure on this set, except possibly for the fact that some variables are considered observed (their values will be given, when using the model) and other unobserved (their values will need to be predicted using the model), and this partitioning may be known in advance.

In many cases, there is more structure in the variables. A typical case is when a dynamic process is described: the values of the variables change over time, and we can talk about the value of a variable V_i at timepoint $0, 1, 2, \ldots$. Let us denote with $V_i^{(t)}$ the variable that represents the value of variable V_i at time point t; we assume discrete time points $t \geq 0$. While the state of the system changes over time, its dynamics are constant: for instance, the value of $V_i^{(t)}$ depends on $V_i^{(t-1)}$ (or more generally on $V_j^{(t-k)}$ for any k) in the same way, for any $t > 0$.

As an example of this, consider a hidden Markov model (HMM). Such a model describes a system that at any time is in one particular state, from a given set of states S. Its current state depends probabilistically on its previous state; that is, for any pair of states (s, s'), there is a fixed probability that a system that is in state s at time t, will be in state s' at time $t+1$. Further, while the state itself is unobservable, at each time point the value of a particular variable X can be observed, and the value of X depends probabilistically on the current state.

Keeping our earlier convention of using $\mathbf{X}$ for the set of observed variables, and $\mathbf{Y}$ for the set of unobserved variables, we use $X^{(t)}$ to denote the output value at time t and $Y^{(t)}$ to denote the state at time t; we then have $\mathbf{X} = \{X^{(0)}, X^{(1)}, X^{(2)}, \ldots\}$ and $\mathbf{Y} = \{Y^{(0)}, Y^{(1)}, Y^{(2)}, \ldots\}$. We can express the assumptions of a HMM in a graphical model by stating that, for all $t > 0$, $Y^{(t)}$ depends on $Y^{(t-1)}$ (and this dependency is the same for all t), and for all $t \geq 0$, $X^{(t)}$ depends on $Y^{(t)}$ (again, in the same way for all t).

Figure 2 shows an example of a hidden Markov model (a), and how it is modeled as an (infinite) Bayesian network (b). Because the dependencies of $Y^{(t)}$ on $Y^{(t-1)}$ and of $X^{(t)}$ on $Y^{(t)}$ are the same, we can express the Bayesian network more compactly by just showing for one particular time slice t how $Y^{(t)}$ and $X^{(t)}$ depend on other variables.

HMMs are a special case of dynamic Bayesian networks (DBNs). In a DBN, we also have variables $V_i^{(t)}$, but there is more flexibility with respect to which variables are observed and which are not, and with respect to the dependencies between variables. Generally, a variable $V_i^{(t)}$ can depend on any variable $V_j^{(t-k)}$ for any $k \geq 0$, as long as the set of dependencies form a sound Bayesian network (i.e., there are no cyclic dependencies). The DBN can be represented by a standard Bayesian network that contains as many variables, spread over as large a time window, as needed.

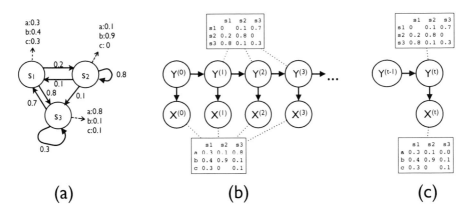

Fig. 2 (a) A schematic representation of a Hidden Markov model. (b) An infinite Bayesian network representing the same model, with the conditional probability tables of $X^{(t)}$ and $Y^{(t)}$ shown. Since each $X^{(t)}$ and $Y^{(t)}$ has the same dependencies, the tables are shown only once. (c) The same network, represented more compactly as a dynamic Bayesian network.

Dynamic Bayesian networks are an example of directed models with a repeating structure. Similarly, undirected models with structure can be defined. A well-known example of such models are conditional random fields (CRFs) [41]. For details on these, we refer to the literature.

Plates. Within the graphical model community, *plates* have been introduced as a way of structuring Bayesian networks [5, 55]. A plate indicates a substructure that is actually repeated multiple times within the graph. More precisely, several isomorphic copies of the substructure occur in the graph, and the factors associated with these copies are the same, modulo renaming of nodes. An example is shown in Figure 3. There is an arc from X to Y, but by drawing a rectangle around Y we indicate that there are actually multiple variables Y_i. The conditional probability distribution of Y_i given X is the same for all Y_i. Note that, because of this, plates are not only a way of writing the graph more compactly; they add expressiveness, as they allow us to impose an additional constraint on the factorization. Without plate notation, one could indicate that $Pr(X, Y_1, Y_2, Y_3) = Pr(X)Pr(Y_1|X)Pr(Y_2|X)Pr(Y_3|X)$, but not that, in addition, $Pr(Y_1|X) = Pr(Y_2|X) = Pr(Y_3|X)$. With a plate model, we can for instance state that the probability of a person having gene X depends on that person's mother having gene X, *and the dependence is exactly the same for all persons.*

Note that plate models are mostly useful when the variables denote properties of different objects, and there can be one-to-many or many-to-many relationships beween these objects. When all variables semantically denote a property of the same object (for instance, length and weight of a person),

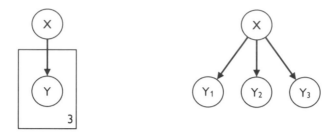

Fig. 3 Left: a simple Bayesian network using plate notation; right: the corresponding Bayesian network without plate notation. The right graph imposes a strictly weaker constraint: it does not indicate that the different Y_i depend on X in exactly the same way.

or denote properties of different objects among which there is a one to one relationship, then plate notation is typically not needed. When there are one-to-many relationships, objects on the "many" side are typically interchangeable, which means that their relationship to the one object must be the same.

The fact that plates imply that the same factor is shared by multiple substructures is not a restriction; when multiple substructures may actually have different factors, it suffices to introduce an additional variable within the plate that indicates a parameter of the factor specification; since that variable may have different values in the different occurrences, the actual factors can be different.

Plates have been introduced *ad hoc* into graphical models. They are mostly defined by illustrations and incomplete definitions examples; a single formal and precise definition for them does not exist, though formal definitions for certain variants have been proposed [28]. While plates are very useful, their expressiveness is limited. For instance, if we would have a sequence of variables X_i, $i = 1, \ldots, n$ where each X_i depends on X_{i-1} in exactly the same way, i.e., $Pr(X_1, \ldots, X_n) = Pr(X_1)Pr(X_2|X_1) \cdots Pr(X_n|X_{n-1}) = Pr(X_1)\prod_{i=2}^{n} Pr(X_i|X_{i-1})$, this cannot be expressed using plate notation. The reason is that a single variable can take two different roles in the same plate (as "parent" in one instantiation of the plate, and as "child" in another instantiation of the same plate). Further, plate models are easy to understand when plates are properly nested, but more difficult when plates can overlap (which is allowed under certain conditions). Heckerman et al. [28] introduce a variant of plates that has additional annotations that lift many of the restrictions; the simplicity of the basic plates models is then lost, however, and the formalism becomes equivalent to entity-relationship models, see later.

Together with plates, Spiegelhalter and his colleagues introduced BUGS [25], a language for defining graphical models with plates. Besides defining the structure of the model itself, the user can also indicate the format of certain factors in some detail; for instance, the conditional probability distribution of one variable given another one can be constrained to a particular family of distributions. Parameter learning is then reduced to learning the parameter of that distribution, rather than learning the conditional probability distribution in tabular format. Thus, the BUGS language is strictly more expressive than plate notation.

Example 5. Imagine that we have two students and three courses; each student has an IQ, each course has a difficulty level (Diff), and the grade (Gr) a student obtains for a course depends on the student's IQ and the course's difficulty. We could simply build a graphical model stating that grade Gr depends on IQ and Diff (Figure 4a), the parameters of which can then be learned by looking at six examples of students getting a grade for some course, but that does not take into account the fact that we know that some of these

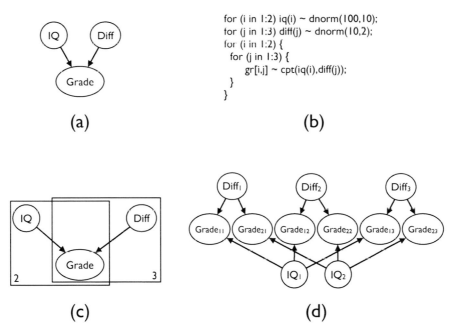

Fig. 4 (a) A graphical model that simply states that a Grade for an exam depends on the IQ (of the student taking that exam) and the Difficulty (of the course being examined). It does not express a certain relational structure, namely that some of these grades are obtained by the same student, or obtained for the same course. (b) a BUGS program that expresses that IQ of students and Difficulty of courses are normally distributed, that the grade obtained at an examination depends on IQ and Difficulty, and that we have 2 students who have taken 3 courses; (c) the corresponding plate model; (d) a corresponding graphical model. From (b) to (d), each consecutive model carries less information than the preceding one.

grades are really about the same student, or about the same course. Figure 4b shows (part of) an example BUGS program that does express this; Figure 4c shows the plate model that corresponds to the BUGS model (but which does not show certain information about the distributions), and Figure 4d shows the ground graphical model that corresponds to this plate model (this ground graphical model does not show that certain conditional distributions must be equal).

5.2 *Approaches in the Relational Database Setting*

Probabilistic Relational Models. Among the best known representation formalisms for statistical relational learning are probabilistic relational models or PRMs [23]. PRMs extend Bayesian networks to the relational representation used in relational databases. They model the joint probability distribution over the non-key attributes in a relational database schema. Each such attribute corresponds to a node and direct dependencies are modeled by directed edges. Such edges can connect attributes from different entity types that are (indirectly) related (such a relationship is called a "slot chain"). Inference in PRMs occurs by constructing a Bayesian network by instantiating the PRM with the data in the database and performing the inference in the latter. To handle 1:N relationships in the Bayesian network, PRMs make use of predefined aggregation functions.

Example 6. Consider again the example with grades obtained depending on the student's IQ and the course's difficulty. Figure 5 shows a graphical representation of a PRM that corresponds to the plate model shown before.

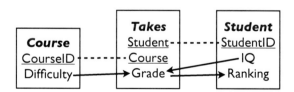

Fig. 5 A probabilistic relational model structure indicating the existence of three classes of objects (Courses, Students, and instances of the Takes relation), what the attributes they have, and how these attributes depend on each other. Dashed lines indicate foreign key relationships. Arrows indicate which attributes depend on which other attributes. As the relationship between students and grades is one to many, the dependency of a student's ranking on her grades is actually a dependency of one variable on multiple variables.

Relational Bayesian Networks. Relational Bayesian Networks [30] are another formalism for statistical relational learning; they were developed independently from PRMs and are similar to them in the sense that they also use the relational model and that the models are essentially bayesian networks. We do not discuss them in more detail here but refer to the relevant literature [30].

Entity-Relationship Probabilistic Models. Heckerman et al. [28] compare the expressiveness of plate models and PRMs and propose a new model, called Entity-Relationship Probabilistic Models, that generalizes both. As the name suggests, these models use the Entity-Relationship model known from relational database theory to specify the structure of probabilistic models. With respect to expressiveness, ERPMs come close to the logic-based formalisms we discuss next, while retaining a graphical, schema-like, flavor. Again, we refer to the literature [28] for a more thorough discussion.

5.3   Approaches in the Logical Setting

The integration of first order logic reasoning with probabilistic inference is a challenging goal, on which research has been conducted for several decades, with clear progress but limited convergence in terms of the practical formalisms that are in use. A particular strength of this type of approaches is that they can rely on a strong theoretical foundation; the combination of logic and probability has been studied more formally than, for instance, the expressiveness of plate models or the annotation of entity-relationship models with probabilistic information.

We first discuss a few general insights about logic and probability; next, we will discuss a number of formalisms.

5.3.1   Probabilistic Logics

While both logical and probabilistic inference are well-understood, their integration is not as straightforward as it may seem. The semantics of probabilistic information in the context of first order logic can be defined in different ways. In this section we consider two dimensions along which the semantics of probabilistic logics may differ.

Type 1 versus Type 2 Semantics. Seminal work on the integration of logic and probability has been conducted by Halpern [27]. To begin with, Halpern points out that there are different types of probabilities that one can make statements about. A statement such as "the probability that a randomly chosen bird flies" is inherently ambiguous, unless we specify what kind of probability we are referring to. In what Halpern calls a Type 1 probabilistic logic, a logical variable has a distribution over its domain associated with it, and we can talk about the probability of that variable x taking a particular value. For instance, if logical variable x has a uniform distribution over the domain $\{Tweety, Oliver, Fred, Larry, Peter\}$, it may hold that $Pr_x(Flies(x)) = 0.4$; this formula states that if we choose a random bird x from this particular distribution, there is a probability of 0.4 that $Flies(x)$ holds.

In Type 2 logics, probabilities about possible worlds are given.[4] Such probabilities are most easily interpreted as a degree of belief. Thus, we might state, for instance, that there is a probability of 0.2 that Tweety flies: $Pr(Flies(Tweety)) = 0.2$. Note that this cannot be expressed (using the same vocabulary of predicates and constants) using a Type 1 logic: in one particular world, Tweety either flies or it does not, so $Pr_x(Flies(Tweety))$ is either 0 or 1.

Which type of logic is most natural in a particular situation depends on the application. If we wish to describe what the probability is that x takes a particular value, given a certain (partially randomized) process for computing it, a Type 1 logic expresses this more directly. However, if we want to express a certain degree of belief that a particular fact is true, a Type 2 logic is more natural. (A Type 1 logic could be used here as well, but this would require introducing a logical variable x that represents a particular world, and talking about $Flies(Tweety, x)$ to indicate whether Tweety flies in a particular world x; this is not a very natural way of expressing things.) The two types of logics can in principle be mixed: for instance, when we talk about the probability that a coin is fair, we talk about the (type 2) probability that the (type 1) probability of obtaining heads is 0.5. Halpern calls such a combined structure a Type 3 logic.

As an example of the kind of reasoning that is possible with a Type 3 logic, consider the following program and query:

```
0.8: Flies(Tweety).
0.2: Flies(Oliver).
0.5: Flies(Fred).
1.0: Flies(Larry).
0.5: Flies(Peter).

Bird(x) -> x ~ Unif({Tweety, Oliver, Fred, Larry, Peter}).
```

The facts are annotated with type 2 probabilities. The rule on line 6 is our way of specifying a distribution over the domain of x when x is of type Bird; it determines type 1 probabilities. Consider the query $?-Pr(Pr_{x|Bird(x)}(Flies(x)) \geq 0.2)$. The query asks: if we select a random world from all possible worlds, what is the (type 2) probability that for this world it holds that the (type 1) probability that a randomly chosen bird flies is at least 0.2? In this case,

---

[4] A "possible world" is an assignment of truth values to all ground facts; for instance, given two propositions p and q, there are four possible worlds: one where p and q are true, one where both are false, and two where exactly one of them is true.

the answer is 1: since Larry flies in each possible world, and when choosing x there is a 0.2 probability that we choose Larry, the probability that a randomly chosen bird flies will always be at least 0.2. A general way to compute $? - Pr(Pr_{x|Bird(x)}(Flies(x)) \leq p)$, for $p > 0.2$, is to compute all possible worlds and their probability, check for each of these worlds whether the mentioned type 1 probability is at least p, and add up the probabilities of all these worlds.

Apart from introducing these logics, Halpern also shows that a complete axiomatization for them is not possible. As a result, for practical uses it is necessary to consider simpler versions of these logics. In the same way that practical logical inference systems (such as Prolog) use subsets of first order logic, practical probabilistic-logical methods will use more restrictive formalisms than Halpern's.

Proof versus Model Semantics. In probabilistic logic learning, two types of semantics are distinguished [15]: the model theoretic semantics and the proof theoretic semantics. Approaches that are based on the model theoretic semantics define a probability distribution over interpretations and extend probabilistic attribute-value techniques, such as Bayesian networks and Markov networks, while proof theoretic semantics approaches define a probability distribution over proofs and upgrade, e.g., stochastic context free grammars.

Example 7. Consider the case where each example is a sentence in natural language. In this example, a model theoretic approach would define a probability distribution directly over sentences. A proof theoretic approach would define a probability distribution over "proofs", in this case possible parse trees of the sentence (each sentence may have several possible parse trees). Note that the proof theoretic view is more general, in the sense that the distribution over sentences can be computed from the distribution over proofs.

5.3.2 Examples of Formalisms

Next, we will look at a number of different formalisms in which probabilistic logical models can be written down. They will mostly be illustrated with examples. As these formalisms sometimes express quite different types of knowledge, the corresponding examples also differ; it is difficult to define a single running example to compare all the formalisms because an example for one formalism is not necessarily suitable for illustrating the other.

Another point is that, in practice, in all these formalisms, probabilities can, but need not, be stated by the programmer; where not stated, they can be learned from data when necessary. A more difficult challenge is the learning of the program structure from data; this typically requires a search through the space of all possible model structures. Methods for learning parameters and structure have been proposed for most formalisms we mention here, but we will not go into detail about them.

All the examples below are examples of what is called *knowledge based model construction* (KBMC). This term was first introduced by Haddawy [26] and refers to the fact that a probabilistic model is constructed by means of a "program" that defines how the model should be constructed from certain knowledge available in the domain. That knowledge can be stated declaratively, or it can be hard-coded in an imperative program; we will see examples of both. In most cases, programming the model requires an understanding of the type of model that is being programmed; for instance, in formalisms based on Bayesian networks, the user is expected to know what a Bayesian network represents. Perhaps the most notable exception to this rule is Markov Logic, where the intuitive meaning of a program is relatively independent from the underlying Markov network.

Stochastic Logic Programs. Stochastic logic programs (SLPs) [46] follow the proof theoretic view and upgrade stochastic context free grammars to first order logic. SLPs are logic programs with probabilities attached to the clauses such that the probabilities of clauses with the same head sum to 1.0. These numbers indicate the probability that upon a call to a predicate, this particular clause is used to resolve the calling literal. (This is similar to how in a stochastic grammar a rule $S \rightarrow A$ is annotated with the probability that, given that a term S is encountered, this particular rule is used to rewrite it.) The probability of one particular inference chain is then computed as the product of the probabilities of the clauses that are used in the proof.

Example 8. The following SLP simulates the tossing of coins a variable number of times.

```
0.7: series([X|Y]) :- toss(X), series(Y).
0.3: series([]).
0.5: toss(heads).
0.5: toss(tails).
```

When the query ?-series(X) is called, there is a 30% chance that it results in X=[] (i.e., the second of the two clauses for series is used to answer the query). There is a 70% chance that the first clause is used to answer the query, instead of the second one; in this case, the first element of the list is instantiated with heads or tails with 50% probability each, after which a recursive call repeats the process. Repeatedly calling the same query is equivalent to random sampling from a process that in the end will give the following distribution:

```
X=[] : 0.3
X=[heads]: 0.105
X=[tails]: 0.105
X=[heads,heads]: 0.03675
X=[heads,tails]: 0.03675
X=[tails,heads]: 0.03675
```

```
X=[tails,tails]: 0.03675
X=[heads,heads,heads]: 0.0128625
. . .
```

Note that, in general, an inference chain may also fail (this was not the case in the above example), in which case no instantiation is returned.

SLPs provide an elegant way of describing stochastic processes; executing them amounts to random sampling using these processes. When failing inference chains exist, the SLP can also be used to estimate the probability of a literal being true using Monte Carlo sampling.

SLPs can easily be used to express Type 1 probabilities. For instance, the SLP

```
0.2: bird(tweety).
0.2: bird(oliver).
0.2: bird(fred).
0.2: bird(larry).
0.2: bird(peter).
```

ensures that when the query ?- bird(X) is called, X is instantiated to tweety or to other constants, each with a probability of 0.2.

Note that the numbers annotating the facts are not type 2 probabilities; the meaning of 0.2: bird(tweety) is not that $Pr(Bird(Tweety)) = 0.2$, but $Pr(x = Tweety|Bird(x)) = 0.2$, which is a Type 1 probability. More specifically, the SLP defines the distribution associated with the variable X when ?- bird(X) is called. Type 2 probabilities are difficult to write down in an SLP.

Prism. The PRISM system [54] follows an approach that is somewhat similar to the SLP approach. Here, no probabilities are associated with clauses, but there is a so-called choice predicate that indicates that one of a number of alternative instantiations is chosen. In the case of PRISM the choice predicate is called msw, for multi-valued switch, and it instantiates a variable with a particular constant according to a given distribution, which is defined by a set_sw predicate. Thus, the following PRISM program and query

```
values(bird, [tweety,oliver,fred,larry,peter]).
set_sw(bird, [0.2, 0.2, 0.2, 0.2, 0.2]).
```

holds exactly the same information as the SLP shown above, and the query ?- bird(X). instantiates the variable X with tweety in 20% of the cases.

PRISM is among the most elaborated probabilistic-logical learning systems, with a clearly defined distribution-based semantics and efficient built-in procedures for inference and learning. Further details on it can be found at the PRISM website, sato-www.cs.titech.ac.jp/prism/, which also contains pointers to the extensive literature.

Bayesian Logic Programs and Logical Bayesian Networks. Bayesian
Logic Programs (BLPs) [34] aim at combining the inference power of Bayesian
networks with that of first-order logic reasoning. Similar to PRMs, the semantics of a BLP is defined by translating it to a Bayesian network. Using
this network, the probability of a given interpretation or the probability that
a given query yields a particular answer can be computed.

Logical Bayesian networks [19] are a variant of BLPs in which a cleaner
distinction is made between the definition of which stochastic variables exist (that is, which objects exist in the domain of discourse, and which objects have what properties), and the probabilistic inference. In other words,
probabilistic inference always happens in a deterministically defined network,
whereas in BLPs the network's definition can itself be a result of probabilistic
inference.

The following (taken from Fierens et al. [19]) is an example of an LBN:

```
/* definition of the random variables */
random(iq(S)) <- student(S).
random(ranking(S)) <- student(S).
random(diff(C)) <- course(C).
random(grade(S,C)) <- takes(S,C).

/* definition of dependencies */
ranking(S) | grade(S,C) <- takes(S,C).
grade(S,C) | iq(S), diff(C).

/* definition of the universe */
student(john).  student(pete).
course(ai).  course(db).
takes(john,ai).  takes(john,db).  takes(pete,ai).
```

The LBN states in its first part that for each student s, a stochastic variable
$iq(s)$ and another stochastic variable $ranking(s)$ is defined; for each course
c, a variable $diff(c)$ is defined; and for each student-course combination a
variable $grade(s, c)$ is defined. The second part states under what conditions
there are direct dependencies among stochastic variables.

The first two parts of this example program together define a function F
that maps interpretations to Bayesian networks. The third part states some
knowledge about the world; here it is a set of ground facts, but it could be
any logic program. The minimal model of this program is the interpretation
that serves as an input to the function F.

Compared to PRMs, the first two parts (defining the variables and dependencies) play the role of the relational database schema. The LBN specification is less rigid in the sense that any logic program can be used to define the
"schema"; for instance, if we wanted to express in the LBN that a student
is only graded for a course if she is registered for examination this term, we
can simply change one rule into

```
random(grade(S,C)) <- takes(S,C), registered(S).
```

Standard schema definition languages for relational databases, such as SQL, do not support such complicated schema definitions; also for PRMs it is not clear how this can be done. In this sense, LBNs are more expressive than PRMs.

In a BLP, the above program would be written as follows:

```
iq(S) | student(S).
ranking(S) | student(S).
diff(C) | course(C).
grade(S,C) | takes(S,C).
grade(S,C) | iq(S), diff(C), takes(S,C).
ranking(S) | grade(S,C), takes(S,C).
```

Note that the BLP has a simpler structure, but does not make certain details explicit. There is an essential difference between structure-determining predicates, such as *takes*, and stochastic variables, such as *grade*. The network resulting from this should contain a stochastic variable *grade(john, ai)*, but should not contain a stochastic variable *takes(john, ai)*; rather, the *takes* predicate defines the conditions under which there should be an edge between other variables. The role of `takes(S,C)` in the BLP is very different from that of `grade(S,C)` or `iq(S)`, but this is not visible in the clauses. To compensate for this, BLPs come with a graphical interface, Balios [35], in which these additional constraints can be specified in the graphical representation of the network. BLPs as defined in Balios are more similar to LBNs. For a more extensive comparison of LBNs with PRMs and BLPs, we refer to Fierens et al. [19].

Markov Logic Networks. Markov networks, also known as Markov random fields, are undirected probabilistic graphical models. In these models, there is an edge between two nodes if and only if knowing the value of one node carries information about the other, regardless of what other evidence is given.

Markov Logic Networks (MLNs) [52] can be seen as an upgrade to first order logic of Markov networks. MLNs are defined as sets of weighted first order logic formulas. These formulas do not have to be universally true in order to be valid. They are viewed as "soft" constraints on logical interpretations: an interpretation that violates a formula is not considered impossible, it is simply considered less likely. More specifically, each ground instantiation of a formula that evaluates to false in a particular interpretation reduces the probability of that interpretation with a constant factor.

Example 9. A frequently used toy example in the context of Markov Logic is the following:

```
Friend(x,y) <=> Friend(y,x).
Friend(x,y), Smokes(x) => Smokes(y).
```

The formulas indicate that the Friend relationship is symmetric, and when one person smokes, friends of this person smoke as well. In standard logic, these clauses would not be useful, because strictly speaking they are incorrect: Friend is not perfectly symmetric, and it is not the case that whenever someone smokes, all their friends necessarily smoke. In Markov logic, the logical formulas are not interpreted as universally true, but as statements that "tend to be true", in the sense that they have few exceptions. Each formula will be assigned a weight, which can be learned from a data set; the larger a weight, the more each exception to the formula reduces the overall probability of the model, other things being equal.

Note that, while the underlying inference engine is based on Markov networks, there is nothing in the structure of the Markov logic network that shows this. The user can simply state some logical formulas that he or she believes are usually true; the Markov logic inference engine will determine weights for the formulas reflecting how strongly they hold on a given dataset, and next, be able to tell the user with what probability a certain statement is true.

The Alchemy system[5] implements structure and parameter learning for MLNs. It is among the most popular SRL systems at the time of writing this text.

CP-Logic. CP-logic [67], originally called "Logic programs with annotated disjunctions" [68], differs from the other approaches in that it defines a *causal* probabilistic model. The causality is not just an interpretation that can be given to the model (and which might be correct or incorrect); the causal interpretation is by definition correct, because it is in the semantics of the model. This is very different from, for instance Bayesian networks, and it sets CP-logic apart from the other formalisms discussed here.

A CP-logic rule is of the form

$$h_1 : \alpha_1 \lor h_2 : \alpha_2 \lor \cdots \lor h_k : \alpha_k \leftarrow b_1, b_2, \ldots, b_n$$

where the h_i and b_j are literals and the α_i are reals such that $\forall i : \alpha_i \geq 0$ and $\sum_i \alpha_i \leq 1$. The rule specifies a part of a causal process, and states that whenever at some point the body becomes true for a particular instantiation of the logical variables, an *event* happens that *causes* at most one of the head literals to become true. (If the selected literal was already true, the event has no effect.) More specifically, one literal is drawn from the set of head literals according to the distribution specified by the α_i (if $\sum_i \alpha_i < 1$, there is a probability of $1 - \sum_i \alpha_i$ that nothing is selected), and the value of that literal is set to true, regardless of what it was before. Whenever an event happens that causes one literal to be selected, the outcome of this event is independent of any other such events that occurred earlier. Thus, selections

---

[5] http://alchemy.cs.washington.edu/

in different rules, as well as selections in different instantiations of the same rule, are made independently.

Example 10. The following CP-logic program [43] describes how two people may go shopping and buy particular kinds of food:

```
0.3: buys(john, spaghetti) v 0.7: buys(john, chicken) <- shops(john).
0.4: buys(mary, spaghetti) v 0.6: buys(mary, fish) <- shops(mary).
0.8: shops(john).
0.5: shops(mary).
```

It states that John may decide to go buy some food today, and with a certain probability will buy spaghetti or chicken (and only one of these); similarly, Mary may buy spaghetti or fish. The rules are causal: if anything prevents John from shopping, there will be no chicken tonight.

While the structure of CP-logic programs may seem similar to that of Bayesian networks, there are several important differences. First: a Bayesian network defines a factorization of a joint distribution, not a causal structure; while arcs can be interpreted as causal, this interpretation is not part of the network's semantics. In a CP-logic program, it is. Second, the numbers we find in the conditional probability distributions of a Bayesian network are conditional probabilities; a node Y with parent X is annotated with a table that contains $Pr(Y|X)$. The numbers we find in a CP-logic program are not conditional probabilities; they can be equal to them, or they can be smaller. For instance, when we have a rule $0.5\colon$ y <- x, the conditional probability of y given x is at least 0.5 (because when x is true, this alone already causes y to be true in 50% of the cases), but it can be greater because there may be other events that make y true. Third, a CP-logic program can indicate cyclic causality, while a Bayesian network cannot contain cycles. For instance, when we have two cogwheels A and B that are connected to each other, if an external cause makes cogwheel A turn, then this causes B to turn as well, but also vice versa: if an external cause makes B turn, that causes A to turn. A Bayesian network cannot express such bidirectional causality, while a CP-logic program can simply contain both a <- b and b <- a [67].

BLOG. BLOG [44], which stands for Bayesian Logic, is yet another approach to combining probabilistic with logic inference. Special about this one is that it explicitly aims at reasoning about worlds with unknown objects, or worlds in which it is not known whether two constants actually refer to the same object or not (identity uncertainty).

Example 11. The following example BLOG program is taken from [44]. The program defines a stochastic process where four balls are drawn (with replacement) from an urn; we do not know how many balls are in the urn, but we do know that the balls that have been put in the urn were selected randomly from a population with 50% blue and 50% green balls.

```
type Color; type Ball; type Draw;

random Color TrueColor(Ball);
random Ball BallDrawn(Draw);
random Color ObsColor(Draw);

guaranteed Color Blue, Green;
guaranteed Draw Draw1, Draw2, Draw3, Draw4;

#Ball ~ Poisson[6]();
TrueColor(b) ~ TabularCPD[[0.5,0.5]]();
BallDrawn(d) ~ Uniform({Ball b});

ObsColor(d)
  if (BallDrawn(d) != null) then
    ~ TabularCPD[[0.8, 0.2], [0.2, 0.8]] (TrueColor(BallDrawn(d)));
```

The first line defines three types of objects; the extension of each type is a set of objects, the size of which is not specified at this point. The `random` statements define random variables associated with each object, and define the values that these random variables can take; for instance, with each ball a variable TrueColor is associated, and this variable takes on a value of type Color. Next, the extensions of Color and Draw are specified: we guarantee that there are exactly two colors (Blue and Green), and four draws. The extension of Ball is not known so precisely: its cardinality, i.e., the number of balls, is unknown but a probability distribution is given for it (Poisson distribution with parameter 6). Given a ball, its TrueColor is Blue or Green with a probability of 0.5 each; given a draw, the ball that is drawn is one from the extension of Ball, drawn uniformly. Finally, the observed color of a ball is the same as its true color in 80% of the cases; in 20% of the cases the color is observed incorrectly.

As can be seen in the example, BLOG strives explicitly at defining *sets* of objects (the elements of which may remain anonymous), rather than individual objects, as is the case in for instance LBNs (where the universe is explicitly defined and each object gets its own name). While most methods work with a fixed universe, and define a distribution over interpretations for this universe, a BLOG model extends this principle by defining a distribution over universes.

ProbLog. By now the reader will be convinced that many different approaches to statistical relational learning exist, and he or she may wonder why this is the case. Part of the answer probaby lies in the fact that (on the one hand) a need is felt for the kind of expressiveness that these formalisms offer, but (on the other hand) this need is application-driven, and thus most researchers have developed a formalism that is suitable for the kind of applications *they* were thinking about. This has led to a variety of formalisms that share many properties, but also have their own specificity, which is often desired for the context they are used in. This is also visible in the examples chosen to illustrate the formalisms: these differ quite strongly.

Seeing the variety as well as the commonalities in all these formalisms, De Raedt et al. [13] argue that there is a need for an underlying programming language, in which the other formalisms could be implemented, but which would itself offer important functionality for efficient probabilistic inference. To this aim, they propose the probabilistic-logical programming language ProbLog. ProbLog is a conceptually very simple extension of the well-known logic programming language Prolog. A ProbLog program consists of a set of definite Horn Clauses, just like a standard Prolog program, but each fact is additionally (and optionally) annotated with a number that expresses the probability that this fact is true.

The following is an example of a ProbLog program:

```
path(X,Y) :- edge(X,Z), path(Z,Y).
edge(1,2).
0.4: edge(2,3).
0.2: edge(3,4).
0.3: edge(1,3).
```

The query ?- path(1,4) results in a probability distribution over yes and no, with yes having a probability of (we shorten path and edge to p and e):

$$Pr(p(1,4)) = Pr(p(1,3))Pr(e(3,4))$$

with

$$
\begin{aligned}
Pr(p(1,3)) &= Pr(e(1,2))Pr(e(2,3)) + Pr(e(1,3)) - Pr(e(1,2))Pr(e(2,3))Pr(e(1,3)) \\
&= Pr(e(2,3)) + Pr(e(1,3)) - Pr(e(2,3))Pr(e(1,3)) \quad (\text{as } Pr(p(1,2)) = 1) \\
&= Pr(e(2,3)) + (1 - Pr(e(2,3)))Pr(e(1,3)) \\
&= 0.4 + 0.6 * 0.3 = 0.58.
\end{aligned}
$$

When the graph becomes more complex than this example graph, the complexity of the calculations rises quickly. This is in part due to the inclusion-exclusion principle, which states that

$$Pr(\bigcup_i A_i) = \sum_i Pr(A_i) - \sum_{\neq i,j} Pr(A_i \cap A_j) + \sum_{\neq i,j,k} Pr(A_i \cap A_j \cap A_k) - \cdots \pm Pr(\bigcap_i A_i)$$

which in the case of independent A_i becomes

$$Pr(\bigcup_i A_i) = \sum_i Pr(A_i) - \sum_{\neq i,j} Pr(A_i)Pr(A_j) + \sum_{\neq i,j,k} Pr(A_i)Pr(A_j)Pr(A_k) - \cdots \pm \prod_i Pr(A_i)$$

The size of this formula increases exponentially with the number of A_i, and computing path probabilities in graphs is a generalization of the above calculation. Thus, computing this probability easily becomes intractable for large graphs. However, much effort has been spent on compiling such structures into more efficient representations, many of which are variants of the so-called binary decision diagrams (BDDs). ProbLog uses a similar translation to render the computation of probabilities more tractable. Besides this, the ProbLog engine implements many other ideas that improve efficiency. All together, they make it possible to answer queries as efficiently as possible. The idea behind ProbLog is that it could be used to implement engines for other formalisms, which will then automatically inherit all these efficient implementations.

For further details about ProbLog we refer to the literature [36, 4].

5.4   Other Approaches

More specific statistical learning techniques, such as Naïve Bayes and Hidden Markov Models, have also been upgraded to the relational setting [22, 59]. While the above formalisms are mostly logic or relational database oriented, other types of languages have been proposed; IBAL [49], for instance, is a functional probabilistic language, while CLP(BN) [53] makes use of constraint logic programming to integrate probabilistic with logical inference. A much more complete and in-depth overview of approaches is presented by Getoor and Taskar [24].

6   General Remarks and Challenges

Many challenges remain in the area of statistical relational learning. We list a few that are currently attracting a significant amount of interest from researchers.

6.1   Understanding Commonalities and Differences

A large number of formalisms for statistical relational learning exist. This situation is sometimes referred to as "the alphabet soup": we have BLOG, BLPs, BUGS, CLP(BN), CP-logic, IBAL, ICL, LBNs, MLNs, PRISM, PRMs, ProbLog, RBNs, SLPs, ... It seems a natural goal to try to merge all these formalisms into a single one (or at least a few ones) that subsumes all of them. Yet, this goal seems difficult to achieve, and the validity of the goal has been challenged: it is possible that depending on the task, one formalism is much more suitable than another, so why try to merge them into one? Still, even if such a grand unification is not necessarily desirable, it seems useful to understand better the differences between all these approaches. Currently, such an understanding exists to some extent, but it remains largely anecdotal and incomplete. Several researchers have shown how to translate programs from one formalism to another, showing equivalence or subsumption between some formalisms. Yet, such translations have to be interpreted with caution, as there are several levels of "equivalence": one can define equivalence in terms of the actual models that can be learned, in terms of the model structures, or in terms of the sets of model structures that can be given by the user as a bias for the learner; models can be interpreted as (constraints over) joint distributions, or as functions that map a logical interpretation onto such a (constraint over a) distribution. A single well-understood framework for comparing formalisms does not appear to exist at this moment. Finally, there is the issue of user-friendliness: how easily can a user write down certain background knowledge, and how easily can that user interpret the models? Again, "interpretation" is a broad term here. For instance, in Markov Logic, clauses are given weights that tell us how likely the clauses are to be violated, but there is no simple connection between these weights and probabilities, which contrasts with the situation in, for instance, BLPs. On the other hand, probabilities of specific facts can of course be inferred by the model whenever necessary.

Earlier in this chapter we have discussed differences in relational learners in terms of the features that they implicitly construct. A comparison between SRL systems from this point of view would be one way in which additional insight can be gained. Such a comparison would indicate to what extent the different SRL formalisms are truly relational; as said before, a relational learner that constructs only a small set of features could be said to be "less relational", and is in that sense less expressive.

There has been a number of practical comparisons of SRL systems. A large number of approaches is compared, from a user's point of view (how easy is it to model a particular problem using a particular SRL approach), by Bruynooghe et al. [3] and Taghipour et al. [58]. In both papers the authors conclude that simple problems can still be hard to model with even the most advanced SRL approaches available today.

6.2 Parameter Learning and Structure Learning

The learning of probabilistic graphical models involves two distinguished tasks: parameter learning and structure learning. *Parameter learning* is the easiest task. Given a model structure (i.e., a directed or undirected graph, in the case of Bayesian or Markov networks; a set of first order logic clauses in the case of Markov logic; etc.), the task is to fit the model to the data, i.e., determine the parameter settings for which an optimal fit is obtained. *Structure learning* is more difficult. It involves determining an optimal model structure (i.e., determining the optimal graph structure, determining the optimal set of clauses, etc.). This in itself often involves a search through the model space, where each model is individually evaluated by fitting it to the data (i.e., via parameter learning) and measuring how good the fit is. However, the structure learning step may exploit additional background knowledge that the user has about the likely structure of the model.

Parameter learning is a relatively standard task by now. Structure learning, on the other hand, needs to be implemented differently depending on the formalism that is being used; for instance, since the syntax of CP-logic programs is quite different from that of Markov logic networks, quite different structure learning approaches are required. For Markov networks, Richardson and Domingos [52] show how the structure can be determined by making use of the ILP system Claudien [12] as an auxiliary system. Meert and Blockeel [43] show how the structure of acyclic CP-logic programs can be learned by turning them into equivalent Bayesian networks that contain one latent variable per CP-logic rule and the structure of which is constrained in a particular way; they then show how standard techniques for learning the structure of Bayesian networks can be adapted to ensure the resulting networks obey these structural constraints. Structure learning methods have been proposed for many other formalisms as well [24, 20].

6.3 Scalability

Probabilistic inference in graphical models is, in the general case, NP-hard: roughly speaking, its computational complexity increases exponentially in the size of the network. While this has always been an issue in probabilistic inference, it is even more so in statistical relational learning. The size of the network generally depends on the size of the domain, which may be very large. For instance, consider the "Friends and Smokers" example from the discussion on Markov logic. There is a stochastic variable for each pair of persons (x, y); when we are talking about a few thousand persons, this means there will be millions of variables in the ground network, interconnected in complex ways. Clearly, inference in this ground network will be challenging. An important approach towards alleviating this problem is the concept of *lifted inference*.

The term "lifted inference" refers to performing inference on a higher level of abstraction than the ground network. A crucial property that lifted inference methods rely on is that of indistinguishability. Sometimes two stochastic variables are exactly equivalent with respect to what is known about them. This can be true in ordinary networks, but it is much more often the case in ground models that have been generated by first-order models. For instance, in the Markov logic network for Friends and Smokers, suppose Bart has one friend who smokes and one who does not, and these do not know each other; and suppose Lisa is in exactly the same situation. If nothing else is known, then whatever we can infer about Bart we can also infer about Lisa: all probabilites will be equal.

The idea of lifted inference is similar to what is used in logic programming: in the Prolog programming language, for instance, inference is done on the level of variables, rather than constants, insofar possible. Given the rule `q(X) :- p(X)`, a Prolog engine can infer from `p(a)` that `q(a)` holds (i.e., apply the rule to a specific case), but it can also infer from `p(X)` that `q(X)` holds (regardless of what X is). Inference is done on the level of universally quantified variables where possible, and on the ground level where needed. Constraint logic programming provides a middle ground by allowing inference that keeps track of sets of ground instantiations for which the current inference could be made (and these sets may be defined extensionally, by listing their elements, or intensionally, using constraints). This allows for much more efficient inference, compared to reasoning only on the ground level.

The same principle can be used in probabilistic-logical models. If we write

```
0.5: p(a).
0.5: p(b).
0.5: p(c).

0.5: q(X) <- p(X).
r(X,Y) <- p(X),q(Y).
```

then it is clear that $q(a)$, $q(b)$ and $q(c)$ must all have the same probability of being true (0.25), and similarly, $Pr(r(X,Y)) = 0.125$ whenever $X \in \{a, b, c\}$ and $Y \in \{a, b, c\}$. Clearly, we can compute probabilities on the level of (extensionally or intensionally defined) sets of ground literals, rather than on the level of individual literals.

Probabilistic inference is much more complicated than pure logical inference, however, and lifting it to the first-order context is a challenge that is still far from solved. It does have a large potential towards more efficient inference in statistical relational learning. Seminal work in this area was performed by Poole [50], and in the ensuing decade many other authors took up the challenge.

7   Recommended Reading

Starting out from a general description of relational learning and how it differs from standard learning, we have discussed neural-network based approaches to relational learning, and statistical relational learning. A much more detailed treatment of all these topics is available in several reference works. Directly relevant references to the literature include the following. A comprehensive introduction to ILP can be found in De Raedt's book [10] on logical and relational learning, or in the collection edited by Džeroski and Lavrač [18] on relational data mining. Learning from graphs is covered by Cook and Holder [7]. Džeroski and Lavrač [18] is also a good starting point for reading about multi-relational data mining, together with research papers on multi-relational data mining systems. Statistical relational learning in general is covered in the collection edited by Getoor and Taskar [24], while De Raedt and Kersting [15] and De Raedt et al. [14] present overviews of approaches originating in logic-based learning.

Acknowledgements. This chapter builds on earlier publications written in close collaboration with Werner Uwents, Jan Struyf, and Maurice Bruynooghe. The author thanks Daan Fierens and an anonymous reviewer for valuable comments.

References

1. Blockeel, H., Bruynooghe, M.: Aggregation versus selection bias, and relational neural networks. In: IJCAI 2003 Workshop on Learning Statistical Models from Relational Data, SRL 2003, Acapulco, Mexico, August 11 (2003)
2. Bratko, I.: Prolog Programming for Artificial Intelligence. Addison-Wesley (1986)
3. Bruynooghe, M., De Cat, B., Drijkoningen, J., Fierens, D., Goos, J., Gutmann, B., Kimmig, A., Labeeuw, W., Langenaken, S., Landwehr, N., Meert, W., Nuyts, E., Pellegrims, R., Rymenants, R., Segers, S., Thon, I., Van Eyck, J., Van den Broeck, G., Vangansewinkel, T., Van Hove, L., Vennekens, J., Weytjens, T., De Raedt, L.: An exercise with statistical relational learning systems. In: Proceedings of the 6th International Workshop on Statistical Relational Learning (2009)
4. Bruynooghe, M., Mantadelis, T., Kimmig, A., Gutmann, B., Vennekens, J., Janssens, G., De Raedt, L.: Problog technology for inference in a probabilistic first order logic. In: Coelho, H., Studer, R., Wooldridge, M. (eds.) ECAI. Frontiers in Artificial Intelligence and Applications, vol. 215, pp. 719–724. IOS Press (2010)
5. Buntine, W.: Operations for learning with graphical models. Journal of Artificial Intelligence Research 2, 159–225 (1994)
6. Cook, D.J., Holder, L.B.: Substructure discovery using minimum description length and background knowledge. J. Artif. Intell. Res. (JAIR) 1, 231–255 (1994)
7. Cook, D.J., Holder, L.B.: Mining Graph Data. Wiley (2007)
8. Garcez, A.S.d., Zaverucha, G.: The connectionist inductive learning and logic programming system. Appl. Intell. 11(1), 59–77 (1999)
9. De Raedt, L.: Logical settings for concept learning. Artificial Intelligence 95, 187–201 (1997)
10. De Raedt, L.: Logical and Relational Learning. Springer (2008)
11. De Raedt, L., Dehaspe, L.: Clausal discovery. Machine Learning 26, 99–146 (1997)
12. De Raedt, L., Dehaspe, L.: Clausal discovery. Machine Learning 26(2-3), 99–146 (1997)
13. De Raedt, L., Demoen, B., Fierens, D., Gutmann, B., Janssens, G., Kimmig, A., Landwehr, N., Mantadelis, T., Meert, W., Rocha, R., Santos Costa, V., Thon, I., Vennekens, J.: Towards digesting the alphabet-soup of statistical relational learning. In: Proceedings of the NIPS*2008 Workshop Probabilistic Programming, pp. 1–3 (2008)
14. De Raedt, L., Frasconi, P., Kersting, K., Muggleton, S.H. (eds.): Probabilistic Inductive Logic Programming. LNCS (LNAI), vol. 4911. Springer, Heidelberg (2008)

15. De Raedt, L., Kersting, K.: Probabilistic logic learning. SIGKDD Explorations 5(1), 31–48 (2003)
16. Dehaspe, L., Toivonen, H.: Discovery of frequent datalog patterns. Data Mining and Knowledge Discovery 3(1), 7–36 (1999)
17. Driessens, K.: Relational reinforcement learning. Mach. Learn. 43, 7–52 (2001)
18. Džeroski, S., Lavrač, N. (eds.): Relational Data Mining. Springer (2001)
19. Fierens, D., Blockeel, H., Bruynooghe, M., Ramon, J.: Logical bayesian networks and their relation to other probabilistic logical models. In: Kramer, Pfahringer [38], pp. 121–135
20. Fierens, D., Ramon, J., Bruynooghe, M., Blockeel, H.: Learning Directed Probabilistic Logical Models: Ordering-Search Versus Structure-Search. In: Kok, J.N., Koronacki, J., Lopez de Mantaras, R., Matwin, S., Mladenič, D., Skowron, A. (eds.) ECML 2007. LNCS (LNAI), vol. 4701, pp. 567–574. Springer, Heidelberg (2007)
21. Finn, P., Muggleton, S., Page, D., Srinivasan, A.: Pharmacophore discovery using the inductive logic programming system Progol. Mach. Learn. 30, 241–270 (1998)
22. Flach, P.A., Lachiche, N.: Naive bayesian classification of structured data. Machine Learning 57(3), 233–269 (2004)
23. Friedman, N., Getoor, L., Koller, D., Pfeffer, A.: Learning probabilistic relational models. In: Dean, T. (ed.) IJCAI, pp. 1300–1309. Morgan Kaufmann (1999)
24. Getoor, L., Taskar, B.: Introduction to Statistical Relational Learning. MIT Press (2007)
25. Gilks, W.R., Thomas, A., Spiegelhalter, D.J.: A language and program for complex bayesian modelling. The Statistician 43, 169–178 (1994)
26. Haddawy, P.: Generating bayesian networks from probablity logic knowledge bases. In: de Mántaras, R.L., Poole, D. (eds.) UAI, pp. 262–269. Morgan Kaufmann (1994)
27. Halpern, J.Y.: An analysis of first-order logics of probability. Artificial Intelligence 46, 311–350 (1990)
28. Heckerman, D., Meek, C., Koller, D.: Probabilistic entity-relationship models, prms, and plate models. In: Introduction to Statistical Relational Learning, pp. 201–238. MIT Press (2007)
29. Horváth, T., Ramon, J., Wrobel, S.: Frequent subgraph mining in outerplanar graphs. In: Proc. of the 12th ACM SIGKDD Int'l Conf. on Knowledge Discovery and Data Mining, pp. 197–206 (2006)
30. Jaeger, M.: Relational bayesian networks. In: UAI 1997: Proceedings of the Thirteenth Conference on Uncertainty in Artificial Intelligence, Brown University, Providence, Rhode Island, USA, August 1-3, pp. 266–273. Morgan Kaufmann (1997)
31. Jensen, D., Neville, J.: Linkage and autocorrelation cause feature selection bias in relational learning. In: Proc. of the 19th Int'l Conf. on Machine Learning, pp. 259–266 (2002)
32. Jensen, D., Neville, J., Gallagher, B.: Why collective inference improves relational classification. In: Proc. of the 10th ACM SIGKDD Int'l Conf. on Knowledge Discovery and Data Mining, pp. 593–598 (2004)
33. Karalič, A., Bratko, I.: First order regression. Machine Learning 26, 147–176 (1997)
34. Kersting, K.: An Inductive Logic Programming Approach to Statistical Relational Learning. IOS Press (2006)

35. Kersting, K., Dick, U.: BALIOS – The Engine for Bayesian Logic Programs. In: Boulicaut, J.-F., Esposito, F., Giannotti, F., Pedreschi, D. (eds.) PKDD 2004. LNCS (LNAI), vol. 3202, pp. 549–551. Springer, Heidelberg (2004)

36. Kimmig, A., De Raedt, L.: Local query mining in a probabilistic prolog. In: Boutilier, C. (ed.) IJCAI, pp. 1095–1100 (2009)

37. Koller, D., Friedman, N., Getoor, L., Taskar, B.: Graphical models in a nutshell. In: Introduction to Statistical Relational Learning, pp. 13–55. MIT Press (2007)

38. Kramer, S., Pfahringer, B. (eds.): ILP 2005. LNCS (LNAI), vol. 3625. Springer, Heidelberg (2005)

39. Krogel, M.-A., Rawles, S., Železný, F., Flach, P.A., Lavrač, N., Wrobel, S.: Comparative Evaluation of Approaches to Propositionalization. In: Horváth, T., Yamamoto, A. (eds.) ILP 2003. LNCS (LNAI), vol. 2835, pp. 197–214. Springer, Heidelberg (2003)

40. Krogel, M.-A., Wrobel, S.: Transformation-Based Learning Using Multirelational Aggregation. In: Rouveirol, C., Sebag, M. (eds.) ILP 2001. LNCS (LNAI), vol. 2157, pp. 142–155. Springer, Heidelberg (2001)

41. Lafferty, J.D., McCallum, A., Pereira, F.C.N.: Conditional random fields: Probabilistic models for segmenting and labeling sequence data. In: Brodley, C.E., Danyluk, A.P. (eds.) ICML, pp. 282–289. Morgan Kaufmann (2001)

42. Lloyd, J.W.: Logic for Learning. Springer (2003)

43. Meert, W., Struyf, J., Blockeel, H.: Learning ground CP-Logic theories by leveraging bayesian network learning techniques. Fundam. Inform. 89(1), 131–160 (2008)

44. Milch, B., Marthi, B., Russell, S.J., Sontag, D., Ong, D.L., Kolobov, A.: Blog: Probabilistic models with unknown objects. In: Kaelbling, L.P., Saffiotti, A. (eds.) IJCAI, pp. 1352–1359. Professional Book Center (2005)

45. Muggleton, S.: Inverse entailment and Progol. New Generation Computing, Special issue on Inductive Logic Programming 13(3-4), 245–286 (1995)

46. Muggleton, S.: Stochastic logic programs. In: De Raedt, L. (ed.) Advances in Inductive Logic Programming, pp. 254–264. IOS Press (1996)

47. Nijssen, S., Kok, J.N.: The gaston tool for frequent subgraph mining. Electr. Notes Theor. Comput. Sci. 127(1), 77–87 (2005)

48. Perlich, C., Provost, F.J.: Aggregation-based feature invention and relational concept classes. In: Getoor, L., Senator, T.E., Domingos, P., Faloutsos, C. (eds.) KDD, pp. 167–176. ACM (2003)

49. Pfeffer, A.: The design and implementation of ibal: A general-purpose probabilistic programming language. Technical Report TR-12-05, Harvard University (2005)

50. Poole, D.: First-order probabilistic inference. In: Gottlob, G., Walsh, T. (eds.) IJCAI, pp. 985–991. Morgan Kaufmann (2003)

51. Quinlan, J.R.: Learning logical definitions from relations. Machine Learning 5, 239–266 (1990)

52. Richardson, M., Domingos, P.: Markov logic networks. Mach. Learn. 62(1-2), 107–136 (2006)

53. Santos Costa, V., Page, D., Qazi, M., Cussens, J.: Clp(bn): Constraint logic programming for probabilistic knowledge. In: Meek, C., Kjærulff, U. (eds.) UAI, pp. 517–524. Morgan Kaufmann (2003)

54. Sato, T., Kameya, Y.: PRISM: A symbolic-statistical modeling language. In: Proceedings of the 15th International Joint Conference on Artificial Intelligence (IJCAI 1997), pp. 1330–1335 (1997)

55. Spiegelhalter, D.J.: Bayesian graphical modelling: a case-study in monitoring health outcomes. Applied Statistics 47, 115–134 (1998)
56. Struyf, J., Blockeel, H.: Relational learning. In: Sammut, C., Webb, G. (eds.) Encyclopedia of Machine Learning, pp. 851–857. Springer (2010)
57. Tadepalli, P., Givan, R., Driessens, K.: Relational reinforcement learning: An overview. In: Proc. of the ICML 2004 Wshp. on Relational Reinforcement Learning, pp. 1–9 (2004)
58. Taghipour, N., Fierens, D., Blockeel, H.: Probabilistic logical learning for biclustering: A case study with surprising results. CW Reports CW597, Department of Computer Science, K.U.Leuven (October 2010)
59. Thon, I., Landwehr, N., De Raedt, L.: A Simple Model for Sequences of Relational State Descriptions. In: Daelemans, W., Goethals, B., Morik, K. (eds.) ECML PKDD 2008, Part II. LNCS (LNAI), vol. 5212, pp. 506–521. Springer, Heidelberg (2008)
60. Towell, G.G., Shavlik, J.W.: Knowledge-based artificial neural networks. Artif. Intell. 70(1-2), 119–165 (1994)
61. Uwents, W.: Learning complex aggregate features with relational neural networks. PhD thesis, Katholieke Universiteit Leuven (2011) (forthcoming)
62. Uwents, W., Blockeel, H.: Classifying relational data with neural networks. In: Kramer, Pfahringer [38], pp. 384–396
63. Uwents, W., Blockeel, H.: A Comparison between Neural Network Methods for Learning Aggregate Functions. In: Boulicaut, J.-F., Berthold, M.R., Horváth, T. (eds.) DS 2008. LNCS (LNAI), vol. 5255, pp. 88–99. Springer, Heidelberg (2008)
64. Uwents, W., Blockeel, H.: Learning Aggregate Functions with Neural Networks Using a Cascade-Correlation Approach. In: Železný, F., Lavrač, N. (eds.) ILP 2008. LNCS (LNAI), vol. 5194, pp. 315–329. Springer, Heidelberg (2008)
65. Uwents, W., Monfardini, G., Blockeel, H., Gori, M., Scarselli, F.: Neural networks for relational learning: An experimental comparison. Machine Learning 82, 315–349 (2011)
66. Van Assche, A., Vens, C., Blockeel, H., Džeroski, S.: First order random forests: Learning relational classifiers with complex aggregates. Machine Learning 64(1-3), 149–182 (2006)
67. Vennekens, J., Denecker, M., Bruynooghe, M.: Cp-logic: A language of causal probabilistic events and its relation to logic programming. TPLP 9(3), 245–308 (2009)
68. Vennekens, J., Verbaeten, S., Bruynooghe, M.: Logic Programs with Annotated Disjunctions. In: Demoen, B., Lifschitz, V. (eds.) ICLP 2004. LNCS, vol. 3132, pp. 431–445. Springer, Heidelberg (2004)
69. Vens, C., Ramon, J., Blockeel, H.: Refining Aggregate Conditions in Relational Learning. In: Fürnkranz, J., Scheffer, T., Spiliopoulou, M. (eds.) PKDD 2006. LNCS (LNAI), vol. 4213, pp. 383–394. Springer, Heidelberg (2006)
70. Washio, T., Motoda, H.: State of the art of graph-based data mining. SIGKDD Explorations 5(1), 59–68 (2003)
71. Yan, X., Han, J.: gspan: Graph-based substructure pattern mining. In: ICDM, pp. 721–724. IEEE Computer Society (2002)
72. Yin, X., Han, J., Yang, J., Yu, P.S.: Efficient classification across multiple database relations: A CrossMine approach. IEEE Trans. Knowl. Data Eng. 18(6), 770–783 (2006)

Chapter 9
Kernel Methods for Structured Data

Andrea Passerini

Kernel methods are a class of non-parametric learning techniques relying on kernels. A kernel generalizes dot products to arbitrary domains and can thus be seen as a similarity measure between data points with complex structures. The use of kernels allows to decouple the representation of the data from the specific learning algorithm, provided it can be defined in terms of distance or similarity between instances. Under this unifying formalism a wide range of methods have been developed, dealing with binary and multiclass classification, regression, ranking, clustering and novelty detection to name a few. Recent developments include statistical tests of dependency and alignments between related domains, such as documents written in different languages. Key to the success of any kernel method is the definition of an appropriate kernel for the data at hand. A well-designed kernel should capture the aspects characterizing similar instances while being computationally efficient. Building on the seminal work by D. Haussler on convolution kernels, a vast literature on kernels for structured data has arisen. Kernels have been designed for sequences, trees and graphs, as well as arbitrary relational data represented in first or higher order logic. From the representational viewpoint, this allowed to address one of the main limitations of statistical learning approaches, namely the difficulty to deal with complex domain knowledge. Interesting connections between the complementary fields of statistical and symbolic learning have arisen as one of the consequences. Another interesting connection made possible by kernels is between generative and discriminative learning. Here data are represented with generative models and appropriate kernels are built on top of them to be used in a discriminative setting.

In this chapter we revise the basic principles underlying kernel machines and describe some of the most popular approaches which have been developed. We give an extensive treatment of the literature on kernels for structured data and suggest some basic principles for developing novel ones. We finally discuss kernel methods

Andrea Passerini
Dipartimento di Ingegneria e Scienza dell'Informazione,
Università degli Studi di Trento, Italy
e-mail: passerini@disi.unitn.it

M. Bianchini et al. (Eds.): *Handbook on Neural Information Processing*, ISRL 49, pp. 283–333.
DOI: 10.1007/978-3-642-36657-4_9   © Springer-Verlag Berlin Heidelberg 2013

for predicting structures. These algorithms deal with structured-output prediction, a learning setting in which the output is itself a structure which has to be predicted from the input one.

1 A Gentle Introduction to Kernel Methods

In the typical statistical learning framework a supervised learning algorithm is given a training set of input-output pairs $\mathscr{D} = \{(x_1, y_1), \ldots, (x_m, y_m)\}$, with $x_i \in \mathscr{X}$ and $y_i \in \mathscr{Y}$, sampled identically and independently from a fixed but unknown probability distribution ρ. The set $\mathscr{X}$ is called the input (or instance) space and can be any set. The set $\mathscr{Y}$ is called the output (or target) space. For instance, in the case of binary classification $\mathscr{Y} = \{-1, 1\}$ while the case of regression $\mathscr{Y}$ is the set of real numbers. The learning algorithm outputs a function $f : \mathscr{X} \mapsto \mathscr{Y}$ that approximates the probabilistic relation ρ between inputs and outputs. The class of functions that is searched is called the *hypothesis space*.

Intuitively, f should assign to a novel input x the same (or a similar) y of similar inputs already observed in the training set. A *kernel* is a function : $\mathscr{X} \times \mathscr{X} \mapsto$ IR measuring the similarity between pairs of inputs. For example, the similarity between a pair of sequences could be the number of common subsequences of length up to a certain m. The kernel should satisfy the following equation:

$$k(x, x') = \langle \Phi(x), \Phi(x') \rangle.$$

It thus corresponds to mapping examples to a (typically high dimensional) *feature space* $\mathscr{H}$ and computing the dot product in that space. In the sequence example, $\mathscr{H}$ is made of vectors of booleans, with an entry for each possible sequence of length up to m given the alphabet. $\Phi(x)$ maps x to a vector with one for entries occurring in x and zero otherwise. However, the kernel function does not need to explicitly do the mapping (which can be even infinite dimensional) in computing the similarity.

Fig. 1 A simple binary classifier in feature space: μ_+ and μ_- are the mean vectors of positive (circles) and negative (squares) examples respectively. The algorithm assigns a novel example to the class with the nearer mean. The decision boundary is a hyperplane (solid line) half-way down between the line linking the means (the dotted line).

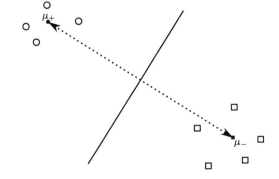

Kernels allow to construct algorithms in dot product spaces and apply them to data with arbitrarily complex structures. Consider a binary classification task ($\mathcal{Y} = \{-1, 1\}$). A simple similarity-based classification function [68] could assign to x the y of the examples which on average are more similar to it (see Figure 1). The algorithm starts by computing the means of the training examples for the two classes in feature space:

$$\mu_+ = \frac{1}{n_+} \sum_{i:y_i=+1} \Phi(x_i) \qquad \mu_- = \frac{1}{n_-} \sum_{i:y_i=-1} \Phi(x_i)$$

where n_+ and n_- are the number of positive and negative examples respectively. It then assigns a novel example to the class of the closer mean:

$$f(x) = \text{sgn}\left(\langle \mu_+, \Phi(x) \rangle - \langle \mu_-, \Phi(x) \rangle\right)$$

where $\text{sgn}(z)$ returns the sign of the argument and we assumed for simplicity that the two means have the same distance from the origin (otherwise a bias term should be included). The decision boundary is represented by a hyperplane which is half way down on the line linking the means and orthogonal to it. By replacing the formulas for the means, we obtain:

$$f(x) = \text{sgn}\left(\frac{1}{n_+} \sum_{i:y_i=+1} \langle \Phi(x_i), \Phi(x) \rangle - \frac{1}{n_-} \sum_{i:y_i=-1} \langle \Phi(x_i), \Phi(x) \rangle\right).$$

Feature space mappings $\Phi(\cdot)$ only appear in dot products and can thus be replaced by kernel functions. This is commonly known as the *kernel trick*. The resulting f can be compactly written as:

$$f(x) = \text{sgn}\left(\sum_i c_i k(x_i, x)\right),$$

where $c_i = y_i / n_{y_i}$. The unthresholded version of f (i.e. before applying sgn) is a linear combination of kernel functions "centered" on training examples x_i. This is a common aspect characterizing (with minor variations) kernel machines, as we will see in the rest of the chapter.

2   Mathematical Foundations

The kernel trick allows to implicitly compute a dot product between instances in a possibly infinite feature space. In this section we will treat in more detail the theory underlying kernel functions, showing how to verify if a given function is actually a valid kernel, and given a valid kernel how to generate a feature space such that the kernel computes a dot product in that space. We will then highlight the connections between kernel machines and regularization theory, showing how most supervised kernel machines can be seen as instances of regularized empirical risk minimization. We will focus on real valued functions but results can be extended to complex ones as well. Details and proofs of the reported results can be found in [4, 6, 64, 68].

2.1 Kernels

Let's start by providing some geometric structure to our feature spaces.

Definition 1 (Inner Product)
Given a vector space $\mathscr{X}$, an inner product is a map $\langle \cdot, \cdot \rangle : \mathscr{X} \times \mathscr{X} \to \mathbb{R}$ such that for every $x, x', x'' \in \mathscr{X}, \alpha \in \mathbb{R}$:

1. $\langle x, x' \rangle = \langle x', x \rangle$ (*symmetry*)
2. $\langle x + x'', x' \rangle = \langle x, x' \rangle + \langle x'', x' \rangle$, $\langle \alpha x, x' \rangle = \alpha \langle x, x' \rangle$
   $\langle x, x' + x'' \rangle = \langle x, x' \rangle + \langle x, x'' \rangle$, $\langle x, \alpha x' \rangle = \alpha \langle x, x' \rangle$ (*bilinearity*)
3. $\langle x, x \rangle \geq 0$ (*positive definiteness*)

If in condition 3. equality only holds for $x = 0_{\mathscr{X}}$ the inner product is *strict*.

Inner products are also known as dot or scalar products. A vector space endowed with an inner product is called an *inner product space*. As a simple example, the standard dot product in the space of n-dimensional real vectors $\mathbb{R}^n$ is

$$\langle x, x' \rangle = \sum_{i=1}^{n} x_i x_i'.$$

A norm can be defined as $||x||_2 = \sqrt{\langle x, x \rangle}$ and a distance as $d(x, x') = ||x - x'||_2 = \sqrt{\langle x, x \rangle - 2\langle x, x' \rangle + \langle x', x' \rangle}$. A *Hilbert* space is an inner product space with two additional properties (completeness and separability) guaranteeing that it is isomorphic to some standard spaces ($\mathbb{R}^n$ or its infinite dimensional analogue L_2, the set of square convergent real sequences). Hilbert spaces are often infinite dimensional.

Feature maps $\Phi : \mathscr{X} \to \mathscr{H}$ map instances into a Hilbert space. In the following we will provide conditions guaranteeing that a kernel function k acts as a dot product in the Hilbert space of a certain map.

Definition 2 (Gram Matrix). Given a function $k : \mathscr{X} \times \mathscr{X} \to \mathbb{R}$ and patterns $x_1, \ldots, x_m$, the $m \times m$ matrix K such that

$$K_{ij} = k(x_i, x_j)$$

is called the *Gram matrix* of k with respect to $x_1, \ldots, x_m$.

Definition 3 (Positive Definite Matrix). A symmetric $m \times m$ matrix K is *positive definite* if

$$\sum_{i,j=1}^{m} c_i c_j K_{ij} \geq 0, \quad \forall \mathbf{c} \in \mathbb{R}^m.$$

If equality only holds for $c = \mathbf{0}$, the matrix is *strictly positive definite*.

Alternative conditions for positive definiteness are that all its eigenvalues are non-negative, or that there exists a matrix B such that $K = B^T B$.

Definition 4 (Positive Definite Kernel). A function $k : \mathcal{X} \times \mathcal{X} \to \mathbb{R}$ such that $\forall m \in \mathbb{N}$ and $\forall x_1, \dots, x_m \in \mathcal{X}$ it gives rise to a positive definite Gram matrix is called a *positive definite kernel*.[1]

Theorem 1 (Valid Kernels)
A kernel function $k : \mathcal{X} \times \mathcal{X} \to \mathbb{R}$ corresponds to a dot product in a Hilbert space $\mathcal{H}$ obtained by a feature map Φ:

$$k(x, x') = \langle \Phi(x), \Phi(x') \rangle \tag{1}$$

if and only if it is positive definite.

Proof. The 'if' implication can be proved by building a map from $\mathcal{X}$ into a space where k acts as a dot product. We will actually build a map into a feature space of functions:

$$\Phi : \mathcal{X} \to \mathbb{R}^{\mathcal{X}} \mid \Phi(x) - k(\cdot, x).$$

Φ maps an instance into a kernel function "centered" on the instance itself. In order to turn this space of functions into a Hilbert space, we need to make it a vector space and provide a dot product. A vector space is obtained taking the span of kernel k, that is all functions

$$f(\cdot) = \sum_{i=1}^{m} \alpha_i k(\cdot, x_i)$$

for all $m \in \mathbb{N}$, $\alpha_i \in \mathbb{R}$, $x_i \in \mathcal{X}$. A dot product in such space between f and another function

$$g(\cdot) - \sum_{j=1}^{m'} \beta_j k(\cdot, x'_j)$$

can be defined as

$$\langle f, g \rangle = \sum_{i=1}^{m} \sum_{j=1}^{m'} \alpha_i \beta_j k(x_i, x'_j). \tag{2}$$

Note that in order for eq. (2) to satisfy the positive definiteness property of an inner product (see Definition 1) the kernel $k(x, x')$ needs to be positive definite. For each given function f, it holds that

---

[1] Note that part of the literature calls such kernels and matrices positive semi-definite, indicating with positive definite the strictly positive definite case.

$$\langle k(\cdot,x), f(\cdot) \rangle = f(x). \tag{3}$$

In particular, for $f = k(\cdot,x')$ we have:

$$\langle k(\cdot,x), k(\cdot,x') \rangle = k(x,x').$$

By satisfying equation (1) we showed that each positive definite kernel can be seen as a dot product in another space. In order to show that the converse is also true, it suffices to prove that given a map Φ from $\mathscr{X}$ to a product space, the corresponding function $k(x,x') = \langle \Phi(x), \Phi(x') \rangle$ is a positive definite kernel. This can be proved by noting that for all $m \in \mathbb{N}$, $\mathbf{c} \in \mathbb{R}^m$ and $x_1, \ldots, x_m \in \mathscr{X}$ we have

$$\sum_{i,j=1}^{m} c_i c_j k(x_i, x_j) = \left\langle \sum_{i=1}^{m} c_i \Phi(x_i), \sum_{j=1}^{m} c_j \Phi(x_j) \right\rangle = \left\| \sum_{i=1}^{m} c_i \Phi(x_i) \right\|^2 \geq 0.$$

The existence of a map to a dot product space satisfying (1) is therefore an alternative definition for a positive definite kernel.   □

A kernel satisfying equation (3) is said to have the *reproducing property*. The resulting space is named *Reproducing Kernel Hilbert Space* (RKHS). This is the hypothesis space we will deal with when developing supervised learning algorithms based on kernels. Note that while for a positive definite kernel k there is always a corresponding feature space of functions constructed as described in the proof[2] (and vice versa), there can be other (possibly finite dimensional) spaces working as well. A constructive example was presented in Section 1 for the common subsequence kernel. Other examples will be shown when describing kernels on structured data (Section 4).

2.2 Supervised Learning with Kernels

A key problem in supervised learning is defining an appropriate measure for the quality of the predictions. This is achieved by a *loss* function $V : \mathscr{Y} \times \mathscr{Y} \to [0, \infty)$, a non-negative function measuring the error between the actual and predicted output for a certain input $V(f(x), y)$. A simple example is the misclassification loss, which outputs one for an incorrect classification and zero otherwise. Learning aims at producing a function with the smallest possible expected risk, i.e. the probability of committing an error according to the data distribution ρ. Unfortunately, this distribution is usually unknown and one has to resort to the empirical risk, i.e. the average error on the training set $\mathscr{D}_m$:

$$R_{emp}[f] = \frac{1}{m} \sum_{i=1}^{m} V(f(x_i), y_i).$$

---

[2] An alternative way of constructing a feature space corresponding to a positive definite kernel is provided by Mercer's Theorem [53].

In order to prevent overfitting of training data, one has to impose some constraints on the possible hypotheses. The typical solution in machine learning is that of tending to prefer *simpler* hypotheses, by restricting the hypothesis space, biasing the learning algorithm for favouring them, or both. Most kernel machines rely on Tikhonov regularization [76], in which a regularization term $\Omega[f]$ is added to the empirical risk in order to bias learning towards more stable solutions:

$$R_{reg}[f] = R_{emp}[f] + \lambda \Omega[f].$$

The regularization parameter $\lambda > 0$ trades the effect of training errors with the complexity of the function. By choosing Ω to be convex, and provided $R_{emp}[f]$ is also convex, the problem has a unique global minimum. A common regularizer is the squared norm of the function, i.e. $\Omega[f] = ||f||^2$.

When the hypothesis space is a reproducing kernel Hilbert space $\mathscr{H}$ associated to a kernel k, the *representer theorem* [41] gives an explicit form of the minimizers of $R_{reg}[f]$.

Theorem 2 (Representer Theorem). *Let* $D_m = \{(x_i, y_i) \in \mathscr{X} \times \mathbb{R}\}_{i=1}^m$ *be a training set,* V *a convex loss function,* $\mathscr{H}$ *a RKHS with norm* $||\cdot||_{\mathscr{H}}$*. Then the general form of the solution of the regularized risk*

$$\frac{1}{m}\sum_{i=1}^m V(f(x_i), y_i) + \lambda ||f||^2_{\mathscr{H}}$$

is

$$f(x) = \sum_{i=1}^m c_i k(x_i, x). \tag{4}$$

The proof is omitted for brevity and can be found in [41]. The theorem states that regardless of the dimension of the RKHS $\mathscr{H}$, the solution lies on the span of the m kernels centered on the training points. Generalization of the representer theorem have been proved [68] for arbitrary cumulative loss functions $V((x_1, y_1, f(x_1)), \ldots, (x_m, y_m, f(x_m)))$ and strictly monotonic regularization functionals. A semi-parametric version of the theorem accounts for slightly more general solutions, including for instance a constant bias term.

3   Kernel Machines for Structured Input

Most supervised kernel machines can be seen as instantiations of the Tikhonov regularization framework for a particular choice of the loss function V. Kernel ridge regression [65, 59] employs the quadratic loss $V(f(x), y) = (f(x) - y)^2$ and is used for both regression and classification tasks. Kernel logistic regression [39] uses the negative log-likelihood of the probabilistic model, i.e. $\log(1 + \exp(-yf(x)))$ for binary classification. Support Vector Machines [13] (SVM) are the most popular class of kernel methods. Initially introduced for binary classification [9], they have been extended to deal with different tasks such as regression and multiclass classification. The common rationale of SVM algorithms is the use of a loss function forcing *sparsity* in the solution. That is, only a small subset of the c_i coefficients in eq. (4) will be non-zero. The corresponding training examples are termed *Support Vectors* (SV). In the following we detail SVM for binary classification and regression. SVM for multiclass classification will arise as a special case of structured-output prediction (see Section 6). We will then discuss a SV approach for novelty detection based on the estimation of the smallest enclosing hypersphere. Finally we will introduce kernel Principal Component Analysis [67] for non-linear dimensionality reduction. Additional algorithms can be found e.g. in [71].

3.1 SVM for Binary Classification

SVM for binary classification employ the so-called *hinge loss*:

$$V(f(x),y) = |1 - yf(x)|_+ = \begin{cases} 0 & \text{if } yf(x) \geq 1 \\ 1 - yf(x) & \text{otherwise} \end{cases}$$

As shown in Fig.2(a), a linear cost is paid in case the confidence in the correct class is below a certain threshold. By plugging the hinge loss in the Tikhonov regularization functional we obtain the optimization problem addressed by SVM:

$$\min_{f \in \mathcal{H}} \frac{1}{m} \sum_{i=1}^{m} |1 - y_i f(x_i)|_+ + \lambda ||f||_{\mathcal{H}}^2.$$

Slack variables $\xi_i = |1 - y_i f(x_i)|_+$ can be used to represent the cost paid for each example, giving the following quadratic optimization problem:

$$\min_{f \in \mathcal{H}, \boldsymbol{\xi} \in \mathbb{R}^m} \frac{1}{m} \sum_{i=1}^{m} \xi_i + \lambda ||f||_{\mathcal{H}}^2$$

$$\text{subject to: } y_i f(x_i) \geq 1 - \xi_i \quad i = 1, \ldots, m$$

$$\xi_i \geq 0 \quad i = 1, \ldots, m.$$

To see that this corresponds to the hinge loss, note that as we minimize over slack variables $\boldsymbol{\xi}$, ξ_i will be zero (it must be non-negative) if $y_i f(x_i) \geq 1$ (hard constraint satisfied) and $1 - y_i f(x_i)$ otherwise. By the representer Theorem we know that the solution of the above problem is given by eq. (4). As for the simple classification algorithm seen in Section 1, the decision function takes the sign of f to predict labels and the decision boundary is a separating hyperplane in the feature space. To see this, let $\Phi(\cdot)$ be a feature mapping associated with kernel k. Function f can be rewritten as:

$$f(x) = \sum_{i=1}^{m} c_i \langle \Phi(x_i), \Phi(x) \rangle = \langle \sum_{i=1}^{m} c_i \Phi(x_i), \Phi(x) \rangle = \langle \boldsymbol{w}, \Phi(x) \rangle.$$

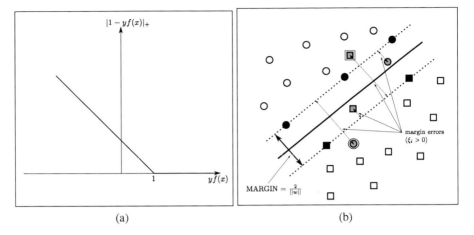

(a)   (b)

Fig. 2 SVM for binary classification: (a) Hinge loss. (b) Classification function. The solid line represents the separating hyperplane, while dotted lines are hyperplanes with confidence margin equal to one. Black points are unbound SVs, grey points are bound SVs and extra borders indicate bound SVs which are also training errors. All other points do not contribute to the function to be minimized. Dotted lines indicate the margin error ξ_i for bound SVs.

The decision boundary $\langle w, \Phi(x) \rangle = 0$ is a hyperplane of points orthogonal to w. Note that w can be explicitly computed only if the feature mapping $\Phi(\cdot)$ is finite-dimensional. From the definition of the dot product in the RKHS $\mathcal{H}$ (see eq. (2)) we can compute the (squared) norm of f:

$$\|f\|_{\mathcal{H}}^2 = \langle f, f \rangle = \sum_{i=1}^{m} \sum_{j=1}^{m} c_i c_j k(x_i, x_j) = \sum_{i=1}^{m} \sum_{j=1}^{m} c_i c_j \langle \Phi(x_i), \Phi(x_j) \rangle$$

$$= \langle \sum_{i=1}^{m} c_i \Phi(x_i), \sum_{j=1}^{m} c_j \Phi(x_j) \rangle = \langle w, w \rangle = \|w\|^2.$$

The minimization problem can be rewritten as:

$$\min_{w \in \mathcal{H}, \xi \in \mathbb{R}^m} C \sum_{i=1}^{m} \xi_i + \frac{1}{2} \|w\|^2$$

$$\text{subject to: } y_i \langle w, \Phi(x_i) \rangle \geq 1 - \xi_i \quad i = 1, \ldots, m$$

$$\xi_i \geq 0 \quad i = 1, \ldots, m,$$

where we replaced $C = 2/\lambda m$ for consistency with most literature on SVM. Hyperplanes $\langle w, \Phi(x) \rangle - 1 = 0$ and $\langle w, \Phi(x) \rangle + 1 = 0$ are "confidence" boundaries for not paying a cost in predicting class $+1$ and -1 respectively. The distance between them $2/\|w\|$ is called geometric margin. By minimizing $\|w\|^2$, this margin is maximized, while slack variables ξ_i account for margin errors. The minimizer thus trades off margin maximization and fitting of training data. Constraints in the optimization problem can be included in the minimization functional using Lagrange multipliers:

$$L(w, \alpha, \beta) = C \sum_{i=1}^{m} \xi_i + \frac{1}{2} ||w||^2 - \sum_{i=1}^{m} \alpha_i (y_i \langle w, \Phi(x_i) \rangle - 1 + \xi_i) - \sum_{i=1}^{m} \beta_i \xi_i$$

where $\alpha_i, \beta_i \geq 0$ for all i. The Wolfe dual formulation for the Lagrangian amounts at maximizing it over α_i, β_i subject to the vanishing of the gradient of w and ξ, i.e.:

$$\frac{\partial L}{\partial w} = w - \sum_{i=1}^{m} \alpha_i y_i \Phi(x_i) = 0 \rightarrow w = \sum_{i=1}^{m} \alpha_i y_i \Phi(x_i) \qquad (5)$$

$$\frac{\partial L}{\partial \xi_i} = C - \alpha_i - \beta_i = 0 \rightarrow \alpha_i \in [0, C]. \qquad (6)$$

The second implication comes from the non-negativity of both α_i and β_i. Substituting into the Lagrangian we obtain:

$$\max_{\alpha \in \mathbb{R}^m} \quad -\frac{1}{2} \sum_{i=1}^{m} \sum_{j=1}^{m} \alpha_i y_i \alpha_j y_j \langle \Phi(x_i), \Phi(x_j) \rangle + \sum_{i=1}^{m} \alpha_i$$

$$\text{subject to: } \alpha_i \in [0, C] \quad i = 1, \dots, m.$$

The problem can be solved using off-the-shelf quadratic programming tools. However, a number of ad-hoc algorithms have been proposed which exploit the specific characteristics of this problem to achieve substantial efficiency improvements [37, 58]. Note that replacing $\langle \Phi(x_i), \Phi(x_j) \rangle = k(x_i, x_j)$ we recover the kernel-based formulation where the feature mapping is only implicitly done. The general form for f (eq. (4)) can be recovered setting $c_i = \alpha_i y_i$ (see eq. (5)).

The Karush−Kuhn−Tucker (KKT) complementary conditions require that the optimal solution satisfies:

$$\alpha_i (y_i \langle w, \Phi(x_i) \rangle - 1 + \xi_i) = 0 \qquad (7)$$

$$\beta_i \xi_i = 0 \qquad (8)$$

for all i. Eq. (7) implies that $\alpha_i > 0$ only for examples where $y_i \langle w, \Phi(x_i) \rangle \leq 1$. These are the support vectors, all other examples do not contribute to the decision function f. If $\alpha_i < C$, equations (8) and (6) imply that $\xi_i = 0$. These are called *unbound* support vectors and lay on the confidence one hyperplanes. *Bound* support vectors ($\alpha_i = C$) are margin errors ($\xi_i > 0$). Figure 2(b) shows an example highlighting hyperplanes and support vectors.

Variants of SVM for binary classification have been developed in the literature. Most approaches include a bias term b to the classification function f. This can be obtained simply setting $k'(x, x') = k(x, x') + 1$. Linear penalties can be replaced with quadratic ones (ξ_i^2) in the minimization functional. The ν-SVM [70] allows to explicitly upper bound the number of margin errors. Further details can be found in several textbooks (see e.g. [13]).

3.2 SVM for Regression

SVM for regression enforce sparsity in the solution by tolerating small deviations from the desired target. This is achieved by the $\varepsilon - insensitive$ loss (see fig.3(a)):

$$V(f(x),y) = |y - f(x)|_{\varepsilon} = \begin{cases} 0 & \text{if } |y - f(x)| \leq \varepsilon \\ |y - f(x)| - \varepsilon & \text{otherwise} \end{cases}$$

which doesn't penalize deviations up to ε from the target value (the so-called ε-tube), and gives a linear penalty to further deviations. By introducing slack variables for penalties, we obtain the following minimization problem:

$$\min_{f \in \mathcal{H}, \xi, \xi^* \in \mathbb{R}^m} \frac{1}{m} \sum_{i=1}^{m} (\xi_i + \xi_i^*) + \lambda \|f\|_{\mathcal{H}}^2$$
$$\text{subject to: } f(x_i) - y_i \leq \varepsilon + \xi_i \quad i = 1, \ldots, m$$
$$y_i - f(x_i) \leq \varepsilon + \xi_i^* \quad i = 1, \ldots, m$$
$$\xi_i, \xi_i^* \geq 0 \quad i = 1, \ldots, m.$$

As for the binary classification case, we can rewrite the problem in terms of weight vector and feature mapping. The Lagrangian is obtained as ($C = 2/\lambda m$):

$$L(\mathbf{w}, \xi, \xi^*, \alpha, \alpha^*, \beta, \beta^*) = \frac{1}{2}\|\mathbf{w}\|^2 + C \sum_{i=1}^{m} (\xi_i + \xi_i^*) - \sum_{i=1}^{m} (\beta_i \xi_i + \beta_i^* \xi_i^*) - \sum_{i=1}^{m} \alpha_i (\varepsilon + \xi_i + y_i - \langle \mathbf{w}, \Phi(x_i) \rangle)$$
$$- \sum_{i=1}^{m} \alpha_i^* (\varepsilon + \xi_i^* - y_i + \langle \mathbf{w}, \Phi(x_i) \rangle) \quad (9)$$

with $\alpha_i, \alpha_i*, \beta_i, \beta_i^* \geq 0, \forall i \in [1, m]$. By vanishing the derivatives of L with respect to the primal variables we obtain:

$$\frac{\partial L}{\partial \mathbf{w}} = \mathbf{w} - \sum_{i=1}^{m} (\alpha_i^* - \alpha_i) \Phi(x_i) = 0 \rightarrow \mathbf{w} = \sum_{i=1}^{m} (\alpha_i^* - \alpha_i) \Phi(x_i) \quad (10)$$

$$\frac{\partial L}{\partial \xi_i} = C - \alpha_i - \beta_i = 0 \rightarrow \alpha_i \in [0, C] \quad (11)$$

$$\frac{\partial L}{\partial \xi_i^*} = C - \alpha_i^* - \beta_i^* = 0 \rightarrow \alpha_i^* \in [0, C].$$

Finally, substituting into the Lagrangian we derive the dual problem:

$$\max_{\alpha \in \mathbb{R}^m} \quad -\frac{1}{2}\sum_{i,j=1}^{m}(\alpha_i^* - \alpha_i)(\alpha_j^* - \alpha_j)\langle \Phi(x_i), \Phi(x_j)\rangle - \varepsilon \sum_{i=1}^{m}(\alpha_i^* + \alpha_i) + \sum_{i=1}^{m}y_i(\alpha_i^* - \alpha_i),$$

subject to: $\alpha_i, \alpha_i^* \in [0, C], \quad \forall i \in [1, m].$

The kernel-based formulation is again recovered setting $\langle \Phi(x_i), \Phi(x_j)\rangle = k(x_i, x_j)$, while the general form for f (eq. (4)) is obtained for $c_i = \alpha_i - \alpha_i^*$. (see eq. (10)). The KKT complementary conditions require that the optimal solution satisfies:

$$\alpha_i(\varepsilon + \xi_i + y_i - \langle \mathbf{w}, \Phi(x_i)\rangle) = 0$$
$$\alpha_i^*(\varepsilon + \xi_i^* - y_i + \langle \mathbf{w}, \Phi(x_i)\rangle) = 0$$
$$(C - \alpha_i)\xi_i = 0$$
$$(C - \alpha_i^*)\xi_i^* = 0$$

These conditions enlighten some interesting analogies to the classification case:

- All patterns within the ε-tube, for which $|f(\mathbf{x}_i) - y_i| < \varepsilon$, have $\alpha_i, \alpha_i^* = 0$ and thus don't contribute to the estimated function f.
- Patterns for which either $0 < \alpha_i < C$ or $0 < \alpha_i^* < C$ are on the border of the ε-tube, that is $|f(\mathbf{x}_i) - y_i| = \varepsilon$. They are the unbound support vectors.
- The remaining training patterns are margin errors (either $\xi_i > 0$ or $\xi_i^* > 0$), and reside out of the ε-insensitive region. They are bound support vectors, with corresponding $\alpha_i = C$ or $\alpha_i^* = C$.

Figures 3(b),3(c),3(d) show examples of SVM regression for decreasing values of ε. In order to highlight the effect of the parameter on the approximation function, we focus on 1D regression and report the form of the function in the input (rather than feature) space. Note the increase in the number of support vectors when requiring tighter approximations. A Gaussian kernel (see Section 4.1) was employed in all cases.

3.3   Smallest Enclosing Hypersphere

A support vector algorithm has been proposed in [75, 66] in order to characterize a set of data in terms of support vectors, thus allowing to compute a set of contours which enclose the data points. The idea is finding the smallest hypersphere which encloses the points in the feature space. Outliers can be dealt with by relaxing the enclosing constraint and allowing some points to stay out of the sphere in feature space. The algorithm can be readily employed for novelty detection, by predicting whether a test instance lays outside of the enclosing hypersphere.

Given a set of m examples $x_i \in \mathcal{X}, i = 1, \dots, m$ and a feature mapping Φ, we can define the problem of finding the smallest enclosing sphere of radius R in the feature space as follows:

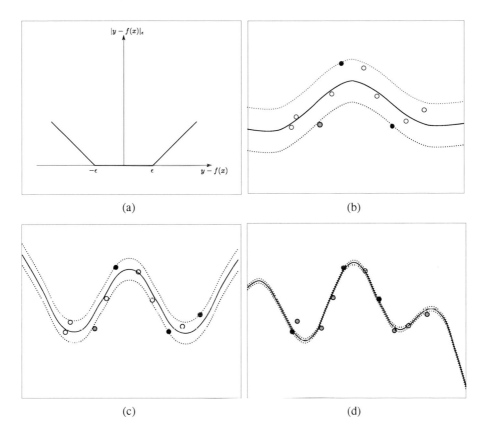

Fig. 3 SVM for regression: (a) Epsilon insensitive loss. (b)(c)(d) Regression functions for decreasing values of ε. Solid lines represent the regression function, dotted lines represent the ε-insensitive tube around the regression function. All points within this tube are considered correctly approximated. Black points are unbound SVs laying on the borders of the tube, gray points are bound SVs laying outside of it. All other points do not contribute to the regression function. Note the increase in the complexity of the function and the number of support vectors for decreasing values of ε.

$$\min_{R\in\mathbb{R},\mathbf{o}\in\mathscr{H},\boldsymbol{\xi}\in\mathbb{R}^m} \quad R^2 + C\sum_{i=1}^{m}\xi_i$$

$$\text{subject to} \quad ||\boldsymbol{\Phi}(x_i) - \mathbf{o}||^2 \le R^2 + \xi_i, \quad i = 1,\ldots,m$$

$$\xi_i \ge 0, \quad i = 1,\ldots,m$$

where $\mathbf{o}$ is the center of the sphere, ξ_i are slack variables allowing for soft constraints and C is a cost parameter balancing the radius of the sphere versus the number of outliers. We consider the Lagrangian

$$L(R,\mathbf{o},\boldsymbol{\xi},\boldsymbol{\alpha},\boldsymbol{\beta}) = R^2 + C\sum_{i=1}^{m}\xi_i - \sum_{i=1}^{m}\alpha_i(R^2 + \xi_i - ||\Phi(x_i) - \mathbf{o}||^2) - \sum_{i=1}^{m}\beta_i\xi_i,$$

with $\alpha_i \geq 0$ and $\beta_i \geq 0$ for all $i \in [1,m]$, and by vanishing the derivatives with respect to the primal variables R, $\mathbf{o}$ and ξ_i we obtain

$$\frac{\partial L}{\partial R} = 1 - \sum_{i=1}^{m}\alpha_i = 0 \rightarrow \sum_{i=1}^{m}\alpha_i = 1$$

$$\frac{\partial L}{\partial \mathbf{o}} = \mathbf{0} - \sum_{i=1}^{m}\alpha_i\Phi(x_i) = 0 \rightarrow \mathbf{o} = \sum_{i=1}^{m}\alpha_i\Phi(x_i) \qquad (12)$$

$$\frac{\partial L}{\partial \xi_i} = C - \alpha_i - \beta_i = 0 \rightarrow \alpha_i \in [0,C].$$

Substituting into the Lagrangian we derive the Wolf dual problem

$$\max_{\boldsymbol{\alpha}\in\mathbb{R}^m} \sum_{i=1}^{m}\alpha_i\Phi(x_i)^2 - \sum_{i=1}^{m}\sum_{j=1}^{m}\alpha_i\alpha_j\langle\Phi(x_i),\Phi(x_j)\rangle, \qquad (13)$$

$$\text{subject to } \sum_{i=1}^{m}\alpha_i = 1, \quad 0 \leq \alpha_i \leq C, \quad i = 1,\ldots,m.$$

The distance of a given point x from the center of the sphere

$$R^2(x) = ||\Phi(x) - \mathbf{o}||^2$$

can be written using (12) as

$$R^2(x) = \langle\Phi(x),\Phi(x)\rangle - 2\sum_{i=1}^{m}\alpha_i\langle\Phi(x),\Phi(x_i)\rangle + \sum_{i=1}^{m}\sum_{j=1}^{m}\alpha_i\alpha_j\langle\Phi(x_i),\Phi(x_j)\rangle. \quad (14)$$

As for the other SV algorithms, both (13) and (14) contain only dot products in the feature space, which can be substituted by a kernel function k. The KKT conditions [17] imply that at the saddle point of the Lagrangian

$$\beta_i\xi_i = 0$$
$$\alpha_i(R^2 + \xi_i - ||\Phi(x_i) - \mathbf{o}||^2) = 0$$

showing the presence of (see fig. 4):

- Unbound support vectors ($0 < \alpha_i < C$), whose images lie on the surface of the enclosing sphere.
- Bound support vectors ($\alpha_i = C$), whose images lie outside of the enclosing sphere, which correspond to outliers.
- All other points ($\alpha = 0$) with images inside the enclosing sphere.

Fig. 4 Novelty detection by smallest enclosing hypersphere. The solid line is the border of the hypersphere enclosing most of the data and represents the decision boundary. Black points are unbound SV laying on the hypersphere border. Grey points are outliers which are left outside of the hypersphere. All other points do not contribute to the decision function.

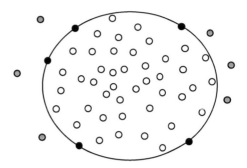

The radius R^* of the enclosing sphere can be computed by (14) provided x is an unbound support vector. A decision function for novelty detection would predict a point as positive if it lays outside of the sphere and negative otherwise, i.e.:

$$f(x) = \mathrm{sgn}\left(R^2(x) - (R^*)^2\right)$$

3.4 Kernel Principal Component Analysis

Principal Component Analysis (PCA) is a standard technique for linear dimensionality reduction which consists of projecting examples onto directions of maximal variance, thus retaining most of their information content. We will introduce it considering explicit feature mappings $\Phi(x)$, and then show how these can be computed only implicitly via kernels (see [71] for further details).

Given a set of orthonormal vectors $u_1, \ldots, u_k$, the orthogonal projection of a point $\Phi(x)$ into the subspace V spanned by them is computed as:

$$P_V(\Phi(x)) = \left(\langle u_i, \Phi(x)\rangle\right)_{i=1}^{k} \tag{15}$$

where each dot product computes the length of the projection in the corresponding direction. Let X be a matrix representation of a set of points $\{\Phi(x_1), \ldots, \Phi(x_m)\}$, with row i representing point $\Phi(x_i)^T$. Let's assume that the points are centered around the origin, i.e. their mean is zero. This can always be obtained subtracting the mean from each point. The covariance matrix of the points C is computed as:

$$C = \frac{1}{m}\sum_{i=1}^{m}\sum_{j=1}^{m}\Phi(x_i)\Phi(x_j)^T = \frac{1}{m}X^T X.$$

The directions of maximal variance are the eigenvectors of C with maximal eigenvalue. PCA basically projects points onto the first k eigenvectors of the covariance matrix, where k is the dimension of the reduced space. The dimension k can be set a priori or chosen according to the amount of variance captured, measured as the sum of the eigenvalues.

Standard PCA requires to explicitly use mappings $\Phi(x)$, both in computing covariance matrix C and in doing the projection (Eq. (15)). However, it is possible to avoid this thanks to the relationship between the eigen-decomposition of covariance and kernel matrices. Let $K = XX^T$ be the kernel matrix of the data. Let (λ, v) be an eigenvalue-eigenvector pair of K. It holds that:

$$C(X^T v) = \frac{1}{m}X^T XX^T v = \frac{1}{m}X^T Kv = \frac{\lambda}{m}X^T v$$

showing that $(\lambda/m, X^T v)$ is an eigenvalue-eigenvector pair of C. The norm of the eigenvector is:

$$\|X^T v\|^2 = v^T XX^T v = v^T Kv = \lambda v^T v = \lambda$$

where the last equality follows from the orthonormality of v. The normalized eigenvector is thus $u = 1/\sqrt{\lambda}X^T v$. Projecting $\Phi(x)$ onto this direction can be computed as:

$$P_u(\Phi(x)) = \langle 1/\sqrt{\lambda}X^T v, \Phi(x)\rangle = 1/\sqrt{\lambda}v^T X\Phi(x) = 1/\sqrt{\lambda}\sum_{i=1}^{m}v_i\langle\Phi(x_i),\Phi(x)\rangle = 1/\sqrt{\lambda}\sum_{i=1}^{m}v_ik(x_i,x)$$

where k is a kernel corresponding to the feature map $\Phi(\cdot)$. Note that v is computed as eigenvector of the kernel matrix K, thus we never need to explicitly compute $\Phi(x)$. Centering of the data in feature space can also be addressed simply using a modified kernel:

$$\hat{k}(x,x') = \langle \Phi(x) - \frac{1}{m}\sum_{i=1}^{m}\Phi(x_i), \Phi(x') - \frac{1}{m}\sum_{i=1}^{m}\Phi(x_i)\rangle$$

$$= k(x,x') - \frac{1}{m}\sum_{i=1}^{m}k(x_i,x') - \frac{1}{m}\sum_{i=1}^{m}k(x,x_i) + \frac{1}{m^2}\sum_{i=1}^{m}\sum_{j=1}^{m}k(x_i,x_j).$$

A number of well-known dimensionality reduction techniques, like Locally Linear Embedding and Laplacian Eigenmaps, can be seen [28] as special cases of kernel PCA. Kernel PCA has been used for addressing a variety of problems, from novelty detection [30] to image denoising [40].

4   Kernels on Structured Data

Kernel design deals with the problem of choosing an appropriate kernel for the task at hand, that is a similarity measure of the data capable of best capturing the available information. We start by introducing a few basic kernels commonly used in practice, and show how to realize complex kernels by combination of simpler ones, allowing to treat different parts or characteristics of the input data in different ways. We will introduce the notion of kernels on discrete structures, providing examples of kernels for strings, trees and graphs. We will then discusses two classes of hybrid kernels, based on probabilistic generative models and logical formalisms respectively. For extensive treatments of kernels for structured data see also [22, 51].

4.1 Basic Kernels

Let's start with some basic kernels on inner product spaces. While they cannot be directly applied to structured objects, they will turn useful when defining complex kernels as combinations of simpler ones. The standard dot product is called linear kernel:

$$k(x,x') = \langle x,x' \rangle.$$

Its normalized version computes the cosine of the angle between the two vectors:

$$k_{norm}(x,x') = \frac{\langle x,x' \rangle}{\sqrt{\langle x,x \rangle \langle x',x' \rangle}}. \tag{16}$$

The *polynomial* kernel:

$$k_d(x,x') = (\langle x,x' \rangle + c)^d \tag{17}$$

with $d \in \mathrm{IN}$ and $c \in \mathrm{IR}_0^+$, allows to combine individual features taking their products. The corresponding feature space contains all possible monomials of degree up to d. Gaussian kernels are defined as:

$$k_\sigma(x,x') = \exp\left(-\frac{||x-x'||^2}{2\sigma^2}\right) = \exp\left(-\frac{\langle x,x \rangle - 2\langle x,x' \rangle + \langle x',x' \rangle}{2\sigma^2}\right) \tag{18}$$

with $\sigma > 0$. They are an example of *Universal* kernels [54], a class of kernels which can uniformly approximate any arbitrary continuous target function. Note that the smallest the variance σ^2, the most the prediction for a certain point will depend only on its nearest (training) neighbours, eventually leading to orthogonality between any pair of points and poor generalization. Tuning this *hyperparameter* according to the complexity of the problem at hand is another mean to control overfitting (see also Section 5).

The simplest possible kernel for arbitrary domains is the *matching* or *delta* kernel:

$$k_\delta(x,x') = \delta(x,x') = \begin{cases} 1 \text{ if } x = x' \\ 0 \text{ otherwise.} \end{cases} \tag{19}$$

While it clearly does not allow any generalization if used alone, it is another useful component for building more complex kernels.

4.2   Kernel Combination

The class of kernels has a few interesting closure properties useful for combinations. It is closed under addition, product, multiplication by a positive constant and pointwise limits [6], that is they form a closed convex cone. Note that the addition of two kernels corresponds to the concatenation of their respective features:

$$
\begin{aligned}
(k_1 + k_2)(x,x') &= k_1(x,x') + k_2(x,x') \\
&= \langle \Phi_1(x), \Phi_1(x') \rangle + \langle \Phi_2(x), \Phi_2(x') \rangle \\
&= \langle \Phi_1(x) \odot \Phi_2(x), \Phi_1(x') \odot \Phi_2(x') \rangle
\end{aligned}
$$

where $\odot$ denotes vector concatenation. Taking the product of two kernels amounts at taking the Cartesian product between their respective features:

$$
\begin{aligned}
(k_1 \times k_2)(x,x') &= k_1(x,x')k_2(x,x') \\
&= \sum_{i=1}^{n} \Phi_{1i}(x)\Phi_{1i}(x') \sum_{j=1}^{m} \Phi_{2j}(x)\Phi_{2j}(x') \\
&= \sum_{i=1}^{n}\sum_{j=1}^{m} (\Phi_{1i}(x)\Phi_{2j}(x))(\Phi_{1i}(x')\Phi_{2j}(x')) \\
&= \sum_{k=1}^{nm} \Phi_{12k}(x)\Phi_{12j}(x') = \langle \Phi_{12}(x), \Phi_{12}(x') \rangle
\end{aligned}
$$

where $\Phi_{12}(x) = \Phi_1(x) \times \Phi_2(x)$. Such properties are still valid in the case that the two kernels are defined on different domains [29]. If k_1 and k_2 are kernels defined respectively on $\mathscr{X}_1 \times \mathscr{X}_1$ and $\mathscr{X}_2 \times \mathscr{X}_2$, then their *direct sum* and *tensor product*:

$$
\begin{aligned}
(k_1 \oplus k_2)((x_1,x_2),(x_1',x_2')) &= k_1(x_1,x_1') + k_2(x_2,x_2') \\
(k_1 \otimes k_2)((x_1,x_2),(x_1',x_2')) &= k_1(x_1,x_1')k_2(x_2,x_2')
\end{aligned}
$$

are kernels on $(\mathscr{X}_1 \times \mathscr{X}_2) \times (\mathscr{X}_1 \times \mathscr{X}_2)$, with $x_1,x_1' \in \mathscr{X}_1$ and $x_2,x_2' \in \mathscr{X}_2$. These combinations allow to treat in a diverse way parts of an individual which have different meanings. Finally, if k is defined on $\mathscr{S} \times \mathscr{S}$ with $\mathscr{S} \subset \mathscr{X}$, a *zero extension* kernel on $\mathscr{X} \times \mathscr{X}$ can be obtained setting $k(x,x') = 0$ whenever x or x' do not belong to $\mathscr{S}$.

   These concepts are at the basis of the so called *convolution* kernels [29, 85] for discrete structures. Suppose $x \in \mathscr{X}$ is a composite structure made of "parts" $x_1,\dots,x_D$ such that $x_d \in \mathscr{X}_d$ for all $i \in [1,D]$. This can be formally represented by a relation R on $\mathscr{X}_1 \times \cdots \times \mathscr{X}_D \times \mathscr{X}$ such that $R(x_1,\dots,x_D,x)$ is true iff $x_1,\dots,x_D$ are the parts of x. For example if $\mathscr{X}_1 = \cdots = \mathscr{X}_D = \mathscr{X}$ are sets containing all finite strings over a finite alphabet $\mathscr{A}$, we can define a relation $R(x_1,\dots,x_D,x)$ which is true iff $x = x_1 \circ \cdots \circ x_D$, with $\circ$ denoting concatenation of strings. Note that in this example x can be decomposed in multiple ways. If their number is finite, the relation is said to be finite. Let $R^{-1}(x) = \{x_1,\dots,x_D : R(x_1,\dots,x_D,x)\}$ return the set of decompositions for x. Given a set of kernels $k_d : \mathscr{X}_d \times \mathscr{X}_d \to \mathbb{R}$, one for each of the parts of x, the *R-convolution* kernel is defined as

$$(k_1 \star \cdots \star k_D)(x,x') = \sum_{(x_1,\ldots,x_D) \in R^{-1}(x)} \sum_{(x'_1,\ldots,x'_D) \in R^{-1}(x')} \prod_{d=1}^{D} k_d(x_d,x'_d) \qquad (20)$$

where the sums run over all the possible decompositions of x and x'. For finite relations R, this can be shown to be a valid kernel [29]. Let X,X' be sets and let $R(\xi,X)$ be the membership relation, i.e. $\xi \in R^{-1}(X) \iff \xi \in X$. The *set kernel* is defined as:

$$k_{set}(X,X') = \sum_{\xi \in X} \sum_{\xi' \in X'} k_{member}(\xi,\xi') \qquad (21)$$

where k_{member} is a kernel on set elements. Simple examples of set kernels include the *intersection kernel*:

$$k_\cap(X,X') = |X \cap X'|$$

and the *Tanimoto kernel*:

$$k_{Tanimoto}(X,X') = \frac{|X \cap X'|}{|X \cup X'|}.$$

A more complex type of *R*-convolution kernel is the so called *analysis of variance* (ANOVA) kernel [83]. Let $\mathscr{X} = \mathscr{S}^n$ be the set of n-sized tuples built over elements of a certain set $\mathscr{S}$. Let $k_i : \mathscr{S} \times \mathscr{S} \to \mathbb{R}$, $i \in [1,n]$ be a set of kernels, which will typically be the same function. For $D \in [1,n]$, the ANOVA kernel of order D, $k_{Anova} : \mathscr{S}^n \times \mathscr{S}^n \to \mathbb{R}$, is defined by

$$k_{Anova}(x,x') = \sum_{1 \leq i_1 < \cdots < i_D \leq n} \prod_{d=1}^{D} k_{i_d}(x_{i_d},x'_{i_d}).$$

Note that the sum ranges over all possible subsets of cardinality D. For $D = n$, the sum consists only of the term for which $(i_1 = 1,\ldots,i_D = n)$, and k becomes the tensor product $k_1 \otimes \cdots \otimes k_n$. Conversely, for $D = 1$, each product collapses to a single factor, while i_1 ranges from 1 to n, giving the direct sum $k_1 \oplus \cdots \oplus k_n$. By varying D we can run between these two extremes. In order to reduce the computational cost of kernel evaluations, recursive procedures are usually employed [80].

Mapping kernels [72] were recently introduced as a generalization of R-convolution kernels, in which the kernel sums over a subset of all possible decompositions of x and x'. If the *mapping system* $M_{x,x'}$ selecting the subset is *transitive*, i.e. $(x_d,x'_d) \in M_{x,x'} \wedge (x'_d,x''_d) \in M_{x',x''} \Rightarrow (x_d,x''_d) \in M_{x,x''}$, then the resulting mapping kernel is positive definite.

Finally, once a kernel k on an arbitrary type of data has been defined, it can readily be composed with basic kernels on inner-product spaces, like the ones seen in the previous section (just set $\langle x,x' \rangle = k(x,x')$). This will produce an overall feature mapping which is the composition of the mappings $\Phi(\cdot)$ and $\Phi'(\cdot)$ associated to the two kernels:

$$\Phi^* : \mathcal{X} \to \mathcal{H}' \mid \Phi^* = \Phi' \circ \Phi.$$

Polynomial (eq. 17) and Gaussian (eq. 18) kernels are commonly employed to allow for nonlinear combinations of features in the mapped space of the first kernel. Cosine normalization (eq. 16) is often used to reduce the dependence on the size of the objects. In the case of set kernels, an alternative is that of dividing by the product of the cardinalities of the two sets, thus computing the mean value between pairwise comparisons:

$$k_{mean}(X,X') = \frac{k_{set}(X,X')}{|X||X'|}.$$

4.3   Kernels on Discrete Structures

R-convolution and mapping kernels are very general classes of kernels, which can be used to model similarity between objects with discrete structures, such as strings, trees and graphs. A large number of kernels on structures have been defined in the literature, mostly as instantiations of these kernel classes. A very common approach consists of defining similarity in terms of counts of common substructures. However, efficiency is a key issue in order to develop kernels of practical utility. A kernel can also be thought of as a procedure efficiently implementing a given dot product in feature space. In the following, we will report a series of kernels developed for efficiently treating objects with discrete structure. We do not aim at providing an exhaustive enumeration of all kernels on structured data developed in the literature. We will focus on some representative approaches, whose description should help in figuring out how kernel design works, while providing pointers to additional literature. In most of the cases, we will explicitly show a feature space corresponding to the kernel. This is one of the most common ways of proving that a kernel is positive definite. An alternative is showing that it belongs to a class of known valid kernels, like the up-mentioned R-convolution and mapping ones.

4.3.1   Strings

Strings allow to represent data consisting of sequences of discrete symbols. They account for variable length objects in which the ordering of the elements matters. Biological sequences, for instance, can be represented as strings of symbols, amino-acids for proteins or nucleotides for DNA and RNA. Text documents can be represented as strings of characters. We introduce some notation before describing a number of common kernels for strings.

Consider a finite alphabet $\mathcal{A}$. A string s is a finite sequence of (possibly zero) characters from $\mathcal{A}$. We define by $|s|$ the length of string s, $\mathcal{A}^n$ the set of all strings of length n, and

$$\mathcal{A}^* = \bigcup_{n=0}^{\infty} \mathcal{A}^n$$

the set of all strings. Concatenation between strings s and t is simply represented as st. A (possibly non-contiguous) subsequence u of s is defined as $u := s(\mathbf{i}) := s(i_1)\ldots s(i_{|u|})$, with $1 \leq i_1 < \cdots < i_{|u|} \leq |n|$ and $s(i)$ the ith element of s. The length $l(\mathbf{i})$ of the subsequence u in s is $i_{|u|} - i_1 + 1$. Note that if $\mathbf{i}$ is not contiguous, $l(\mathbf{i}) > |u|$.

The *spectrum kernel* [46] is a simple string kernel originally introduced for protein classification. The *k-spectrum* of a string is the set of all its *k-mers*, i.e. contiguous substrings of length k. The feature space $\mathcal{H}_k = \mathbb{R}^{|\mathscr{A}|^k}$ of the spectrum kernel has a coordinate for each possible k-length sequence given the alphabet $\mathscr{A}$. Its corresponding feature map is:

$$\Phi(s) = (\phi_u(s))_{u \in \mathscr{A}^k}$$

where

$$\phi_u(s) = \text{number of times in which } u \text{ occurs in } s$$

giving a weighted representation of the k-spectrum. The k-spectrum kernel k_k is the inner product in this feature space:

$$k_k(s,t) = \langle \Phi(s), \Phi(t) \rangle = \sum_{u \in \mathscr{A}^k} \phi_u(s)\phi_u(t).$$

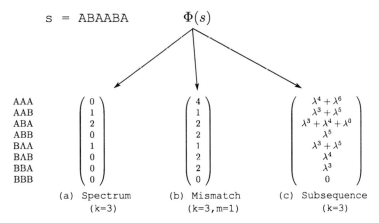

Fig. 5 Feature mappings for different types of string kernels on a toy example with an alphabet made of two symbols ($\mathscr{A} = \{A, B\}$). The feature space is indexed by all substrings of length $k = 3$ in all cases. (a) Mapping for spectrum kernel. Each entry is the number of occurrences of the corresponding 3-mer in s. (b) Mapping for mismatch string kernel ($m = 1$). Note that each entry is equal to the sum of the entries of the spectrum kernel mapping for all 3-mers with one mismatch (e.g. AAA,AAB,ABA,BAA for entry AAA). (c) Mapping for the string subsequence kernel. Each entry is the number of times the corresponding substring is found in s, possibly with gaps, weighted according to the length of the match. Substring ABA for instance has matches ABA,ABAA,ABAABA with zero, one and three gaps respectively.

Figure 5(a) shows the feature mapping corresponding to a 3-spectrum kernel for a toy example with a simple alphabet of two symbols. For real-world cases, the feature space will have a very large dimension and feature vectors for strings will be typically extremely sparse. A very efficient procedure to compute the kernel can be devised using *suffix trees* [27]. A suffix tree for a given string s of length n is a tree with exactly n leaves, where each path from the root to a leaf is a suffix of the string s. The suffix tree for the string s can be constructed in $O(n)$ time using Ukkonen's algorithm [79]. In our case, a suffix tree can be used to identify all k-mers contained in the given sequence, simply following all the possible paths of size k starting from the root of the tree. Moreover, the problem of calculating the number of occurrences of each k-mer can be solved just counting the number of leaves in the subtree that starts at the end of the corresponding path. Given that the number of leaves of the tree is simply the size of the represented string, we have a linear-time method to calculate the k-spectrum of a string. Further modifications are needed to avoid the need of directly calculating the scalar product of the two feature vectors for the computation of the kernel. A generalized suffix tree is a suffix tree constructed using more than one string [27]. Given a set of strings there exists a variant of Ukkonen's algorithm that can build the corresponding generalized suffix tree in a time linear in the sum of the sizes of all the strings. A generalized suffix tree can be used to calculate the k-spectrum kernel of two strings at once, just traveling the tree in a depth first manner and summing up the products of the number of occurrences of every k-mer in the two strings. The procedure can also be used to compute a whole Gram matrix at once.

Spectrum kernels can be generalized to consider weighted combinations of substrings of arbitrary length:

$$k(s,t) = \sum_{u \in \mathscr{A}^*} w_u \phi_u(s) \phi_u(t) \tag{22}$$

where the non-negative coefficients w_u can be used for instance to give different weights according to the substring length. The k-spectrum kernel, for instance, is recovered setting $w_u = 1$ if $|u| = k$ and zero otherwise. Viswanathan and Smola [82] devised an efficient procedure for computing these types of kernels which exploits feature sparsity. The basic idea is to use the suffix tree representation for sorting all non-zero entries in $(\phi_u(s))_{u \in \mathscr{A}^*}$ and $(\phi_u(t))_{u \in \mathscr{A}^*}$ for s and t and evaluate only the matching ones.

The *mismatch string kernel* [47] is a variant of the spectrum kernel allowing for approximate matches between k-mers. Let the $(k, m) - neighbourhood$ of a k-mer u the set of all k-mers v which differ from u by at most m mismatches. Let $N_{(k,m)}(u)$ indicate this neighbourhood. The feature map of a k-mer u is:

$$\phi_{k,m}(u) = (\phi_v(u))_{v \in \mathscr{A}^k}$$

where $\phi_v(u) = 1$ if $v \in N_{(k,m)}(u)$ and zero otherwise. The feature map of a string s is computed summing the feature vectors of all its k-mers:

$$\Phi(s) = \sum_{\text{k-mers } u \text{ in } s} \phi_{k,m}(u).$$

Figure 5(b) shows the feature mapping corresponding to a mismatch string kernel with $k = 3$ and $m = 1$, on the same toy example used for the spectrum kernel. Note the increase in density of the feature vector. On real-world cases with large alphabets this will nonetheless be still very sparse. A (k,m)-mismatch tree is a data structure similar to a suffix tree. At each depth-k node, it allows to compute the number of k-mers in the string which have at most m mismatches with the one along the path from the root to the node. Generalized (k,m)-mismatch trees can be created as for the suffix tree in order to directly compute the kernel for pairs of strings or full Gram matrices.

The *string subsequence kernel* (SSK) [49] is an alternative string kernel also considering gapped substrings in computing similarities. The feature map Φ for a string s is defined as:

$$\Phi(s) = (\phi_u(s))_{u \in \mathscr{A}^k} = \left(\sum_{i:s(i)=u} \lambda^{l(i)} \right)_{u \in \mathscr{A}^k}$$

where $0 < \lambda \leq 1$ is a weight decay penalizing gaps. Such feature measures the number of occurrences of u in s, weighted according to their lengths. Note that the longer the occurrence, the more gaps in the alignment between u and s. Figure 5(c) shows the feature mapping obtained for $k = 3$ and arbitrary λ on the toy example already discussed, highlighting the contribution of the different occurrences of each substring. The inner product between strings s and t is computed as:

$$k_k(s,t) = \sum_{u \in \mathscr{A}^k} \phi_u(s)\phi_u(t) = \sum_{u \in \mathscr{A}^k} \sum_{i:s(i)=u} \sum_{j:t(j)=u} \lambda^{l(i)+l(j)}.$$

In order to make this product computationally efficient, we first introduce the auxiliary function

$$k'_i(s,t) = \sum_{u \in \mathscr{A}^i} \sum_{i:s(i)=u} \sum_{j:t(j)=u} \lambda^{|s|+|t|-i_1-j_1+2}$$

for $i = 1,\dots,k-1$, counting the length from the beginning of the substring match to the end of s and t instead of $l(i)$ and $l(j)$. The SSK can now be computed by the following recursive procedure, where $a \in \mathscr{A}$:

$$k_0(s,t) = 1 \quad \forall s,t \in \mathscr{A}^*$$
$$k'_i(s,t) = 0 \quad \text{if } \min(|s|,|t|) < i$$
$$k_i(s,t) = 0 \quad \text{if } \min(|s|,|t|) < i$$
$$k'_i(sa,t) = \lambda k'_i(s,t) + \sum_{j:t(j)=a} k'_{i-1}(s,t[1,\dots,j-1])\lambda^{|t|-j+2}, \forall i \in [1,k-1]$$
$$k_k(sa,t) = k_k(s,t) + \sum_{j:t(j)=a} k'_{k-1}(s,t[1,\dots,j-1])\lambda^2. \tag{23}$$

To prove the correctness of the procedure note that $k_k(sa,t)$ is computed by adding to $k_k(s,t)$ all the terms resulting by the occurrences of substrings terminated by a, matching t anywhere and sa on its right terminal part. In fact, in the second term of the recursion step for k_k, k'_{k-1} will count any matching substring found in s as if it finished at $|s|$, and the missing λ for the last element a is added for both s and t.

This kernel can be readily expanded to consider substrings of different lengths, i.e. by using a linear combination like

$$k(s,t) = \sum_k c_k k_k(s,t)$$

with $c_k \geq 0$. In such case, we simply compute k'_i for all i up to one less than the largest k required, and then apply the last recursion in (23) for each k such that $c_k > 0$, using the stored values of k'_i.

A number of fast kernels for inexact string matching have been proposed in [48]. Alternative types of string kernels will be discussed in Section 4.4.

4.3.2 Trees

Trees allow to represent structured objects which include hierarchical relationships (without cycles). Phylogenetic trees, for instance, are commonly used in biology to represent the evolutionary relationships among different species or biological sequences. Parse trees are a standard way to represent the syntactic structure of a string in a certain language. Let's start with some definitions useful for describing examples of kernels on trees.

A tree is a connected graph without cycles. A *rooted* tree is a tree where one node is chosen as the root. A natural orientation arises in rooted trees, moving along paths starting from the root. The nodes on the path from the root to v are called *ancestors* of v, with the last one being its *parent*. The nodes on the path leaving v are its *descendants*. The direct descendants of v are its *children*. A leaf is a node with no children. An *ordered* tree is a tree with a total order relationship among the children of each node. A *labelled* tree is a tree where each node is labelled with a symbol from an alphabet $\mathscr{A}$. Let $l(v)$ return the label of node v. A *subtree t'* of t is a tree made of a subset of nodes and edges in t. A *proper subtree t'* of t is a tree made of a node a and all its descendants in t. A *subset tree* is a subtree with more than one node, having either all children of a node or none of them. Figure 6 shows examples of different types of subtrees.

First of all, note that a generic string kernel can be applied to trees provided they are turned into a suitable string representation. Viswanathan and Smola [82] show how to employ the weighted substring kernel of eq. (22) to develop a kernel matching arbitrary subtrees. First, a procedure $tag(v)$ encodes a tree rooted at v in a string as follows:

- if v is an unlabelled leaf then $tag(v) = [\,]$;
- if v is a labelled leaf then $tag(v) = [l(v)]$;

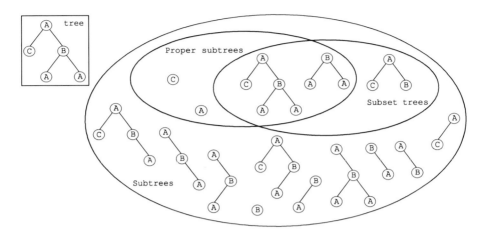

Fig. 6 Examples of different types of subtrees for a rooted ordered labelled tree. Tree kernels typically construct feature spaces based on these kind of fragments.

- if v is an unlabelled node with children $v_1, \ldots, v_m$ then $tag(v) = [tag(v_1) \cdots tag(v_m)]$;
- if v is a labelled node with children $v_1, \ldots, v_m$ then $tag(v) = [l(v)tag(v_1) \cdots tag(v_m)]$.

If the tree is unordered, the tags of the children are sorted in lexicographic order. The top leftmost tree in Fig. 6, for instance, would be encoded as: [A[C][B[A][A]]].

The resulting string is fed to the weighted substring kernel, again relying on its suffix tree representation to exploit feature sparsity. Different choices for the w_s lead to different subtree features. Only proper subtrees, for instance, can be used by setting $w_s = 0$ for substrings not starting and ending with balanced brackets.

Collins and Duffy [10, 11] introduced a *subset tree kernel* based on subset tree matches in the field of Natural Language Processing (NLP). Here parse trees are rooted ordered labelled trees representing the syntactic structure of a sentence according to an underlying (stochastic) context free grammar (CFG). Each subtree consisting of a node and the set of its children is a production rule of the grammar. Only subset trees are considered as valid features for the kernel. The rationale behind this choice is not splitting production rules in defining subtrees.

Let $\mathscr{T}$ be a set of rooted ordered labelled trees. The feature space has a coordinate for each possible subset tree in $\mathscr{T}$. Given by M the number of these fragments, a tree t is mapped to:

$$\Phi(t) = (\phi_i(t))_{i \in [1,M]}$$

where:

$$\phi_i(t) = \text{number of times in which the } i^{th} \text{ tree fragment occurs in } t$$

We define the set of nodes in t_1 and t_2 as N_1 and N_2 respectively. We further define an indicator function $I_i(n)$ to be 1 if subset tree i is seen rooted at node n and 0 otherwise. The kernel between t_1 and t_2 can now be written as

$$k(t_1,t_2) = \sum_{i=1}^{M} \phi_i(t_1)\phi_i(t_2) = \sum_{i=1}^{M} \sum_{n_1 \in N_1} I_i(n_1) \sum_{n_2 \in N_2} I_i(n_2) = \sum_{n_1 \in N_1} \sum_{n_2 \in N_2} C(n_1,n_2)$$

where we define $C(n_1,n_2) = \sum_{i=1}^{M} I_i(n_1)I_i(n_2)$, that is the number of common subset trees rooted at both n_1 and n_2. Given two nodes n_1 and n_2, we say that they *match* if their have the same label, same number of children and each child of n_1 has the same label of the corresponding child of n_2. The following recursive definition permits to compute $C(n_1,n_2)$ in polynomial time:

- If n_1 and n_2 don't match $C(n_1,n_2) = 0$.
- if n_1 and n_2 match, and they are both pre-terminals[3] $C(n_1,n_2) = 1$.
- Else

$$C(n_1,n_2) = \prod_{j=1}^{nc(n_1)} (1 + C(ch(n_1,j),ch(n_2,j))) \qquad (24)$$

where $nc(n_1)$ is the number of children of n_1 (equal to that of n_2 for the definition of match) and $ch(n_1,j)$ is the jth child of n_1.

To prove the correctness of (24), note that each child of n_1 contributes exactly $1 + C(ch(n_1,j),ch(n_2,j))$ common subset trees for n_1, n_2, the first with the child alone, and the other $C(ch(n_1,j),ch(n_2,j))$ with the common subset trees rooted at the child itself. The product in (24) considers all possible combinations of subset trees contributed by different children.

Given the large difference in size of trees to be compared, it is usually convenient to employ a normalized version of the kernel (see eq. (16)). Moreover, the kernel tends to produce extremely large values for very similar trees, thus making the algorithm behave like a one-nearest neighbour rule. This effect can be reduced by restricting the depth of the allowed subset trees to a fixed value d, or by scaling their relative importance with their size. To this extent we can introduce a parameter $0 < \lambda \leq 1$, turning the last two points of the definition of C into:

- if n_1 and n_2 match, and they are both pre-terminals $C(n_1,n_2) = \lambda$.
- Else

$$C(n_1,n_2) = \lambda \prod_{j=1}^{nc(n_1)} (1 + C(ch(n_1,j),ch(n_2,j))).$$

This corresponds to a modified inner product

---

[3] A pre-terminal is the parent of a leaf.

$$k(t_1,t_2) = \sum_{i=1}^{M} \lambda^{size_i} \phi_i(t_1)\phi_i(t_2)$$

where $size_i$ is the number of nodes of the corresponding subset tree.

Most tree kernels developed in the literature are extensions of the subset tree kernel. Moschitti [55] proposed the *partial tree kernel* in which arbitrary subtrees are used as fragments in place of subset trees. The kernel is obtained replacing equation (24) with:

$$C(n_1,n_2) = 1 + \sum_{J_1,J_2:|J_1|=|J_2|} \prod_{i=1}^{|J_1|} C(ch(n_1,J_{1i}), ch(n_2,J_{2i})))$$

where $J_1 = (J_{11},J_{12},\ldots,J_{1|J_1|})$ and $J_2 = (J_{21},J_{22},\ldots,J_{2|J_2|})$ and index sequences associated with ordered child sequences of n_1 and n_2 respectively. The *elastic tree kernel* extends the subset tree kernel by allowing: 1) matches between nodes with different number of children, provided the comparison still follows their left-to-right ordering; 2) approximate label matches, by introducing a similarity measure between them; 3) elastic matching between subtrees, where subtrees can be "stretched" in a tree provided the relative positions of the nodes in the subtree are preserved. Aiolli et al. [1] show that kernels defined on routes between tree nodes provide competitive discriminative power at reduced computational complexity with respect to most tree kernels. For a detailed treatment of the literature on tree kernels see [50].

4.3.3 Graphs

Graphs are a natural and powerful way to represent structured objects in a variety of domains, ranging from chemo- and bio-informatics to the World Wide Web. Here we will focus on graphs representing individual objects, like a chemical compound with atoms as vertices and bonds as edges. A different problem is that of treating objects that are related to each other by a graph structure. For a description of kernels on this type of data see [42, 21, 20, 69]. Let's start with some useful definitions concerning graphs.

A graph $G = (\mathcal{V},\mathcal{E})$ is a finite set of vertices (also called nodes) $\mathcal{V}$ and edges $\mathcal{E} \in \mathcal{V} \times \mathcal{V}$. In *directed* graphs edges (v_i,v_j) are oriented from initial node v_i to terminal node v_j. In *undirected* graphs edges have no orientation. This can be represented for instance by letting $(v_i,v_j) \in \mathcal{E} \iff (v_j,v_i) \in \mathcal{E}$. A *labelled* graph is a graph with a set of labels $\mathcal{L}$ and a function $long(\cdot)$ assigning labels to nodes (*node-labelled* graph), edges (*edge-labelled* graph) or both (*fully-labelled* graph). In the following we will call node-labelled graphs simply labelled graphs unless otherwise specified. A labelled graph can also be represented by its *adjacency* and label matrices A and L. The adjacency matrix is such that $A_{ij} = 1$ if $(v_i,v_j) \in \mathcal{E}$ and zero otherwise. The node-label matrix L is such that $L_{ij} = 1$ if $l(v_j) = \ell_i$ and zero otherwise. A *walk* in a graph is a sequence of vertices $(v_1,\ldots,v_{n+1})$ with $v_i \in \mathcal{V}$ and $(v_i,v_{i+1}) \in \mathcal{E}$ for all i. The length of a walk is the number of its edges (n in the example before). Let $W_n(G)$ return all n-length walks in graph G. A *path* is a walk such that $v_i \neq v_j \iff i \neq j$.

A *cycle* is a path such that $(v_{n+1}, v_1) \in \mathcal{E}$. A walk can contain an arbitrary number of cycles. A *connected* graph is a graph having at least one undirected path for each pair of its nodes. The *distance* between two nodes is the length of the minimal undirected path between them (if any). Given a graph G a *subgraph* G' is a graph made of a subset of the nodes and edges of G. Given a set of nodes $\mathcal{V}' \subset \mathcal{V}$, the subgraph *induced* by $\mathcal{V}'$ is made of nodes $\mathcal{V}'$ and all edges $\mathcal{E}' \subset \mathcal{E}$ connecting them. Two graphs G and G' are *isomorphic* if there is an isomorphism, a bijection ϕ such that for any two nodes $v_1, v_2 \in \mathcal{V}$ there is an edge $(v_1, v_2) \in \mathcal{E} \iff (\phi(v_1), \phi(v_2)) \in \mathcal{E}'$. For labelled graphs the isomorphism must also preserve label information, i.e. $l(v) = l(\phi(v))$. An isomorphism basically defines equivalence classes for graphs.

First, note that as discussed in the case trees, string kernels can be applied to the string encoding of a graph. The Simplified Molecular Input Line Entry Specification (SMILES), for instance, is a popular string notation for chemical molecules. Kernels for SMILES strings are described in [74], among other kernels for 2D and 3D molecular representations.

In defining kernels on graphs, we of course do not want to distinguish among isomorphic graphs. Given the set $\mathcal{G}$ of all graphs, the subgraph feature space is the space of all possible subgraphs of $\mathcal{G}$ modulo isomorphism, i.e. any two isomorphic subgraph are mapped to the same coordinate. Gärtner et al. [23] proved that no polynomial time algorithm can be devised for computing an inner product in such space (unless $P = NP$). The same holds if we restrict the feature space to consider paths only. Most existing kernels on generic graphs either use features based on graph walks, or employ some strategy to limit the set of valid subgraphs.

Examples of the first approach can be found in [19]. Consider the case of undirected node-labelled graphs. The simplest examples compute similarities in terms of walks in the two graphs which start and end with the same labels, i.e. $(v_1, \ldots, v_{n+1}) \in W_n(G)$ and $(v_1', \ldots, v_{m+1}') \in W_m(G')$ for which $l(v_1) = l(v_1')$ and $l(v_{n+1}) = l(v_{m+1}')$. The feature space of the kernel is defined in terms of features for label pairs:

$$\Phi(G) = \left(\phi_{\ell_i, \ell_j}(G) \right)_{\ell_i, \ell_j \in \mathcal{L}}$$

where:

$$\phi_{\ell_i, \ell_j}(G) = \sum_{n=1}^{\infty} \lambda_n |\{(v_1, \ldots, v_{n+1}) \in W_n(G) : l(v_1) = \ell_1 \wedge l(v_{n+1}) = \ell_j\}|.$$

and λ is a sequence of non-negative weights ($\lambda_n \in \mathbb{R}_0^+$ for all $n \in \mathbb{N}$). The feature map for a label pair ℓ_i, ℓ_j returns a weighted sum of the number of walks of length n starting with ℓ_i and ending with ℓ_j for all possible lengths n.

The feature mapping can be computed by operations on the graph matrices. The adjacency matrix has the useful property that its n power A^n extends the adjacency concept to walks of length n, that is $(A^n)_{ij}$ is the number of walks of length n from v_i to v_j. By introducing the label matrix we obtain that $(LA^nL^T)_{ij}$ is the number of walks of length n which start and end with labels ℓ_i and ℓ_j respectively. Thus:

$$\phi_{\ell_i,\ell_j}(G) = \left(\sum_{n=1}^{\infty} \lambda_n L A^n L^T \right)_{\ell_i,\ell_j}$$

and the corresponding kernel is:

$$k(G,G') = \langle L \left(\sum_{i=1}^{\infty} \lambda_i A^i \right) L^T , L' \left(\sum_{j=1}^{\infty} \lambda_j A'^j \right) L'^T \rangle$$

where the dot product between two matrices M, M' is is defined as:

$$\langle M,M' \rangle = \sum_{i,j} M_{ij} M'_{ij}. \tag{25}$$

Note that the kernel has to be absolute convergent in order to be a valid positive definite kernel. A sufficient condition for *absolute convergent graph kernels* can be found in [19]. An example of valid kernel is the *exponential graph kernel* defined as:

$$k_{exp}(G,G') = \langle L e^{\beta A} L^T , L' e^{\beta A'} L'^T \rangle$$

where $\beta \in \mathbb{R}$ is a parameter and the exponential of a matrix M is defined as

$$e^{\beta M} = \lim_{n \to \infty} \sum_{i=0}^{n} \frac{(\beta M)^i}{i!}.$$

Feasible matrix exponentiation usually requires diagonalizing the matrix. If we can diagonalize A such that $A = T^{-1}DT$, we can easily compute any power of A as $A^n = (T^{-1}DT)^n = T^{-1}D^nT$, where the power of the diagonal matrix D is calculated component-wise $[D^n]_{ij} = [D_{ij}]^n$. Therefore we have

$$e^{\beta A} = T^{-1} e^{\beta D} T$$

where $e^{\beta D}$ is calculated component-wise.

It's straightforward to extend this kernel to graphs with weighted edges by setting $A_{ij} = weight(v_i, v_j)$. The feature space of these kernels has a dimension equal to the square of the number of labels. When there are few possible labels, this can prevent the realization of an informative similarity measure. Extensions to this type of kernels consider the labels along the entire walk instead of only those of the terminal nodes. The feature space in this case is indexed by strings u, i.e.:

$$\Phi(G) = (\phi_u(G))_{u \in \mathscr{A}*}$$

where:

$$\phi_u(G) = \lambda_n |(v_1,\ldots,v_{n+1}) \in W_n(G) : n = |u| - 1 \wedge l(v_1) = u_1 \wedge \cdots \wedge l(v_{n+1}) = u_{n+1}\}|.$$

Note that it is straightforward to consider edge-labelled or fully-labelled graphs by replacing or adding edge labels in the comparisons. The kernel can be computed based on the powers of the adjacency matrix of the direct product [4] of the two graphs. A further extension accounts for up to $m \geq 0$ mismatches in the walk labels. The kernel can be computed by a combination of the direct product of the labelled and unlabelled graphs respectively. See [22] for the details of these kernels.

A large number of kernels have been developed relying on possibly domain-inspired strategies for choosing which subgraphs to include in the feature space. The *cyclic pattern kernel* [32] extracts features consisting of cycle and tree patterns. Cycles are directly extracted from the graph. A set of trees (called a forest) is obtained by removing from the graph all edges of all cycles. A canonical string representation for cycles and trees is employed in order to map each of them to a distinct feature coordinate modulo isomorphism (the pattern). Note that cyclic pattern kernels still cannot be computed in polynomial time in general [32], but it's sufficient to limit the computation to a subset of well-behaved graphs with a small enough number of cycles. The *shortest-path kernel* [8] considers the shortest-path between pairs of nodes. The *graph fragment kernel* [84] considers all connected subgraphs up to a given number of edges.

Note that kernels on graphs are not limited to exact or approximate matches between substructures. The *weighted decomposition kernel* (WDK) [52], for instance, compares two graphs by a combination of kernels between node pairs together to their contexts. Recalling the general form for R-convolution kernels (see eq. (20)), let $R^{-1}(G)$ be a decomposition of the graph into a node v and its *context V* in the graph. A possible context is the subgraph induced by all nodes at distance at most d from v (called its n-neighbourhood subgraph). The WDK is defined as:

$$k(G,G') = \sum_{(v,V)\in R^{-1}(G)} \sum_{(v',V')\in R^{-1}(G')} \delta(l(v),l(v'))k_{neigh}(V,V')$$

where δ is the matching kernel (see eq.(19)) and k_{neigh} is a kernel on the neighbourhood subgraph. Further details on graph kernels can be found e.g. in [22, 7].

4.4 Kernels from Generative Models

Generative models such as Hidden Markov Models [61] are a principled way to represent the probability distribution underlying the generation of data, and allow to treat aspects like uncertainty and missing information under a unifying formalism. On the other hand, discriminative methods such as kernel machines are an effective way to build decision boundaries, and often outperform generative models in prediction tasks. It would thus be desirable to have a learning method able to combine these complementary approaches. In the following we will present some examples of kernels derived from generative models, by directly modeling joint probability distributions [85, 29], defining a similarity measure between the models

---

[4] The direct product of two graphs G, G' has nodes $(v,v') \in \mathcal{V} \times \mathcal{V}' : l(v) = l(v')$ and edges $((v,v'),(u,u')) \in (\mathcal{V} \times \mathcal{V}')^2 : (v,u) \in \mathcal{E} \wedge (v',u') \in \mathcal{E}' \wedge l(v,u) = l(v',u')$. Nodes and edges in the direct product inherit the labels of the corresponding nodes and edges in two graphs.

underlying two examples [35, 34], or defining arbitrary kernels over observed and hidden variables and marginalizing over the hidden ones [78, 38]. For an additional general class of kernels from probability distributions see [36].

4.4.1 Dynamic Alignment Kernels

Joint probability distributions are a natural way of representing relationships between objects. The similarity of two objects can be modeled as a joint probability distribution that assigns high probabilities to pairs of related objects and low probabilities to pairs of unrelated objects. These considerations have been used in [85] to propose a kernel based on joint probability distributions. An analogous kernel was independently presented in [29] as a special case of convolution kernel (see section 4.2).

Definition 5. A joint probability distribution is *conditionally symmetrically independent* (CSI) if it is a mixture of a finite or countable number of symmetric conditionally independent distributions.

In order to show that a CSI joint p.d. is a positive definite kernel, let's write it as a dot product. Let X, Z, C be three discrete random variables such that

$$p(x,z) = P(X = x, Z = z) = p(z,x)$$

and

$$p(x,z|c) = P(X = x, Z = z|C = c) = p(x|c)p(z|c)$$

for all possible realizations of X, Z, C. We can thus write

$$p(x,z) = \sum_c p(x|c)p(z|c)p(c) = \sum_c \left(p(x|c)\sqrt{p(c)} \right) \left(p(z|c)\sqrt{p(c)} \right)$$

where the sum is over all possible realizations $c \in \mathscr{C}$ of C. This corresponds to a dot product with feature map

$$\Phi(x) = \{ p(x|c)\sqrt{p(c)} \mid c \in \mathscr{C} \}.$$

For a more general proof see [85].

A joint p.d. for a finite symbol sequence can be defined with a pair Hidden Markov Model. Such models generate two symbol sequences simultaneously, and are used in bioinformatics to align pairs of protein or DNA sequences [16]. A PHMM can be defined as follows, where A, B represent the two sequences modeled.

- A finite set S of states, given by the disjoint union of:

    S^{AB} - states that emit one symbol for A and one for B,
    S^A - states that emit one symbol only for A,
    S^B - states that emit one symbol only for B,
    a starting state START and an ending state END, which don't emit symbols.

- An $|S| \times |S|$ state transition probability matrix T.
- An alphabet $\mathcal{A}$.
- For states emitting symbols:

  - for $s \in S^{AB}$ a probability distribution over $\mathcal{A} \times \mathcal{A}$,
  - for $s \in S^A$ or $s \in S^B$ a probability distribution over $\mathcal{A}$.

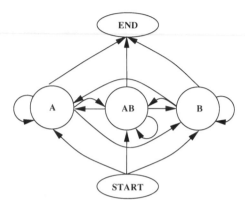

Fig. 7 State diagram for PHMM modeling pairs of sequences AB. The state AB emits common or similar symbols for both sequences, while the states A and B model insertions in sequence A and B respectively.

The state diagram for this PHMM is represented in figure 7. The state AB emits matching or nearly matching symbols for both sequences, while states A an B model insertions, that is symbols found in one sequence but not in the other. The joint p.d. for two sequences is given by the combination of all possible paths from START to END, weighted by their probabilities. This can be efficiently computed by well known dynamic programming algorithms [61]. Sufficient conditions for a PHMM to be CSI can be found in [85].

4.4.2 Fisher Kernel

The basic idea of the Fisher Kernel [35, 34] is that of representing the generative processes underlying two examples into a metric space, and compute a similarity measure in such space. Given a generative probability model $P(X|\theta)$ parametrized by $\theta = (\theta^1, \ldots, \theta^r)$, the gradient of its loglikelihood with respect to θ, $V_\theta(X) := \nabla_\theta \log P(X|\theta)$, is called *Fisher score*. It indicates how much each parameter θ^i contributes to the generative process of a particular example. The gradient is directly related to the expected *sufficient statistics* for the parameters. In the case that the generative model is an HMM, such statistics come as a by product of the forward backward algorithm [61] used to compute $P(X|\theta)$, without any additional cost. The derivation of the gradient for HMM and its relation to sufficient statistics is described in [33].

A class of models $P(X|\theta)$, $\theta \in \Theta$ defines a Riemannian manifold M_Θ (see [2, 3]), with metric tensor given by the covariance of the Fisher score, called *Fisher information matrix* and computed as

$$F := E_p[V_\theta(X)V_\theta(X)^T]$$

where the expectation is over $P(X|\theta)$. The direction of steepest ascent of the log-likelihood along the manifold is given by the *natural gradient* $\tilde{V}_\theta(X) = F^{-1}V_\theta(X)$ (see [3] for a proof). The inner product between such natural gradients relative to the Riemannian metric,

$$k(X,X') = \tilde{V}_\theta(X)^T F\tilde{V}_\theta(X) = V_\theta(X)^T F^{-1} V_\theta(X)$$

is called *Fisher kernel*. When the Fisher information matrix is too difficult to compute, it can be approximated by $F \approx \sigma^2 I$, where I is the identity matrix and σ a scaling parameter. Moreover, as $V_\theta(X)$ maps X to a vectorial feature space, we can simply use the dot product in such space, giving rise to the *plain* kernel

$$k(X,X') = V_0(X)^T V_0(X).$$

The Fisher kernel has been successfully employed for instance for detecting remote protein homologies [33], where the generative model is chosen to be an HMM representing a given protein family.

4.4.3 Marginalized Kernels

Marginalized kernels [78] are a hybrid class of kernels combining probabilistic models and arbitrary kernels over structures. Assume that a reasonable probabilistic model for the examples should include both observed variables x and hidden ones h. For instance, a set of images of handwritten characters could come from a number of different writers. Let $p(h|x)$ be the posterior probability of hidden variables given observed ones. Let $z = (x, h)$ and let $k_z(z, z')$ be a *joint* kernel over both observed and hidden variables. A *marginalized* kernel is obtained taking the expectation of the joint kernel with respect to the hidden variables, i.e.:

$$k(x,x') = \sum_h \sum_{h'} p(h|x)p(h'|x')k_z(z,z').$$

A first example of this type of kernel is the marginalized count kernel [78] for strings. Let x be a string of symbols from an alphabet $\mathscr{A}_x$. A very simple kernel is the 1-mer spectrum kernel (see Section 4.3.1), based on the counts of symbol co-occurrences:

$$k(x,x') = \sum_{u \in \mathscr{A}_x} \phi_u(x)\phi_u(x')$$

where $\phi_u(x)$ counts the number of occurrences of symbol u in x. This kernel treats all elements of the string the same, i.e. as if they were coming from the same distribution. Knowledge of the domain can suggest us that considering a number of different distributions should be more appropriate. For instance, in treating protein sequences we could distinguish between residues which are exposed at the surface and those that are buried in the protein core. We would thus like to compute separate counts for residues in the two conditions. As this information is not directly available from the sequence, we will model it using hidden variables. Let h be a string of hidden variables from an alphabet $\mathscr{A}_h$ ($\mathscr{A}_h = \{E,B\}$ in the protein example), with $|h| = |x|$. The marginalized count kernel is defined as:

$$k(x,x') = \sum_h \sum_{h'} p(h|x) p(h'|x') \sum_{u_x \in \mathscr{A}_x} \sum_{u_h \in \mathscr{A}_h} \phi_{u_x,u_h}(z) \phi_{u_x,u_h}(z')$$

where

$\phi_{u_x,u_h}(z)$ = number of times in which u_x and u_h appear in the same position in x and h respectively.

The kernel can be written in terms of marginalized counts $\hat{\phi}_{u_x,u_h}$:

$$k(x,x') = \sum_{u_x \in \mathscr{A}_x} \sum_{u_h \in \mathscr{A}_h} \underbrace{\sum_h p(h|x) \phi_{u_x,u_h}(z)}_{\hat{\phi}_{u_x,u_h}(x)} \underbrace{\sum_{h'} p(h'|x') \phi_{u_x,u_h}(z')}_{\hat{\phi}_{u_x,u_h}(x')}.$$

Note that counts can be easily generalized to k-mers with $k > 1$. Marginalized kernels have been defined for graphs [38] by defining transition probabilities between nodes and computing kernels in terms of random walks. These are tightly related to the walk-based kernels described in Section 4.3.3. See [81] for a unifying framework.

4.5 Kernels on Logical Representations

Logic representation formalisms allow to naturally express complex domain knowledge and perform reasoning on it. Developing kernels capable of handling this type of representation can greatly enhance their expressive power, allowing them to incorporate the semantics of the domain under consideration. In this section we will describe a generic class of kernels on logic terms. We will then show how to employ it to define kernels based on logic proofs. We will focus on the widespread Prolog programming language [73], which is based on first order logic enriched with partial support for arithmetic operations and some higher order structures like sets. Further details and additional examples of logic kernels can be found in [18]. For a general treatment of kernels on higher order logic representations see [24].

Let's first introduce some definitions and notation. A *definite clause* is an expression of the form $h \leftarrow b_1,...,b_n$, where h and the b_i are atoms and commas indicate logical conjunctions. Atoms are expressions of the form $p(t_1,...,t_n)$ where $p^{/n}$ is a predicate symbol of arity n and the t_i are *terms*. Terms are constants (denoted by lower case), variables (denoted by upper case), or structured terms. Structured terms are expressions of the form $f(t_1,...,t_k)$, where $f^{/k}$ is a functor symbol of arity k and $t_1,...,t_k$ are terms. A term is *ground* if it contains no variable. The atom h is also called the head of the clause, and $b_1,...,b_n$ its body. Intuitively, a clause represents that the head h will hold whenever the body $b_1,...,b_n$ holds. A clause with an empty body is called a *fact*. A set of clauses forms a *knowledge base* $\mathcal{B}$. When representing a domain of interest, we typically distinguish between *extensional* knowledge modeling single individuals of the domain (typically as a set of ground facts) and *intensional* knowledge providing general rules. A *substitution* $\theta = (V_1/t_1,...,V_n/t_n)$ is an assignment which applied to a formula F (written as $F\theta$, where F is a clause, atom or term) replaces each occurrence of variable V_i in h with term t_i, for all i. Queries can be used in order to derive novel information from a knowledge base. A query is a special clause (called *goal*) with an empty head. A goal $\leftarrow g$ is *entailed* by a knowledge base $\mathcal{B}$ (written as $\mathcal{B} \models g$) if and only if can be proved using clauses in $\mathcal{B}$ (otherwise $\mathcal{B} \not\models g$). A correct answer for g is a (possibly empty) substitution θ used to prove the goal.

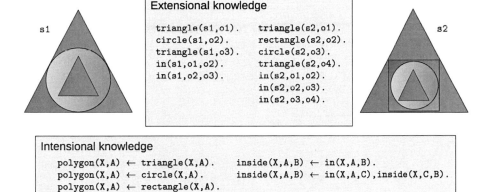

Fig. 8 Example of Prolog-based representation for an artificial domain. Top: extensional knowledge for two sample scenes. Bottom: intensional knowledge representing generic polygons and nesting in containment. The first argument of all predicates indicates the scene, the other arguments are objects within the scene.

Figure 8 shows an artificial domain with scenes containing nested polygons, highlighting extensional and intensional knowledge. The `polygon/2` predicate models generic polygons, while `inside/3` models the concept of nesting in containment. Novel facts can be derived from the knowledge base. Substitution $\theta = (X/bong1,A/o3)$, for instance, is a correct answer for goal $\leftarrow$ `polygon(X,A)`, while `inside(bong2,o2,o4)` can be proved from goal $\leftarrow$ `inside(X,A,B)` with answer $\theta = (X/bong2,A/o2,B/o4)$.

4.5.1 Kernels on Ground Terms

Complex objects can be represented as structured terms. In the *individual-as-term* representation an entire individual in the domain (e.g. a molecule) is encoded as a single structured term. Defining kernels between terms allows to apply kernel methods on this type of representation.

Let $\mathscr{C}$ be a set of constants and $\mathscr{F}$ a set of functors, and denote by $\mathscr{U}$ the corresponding Herbrand universe (the set of all ground terms that can be formed from constants in $\mathscr{C}$ and functors in $\mathscr{F}$). Let's assume that a type syntax exists. Types of constants are indicated with single characters (e.g. τ). Structured terms $f(t_1, ..., t_k)$ have type signatures $\tau_1 \times, ..., \times \tau_k \mapsto \tau'$, where τ' is the type of the term and $\tau_1 \times, ..., \times \tau_k$ the types of its arguments. We write $s : \tau$ to indicate that s is of type τ. The kernel between two ground typed terms t and s is a function $k : \mathscr{U} \times \mathscr{U} \to \mathbb{R}$ defined inductively as follows:

- if $s \in \mathscr{C}, t \in \mathscr{C}, s : \tau, t : \tau$ then

$$k(s,t) = \kappa_\tau(s,t)$$

  where $\kappa_\tau : \mathscr{C} \times \mathscr{C} \mapsto \mathbb{R}$ is a valid kernel on constants of type τ;
- else if s and t are structured terms that have the same type but different functors or signatures, i.e., $s = f(s_1, ..., s_n)$ and $t = g(t_1, ..., t_m)$, $s : \sigma_1 \times, ..., \times \sigma_n \mapsto \tau'$, $t : \tau_1 \times, ..., \times \tau_m \mapsto \tau'$, then

$$k(s,t) = \iota_{\tau'}(f^{/n}, g^{/m})$$

  where $\iota_{\tau'} : \mathscr{F} \times \mathscr{F} \mapsto \mathbb{R}$ is a valid kernel on functors that construct terms of type τ'
- else if s and t are structured terms and have the same functor and type signature, i.e., $s = f(s_1, ..., s_n)$, $t = f(t_1, ..., t_n)$, and $s, t : \tau_1 \times, ..., \times \tau_n \mapsto \tau'$, then

$$k(s,t) = \iota_{\tau'}(f^{/n}, f^{/n}) + \sum_{i=1}^{n} k(s_i, t_i) \tag{26}$$

- in all other cases $k(s,t) = 0$.

As a simple example consider data structures intended to describe scientific references:

$r = $ `article("Kernels on Gnus and Gnats",journal(ggj,2004))`
$s = $ `article("The Logic of Gnats",conference(icla,2004))`
$t = $ `article("Armadillos in Hilbert space",journal(ijaa,2004))`

Using $\kappa_\tau(x,z) = \delta(x,z)$ for all τ and $x, z \in \mathscr{C}$ and $\iota_{\tau'}(x,z) = \delta(x,z)$ for all τ' and $x, z \in \mathscr{F}$, we obtain $k(r,s) = 1, k(r,t) = 3$, and $k(s,t) = 1$. The fact that all papers are published in the same year does not contribute to $k(r,s)$ or $k(s,t)$ since these pairs have different functors describing the venue of the publication; it does contribute to

$k(r,t)$ as they are both journal papers. Note that strings have been treated as constants. A more informed similarity measure can be obtained employing a string kernel for comparing constants of type string.

Positive semi-definiteness of kernels on ground terms follows from their being special cases of decomposition kernels (see [57] for details). Variants where direct summations over sub-terms are replaced by tensor products are also possible. These kernels can be generalized to deal with higher order logic representations, see [24] for a detailed treatment.

4.5.2 Kernels on Proof Trees

In proving goals, an interpreter is required to recursively prove a number of subgoals according to a certain ordering (top-down for different clauses in the knowledge base and left-to-right in the clause body for Prolog). The proof has a structure which contains relevant information for the reasons for the final outcome. It would be desirable to be able to exploit this additional information in computing similarity between examples. This can be achieved by defining kernels directly on proofs [57], instead of simply on their outcomes in terms of goal satisfaction and goal variable substitutions. Given a knowledge-base $\mathscr{B}$ and a goal g, the proof tree for g is empty if $\mathscr{B} \not\models g$ or, otherwise, it is a tree t recursively defined as follows:

- if there is a fact f in $\mathscr{B}$ and a substitution θ such that $g\theta = f\theta$, then $g\theta$ is a leaf of t.
- otherwise there must be a clause $h \leftarrow b_1, ..., b_n \in \mathscr{B}$ and a substitution θ' such that $h\theta' = g\theta'$ and $\mathscr{B} \models b_j\theta' \,\forall j$, $g\theta'$ is the root of t and there is a subtree of t for each $b_j\theta'$ that is a proof tree for $b_j\theta'$.

Each internal node contains a clause head and its ordered set of children is the ordered set of atoms in its body. A simple bottom-up recursive procedure to turn a proof tree into a structured term consists of appending after the clause arguments the term representation of each of its children. A kernel on Prolog ground terms like the ones defined in the previous section can be applied to this representation. Note that a goal g can often be proved in multiple ways (if it is satisfiable), leading to a set of proof trees. A set kernel (eq. 21) can be used to compare two sets of proof trees by combining each pairwise comparison between proofs.

Let's recall the artificial example in Figure 8. Assume that positive scenes contain two triangles nested into one another with exactly n objects (possibly triangles) in between. The scene on the left would be positive for $n = 1$, the one on the right for $n = 2$. Having hints on the target concept, one could define a predicate (let's call it *visit*) looking for two polygons contained one into the other:

```
visit(X) ← inside(X,A,B),polygon(X,A),polygon(X,B)
```

Figure 9 shows the proofs trees obtained running such a visitor on the first scene in Figure 8, plus their representation as terms. Note that all the information required to identify the target concept is buried in the tree/term structure. A very simple kernel capable of exploiting this information would employ product (instead of sum)

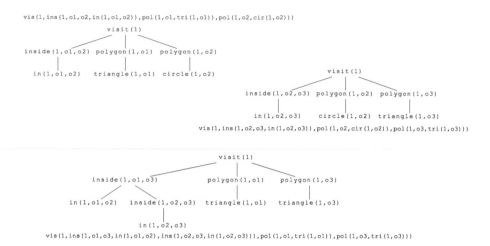

Fig. 9 Proof trees obtained by running the visitor on the first scene in Fig. 8. Representation of proof trees as ground structured terms (functor name abbreviated).

between terms (see eq. (26)), delta kernel (see eq. (19)) between functors, and a kernel between constants always evaluating to one. Applied to a pair of proof trees, this kernel evaluates to one if they both contain the same pair of polygons (e.g. a triangle and a circle) nested one into the other with the same number of objects in between, and zero otherwise. For any value of n, such a kernel maps the examples into a feature space where there is a single feature discriminating between positive and negative examples. Simply using extensional knowledge would not allow to produce effective similarities for this task. See [57] for a more extensive treatment of proof tree kernels and real-world applications.

5 Learning Kernels

Choosing the most appropriate kernel for the problem at hand is usually a rather hard task. Standard approaches consist of trying a number of candidate alternatives (e.g. Gaussian kernels with varying width, spectrum kernels with varying k) and selecting the best one according to some cross validation procedure. Measures of kernel quality can also be used to this aim. *Kernel target alignment* [14], for instance, computes the alignment between a kernel and the "target" kernel, $k_t(x_i, x_j) = y_i y_j$ for the binary classification case. Given a Gram matrix K and a target matrix $Y = yy^T$, the kernel target alignment is computed as:

$$A(K,Y) = \frac{\langle K,Y \rangle}{\sqrt{\langle K,K \rangle \langle Y,Y \rangle}} \tag{27}$$

where the dot product is defined as in Eq. (25). The selection procedure can be problematic in the common situation in which multiple kernels provide useful complementary features, or when there is no clue on which kernels to try. An effective alternative consists in jointly learning the kernel and the kernel machine based on it (i.e. the coefficients c_i of the kernel combination). In the following we present two rather complementary approaches to this aim: learning a sparse convex combination of basic kernels and learning a logic kernel relying on Inductive Logic Programming techniques.

5.1 Learning Kernel Combinations

It is often the case that multiple different kernels can be defined on a certain domain. Using the closure properties of the kernel class, it is always possible to combine all of them on an equal basis, for instance by direct summation or product. However, this is not necessarily the best choice in case many of them actually provide noisy or redundant features. Learning functions based on a large set of kernels has drawbacks also from an interpretation viewpoint, as it becomes difficult to identify the most relevant features for the discrimination. Recent approaches to kernel learning try to overcome these issues by learning weighted combinations of kernels and forcing sparsity in the weights. The overall kernel becomes a convex combination of *basis* kernels:

$$k(x,x') = \sum_{k=1}^{K} d_k k_k(x,x') \tag{28}$$

where K is the total number of kernels, $d_k \geq 0$ for all m and $\sum_{k=1}^{K} d_k = 1$. Multiple kernel learning (MKL) amounts at jointly learning the coefficients c_i of the overall function f and the weights d_k of the kernel combination. A sparse combination, in which only few d_k are different from zero, is typically enforced using some sparsity-inducing norm like the one-norm. MKL was initially addressed using semidefinite programming techniques [43]. A number of alternative solutions have been later proposed in the literature, especially for boosting efficiency towards large-scale applicability. Here we report a simple and efficient formulation named *SimpleMKL* [62].

Let $f(x) - \sum_{k=1}^{K} f_k(x)$, where each f_k belongs to a different RKHS $\mathscr{H}_k$ with associated kernel k_k. The SimpleMKL formulation addresses the following constrained optimization problem:

$$\min_{d} \ J(d)$$

$$\text{subject to: } \sum_{k=1}^{K} d_k = 1$$

$$d_k \geq 0 \quad k = 1,\ldots,K$$

where:

$$J(d) = \begin{cases} \min_{f,\xi} & \frac{1}{2}\sum_{k=1}^{K}\frac{1}{d_k}||f_k||_{\mathscr{H}_k}^2 + C\sum_{i=1}^{m}\xi_i \\ \text{subject to: } & y_i\sum_{k=1}^{K}f_k(x_i) \geq 1-\xi_i \quad i=1,\ldots,m \\ & \xi_i \geq 0 \quad i=1,\ldots,m. \end{cases}$$

Solving $J(d)$ for a particular value of d amounts at solving a standard SVM classification problem with a convex combination kernel as in equation (28). To see this, let's compute the Lagrangian and derive the dual formulation as in Section 3.1. The Lagrangian is given by:

$$L(f,\alpha,\beta) = C\sum_{i=1}^{m}\xi_i + \frac{1}{2}\sum_{k=1}^{K}\frac{1}{d_k}\|f_k\|_{\mathcal{H}_k}^2 - \sum_{i=1}^{m}\alpha_i(y_i\sum_{k=1}^{K}f_k(x_i)-1+\xi_i) - \sum_{i=1}^{m}\beta_i\xi_i$$

where $\alpha_i,\beta_i \geq 0$ for all i are the Lagrange multipliers. Zeroing the gradient with respect to the primal variables gives:

$$\frac{\partial L}{\partial f_k(\cdot)} = \frac{1}{d_k}f_k(\cdot) - \sum_{i=1}^{m}\alpha_i y_i k(\cdot,x_i) \quad k = 1,\ldots,K$$

$$\frac{\partial L}{\partial \xi_i} = C - \alpha_i - \beta_i = 0 \quad i = 1,\ldots,m$$

where we used the reproducing property (see eq. (3)) to derive $\frac{\partial f_k(x_i)}{\partial f_k(\cdot)} = \frac{\partial \langle k(\cdot,x_i),f_k(\cdot)\rangle}{\partial f_k(\cdot)}$ $= k(\cdot,x_i)$. Substituting into the Lagrangian we obtain:

$$\max_{\alpha\in\mathbb{R}^m} \quad -\frac{1}{2}\sum_{i=1}^{m}\sum_{j=1}^{m}\alpha_i y_i \alpha_j y_j \sum_{k=1}^{K}d_k k_k(x_i,x_j) + \sum_{i=1}^{m}\alpha_i$$

$$\text{subject to: } \alpha_i \in [0,C] \quad i = 1,\ldots,m$$

which is the standard SVM dual formulation for $k(x_i,x_j)$ as in eq. (28) and can be solved with one of the available SVM solvers. Differentiating J with respect to the weights d gives:

$$\frac{\partial J}{\partial d_k} = -\frac{1}{2}\sum_{i=1}^{m}\sum_{j=1}^{m}\alpha_i^* y_i \alpha_j^* y_j k_k(x_i,x_j) \quad k = 1,\ldots,K$$

where α^* are the solutions of the inner dual maximization problem. This gradient is used to update d along a gradient descent direction which retains its equality and non-negativity constraints. For a detailed description of the optimization algorithm see [62].

5.2   Learning Logical Kernels

Inductive Logic Programming [56] (ILP) amounts at learning a logic hypothesis capable of explaining a set of observations. In the most common setting of binary classification, the hypothesis should cover positive examples and not cover negative ones. More formally, let $\mathscr{B}$ be a logic knowledge base. Let $\mathscr{D} = \{(x_1,y_1),\ldots,$ $(x_m,y_m)\}$ be a dataset of input-output pairs, where inputs are identifiers for entities in the knowledge base. A hypothesis H is a set of definite clauses. Let $\mathscr{H}$ be the space of all hypotheses which can be constructed from a language $\mathscr{L}$. A generic

ILP algorithm aims at addressing the following maximization problem:

$$\max_{H \in \mathcal{H}} S(H, \mathcal{D}, \mathcal{B})$$

where $S(H, \mathcal{D}, \mathcal{B})$ is an appropriate scoring function for evaluating the quality of the hypothesis, e.g. accuracy of classification. The hypothesis space is structured by a (partial) generality relation $\preceq$: a hypothesis H_1 is more general than a hypothesis H_2 ($H_1 \preceq H_2$) if and only if any example covered by H_2 is also covered by H_1. Search in the hypothesis space is conducted in a general-to-specific or specific-to-general fashion, using an appropriate *refinement operator* [56, 15]. The computational cost of searching in a discrete space typically forces one to resort to heuristic search algorithms, such as (variations of) greedy search (see [15] for more details).

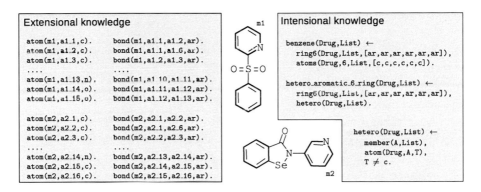

Fig. 10 Example of logical representations for the NCI-HIV dataset of compounds. Extensional knowledge encodes the atom-bond representation of molecules: atom(m,a,t) indicates that molecule m has atom a which is a t chemical element; bond(m,a1,a2,t) indicates a chemical bond of type t (e.g. ar for aromatic bond) between atoms a1 and a2 of molecule m. Intensional knowledge encodes functional groups such as benzene and other aromatic rings. An aromatic ring is a ring of atoms connected by aromatic bonds. A benzene is an aromatic ring of six carbon atoms. A hetero aromatic ring is an aromatic ring containing at least one non-carbon atom.

Figure 10 shows a pair of chemical compounds from the NCI-HIV database which were measured active in inhibiting HIV. Both extensional and intensional knowledge are (partially) reported. Clause active(M) ← atom(M,A1,o), bond(M,A1,A2,2), for instance, covers both compounds, while its specialization active(M) ← atom(M,A1,o),bond(M,A1,A2,2),atom(M,A2,s) covers m1 but not m2.

The integration of ILP and statistical learning has the appealing potential of combining the advantages of the respective approaches, namely the expressivity and interpretability of ILP with the effectiveness and efficiency of statistical learning as well as its ability to deal with other tasks than standard binary classification. ILP and kernel machines can be integrated by defining an appropriate kernel based on logic

hypotheses. Given that a hypothesis is a set of clauses, the simplest kernel consists
of counting the number of clauses which cover both examples. This corresponds to
a feature space with one binary feature for each clause.

Consider again the example of Figure 10. Let $H = \{c_1, c_2, c_3\}$ be a hypothesis
consisting of the following three clauses:

```
active(M) ← atom(M,A1,o),bond(M,A1,A2,2),atom(M,A2,s).
active(M) ← hetero_aromatic_6_ring(M,List),member(A,List),atom(M,A,n).
active(M) ← benzene(M,[A1,A2,A3,A4,A5,A6]),bond(M,A5,A10,ar),atom(M,A10,se).
```

Clauses c_1 and c_2 succeed on molecule m1, while clauses c_2 and c_3 succeed on
molecule m2. The resulting feature space representation according to the
up-mentioned kernel is given by:

$$\Phi_H(\text{m1}) = \begin{pmatrix} 1 \\ 1 \\ 0 \end{pmatrix} \qquad \Phi_H(\text{m2}) = \begin{pmatrix} 0 \\ 1 \\ 1 \end{pmatrix}.$$

More complex kernels can be obtained using kernel composition (e.g. polynomial
or Gaussian kernel, see Section 4.2).

Given a kernel over logical hypothesis, we can construct a prediction function as:

$$f(x; H, \mathscr{B}) = \sum_{i=1}^{m} c_i k(x, x_i; H, \mathscr{B}).$$

The generic maximization problem becomes:

$$\max_{H \in \mathscr{H}} \max_{f \in \mathscr{F}_H} S(f, \mathscr{D}, \mathscr{B})$$

where $\mathscr{F}_H$ is the set of all functions that can be generated with hypothesis H. Learning these logic kernel machines [45] amounts at jointly learning the kernel, in terms
of the logic hypothesis H, and the function in the RKHS associated with it. Learning
convex combinations of basic kernel functions, as seen in Section 5.1, can be cast
into constrained optimization problems for which efficient algorithms exist. Conversely, here we face the discrete space of logic hypotheses and heuristic search
algorithms need to be employed.

kFOIL [44] is a simple example of this paradigm, based on an adaptation of the
well-known FOIL algorithm [60]. The algorithm is briefly sketched in Algorithm 1.
It repeatedly searches for clauses that score well with respect to the data set and
the current hypothesis and adds them to the current hypothesis. In the inner loop,
kFOIL greedily searches for a clause that scores well. To this aim, it employs a
general-to-specific hill-climbing search strategy. Let $p^{/n}$ denote the predicate that
is being learned. Then the most general clause, which succeeds on all examples,
is "$p(X_1, ..., X_n) \leftarrow$". The set of all refinements of a clause c is produced by a refinement operator $\rho(c)$. For our purposes, a refinement operator specializes a clause
$h \leftarrow b_1, \cdots, b_k$ by adding new literals b_{k+1} to the clause, though other refinements

---

Algorithm 1. kFOIL learning algorithm

Initialize $H := \emptyset$
repeat
    Initialize $c := p(X_1, \cdots, X_n) \leftarrow$
    repeat
        $c := \mathrm{argmax}_{c' \in \rho(c)} \max_{f \in \mathcal{F}_{H \cup \{c'\}}} S(f, \mathcal{D}, \mathcal{B})$
    until stopping criterion
    $H := H \cup \{c\}$
until stopping criterion
return $\mathrm{argmax}_{f \in \mathcal{F}_H} S(f, \mathcal{D}, \mathcal{B})$

---

have also been used in the literature. Scoring a clause amounts at training a kernel machine using the current hypothesis, including the candidate clause, and returning a measure of its performance (e.g. its accuracy on the training set). Learning in kFOIL is stopped when there is no improvement in score between two successive iterations. Several extensions and improvements have been proposed. Substantial efficiency gains, for instance, can be obtained using a kernel quality measure such as kernel target alignment (see Eq. (27)) to score clauses, instead of training a kernel machine every time. Further details and extensions can be found in [45].

6   Supervised Kernel Machines for Structured Output

Many relevant learning problems require to predict outputs which are themselves structures. Speech recognition or protein secondary structure prediction, for instance, can eventually be formalized as sequential labelling tasks. A key problem in NLP consists in predicting the parse tree of a sentence. Many techniques have been developed in the literature to address this kind of problems, often based on directed or undirected graphical models. A quite general approach consists of learning a function $f(x, y)$ over both input and output which basically evaluates the quality of y as an output for x (e.g. the conditional probability of the output given the input i.e. $P(Y = y|X = x)$). Prediction then amounts at finding the output maximizing this score, i.e.:

$$y^* = \underset{y \in \mathcal{Y}}{\mathrm{argmax}} \, f(x, y). \tag{29}$$

Kernel machines can be generalized to this kind of setting by defining a *joint feature map* $\Phi(x, y)$ over both input and output, with corresponding kernel $k((x, y), (x', y'))$. A common approach consists of defining basic feature maps for input and output components. The joint feature map is then obtained as a combination of tensor products and direct sums (see Section 4.2) involving these basic maps. Multiclass classification can be obtained for instance setting:

$$\Phi(x, y) = \Phi(x) \otimes \hat{\Phi}(y) \tag{30}$$

$$
x = \texttt{ACCCCGGGTTTTAA}
$$
$$
y = \texttt{EEEIIIIEEEIIEE}
$$

$$
\Phi(x,y) = \begin{array}{c} \left(\begin{array}{c} 5 \\ 2 \\ 2 \\ 4 \\ 3 \\ 2 \\ 1 \\ 2 \\ 0 \\ 2 \\ 2 \\ 2 \end{array}\right) \end{array} \begin{array}{c} \text{EE} \\ \text{EI} \\ \text{IE} \\ \text{II} \\ \text{AE} \\ \text{CE} \\ \text{GE} \\ \text{TE} \\ \text{AI} \\ \text{CI} \\ \text{GI} \\ \text{TI} \end{array}
$$

$$
\left.\begin{array}{c}\text{EE}\\\text{EI}\\\text{IE}\\\text{II}\end{array}\right\} \left[\sum_{t=1}^{T-1} \hat{\phi}(y_t) \otimes \hat{\phi}(y_{t+1})\right]
$$

$$
\left.\begin{array}{c}\text{AE}\\\text{CE}\\\text{GE}\\\text{TE}\\\text{AI}\\\text{CI}\\\text{GI}\\\text{TI}\end{array}\right\} \left[\sum_{t=1}^{T} \phi(x_t) \otimes \hat{\phi}(y_t)\right]
$$

Fig. 11 Example of joint feature map for sequential labelling. One-hot encoding is used for feature mappings of both input and output symbols.

where $\hat{\Phi}(y)$ is the one-hot encoding of the class label. A sequential labelling algorithm encoding Markov chain assumptions for dependencies would use a feature mapping like:

$$
\Phi(x,y) = \left[\sum_{t=1}^{T-1} \hat{\phi}(y_t) \otimes \hat{\phi}(y_{t+1})\right] \odot \left[\sum_{t=1}^{T} \phi(x_t) \otimes \hat{\phi}(y_t)\right]. \tag{31}
$$

Figure 11 shows an example in which the sequential labelling task is predicting coding (exons) and non-coding (introns) regions in the genome. The joint feature map is obtained from Eq. (31) using one-hot encoding for the input and output basic mappings. More complex kernels on structures can also be employed in principle, (see Section 4.3). Note however that limitations to the form of the usable kernels are often needed in order to retain efficiency in computing the argmax in Eq. (29).

A generalized version of the representer theorem gives the form of the solution of Tikhonov regularized problems.

Theorem 3 (Generalized Representer Theorem). *Let $D_m = \{(x_i,y_i) \in \mathcal{X} \times \mathcal{Y}\}_{i=1}^{m}$ be a training set, $V(x_i,y_i,f)$ a general loss function, $\mathcal{H}$ a RKHS with norm $||\cdot||_{\mathcal{H}}$. Then the general form of the solution of the regularized risk*

$$
\frac{1}{m}\sum_{i=1}^{m} V(x_i,y_i,f) + \lambda||f||_{\mathcal{H}}^2
$$

is

$$
f(x,y) = \sum_{i=1}^{m}\sum_{y'\in\mathcal{Y}} c_{iy'} k((x_i,y'),(x,y)).
$$

The solution is as a linear combination of kernel functions centered on "augmented" training examples (x_i,y'), where x_i are training inputs and $y' \in \mathcal{Y}$ possible outputs.

Structured-Output Support Vector Machines [77] (SO-SVM) generalize large-margin classification to this setting, by enforcing a large separation between correct and incorrect output assignments. A hinge loss for structured-output prediction can be written as:

$$V(x,y,f) = |1 - (f(x,y) - \operatorname*{argmax}_{y' \neq y} f(x,y'))|_+ .$$

The resulting Tikhonov regularized functional is:

$$\min_{f \in \mathcal{H}} \frac{1}{m} \sum_{i=1}^{m} |1 - (f(x,y) - \operatorname*{argmax}_{y' \neq y} f(x,y'))|_+ + \lambda ||f||^2_{\mathcal{H}} .$$

As for the scalar case, f can be written as a dot product between the feature space representation of the example and a parameter vector, i.e.:

$$f(x,y) = \sum_{i=1}^{m} \sum_{y' \in \mathcal{Y}} c_{iy'} \langle \Phi(x_i, y'), \Phi(x,y) \rangle = \langle \sum_{i=1}^{m} \sum_{y' \in \mathcal{Y}} c_{iy'} \Phi(x_i, y'), \Phi(x,y) \rangle = \langle w, \Phi(x,y) \rangle .$$

By introducing slack ξ_i for the cost paid for each incorrect prediction we obtain the following quadratic optimization problem:

$$\min_{w \in \mathcal{H}, \xi \in \mathbb{R}^m} \quad C \sum_{i=1}^{m} \xi_i + \frac{1}{2} ||w||^2$$

$$\text{subject to:} \quad \langle w, \Phi(x_i, y_i) \rangle - \operatorname*{argmax}_{y' \neq y_i} \langle w, \Phi(x_i, y') \rangle \geq 1 - \xi_i \quad i = 1, \ldots, m$$

$$\xi_i \geq 0 \quad i = 1, \ldots, m$$

where again we replaced $C - 2/\lambda m$ for consistency with most literature on SO-SVM. Multiclass SVM [12] can be seen as a simple instantiation of this optimization problem, with a joint feature map as in Eq. (30). Note that the margin constraint for each training example can be equivalently replaced with a set of constraints, one for each possible output y, all sharing the same slack variable ξ_i. The Langragian of the optimization problem becomes:

$$L(w, \alpha, \beta) = C \sum_{i=1}^{m} \xi_i + \frac{1}{2} ||w||^2 - \sum_{i=1}^{m} \sum_{y' \neq y_i} \alpha_{iy'} (\langle w, \Phi(x_i, y_i) \rangle - \langle w, \Phi(x_i, y') \rangle - 1 + \xi_i) - \sum_{i=1}^{m} \beta_i \xi_i$$

where $\alpha_{iy'}, \beta_i \geq 0$ for all i and y'. Zeroing the derivatives with respect to the primal variables gives:

$$\frac{\partial L}{\partial w} = w - \sum_{i=1}^{m} \sum_{y' \neq y_i} \alpha_{iy'} \underbrace{\Phi(x_i, y_i) - \Phi(x_i, y')}_{\delta \Phi_i(y')} = 0 \rightarrow w = \sum_{i=1}^{m} \sum_{y' \neq y_i} \alpha_{iy'} \delta \Phi_i(y')$$

$$\frac{\partial L}{\partial \xi_i} = C - \sum_{y' \neq y_i} \alpha_i - \beta_i = 0 \rightarrow \sum_{y' \neq y_i} \alpha_i \in [0, C]$$

Algorithm 2. Cutting plane algorithm for SO-SVM.

Initialize $S_i := \emptyset$ for all i
repeat
  for $i = 1, \ldots, m$ **do**
    $H(y) = 1 - \langle w, \delta\Phi_i(y) \rangle$
    where $w = \sum_j \sum_{y' \in S_j} \alpha_{jy'} \delta\Phi_j(y')$
    compute $\hat{y} = \mathrm{argmax}_{y \neq y_i} H(y)$
    compute $\xi_i = \max\{0, \max_{y \in S_i} H(y)\}$
    if $H(\hat{y}) > \xi_i + \varepsilon$ **then**
      $S_i .- S_i \cup \{\hat{y}\}$
      solve problem (32) restricted to variables $S = \bigcup_i S_i$
    end if
  end for
until no S_i has changed during iteration

---

where we replaced $\delta\Phi_i(y') = \Phi(x_i, y_i) - \Phi(x_i, y')$ for compactness. Substituting into the Lagrangian we obtain:

$$\max_{\alpha \in \mathbb{R}^m} \quad -\frac{1}{2} \sum_{i=1}^{m} \sum_{j=1}^{m} \sum_{y' \neq y_i} \sum_{y'' \neq y_j} \alpha_{iy'} \alpha_{jy''} \langle \delta\Phi_i(y'), \delta\Phi_j(y'') \rangle + \sum_{i=1}^{m} \sum_{y' \neq y_i} \alpha_{iy'} \quad (32)$$

$$\text{subject to:} \quad \sum_{y' \neq y_i} \alpha_{iy'} \in [0, C] \quad i = 1, \ldots, m.$$

Note that replacing $\langle \delta\Phi_i(y'), \delta\Phi_j(y'') \rangle = \langle \Phi(x_i, y_i) - \Phi(x_i, y'), \Phi(x_j, y_j) - \Phi(x_j, y'') \rangle$ $= k((x_i, y_i), (x_j, y_j)) - k((x_i, y_i), (x_j, y'')) - k((x_i, y'), (x_j, y_j)) + k((x_i, y'), (x_j, y''))$ we recover the kernel-based formulation where the feature mapping is only implicitly done.

A main problem in optimizing (32) is the number of variables α, which is often exponential in the size of the output. The problem is addressed using a cutting plane algorithm, which iteratively solves larger optimization problems obtained incrementally adding the most violated constraint. Algorithm 2 reports a sketch of the algorithm. Starting from an empty set of constraints, the algorithm repeatedly iterates over training examples. For each training example the most violated constraint according to the current version of f is computed ($H(y)$ is the cost paid for predicting y). Its cost is compared with that of previous constraints involving the same example (or zero if none exists). If the new cost is larger by more than a user-defined tolerance ε, the constraint is added and a new optimization problem is solved. The algorithm terminates when no further constraint is added for any of the training examples.

In both training and classification phases, the argument maximizing f needs to be returned. The efficiency of this computation is crucial to the applicability of the approach. Dynamic programming techniques can be employed in a number of common cases, like the Viterbi algorithm for sequential labelling or its extension to probabilistic context free grammars for predicting parse trees.

A number of variants of the approach described here exist. A common extension consists in adding a loss function $\Delta(y_i, y')$ to the constraints, measuring the loss incurred in predicting y' in place of y. Alternative approaches have also been proposed in the literature, for instance to address cases in which exact computation of argmax is intractable. For a comprehensive treatment of kernel methods for structured-output prediction see [5].

7 Conclusions

In this chapter we provided a comprehensive introduction to kernel machines for structured data. We reviewed the mathematical foundations underlying kernel methods and described a number of popular kernel machines for both supervised and unsupervised learning. We gave an extensive treatment of kernels for structured data, including strings, trees and graphs, as well as kernels based on generative models and logical representations. Techniques for learning kernels from data were also discussed. Finally we introduced kernel machines for predicting complex output structures, a promising research direction combining kernel methods with probabilistic graphical models and search-based approaches.

Research on kernel methods is extremely active and a large number of diverse problems have been tackled under this formalism. This chapter is aimed at providing a clear and detailed explanation on kernel methods and their use for dealing with structured data, rather than a complete survey on all approaches which have been developed. Techniques which were not covered include Gaussian Processes [63] and distribution embeddings. The former take a Bayesian viewpoint and allow to provide predictions in terms of expected value and variance, where uncertainty is reduced in the proximity of observed points. The latter is a recent trend of research aimed at embedding probability distributions in RKHS. This allows to develop effective nonparametric techniques for problems like testing whether two samples come from the same distribution [25] or measuring the strength of dependency between two variables [26]. Additional references to advanced material on kernel methods can be found e.g. in [31].

References

1. Aiolli, F., Da San Martino, G., Sperduti, A.: Route kernels for trees. In: Proceedings of the 26th Annual International Conference on Machine Learning, ICML 2009, pp. 17–24. ACM, New York (2009)
2. Amari, S.I.: Mathematical foundations of neurocomputing. Proceedings of the IEEE 78(9), 1443–1463 (1990)
3. Amari, S.I.: Natural gradient works efficiently in learning. Neural Computation 10, 251–276 (1998)
4. Aronszajn, N.: Theory of reproducing kernels. Trans. Amer. Math. Soc. 686, 337–404 (1950)
5. Bakir, G.H., Hofmann, T., Schölkopf, B., Smola, A.J., Taskar, B., Vishwanathan, S.V.N.: Predicting Structured Data (Neural Information Processing). The MIT Press (2007)

6. Berg, C., Christensen, J.P.R., Ressel, P.: Harmonic Analysis on Semigroups. Springer, New York (1984)
7. Borgwardt, K.M.: Graph Kernels. PhD thesis, Ludwig-Maximilians-University Munich (2007)
8. Borgwardt, K.M., Kriegel, H.-P.: Shortest-path kernels on graphs. In: Proceedings of the Fifth IEEE International Conference on Data Mining, ICDM 2005, pp. 74–81. IEEE Computer Society, Washington, DC (2005)
9. Boser, B.E., Guyon, I.M., Vapnik, V.N.: A training algorithm for optimal margin classifier. In: Proc. 5th ACM Workshop on Computational Learning Theory, Pittsburgh, PA, pp. 144–152 (July 1992)
10. Collins, M., Duffy, N.: Convolution kernels for natural language. In: Dietterich, T.G., Becker, S., Ghahramani, Z. (eds.) Advances in Neural Information Processing Systems 14. MIT Press (2002)
11. Collins, M., Duffy, N.: New ranking algorithms for parsing and tagging: Kernels over discrete structures, and the voted perceptron. In: Proceedings of the 40th Annual Meeting of the Association for Computational Linguistics (ACL 2002), Philadelphia, PA, USA, pp. 263–270 (2002)
12. Crammer, K., Singer, Y.: On the algorithmic implementation of multiclass kernel-based vector machines. J. Mach. Learn. Res. 2, 265–292 (2002)
13. Cristianini, N., Shawe-Taylor, J.: An Introduction to Support Vector Machines. Cambridge University Press (2000)
14. Cristianini, N., Kandola, J., Elisseeff, A., Shawe-Taylor, J.: On kernel-target alignment. In: Advances in Neural Information Processing Systems 14, vol. 14, pp. 367–373 (2002)
15. De Raedt, L.: Logical and Relational Learning. Springer (2008)
16. Durbin, R., Eddy, S., Krogh, A., Mitchison, G.: Biological Sequence Analysis: Probabilistic Models of Proteins and Nucleic Acids. Cambridge University Press (1998)
17. Fletcher, R.: Practical Methods of Optimization, 2nd edn. John Wiley & Sons (1987)
18. Frasconi, P., Passerini, A.: Learning with Kernels and Logical Representations. In: De Raedt, L., Frasconi, P., Kersting, K., Muggleton, S.H. (eds.) Probabilistic Inductive Logic Programming. LNCS (LNAI), vol. 4911, pp. 56–91. Springer, Heidelberg (2008)
19. Gärtner, T.: Exponential and geometric kernels for graphs. In: NIPS Workshop on Unreal Data: Principles of Modeling Nonvectorial Data (2002)
20. Gärtner, T., Flach, P., Kowalczyk, A., Smola, A.J.: Multi-instance kernels. In: Sammut, C., Hoffmann, A. (eds.) Proceedings of the 19th International Conference on Machine Learning, pp. 179–186. Morgan Kaufmann (2002)
21. Gärtner, T., Lloyd, J.W., Flach, P.A.: Kernels for Structured Data. In: Matwin, S., Sammut, C. (eds.) ILP 2002. LNCS (LNAI), vol. 2583, pp. 66–83. Springer, Heidelberg (2003)
22. Gärtner, T.: Kernels for Structured Data. PhD thesis, Universität Bonn (2005)
23. Gärtner, T., Flach, P.A., Wrobel, S.: On Graph Kernels: Hardness Results and Efficient Alternatives. In: Schölkopf, B., Warmuth, M.K. (eds.) COLT/Kernel 2003. LNCS (LNAI), vol. 2777, pp. 129–143. Springer, Heidelberg (2003)
24. Gärtner, T., Lloyd, J.W., Flach, P.A.: Kernels and distances for structured data. Mach. Learn. 57, 205–232 (2004)
25. Gretton, A., Borgwardt, K.M., Rasch, M.J., Schölkopf, B., Smola, A.J.: A kernel method for the two-sample-problem. In: Schölkopf, B., Platt, J., Hoffman, T. (eds.) Advances in Neural Information Processing Systems 19, pp. 513–520. MIT Press, Cambridge (2007)
26. Gretton, A., Fukumizu, K., Teo, C.H., Song, L., Schölkopf, B., Smola, A.J.: A kernel statistical test of independence. In: Platt, J.C., Koller, D., Singer, Y., Roweis, S. (eds.) Advances in Neural Information Processing Systems 20. MIT Press, Cambridge (2008)

27. Gusfield, D.: Algorithms on Strings, Trees, and Sequences: Computer Science and Computational Biology. Cambridge University Press (1997)
28. Ham, J., Lee, D.D., Mika, S., Schölkopf, B.: A kernel view of the dimensionality reduction of manifolds. In: Proceedings of the Twenty-First International Conference on Machine Learning, ICML 2004, p. 47. ACM, New York (2004)
29. Haussler, D.: Convolution kernels on discrete structures. Technical Report UCSC-CRL-99-10, University of California, Santa Cruz (1999)
30. Hoffmann, H.: Kernel pca for novelty detection. Pattern Recogn. 40, 863–874 (2007)
31. Hofmann, T., Schölkopf, B., Smola, A.J.: Kernel methods in machine learning. Annals of Statistics 36(3), 1171–1220 (2008)
32. Horváth, T., Gärtner, T., Wrobel, S.: Cyclic pattern kernels for predictive graph mining. In: Proceedings of the Tenth ACM SIGKDD International Conference on Knowledge Discovery and Data Mining, KDD 2004, pp. 158–167. ACM, New York (2004)
33. Jaakkola, T., Diekhans, M., Haussler, D.: A discriminative framework for detecting remote protein homologies. Journal of Computational Biology 7(1-2), 95–114 (2000)
34. Jaakkola, T., Haussler, D.: Probabilistic kernel regression models. In: Proc. of Neural Information Processing Conference (1998)
35. Jaakkola, T., Haussler, D.: Exploiting generative models in discriminative classifiers. In: Proceedings of the 1998 Conference on Advances in Neural Information Processing Systems II, pp. 487–493. MIT Press, Cambridge (1999)
36. Jebara, T., Kondor, R., Howard, A.: Probability product kernels. J. Mach. Learn. Res. 5, 819–844 (2004)
37. Joachims, T.: Making large-scale SVM learning practical. In: Schölkopf, B., Burges, C., Smola, A. (eds.) Advances in Kernel Methods – Support Vector Learning, ch. 11, pp. 169–185. MIT Press (1998)
38. Kashima, H., Tsuda, K., Inokuchi, A.: Marginalized kernels between labeled graphs. In: Proceedings of the Twentieth International Conference on Machine Learning, pp. 321–328. AAAI Press (2003)
39. Keerthi, S.S., Duan, K.B., Shevade, S.K., Poo, A.N.: A fast dual algorithm for kernel logistic regression. Mach. Learn. 61, 151–165 (2005)
40. Kim, K., Franz, M.O., Schölkopf, B.: Iterative kernel principal component analysis for image modeling. IEEE Transactions on Pattern Analysis and Machine Intelligence 27(9), 1351–1366 (2005)
41. Kimeldorf, G., Wahba, G.: Some results on tchebycheffian spline functions. J. Math. Anal. Applic. 33, 82–95 (1971)
42. Kondor, R.I., Lafferty, J.: Diffusion kernels on graphs and other discrete input spaces. In: Sammut, C., Hoffmann, A. (eds.) Proc. of the 19th International Conference on Machine Learning, pp. 315–322. Morgan Kaufmann (2002)
43. Lanckriet, G.R.G., Cristianini, N., Bartlett, P., El Ghaoui, L., Jordan, M.I.: Learning the kernel matrix with semidefinite programming. J. Mach. Learn. Res. 5, 27–72 (2004)
44. Landwehr, N., Passerini, A., De Raedt, L., Frasconi, P.: kfoil: learning simple relational kernels. In: Proceedings of the 21st National Conference on Artificial Intelligence, vol. 1, pp. 389–394. AAAI Press (2006)
45. Landwehr, N., Passerini, A., Raedt, L., Frasconi, P.: Fast learning of relational kernels. Mach. Learn. 78, 305–342 (2010)
46. Leslie, C., Eskin, E., Noble, W.S.: The spectrum kernel: a string kernel for svm protein classification. In: Proc. of the Pacific Symposium on Biocomputing, pp. 564–575 (2002)
47. Leslie, C., Eskin, E., Weston, J., Noble, W.S.: Mismatch string kernels for svm protein classification. In: Becker, S., Thrun, S., Obermayer, K. (eds.) Advances in Neural Information Processing Systems 15, pp. 1417–1424. MIT Press, Cambridge (2003)

48. Leslie, C., Kuang, R., Eskin, E.: Inexact matching string kernels for protein classificatio. In: Schölkopf, B., Tsuda, K., Vert, J.-P. (eds.) Kernel Methods in Computational Biology, MIT Press (2004) (in press)
49. Lodhi, H., Shawe-Taylor, J., Cristianini, N., Watkins, C.: Text classification using string kernels. In: Advances in Neural Information Processing Systems, pp. 563–569 (2000)
50. Da San Martino, G.: Kernel Methods for Tree Structured Data. PhD thesis, Department of Computer Science, University of Bologna (2009)
51. Menchetti, S.: Learning Preference and Structured Data: Theory and Applications. PhD thesis, Dipartimento di Sistemi e Informatica, DSI, Università di Firenze, Italy (December 2005)
52. Menchetti, S., Costa, F., Frasconi, P.: Weighted decomposition kernels. In: Proceedings of the 22nd International Conference on Machine Learning, ICML 2005, pp. 585–592. ACM, New York (2005)
53. Mercer, J.: Functions of positive and negative type and their connection with the theory of integral equations. Philos. Trans. Roy. Soc. London A 209, 415–446 (1909)
54. Micchelli, C.A., Xu, Y., Zhang, H.: Universal kernels. J. Mach. Learn. Res. 7, 2651–2667 (2006)
55. Moschitti, A.: Efficient Convolution Kernels for Dependency and Constituent Syntactic Trees. In: Fürnkranz, J., Scheffer, T., Spiliopoulou, M. (eds.) ECML 2006. LNCS (LNAI), vol. 4212, pp. 318–329. Springer, Heidelberg (2006)
56. Muggleton, S., De Raedt, L.: Inductive logic programming: Theory and methods. Journal of Logic Programming 19/20, 629–679 (1994)
57. Passerini, A., Frasconi, P., De Raedt, L.: Kernels on prolog proof trees: Statistical learning in the ILP setting. Journal of Machine Learning Research 7, 307–342 (2006)
58. Platt, J.C.: Fast training of support vector machines using sequential minimal optimization. In: Burges, C., Schölkopf, B. (eds.) Advances in Kernel Methods–Support Vector Learning. MIT Press (1998)
59. Poggio, T., Smale, S.: The mathematics of learning: Dealing with data. Notices of the American Mathematical Society 50(5), 537–544 (2003)
60. Quinlan, J.R.: Learning Logical Definitions from Relations. Machine Learning 5, 239–266 (1990)
61. Rabiner, L.R.: A tutorial on hidden markov models and selected applications in speech recognition. Proceedings of the IEEE 77(2), 257–286 (1989)
62. Rakotomamonjy, A., Bach, F.R., Canu, S., Grandvalet, Y.: SimpleMKL. Journal of Machine Learning Research 9, 2491–2521 (2008)
63. Rasmussenand, C.E., Williams, C.K.I.: Gaussian Processes for Machine Learning (Adaptive Computation and Machine Learning). The MIT Press (December 2005)
64. Saitoh, S.: Theory of Reproducing Kernels and its Applications. Longman Scientific Technical, Harlow (1988)
65. Saunders, G., Gammerman, A., Vovk, V.: Ridge regression learning algorithm in dual variables. In: Proc. 15th International Conf. on Machine Learning, pp. 515–521 (1998)
66. Schölkopf, B., Platt, J.C., Shawe-Taylor, J., Smola, A.J., Williamson, R.C.: Estimating the support of a high dimensional distribution. Neural Computation 13, 1443–1471 (2001)
67. Schölkopf, B., Smola, A., Müller, K.R.: Kernel principal component analysis. In: Advances in Kernel Methods–Support Vector Learning, pp. 327–352. MIT Press (1999)
68. Schölkopf, B., Smola, A.J.: Learning with Kernels. The MIT Press, Cambridge (2002)
69. Schölkopf, B., Warmuth, M.K. (eds.): COLT/Kernel 2003. LNCS (LNAI), vol. 2777. Springer, Heidelberg (2003)

70. Schölkopf, B., Smola, A.J., Williamson, R.C., Bartlett, P.L.: New support vector algorithms. Neural Comput. 12, 1207–1245 (2000)
71. Shawe-Taylor, J., Cristianini, N.: Kernel Methods for Pattern Analysis. Cambridge University Press, New York (2004)
72. Shin, K., Kuboyama, T.: A generalization of haussler's convolution kernel: mapping kernel. In: Proceedings of the 25th International Conference on Machine Learning, ICML 2008, pp. 944–951. ACM, New York (2008)
73. Sterling, L., Shapiro, E.: The art of Prolog: advanced programming techniques, 2nd edn. MIT Press, Cambridge (1994)
74. Swamidass, S.J., Chen, J., Bruand, J., Phung, P., Ralaivola, L., Baldi, P.: Kernels for small molecules and the prediction of mutagenicity, toxicity and anti-cancer activity. Bioinformatics 21, 359–368 (2005)
75. Tax, D.M.J., Duin, R.P.W.: Support vector domain description. Pattern Recognition Letters 20, 1991–(1999)
76. Tikhonov, A.N.: On solving ill-posed problem and method of regularization. Dokl. Akad. Nauk USSR 153, 501–504 (1963)
77. Tsochantaridis, I., Joachims, T., Hofmann, T., Altun, Y.: Large margin methods for structured and interdependent output variables. JMLR 6, 1453–1484 (2005)
78. Tsuda, K., Kin, T., Asai, K.: Marginalized kernels for biological sequences. Bioinformatics 18(suppl. 1), S268–S275 (2002)
79. Ukkonen, E.: On-line construction of suffix trees. Algorithmica 14, 249–260 (1995)
80. Vapnik, V.N.: Statistical Learning Theory. Wiley, New York (1998)
81. Vishwanathan, S.V.N., Schraudolph, N.N., Kondor, R., Borgwardt, K.: Graph kernels. Journal of Machine Learning Research 11, 1201–1242 (2010)
82. Vishwanathan, S.V.N., Smola, A.: Fast Kernels for String and Tree Matching. Advances in Neural Information Processing Systems 15 (2003)
83. Wahba, G.: Splines Models for Observational Data. Series in Applied Mathematics, vol. 59. SIAM, Philadelphia (1990)
84. Wale, N., Watson, I.A., Karypis, G.: Comparison of descriptor spaces for chemical compound retrieval and classification. Knowl. Inf. Syst. 14, 347–375 (2008)
85. Watkins, C.: Dynamic alignment kernels. In: Smola, A.J., Bartlett, P., Schölkopf, B., Schuurmans, D. (eds.) Advances in Large Margin Classiers, pp. 39–50. MIT Press (2000)

Chapter 10
Multiple Classifier Systems:
Theory, Applications and Tools

Francesco Gargiulo, Claudio Mazzariello, and Carlo Sansone

In many Pattern Recognition applications, the achievement of acceptable recognition rates is conditioned by the large pattern variability, whose distribution cannot be simply modeled.

This affects the results at each stage of the recognition system so that, once this has been designed, its performance cannot be improved over a certain bound, despite the efforts in refining either the classification or the description method.

Starting from the early nineties, some research groups concentrated the attention on a multiple classifier approach. The rationale of this approach lies in the assumption that, by suitably combining the results of a set of base classifiers, the obtained performance is better than that of any base classifier. In other words, it is claimed that the consensus of a set of classifiers may compensate for the weakness of a single classifier, while each classifier preserves its own strength.

The implementation of a multiple classifier system implies the definition of a combining rule or a strategy for determining the most likely class a sample should be attributed to, on the basis of the class to which it is attributed by each base classifier.

Different combining rules and strategies, independent of the adopted classification model, have been proposed so far in the open literature. In this Chapter we first present a survey of such approaches, by considering both different architectures and combining rules as well as methods for constructing ensembles. Afterward, a taxonomy of the applications where a multiple classifier approach has been successfully applied is organized. Finally, some of the tools currently available for implementing a multiple classifier system are presented.

1 MCS Theory

In the area of Pattern Recognition the idea of combining classifiers has been proposed as a new direction for the improvement of the performance of individual

Francesco Gargiulo · Claudio Mazzariello · Carlo Sansone
Dipartimento di Informatica e Sistemistica
Università degli Studi di Napoli "Federico II", Italy
e-mail: {francesco.grg,cmazzari,carlosan}@unina.it

M. Bianchini et al. (Eds.): *Handbook on Neural Information Processing*, ISRL 49, pp. 335–378.
DOI: 10.1007/978-3-642-36657-4_10   © Springer-Verlag Berlin Heidelberg 2013

classifiers. These classifiers could be based on a variety of classification methodologies, and they could achieve different rates of correctly classified samples. The
goal of classification result integration algorithms is to generate more certain, precise and accurate classification results. Dietterich [37] provides an accessible and
informal reasoning, from statistical, computational and representational viewpoints,
on why ensembles can improve results. His considerations are also reported in the
Kuncheva's book [122], which is now the most comprehensive reference for researchers in the Multiple Classifier Systems field.

1.1   MCS Architectures

The combination of multiple classifiers can be considered as a generic pattern recognition problem in which the input consists of the results of the individual classifiers,
and the output is the combined decision [122].

Combination of multiple classifiers is a fascinating problem that can be considered from many perspectives, and combination techniques can be grouped and analyzed in different ways. In terms of implementation, a categorization of combination
methods can be made by considering the combination topologies or structures employed, as described in [97]. We could have different MCS depending on [79]:

- Types of classifier outputs

    - *Type 1 - abstract level*: The classifier produces only a label without any infor-
      mation about the classification accuracy.
    - *Type 2 - rank level*: The classifier gives an ensemble of possible classes ranked
      in order of importance.
    - *Type 3 - confidence or measurement level*: The classifier gives a vector of
      scores each one associated to a possible class.

- Trainable or not-Trainable combiners.
- Topology.

Lu [86] categorizes *MCS* topologies into three categories: Cascading, Parallel and
Hierarchical.

In a **cascading classifier**, the classification result generated by a classifier is used
as an input to the next classifier. The results obtained through each classifier are
similarly passed onto the next classifier until a result is obtained through the final
classifier in the chain. The main disadvantage of the use of this methodology is the
inability of later classifiers to correct mistakes made by earlier classifiers.

Parallel classifiers integrate the results of all classifiers in a singular location.
The main design decision that has to be made in the implementation of such a configuration is the selection of a representative combination methodology. If the decision process is well designed, the system can reach good performance. However,
the improper selection of a combinatorial strategy could accentuate the influence of
poorly performing classifiers, which could eventually adversely affect the overall
performance.

Hierarchical classifiers combine both parallel and cascading classifier configurations to obtain optimal performance. The use of such a methodology can compensate the disadvantages encountered through the use of a cascading integration. Hierarchical systems could also be used to introduce error checking, which would nullify the influence of poorly performing classifiers.

A more comprehensive categorization of multiple classifier architectures is presented in [80]. This categorization divides topologies into conditional, hierarchical (serial), multiple-parallel and hybrid topologies.

Conditional Topology

This strategy first selects one classifier to perform the task of classification. If this classifier fails to correctly identify the presented data, another classifier is selected, as shown in the figure 1. Most implementations include a primary classifier, which is usually selected as the first classifier to be selected. The selection of the next classifier can either be a static decision or maybe based on the values obtained through the use of the primary classifier. Examples methods for dynamic selection include decision trees. This process can continue for as long as there are classifiers available or the pattern is correctly classified. If the primary classifier is an efficient one, the process is computationally efficient. The queue of selected classifiers could be organized in order for the computationally heavy classifiers to be only selected at the end of the classifier queue. One difficult aspect of such an implementation is the selection of a process by which the failures and successes of a classifier can be evaluated. This method can become overly complicated when the number of available classifiers increases.

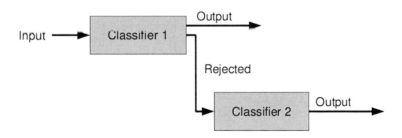

Fig. 1 A Conditional Topology

Hierarchical (Serial) Topology

This topology employs a method where classifiers are applied in succession. Each classifier applied to the data is used to reduce the number of possible classes to which such input data belongs to. As the data passes through the classifiers, the decision becomes more and more focused. The common strategy for the design of the classifier queue is to insert classifiers ordered according to decreasing error values. That is to say the classifier with the highest error is used first, whereas the classifier with the lowest error is used last. Of course, there should be safeguards to ensure that the classes selected by each classifier always include the correct class. If not, the next classifier will not have the option of selecting the correct output class.

In the figure 2 we show an example of hierarchical topology where each base classifier is a binary one, that can distinguish between the true class and all the rest.

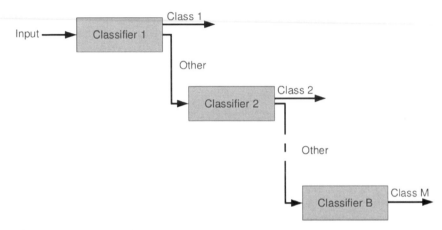

Fig. 2 A Hierarchical Topology

Multiple (Parallel) Topology

This is the most common implementation of a multiple classifier system. All the classifiers first operate in parallel on the input and the results are then pooled to obtain a consensus result. This methodology does incur in a cost as it is computationally heavy, with each classifier having to be executed before the final result is obtained.

Parallel combinations can be implemented using different strategies, and the combination method depends on the types of information produced by the base classifiers (see next Section).

Hybrid Topology

A hybrid topology based system incorporates a mechanism for the selection of the best classifier for a given input. It is obvious that certain classifiers perform better than others on certain data. Thus, the selection of an appropriate classifier would streamline the entire classification process.

This topology could be considered a trade-off between parallel and serial topology. A possible example is shown in fig. 4. The major disadvantage of this architecture is its complexity, even if we can reach better performance with respect to the others topologies.

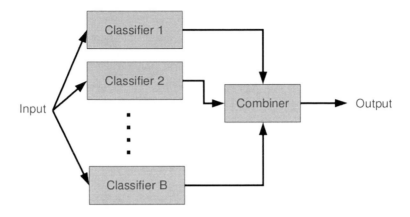

Fig. 3 A Parallel Topology

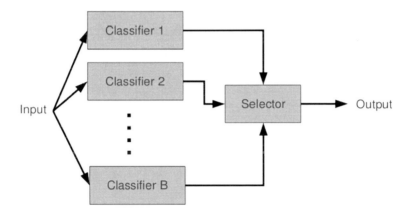

Fig. 4 A Hybrid Topology

Hybrid and parallel topologies are also known as *selection-based* and *fusion-based*, respectively. Until now, classifier selection has not attracted as much as classifier fusion. This might change in the future as classifier selection is probably the better of the two strategies, if trained well [79].

1.2   Combining Rules

The type of the combiner that can be used in the parallel architecture depicted in Fig. 3 depends on the base classifiers' output. If these outputs are of *Type 1*, we can have different kind of combiners such as, for example, *Majority Voting*, *Weighted Majority Voting*, *Behavior-Knowledge Space* and *Bayesian Combination*, while if the base classifiers' output is of *Type 3, Product rule, Sum rule, Max rule, Min rule, Median rule* and the *Dempster-Shafer* approach can be used among others.

Majority Voting

As recalled in [79]: *"Dictatorship and majority vote are perhaps the two oldest strategies for decision making. Their roots can be traced back to the era of ancient Greek city states and the Roman Senate. The majority criterion became established in 1356 for the election of German kings, by 1450 was adopted for elections to the British House of Commons, and by 1500 as a rule to be followed in the House itself"*.

This combiner is then based on a *democratic* method, even used in democratic countries: the Vote. Each classifier gives its own evaluation; the final result will be given from the class with more votes. In this case the combiner has to count only the occurrences of each class, and evaluate which class has the greatest number of votes.

If we want to formalize this concept, following the notation adopted in [79], we can assume that the outputs of the classifier will be denoted with a binary vector of size M,

$$[d_{i,1}, \ldots, d_{i,M}]^T \in \{0,1\}^M, i = \{1, \ldots B\}$$

where B is the number of classifiers involved into the ensemble, M is the number of the possible classes, and where $d_{i,j} = 1$ if the ith classifier votes the class C_j for the actual sample, while $d_{i,j} = 0$ otherwise. So the system will decide for the class C_k if:

$$\sum_{i=1}^{B} d_{i,k} = \max_{j=1}^{M} \sum_{i=1}^{B} d_{i,j} \tag{1}$$

or, in other words, if the number of votes obtained by the C_k class is the maximum of the evaluation obtained from all the possible classes. We can also have a tie: in this case, the sample should be rejected.

Weighted Majority Voting

A variation of the previously described technique is the *weighted majority voting*. In this case, for each classifier, we have also a weight. Obviously this weight will be defined before the decision process. If we want to formalize this method, we can consider the outputs of each classifier as in the previous method. In this case we have to consider another vector b_i, that represents the weights associated to the ith classifier. In this case C_k will be given as output class if:

$$\sum_{i=1}^{B} b_i d_{i,k} = \max_{j=1}^{M} \sum_{i=1}^{B} b_i d_{i,j} \tag{2}$$

It is worth noting that if the weights b_i are all the same, *weighted majority voting* behaves exactly as majority voting.

A good way to choose the weights could be the following one, as demonstrated in [79]:

> *If we consider an ensemble of B independent classifiers, each of them with an his own accuracy, p_i, in which their accuracy will be combined through the weighted majority voting. The accuracy of the combination is maximized put the votes in accord with the following method:*

$$b_i \propto \log \frac{p_i}{1 - p_i} \tag{3}$$

Behavior-Knowledge Space

Behavior-Knowledge Space (BKS) derives the information needed to combine the classifiers from a knowledge space, which can concurrently record the decision of all the classifiers on a suitable set of samples. This means that this space records the behavior of all the classifiers on this set, and it is therefore called the *Behavior-Knowledge Space* [63]. So, a BKS is a N-dimensional space where each dimension corresponds to the decision of a classifier. Given a sample to be assigned to one of m possible classes, the ensemble of the classifiers can in theory provide m^N different decisions. Each one of these decisions constitutes one *unit* of the BKS. In the learning phase each BKS *unit* can record M different values c_i, one for each class. Given a suitably chosen data set, each sample x of this set is classified by all the classifiers and the *unit* that corresponds to the particular classifiers' decision (called *focal unit*) is activated. It records the actual class of x, say j, by adding one to the value of c_j. At the end of this phase, each *unit* can calculate the best representative class associated to it, defined as the class that exhibits the highest value of c_i. It corresponds to the most likely class, given a classifiers' decision that activates that *unit*. In the operating mode, the BKS acts as a look-up table. For each sample x to be classified, the B decisions of the classifiers are collected and the corresponding *focal unit* is selected. Then x is assigned to the best representative class associated to its *focal unit*.

In order to detect *unknown* classes we can consider the use of the following decision rule:

$$C(x) = i$$

when $c_i > 0$ and $\frac{c_i}{T} \geq \lambda$, otherwise x is rejected. T is the total number of samples belonging to that focal unit (i.e. $T = \sum_{k=1}^{M} c_k$), while λ is a suitably chosen threshold $(0 \leq \lambda \leq 1)$ which controls the reliability of the final decision.

Bayesian Combination

The Bayesian Combination rule is based on the *a posteriori* probability, where the conditioning terms are represented by the outputs of individual base classifiers. In fact, to an input sample x it will assign the class that maximizes such probability. If we denote it as $y_i(x)$ (for the sake of simplicity, hereinafter the x will be ignored) the output of the B classifiers involved into the ensemble, and with C_k the generic class, the combiner has to choose the class that maximize the quantity:

$$p(C_k|y_1, y_2, \ldots, y_B) \tag{4}$$

This is the best combination method that we can use to reduce the error probability. The problem in this case regards the knowledge of all the conditional probabilities for the available classes. This information is often unknown. To overcome this problem, it is possible to use some decision rules directly derived from the bayesian formalism, that are an approximation of eq. 4.

The principal combination rules are [71]:

- Product Rule
- Sum rule
- Max rule
- Min rule
- Median rule

Product Rule. If we use the Bayes Rule it is possible to rewrite eq. 4 as:

$$p(C_k|y_1, y_2, \ldots, y_B) = \frac{p(C_k)p(y_1, y_2, \ldots, y_B|C_k)}{p(y_1, y_2, \ldots, y_B)} \tag{5}$$

It is possible to rewrite the denominator as:

$$p(y_1, y_2, \ldots, y_B) = \sum_{l=1}^{M} p(y_1, y_2, \ldots, y_B|C_l)p(C_k) \tag{6}$$

where M is the number of the possible classes. Now, if we assume that the outputs of all the classifiers are **conditionally independent**, we can rewrite the conditional probability as:

$$p(y_1, y_2, \ldots, y_B|C_k) = \prod_{i=1}^{B} p(y_i|C_k) \tag{7}$$

consequently eq. 5 becomes:

$$p(C_k|y_1, y_2, \ldots, y_B) = \frac{p(C_k)^{-B+1} \prod_{i=1}^{B} p(C_k|y_i)}{\sum_{l=1}^{M} \prod_{i=1}^{B} p(y_i|C_l)p(C_k)} \tag{8}$$

To maximize eq. 8, it is necessary to maximize its numerator with respect to k, that is:

$$\max_{k}\{p(C_k)^{-B+1}\prod_{i=1}^{B}p(C_k|y_i)\} \tag{9}$$

Eq. 9 represents the product rule. In fact we try to maximize the product of the conditional probability of each classifier, with respect to all the classes. One of the major problems of this technique is linked to the possibility that one or more classifiers give a result very close to zero. In this case, the product will give us a value very close to zero, too, and the combiner could fail.

Sum Rule. To define the sum rule we have to make the hypothesis that all the *a priori* probabilities and the *a posteriori* probabilities are very close each other:

$$p(C_k|y_i) = p(C_k)(1 + \delta_{i,j}) \qquad with \quad \delta_{i,j} << 1 \tag{10}$$

After this, we can substitute eq. 10 into eq. 9, and we can obtain:

$$p(C_k)^{-B+1}\prod_{i=1}^{B}p(C_k|y_i) - p(C_k) = p(C_k)\prod_{i=1}^{B}(1 + \delta_{i,j}) \tag{11}$$

If we expand the second member product and we do not consider the second order terms, we obtain:

$$\max_{k}\{(1 - B)p(C_k) + \sum_{i=1}^{B}p(C_k|y_i)\} \tag{12}$$

Eq.12 represents the *sum rule*. The limit is in the initial hypothesis which is very restrictive.

Max Rule. This rule is obtained directly from the sum rule. It is obtained as an approximation of the sum with the maximum into the eq. 12:

$$\max_{k}\{(1 - B)p(C_k) + M \max_{i=1}^{B} p(C_k|y_i)\} \tag{13}$$

Min Rule. This rule is obtained starting from eq. 9 with an approximation of the product with the minimum:

$$\max_{k}\{p(C_k)^{-B+1} \min_{i=1}^{B} p(C_k|y_i)\} \tag{14}$$

Median Rule. Finally, the median rule is obtained starting from eq. 14, using the median instead of the minimum:

$$\max_k\{p(C_k)^{-B+1}med_{i=1}^B p(C_k|y_i)\} \tag{15}$$

The Dempster-Shafer Approach

The theory of Dempster and Shafer (D-S theory) has been frequently applied to deal with uncertainty management and incomplete reasoning [55] . Differently from the classical Bayesian theory, D-S theory can explicitly model the absence of information, while in case of absence of information a Bayesian approach attributes the same probability to all the possible events.

The D-S theory could narrow down a hypothesis set with the accumulation of evidence and it allows for a representation of the *ignorance* due to the uncertainty in the evidence. When the ignorance reaches the value zero, the D-S model reduces to the standard Bayesian model. Thus, the D-S theory could be considered as a generalization of the theory of probability.

Some Theoretical Issues. Let θ be a finite, non-empty set consisting of all the possible values of a certain attribute. The set θ serves as our universal set, and it is called the *frame of discernment*. A *mass function*, also called *basic probability assignment*, is a mapping m from the set of all subsets of θ into the closed interval $[0, 1]$ such that

$$m(\varnothing) = 0 \qquad\qquad \sum_{A \subseteq 2^\theta} m(A) = 1 \tag{16}$$

The function value $m(A)$ measures the degree of evidence that is assigned to the subset and (1) reflects that the total evidence is one. The simplest mass function corresponds to the case when there is no available evidence at all (i.e.,*total ignorance*), in this case we set $m(\theta) = 1$ and $m(A) = 0$ for all other subsets of θ.

When assigning a *bpa*, there are some requirements which have to be met. They descend from the fact that the *bpa* is still a probability function, hence has to respect the constraints for mass probability functions. Each *bpa* is such that $m : 2^\theta \rightarrow [0, 1]$, where θ indicates the so called *frame of discernment*. Usually, the frame of discernment θ consists of M mutually exclusive and exhaustive hypotheses A_i, $i = 1, \ldots, M$. A subset $\{A_i, \ldots, A_j\} \subseteq \theta$ represents a new hypothesis. As the number of possible subsets of θ is 2^θ, the generic hypothesis is an element of 2^θ.

For example, if we only consider two hypotheses (classes), namely *Positive*(P) and *Negative*(N); hence, the frame of discernment is $\theta = \{\{P\}, \{N\}\}$ and $2^\theta = \{\{P\}, \{N\}, \{P, N\}\}$, whereas in the Bayesian case only the events $\{\{P\}, \{N\}\}$ would be considered.

$\{P\}$ and $\{N\}$ are referred to as *simple events* or *singletons*, while $\{P, N\}$ is referred to as *composite event*. Furthermore, also the following properties have to hold:

$$m(\varnothing) = 0 \qquad\qquad \sum_{A \subseteq 2^\theta} m(A) = 1$$

The aim of assigning a *bpa* is to describe the reliability of a particular classifier in reporting a specific event. Such a representation is suitable for combination, but as we want to deal with combined results in the same way, we also impose the constraint that the combination of several *bpa* by means of the D-S rule still has to be a *bpa*. The uncertainty in the final decision will be inversely proportional to the extent to which the base classifiers agree. If we have B base classifiers, the combination rule is such that:

$$m(A) = K \sum_{\cap_{i=1}^{B} A_i = A} \prod_{i=1}^{n} m_i(A_i)$$

where:

$$K^{-1} = 1 - \sum_{\cap_{i=1}^{B} A_i = \varnothing} \prod_{i=1}^{B} m_i(A_i)$$

$$= \sum_{\cap_{i=1}^{n} A_i \neq \varnothing} \prod_{i=1}^{n} m_i(A_i)$$

It is worth observing that the normalizing factor K is independent from any specific value of A. The value K can therefore be considered a constant, once the bpa's are fixed.

1.3   Strategies for Constructing a Classifier Ensemble

Another problem to be considered when designing a new MCS regards the strategies that must be chosen for selecting the base classifiers. [65] list eighteen classifier combination schemes which also contain several strategies for building an ensemble; [136] details four methods of combining multiple models: bagging, boosting, stacking and error-correcting output codes whilst [4] covers seven methods of combining multiple learners: among them, error-correcting output codes, bagging, boosting and stacked generalization are reported. Since here the literature is reviewed, with an emphasis on both theoretical and practical advices, the taxonomy from [65] is first provided in Table 1, and then five methods for building an ensemble are reported: bagging, boosting (including AdaBoost), stacked generalization, the random subspace method and error-correcting output codes.

Table 1 Ensemble Methods (taken from [65])

Scheme	Architecture	Trainable	Adaptive	Info-level	Comments
Voting	Parallel	No	No	Abstract	Assumes independent classifiers
Sum, mean, median	Parallel	No	No	Confidence	Robust; Assumes independent confidence estimators
Product, min, max	Parallel	No	No	Confidence	Assumes independent features
Generalized ensemble	Parallel	Yes	No	Confidence	Considers error correlation
Adaptive weighting	Parallel	Yes	Yes	Confidence	Explores local expertise
Stacking	Parallel	Yes	No	confidence	Good utilization of training data
Borda count	Parallel	Yes	No	Rank	Converts ranks into confidences
Logistic regression	Parallel	Yes	No	Rank confidence	Converts ranks into confidences
Class set reduction	Parallel cascading	Yes/No	No	Rank confidence	Efficient
Dempster-Shafer	Parallel	Yes	No	Confidence	Fuses non-probabilistic confidences
Fuzzy integrals	Parallel	Yes	No	Confidence	Fuses non-probabilistic confidences
Mixture of local experts (MLE)	Gated parallel	Yes	Yes	Confidence	Explores local expertise; joint optimization
Hierarchical MLE	Gated parallel hierarchical	Yes	Yes	Confidence	Same as MLE; hierarchical
Associative switch	Parallel	Yes	Yes	Abstract	Same as MLE, but non joint optimization
Bagging	Parallel	Yes	No	confidence	Needs many comparable classifiers
Boosting	Parallel hierarchical	Yes	No	Abstract	Improves margins; unlikely to overtrain, sensitive to mislabels; needs many comparable classifiers
Random subspace	Parallel	Yes	No	Confidence	Needs many comparable classifiers
Neural trees	Hierarchical	Yes	No	confidence	Handles large numbers of classes

Boosting

Boosting was inspired by an on-line learning algorithm called *Hedge(β)* [79]. This algorithm allocates weights to a set of strategies used to predict the outcome of a certain event. Strategies with the correct prediction receive more weight while the weights of the strategies with incorrect predictions are reduced.

Boosting is related to the general problem of producing a very accurate prediction rule by combining rough and moderately inaccurate rules-of-thumb. The general boosting idea is to develop the classifier team D incrementally, adding one classifier at a time. The classifier that joins the ensemble at step k is trained on a data set selectively sampled from the training data set Z. The sampling distribution starts from uniform, and progresses toward increasing the likelihood of "difficult" data points. Thus the distribution is updated at each step, increasing the likelihood of the objects misclassified at step $k - 1$.

The classifiers in D are the trials or events, and the data points in Z are the strategies whose probability distribution we update at each step. The algorithm is called *AdaBoost* which comes from ADAptive BOOSTing. There are two implementation of AdaBoost: with *reweighting* and with *resampling*.

AdaBoost [48] is one of the best-known and best-performing ensemble classifier learning algorithms. It constructs a sequence of base models, where each model is constructed based on the performance of the previous model on the training set. In particular, AdaBoost calls the base model learning algorithm with a training set weighted by a distribution. After the base model is created, it is tested on the training set to see how well it learned.

The figure 1.3 shows AdaBoost's pseudocode. AdaBoost constructs a sequence of base models $h_t \, for \, t \in \{1, 2, \ldots, T\}$, where each model is constructed based on the performance of the previous base model on the training set. In particular, AdaBoost maintains a distribution over the m training examples. The distribution $\mathbf{d}_1$ used in creating the first base model gives equal weight to each example ($d_{1,i} = 1/m \, \forall i \in \{1, 2, \ldots, m\}$). AdaBoost now enters the loop, where the base model learning algorithm L_b is called with the training set and $\mathbf{d}_1$. The returned model h_1 is then tested on the training set to see how well it learned. The total weight of the misclassified examples (ϵ_1) is calculated. The weights of the correctly-classified examples are multiplied by $\epsilon_1/(1\epsilon_1)$ so that they have the same total weight as the misclassified examples. The weights of all the examples are then normalized so that they sum to 1 instead of $2\epsilon_1$. AdaBoost assumes that L_b is a weak learner, i.e., $\epsilon_t < 1/2$ with high probability. Under this assumption, the total weight of the misclassified examples $\epsilon_t < 1/2$ is increased to $1/2$ and the total weight of the correctly classified examples $1\epsilon_t > 1/2$ is decreased to $1/2$. This is done so that, by the weak learning assumption, the next model h_{t+1} will classify at least some of the previously misclassified examples correctly. Returning to the algorithm, the loop continues, creating the T base models in the ensemble. The final ensemble returns, for a new example, the class that gets the highest weighted vote from the base models. Each base model's vote is proportional to its accuracy on the weighted training set used to train it.

$AdaBoost((x_1, y_1), \ldots, (x_m, y_m), L_b, T)$

>   Initialize $\quad d_{1,i} = \frac{1}{m} \quad \forall i \in \{1, 2, \ldots m\}$.
>   **for** $\quad t = 1, 2, \ldots, T$,
> $\qquad h_t = L_b(\{(x_1, y_1), \ldots, (x_m, y_m)\}, \mathbf{d_t})$
> $\qquad$ Calculate the error of $h_t : \epsilon_t = \sum_{i:h_t(x_i) \neq y_i} d_{t,i}$
> $\qquad$ **if** $(\epsilon_t \geq 1/2)$ then,
> $\qquad\qquad$ set $T = t - 1$ and abort this loop.
> $\qquad \beta_t = \frac{\epsilon_t}{1-\epsilon_t}$
> $\qquad$ Calculate distribution d_{t+1}:

$$w_i = d_{t,i} \times \begin{cases} \beta_t, & if\, h_t(x_i) = y_i \\ 1, & otherwise \end{cases}$$

$$d_{t+1,i} = \frac{w_i}{\sum_{i=1}^{m} w_i}$$

>   **return** the final hypothesis:
> $\qquad h_{fin}(x) = argmax_{y \in Y} \sum_{t:h_t(x)=y} \log \frac{1}{\beta_t}$

Fig. 5 The AdaBoost algorithm

Bagging

Bagging is introduced by [21] as an acronym for *Bootstrap AGGregatING*. The idea of bagging is simple and appealing: the ensemble is made of classifiers built on bootstrap replicates of the training set. The classifier outputs are combined by majority voting. The algorithm, which is a special case of model averaging, was originally designed for classification and is usually applied to decision tree models, but it can be used with any type of model for classification or regression. The method uses multiple versions of a training set by using the bootstrap, i.e. sampling with replacement. Each of these data sets is used to train a different model. The outputs of the models are combined by averaging (in the case of regression) or voting (in the case of classification) to create a single output.

Bagging is only effective when using unstable (i.e., a small change in the training set can cause a significant change in the model) non-linear models.

Stacked Generalization

Stacked generalization (or stacking) [138] is a different way of combining multiple models, that introduces the concept of a meta-learner. Although this is an attractive idea, it is less widely used with respect to bagging and boosting. Unlike bagging and boosting, stacking may be (and normally is) used to combine models of different types. The procedure is as follows:

1. Split the training set into two disjoint sets.
2. Train several base learners on the first part.

3. Test the base learners on the second part.
4. Using the predictions from 3) as the inputs, and the correct responses as the outputs, train a higher level learner.

Note that steps 1) to 3) are the same as cross-validation, but instead of using a winner-takes-all approach, the base learners are combined, possibly non-linearly.

Random Subspace Method

The random subspace method (RSM) [62] is a relatively recent method of combining models. Learning machines are trained on randomly chosen sub-spaces of the original input space (i.e. the training set is sampled in the feature space). The outputs of the models are then combined, usually by a simple majority vote.

Error-Correcting Output Codes (ECOC)

It has often been noticed that obtaining a classifier that discriminates between two classes is much easier than one that simultaneously distinguishes among all classes [102]. This observation has motivated substantial research on using an ensemble of binary classifiers to address multiclass problems. To provide the final classification, the outputs of the binary classifiers are combined according to a given rule, usually referred to as *decision* or *reconstruction* rule.

So, given a multiclass problem it is always possible assigns a unique codeword, i.e. a binary string, to each class. If we assume that the string has L bits, the MCS will be composed by L binary classification functions. Given an unknown sample, the base classifiers provide an L-bits string that is compared with the codewords to set the final decision. For example, the input sample can be assigned to the class with the closest codeword according to a distance measure such as the Hamming one. In this framework, in [38] the authors proposed an approach, known as *Error-Correcting Output Codes* (ECOC), where the use of error correcting codes yielded a MCS less sensitive to noise. This result could be achieved via the implementation of an error-recovering capability derived from the coding theory.

2 Applications

In this Section we will present a review of the application fields where Multiple Classifier Systems are profitably used. The solutions and application fields described here are not a complete list of every single work presented by the scientific community; yet, we will thoroughly analyze the proceedings of a well established and acknowledged workshop, which has gained the role of a reference point in the Multiple Classifiers community, and of the Information Fusion and Classification theory communities at large as well. Therefore, we will briefly explore the works published in the proceeding of the Multiple Classifier Systems [1] workshop, which has been held since 2000, and has in 2011 reached its 10th edition.

A breakdown of the application types considered into the workshop papers is reported in Table 2, while the distribution of the various application papers within the ten editions of the workshop is shown in Table 3.

Table 2 Number of published papers per application type

Application type	Number of published papers in MCS workshops
Biometrics	23
Remote-Sensing Data Analysis	12
Document Analysis	19
Figure-and-Ground	3
Medical Diagnosis Support	7
Chemistry & Biology	9
Time series prediction & analysis	3
Image & Video Analysis	8
Computer & Network Security	6
Miscellanea	13

Table 3 Application type per year

Application type	2000	2001	2002	2003	2004	2005	2007	2009	2010	2011
Biometrics	2	3	2	-	4	3	4	1	1	3
Remote-Sensing Data Analysis	4	3	-	-	-	1	3	1	-	-
Document Analysis	4	2	-	3	1	-	2	2	4	1
Figure-and-Ground	1	1	-	-	-	-	-	1	-	-
Medical Diagnosis Support		2	-	1	-	1	2	1	-	-
Chemistry & Biology	-	1	1	-	1	-	-	2	4	-
Time series prediction & analysis	-	-	-	-	-	-	1	1	1	-
Image & Video Analysis	-	-	2	3	-	1	-	2	-	-
Computer & Network Security	-	-	-	1	1	-	-	-	-	4

2.1 Remote-Sensing Data Analysis

In this peculiar application field, researchers mainly have to analyze information sources in order to gain knowledge about earth and ground characterization. Typical sources of information can be sensors which can be installed on airplanes or space stations. Such sensors gather information about electromagnetic fields which are reflected or emitted by the Earth surface, and based on the observation of the variation of such fields allow to infer information about the chemical constitution, and the morphological characteristics of the Earth surface. In [77] the authors propose to recursively decompose a C-class Hyperspectral data analysis problem in C-1 two-class problem, thus defining a tree-structured multiple classifier system.

Several authors investigate the performance of multiple different combination schemes in this field. In [13] several data sources, based on different classification techniques are separately modeled, and consequently several combination schemes, based on consensus and weighted consensus, on consensual neural networks, and on genetic and fuzzy combination rules are applied. Fuzzy combination approaches, in particular, allow to determine a degree of reliability in attributing a class label to a given sample. The authors of [24] propose to combine parametric and nonparametric classifiers for unsupervised retraining in the field of Land-Cover maps classification. The usage of multiple classifiers is proposed in order to improve classification robustness; they mainly focus on Majority voting, Bayesian and Maximum posterior probability combination rules. Supervised, unsupervised and hybrid methods for multiple SOM combination are tested in [129], whereas in [121] majority voting, belief networks and Dynamic Classifier Selection by Simple Partitioning are tested. In particular, Dynamic Classifier Selection uses the average class accuracy to assign each classifier to a class, and checks the individual class assignment of the 10 nearest samples, and selects the output of the locally most accurate classifier. The usage of Multiple Classifier Systems in this application field, is justified in [22] by the need of collecting information from multiple sources, namely by sensors observing different parts of the EM Spectrum, in order to improve the performance of remote sensing. Experiments are carried out by using bagging algorithms, boosting algorithms, and consensus theoretic classifiers. Some focus mainly on base classifier selection and design, such as the authors of [41], which propose an ensemble made of maximum likelihood classifier (MLC), minimum distance classifier (MDC), Mahalanobis distance classifier (MHA), decision tree classifier (DTC) and support vector machine (SVM), and use double fault measure to select three classifiers for further combination, whose independence and diversity are evaluated.

Some authors use multiple classification schemes based on computational intelligence algorithm. In [39] they build a hierarchy of neural network classifiers based on a modification of the error backpropagation mechanism, combining supervised learning with self-organization. An SVM based ensemble is discussed in [131]. In the latter case, diversity is pursued by separating the feature space in several subsets, each classified with an SVM, and then all combined by a further SVM. Results are compared with random forests, bagging and boosting.

The impact of diversity is evaluated in [40], to address the claim that performance of combined classifier is closely related to the selection of member classifiers. General consistency measure, binary prior measure and consistency of errors are evaluated for the optimal selection of the best set of member classifiers. The issue of limited ground truth data is dealt with in [103], considering that a number of factors may cause the spectral signatures of the same class to vary with location and/or time. The authors use the Binary Hierarchical Classifier to propose a knowledge transfer framework that leverages the information gathered from existing labeled data to classify the data obtained from a spatially separate test area. Experiments prove that, when small amounts of labeled data are available from a new area, further improvements can be achieved through semi-supervised learning mechanisms. In [111] different subsets of data are used to train Support Vector Machines; the best different subset of features that are proper for object extraction in LIDAR is selected. Two multiclass SVM methods known as one-against-one and one-against-all are investigated for classification of LIDAR data and then final decision is achieved by majority voting.

2.2 Document Analysis

In this section we will review the works focusing on problems related to the analysis, categorization and classification of document content. We will group here works about handwritten text and numbers recognition, as well as automatic classification of document contents for archive creation and indexing.

Some authors, such as in [81] and [29] work on signature recognition. The former combine responses of three experts, working based on different features and classification techniques, by means of majority voting. The latter propose a multistage architecture, with each stage comprising an 2-class classifier and a decider. The last stage is the actual combiner, evaluating the reliability of the final decision and rejecting unreliable verdicts. In [10] a two-stage off-line signature verification system based on dissimilarity representation is proposed. In the first stage, a set of discrete HMMs are trained with different number of states and/or different codebook sizes, and consequently used to calculate similarity measures that populate new feature vectors. In the second stage, these vectors are employed to train a SVM (or an ensemble of SVMs) that provides the final classification. For the same application, in [11] two new dynamic selection strategies are proposed, both based on the KNORA (K-nearest-oracles) approach. They use the classifier outputs (i.e., the output profile) to find the K-nearest neighbors. To validate the proposed and other dynamic and static selection strategies, a multiple classifier generative-discriminative system is considered. In this system, HMMs are employed as feature extractors followed by SVMs as two-class classifiers.

The authors of [120] define and use a metric called "lexicon density", depending on both the analyzed entries and on the selected recognizer, in order to perform a wise choice of the best recognizer when trying to decode handwritten text (base classifiers selection rather than fusion). They also propose to use lexicon density to choose the best classifiers in a serial chain of multiple base classifiers. Work described in [64] stresses the need for general theory of combination. Their method allows to systematically obtain the best combination, either parallel or serial. In this paper the authors focus on serial methods for handwriting classification.

The most popular problem is handwritten text analysis. In [118] a MCS with self-configuration capabilities, based on genetic algorithms, is described. An innovative approach, contrary to classifier combination, is investigated in [33], where the authors use the outputs of a number of classifiers to come to a combined decision for a given observation. Multiple instances instances generated from the original observation are used. The virtual test sample method is used to improve the performance of a statistical classifier based on Gaussian mixture densities. AdaBoost is used in [56] instead. Here novel boosting methods specifically suited to solve the handwriting recognition problem, such as simple probabilistic boosting, effort-based boosting, effort-based boosting with doubling of training set, simple probabilistic boosting with effort are presented. The authors of [57] combine the development of several classifiers performed completely independent of each other and combined in a last step, together with the creation of several classifiers out of one prototype classifier by using classifier ensemble methods. Arabic handwriting, with all its peculiarities, is the subject of [7], where MCS optimization based on diversity measures is performed. Two approaches are tested, the former selecting the best classifier subset from large classifiers set taking into account different diversity measures, and the latter choosing among the classifier set the one with the best performance and adding it to the selected classifiers subset.In [127] construction of combination functions in identification systems is performed, since the optimal combination functions for identification systems are not known. The authors use neural networks to represent such functions, and explore different training methods in an aim of optimizing performance. The modifications are based on the principle of utilizing best impostors from each training identification trial.

OCR result strings are studied in [133], where an adaptive combination framework and methods for finding suitable parametrizations automatically are studied. Given an image of a text-line, string results are obtained from geometrical decomposition followed by character recognition. Input string synchronization is improved by means of geometric criteria before applying classical voting algorithms like Majority Vote or Borda Count [35] on the character level. The best string candidate is determined by an incomplete graph search.

The authors of [119] exploit input space transformation to exploit the structure of multiple classifiers, as well as the complementarity of different transformation schemes. This paper has a strong focus on diversity. HMM ensembles are generated in [76]. By using a selected clustering validity index, they show that the optimization of HMM codebook size can be selected without training HMM classifiers. In [17] a voting method (ROVER), and two novel statistical methods using MLP as a detector and classifier are discussed.

Cascading is used in [139] for document image classification, performed on two different base classifiers with different feature sets. The first classifier uses image features, learns physical representation of the document, and outputs a set of candidate class labels for the second classifier. The succeeding classifier is a hierarchical classification model based on textual features. The resulting system is applied to tax document classification. Document indexing is performed by the authors of [49], who retrieve all instances of a keyword within a document. Combination of base classifiers is based on a novel type of neural network.

CoBC (Co-Training by Committee), which is basically a single-view variant of Co-Training, is proposed in [58] for handwritten digit recognition. It requires an ensemble of diverse classifiers instead of the redundant and independent views. Furthermore, two novel learning algorithms, namely QBC-then-CoBC and QBC-with-CoBC are presented and discussed. The proposed methods combine the merits of committee-based semi-supervised learning and committee-based active learning.

2.3 Biometrics

Multiple classifier systems are also widely used in the field of biometrics, where multimodal analysis has proven to be quite effective. In fact, research seems to prefer solutions based on the combination of information related to voice, iris, fingerprint, body structure and posture, as well as other features which are of great utility in discriminating among individuals.

A great variety of descriptive models have been proposed so far, trying to identify and expose the peculiarity of each individual. As an example, [50] describes an audio-visual person recognition system based on voice, lip motion and still image. Statistical sensor calibration is necessary prior to deployment, and combination is performed by either multiplying or summing outputs of base classifiers. Multimodal analysis is exploited in [108], where experimental comparison between fixed and trainable combiners is performed. Sum rule, majority voting, Order Statistics operators are applied to the experts' outputs, alongside with Weighted average, and Behavior Knowledge Spaces. BKS is used in [70] as well, together with Decision Templates, in a context involving six intramodal experts exploiting frontal face biometrics. Other authors use speech, face and driving information [44]. Face features are extracted by using PCA, driving behavior is characterized by observing accelerator and brake pedal usage pattern. Fusion is implemented at the matching score level, using "decision" fusion, by a weighted sum with fixed weights. Serial fusion of multiple biometric traits [89] can also be used for personal identity verification. This approach can reduce the verification time for genuine users and the requested degree of user cooperation. Moreover, the use of multiple biometrics can discourage fraudulent attempts to deceive the system. From a security perspective, in fact, a multimodal system appears more protected than its unimodal components. The reason is that, one assumes that an impostor must fake all the fused modalities to be accepted and spoofing multiple modalities is harder than spoofing only one. However, a hacker may fake only a subset of the fused biometric traits. Recently, researchers demonstrated that the existing multimodal systems can be deceived also when only a subset of the fused modalities is spoofed. In [87] it is then demonstrated that, by incorporating a liveness detection algorithm in the combining scheme, the multimodal system results robust in presence of spoof attacks involving only a subset of the combined modalities. In the framework of identity verification a very interesting paper is [72]. The aim of the authors is to establish the relationship between score normalization methods and decision templates in the context of class identity verification. They showed that decision templates correspond to cohort normalization methods. Thanks to this relationship, some of the recent features of cohort score normalization techniques can be adopted by decision templates, with the benefit of noise reduction and the capacity for a distribution drift compensation.

Rather than multimodal analysis, in [75] multichannel input data is used. Decorrelation of input channels is performed based on RGB decomposition, in order to enhance diversity. In [88] an analysis of perceptron-based fusion in presence of stringent requirements in terms of errors is performed; this paper focuses on the peculiar case in which few training samples are available. Linear Discriminant Analysis is optimized in [130] for high dimensional data. In such a case, in fact, LDA often suffers from the small sample size problem and the constructed classifier is biased and unstable. The authors propose an approach to generate multiple Principal Space LDA and Null Space LDA classifiers by random sampling on the feature vector and training set. The two kinds of complementary classifiers are integrated to preserve all the discriminative information in the feature space.

Concentrating more on fingerprint recognition, in [25] the authors propose to combine a structural method, specifically conceived for continuous classification, with one based on a variation of the KL transform. The combination approach chosen for exclusive classification is the majority vote rule, together with a reject option based on confidence evaluation. In the case of continuous classification, instead, The authors combine measures of dissimilarity between fingerprints rather than decisions of base classifiers. The same issue is tackled by the authors of [98], who stress more on classifier selection during combination. The proposed scheme is optimal (in the Neyman-Pearson sense) when sufficient data are available to obtain reasonable estimates of the joint densities of classifier outputs.

Speaker identification is another interesting issue. [61] describe the exploitation of hidden Markov modeling and artificial neural network (ANN). The experiments perform recognition based on spoken digits seven and nine.

More in general, many works propose methods for face detection and recognition, which are mainly based on characterization and detection of particular shapes and significant points occurring in human face. Some researchers work on face movements [68] by coding a set of 44 action units (AUs) and more than 7000 combinations. They define an accurate real-time sequence-based system for representation and recognition of facial AUs, which employs a mixture of HMMs and neural network. Both geometric and appearance-based features are used, and the resulting system is robust to illumination changes and it can represent the temporal information involved in formation of the facial expressions by using appropriate dimension reduction of the dataset. In [51] integration of three simple visual cues for the task of face detection in gray level images is proposed, achieved by a combination of edge orientation matching, hough transform and an appearance based detection method. Several template matchers are combined by using sum and product rule. Multiple images can also be combined to construct a face space [26], thus gaining more robustness with respect to single base classifiers. Another possibility is to combine the outputs of different face detectors (FD). In [36] an approach based on the geometry of the competing face detection results is presented. The main idea of this work lies in finding groups of similar face detection results obtained by several algorithms and further averaging them. The combination result essentially depends on the number of algorithms that have fallen in each of the groups. An experimental evaluation has been performed by using seven FD algorithms on a test dataset of 59888 images, namely, 11677 faces and 48211 non-faces. On these data, the proposed method demonstrated better performance than each of its component algorithms. Color information can be fused as well [110] for the task of face authentication. The confidence level measurement can be calculated for each color subspace. Confidence measures are used within the framework of a gating process in order to select a subset of color space classifiers. The selected classifiers are finally combined using the voting rule for decision making. An interesting proposal for view independent face recognition [43] divides the problem space into several subspaces for the experts, and the outputs of experts are combined by a gating network.

Other works are much more focused on performance, such as [101], where the authors build custom classifiers aimed at enhancing the performance of black box classifiers available off-the-shelf. They use a serial scheme: since the classifiers often have different output types, serial combination allows to overcome such problem. As to fusion, [46] implement a new score fusion technique, based on a form of Bayesian adaptation aimed at deriving the personalized fusion functions from prior user-independent data. Scores are also used as an input to fusion in [117], where SVM score fusion is implemented after a dimension reduction with Bilateral projection-based Two-Dimensional Principal Component Analysis (B2DPCA) for Gabor features. A final fusion is performed combining the SVM scores for the 40 wavelets with a raw average. The authors of [127] investigate the construction of combination functions in identification systems. In contrast to verification systems, the optimal combination functions for identification systems are not known. The combination function is performed here by means of a neural network, and several training methods are explored.

2.4   *Figure and Ground*

Authors performing research in this field mainly work on the identification of figures in a specific background. In well structured application scenarios, template matching is a good solution [78] trying to match a template to a target object. In this work, a 2 dimensional array of figure-and-ground classifiers is used, each observing a point of an image and determining whether the point belongs to the figure (an object) or thew background (ground). A proposal for regular updating of landcover maps is presented by [23]. The authors use temporal series of remote sensing images, thus tackling the temporal dimension of the problem as well. The proposed system is composed of an ensemble of partially unsupervised classifiers integrated in a multiple classifier architecture. The updating problem is formulated under the complex constraint that for some images of the considered multitemporal series no ground-truth information is available. On-line decisions are also taken into account in [100], where an approach based on mixture experts (ME) model is discussed. The main application envisioned is terrain prediction in autonomous outdoor robot navigation. Binary linear models, trained on-line on images seen by the robot at different points in time, are added to a model library as the robot navigates. To predict terrain in a given image, each model in the library is applied to feature data from that image, and the models predictions are combined according to a single-layer (flat) ME approach.

2.5   *Medical Diagnosis Support*

Automated classification, recognition and detection have a very strong and immediate impact in the field of medical diagnosis, since their contribution can dramatically change the way diagnostic supports influence people's lives and doctors' job. In this field, errors are of crucial importance.

Different types of tumor can be diagnosed with the aid of multiple classifier systems, as in [92] and [93]. The former uses boosting to tune classification performance toward the sensitivity and specificity required by the user by means of a cost-sensitive approach. The issue of melanoma detection is tackled, and experiments are carried out on a set of 152 skin lesions described by geometric and colorimetric features. The latter analyzes magnetic resonance spectra of human brain tumors as a part of a decision support system for radiologists. The basic idea is to decompose a complex classification scheme into a sequence of classifiers, each specialized in different classes of tumors and trying to reproduce part of the WHO classification hierarchy. Classifiers with different behavior are combined using a simple voting scheme in order to extract different error patterns. A special label named "unknown" is used when the outcomes of the different classifiers disagree, in order to implement some form of rejection. Mammography is a not invasive diagnostic technique widely used for early detection of breast cancer. One of the main indicants of cancer is the presence of microcalcifications, i.e. small calcium accumulations, often grouped into clusters. Automatic detection and recognition of malignant clusters of microcalcifications are very difficult because of the small size of the microcalcifications and of the poor quality of the mammographic images. In [47] a novel approach using the evidences obtainable from the classification of the single microcalcifications and from the classification of the cluster considered as a whole is proposed. The inherent properties of spectra are also exploited in [99], making possible a domain-based feature selection model on real-life high-dimensional biomedical magnetic resonance (MR) spectra.

ECG and EEG signals are also a subject of study, allowing the early detection of heart diseases or possible pathological conditions related to brain malfunctions. In [82], a novel hybrid kernel machine ensemble is proposed for abnormal ECG beat detection to facilitate long-term monitoring of heart patients. A binary SVM is trained using ECG beats from different patients to adapt to the reference values based on the general patient population. A one-class SVM is trained using only normal ECG beats from a specific patient to adapt to the specific reference value of the patient. Subspace ensembles for classification are explored in [124]. The authors propose to use region partitioning and region weighting to implement effective subspace ensembles. An improved random subspace method that integrates this mechanism is presented for the classification of EEG signals. The authors of [123] propose an extensive and systematical evaluation of the effectiveness and feasibility of bagging, boosting and random subspace ensemble learning methods for electroencephalogram (EEG) signal classification. Experiments are conducted on three BCI subjects with k-nearest neighbor and decision tree as base classifiers.

2.6 Chemistry and Biology

This field is typically characterized by the presence of huge, multidimensional and unlabeled datasets. Several proposals have been made for the study of organ toxicity, based on the analysis of different information sources. Data of biofluids can be used for the automatic classification of Nuclear Magnetic Resonance Spectroscopy with respect to drug induced organ toxicities [83]. Classification is realized by an Ensemble of Support Vector Machines, trained on different subspaces according to a modified version of Random Subspace Sampling. Features most likely leading to an improved classification accuracy are favored by the determination of subspaces, resulting in an improved classification accuracy of base classifiers within the Ensemble. Ensembles of local experts have successfully been applied for the automatic detection of drug-induced organ toxicities based on spectroscopic data as well [84]. For suitable Ensemble composition an expert selection optimization procedure is required that identifies the most relevant classifiers to be integrated. However, it has been observed that Ensemble optimization tends to overfit on the training data. To tackle this problem the authors propose to integrate a stacked classifier optimized via cross-validation that is based on the outputs of local experts. The authors of [54] propose to integrate rule based systems and neural networks. They define chemical structures of fragments responsible for carcinogenicity according to human experts, used to develop specialized rules. Each rule is used to associates a category to each fragment found, then a category to the molecule. Furthermore, an ANN-based expert that uses molecular descriptors in input and predicts carcinogenicity as a numerical value is used. An approach based on hybrid intelligent methods is presented in [14]. Due to the increasing number of different classifiers applied in toxicity prediction, there exist a need to develop tools to integrate various approaches. Neuro-fuzzy networks to provide an improvement in combining the results of five classifiers applied in toxicity of pesticides are discussed here.

The problem of Quantitative Structure-Activity Relationship (QSAR) modeling for pharmaceutical molecules has been tackled by using a quantitative description of a compounds molecular structure to predict that compounds biological activity as measured in an in vitro assay [125] .

The study of genome has also been supported by classification tools. In [105] the authors evaluate the performances of three basic ensemble methods to integrate six different sources of high-dimensional biomolecular data for gene function prediction. The computational genome-wide annotation of gene functions requires the prediction of hierarchically structured functional classes and can be formalized as a multiclass, multilabel, multipath hierarchical classification problem, characterized by very unbalanced classes. In [106] the authors focus on the experimental comparison of the hbayes and tpr hierarchical gene function prediction methods and their cost-sensitive variants.

The authors of [140] investigate and visualize the connectivity of patterns in huge arbitrary shaped clusters. A sequential ensemble, that uses an efficient distance-relatedness based clustering, Mitosis, followed by the centre-based K-means algorithm, is proposed. K-means is used to segment the clusters obtained by Mitosis into a number of subclusters. The ensemble is used to reveal the gradual change of patterns when applied to gene expression sets. High-throughput applications are also popular in biomedicine. [113] presents a methodology to improve classification accuracy in the field of 3D confocal microscopy. A set of 3D cellular images (z-stacks) were taken, each depicting HeLa cells with different mutations of the UCE protein

([Mannose-6-Phosphate] UnCovering Enzyme). This dataset was classified to obtain the mutation class from the z-stacks. 3D and 2D features were extracted, and classifications were carried out with cell by cell and z-stack by z-stack approaches, with 2D or 3D features. Also, a classification approach that combines 2D and 3D features is proposed, which showed interesting improvements in the classification accuracy.

2.7   Time Series Prediction/Analysis

Time series analysis deserves a separate section, since results obtained in this field can be virtually applied to all problems characterized by continuously updating input data.

Time series forecasting is a challenging problem, that has a wide variety of application domains such as in engineering, environment, finance and others. When confronted with a time series forecasting application, typically a number of different forecasting models are tested and the best one is considered. Alternatively, instead of choosing the single best method, a wiser action could be to choose a group of the best models and then to combine their forecasts. In [9] the authors use a Multi-layer perceptron (MLP), Gaussian Processes Regression (GPR) and a Negative Correlation Learning (NCL) model. The authors of [8], instead, investigate the performance of using forecast combination in handling breaks in data series, observing how the performance of prediction strategies varies in presence of discontinuities and "holes" in time series. The problem of missing data is also dealt with in [94]. The authors propose a semi-supervised co-training method. Time series data are transformed to set of labeled and unlabeled data. Different predictors are used to predict the unlabeled data and the most confident labeled patterns are used to retrain the predictors further to and enhance the overall prediction accuracy. By labeling the unknown patterns the missing data is compensated for.

2.8   Image and Video Analysis

Tracking objects in videos is a challenging task, due to varying light conditions, camera angle, and other environmental disturbing factor, as well as occlusion, both partial and total. In [128] the authors propose a novel pairwise diversity measure, that recalls the Fisher linear discriminant, to construct a classifier ensemble for tracking a non-rigid object in a complex environment. A subset of constantly updated classifiers is selected exploiting their capability to distinguish the target from the background and, at the same time, promoting independent errors. Target selection and tracking can also be assimilated to this first type of problems. Multiple Description Coding is used in [6]. The proposal was inspired from the framework of transmitting data over heterogeneous network, especially wireless network.

Due to the increasing amount of media produced, stored, retrieved over the Internet, video and image annotation and segmentation are becoming more and more important, allowing to index and categorize huge amounts of media. In [115] images are annotated with a description of the content in order to facilitate the organization, storage and retrieval of image databases. Several features have been designed and experimentally compared, producing a classifier that can provide a reasonably good performance on a generic photograph database. Movie segmentation is discussed instead in [31] and [114], where combinations of audio and visual features are exploited for effective scene cut and scene type recognition. Image segmentation is also performed, in [85], by using multi-scale features for pixel classification. The link between scale selection and the maximum combination rule from pattern recognition is explored. Automatic annotation is investigated in [66]. Such paper deals, in particular, with the annotation of sports videos. The concept of "cues" allows to attach semantic meaning to low-level features computed on the video and audio. Shots are classified, based on the cues they contain, into the sports they belong to. Diversity is tackled by using the RGB color space, which is prominent both as a color scheme, and a display scheme. The authors of [27], too, claim that the recent advent of multiple classifier systems provides the unique opportunity to exploit the diverse information encapsulated in the different color representations in a systematic fashion. They use information gathered in different color spaces, and subsequently use suitable measures to investigate the diversity of the information infused by the different color spaces.

2.9   Computer and Network Security

Within the field of computer and network security, computational intelligence techniques are gaining more and more momentum. Such techniques are mainly used for novelty detection applied to the problem of zero-day attack detection, as well as typical traffic monitoring and characterization issues. Multiple classifiers are typically employed to overcome the scarce availability of training data. Furthermore, their well known generalization capabilities can be exploited for false alarm and missed detection reduction, in order to improve detection capabilities, which are very critical in this peculiar application field. In [53] a pattern recognition approach to network intrusion detection based on the fusion of multiple classifiers is proposed. A modular Multiple Classifier architecture is designed, where each module takes care of analyzing traffic related to one of the services offered by the protected network. Diversity is obtained by feeding each Multiple Classifier System with different representations of the monitored network traffic. In [30] a serial multi-stage classification system is described instead. Each stage analyzes different features, giving manifold representations of events occurring in the monitored scenario. Decision reliability is estimated at each stage, thus allowing the evaluation of uncertainty, and eventually logging unreliable decisions for further processing. Another multi-stage approach for intrusion detection is proposed in [5]. This has been explicitly devised for the detection of attacks against a web server. In particular, it analyzes the requests received by a web server and is based on a two-stages classification algorithm. In the first stage the structure of the HTTP requests is modeled using several ensembles of HMMs. Then, the outputs of these ensembles are combined using a one-class classification algorithm. In case of intrusion detection, as well as in spam filtering, a malicious adversary may successfully mislead a classifier by *poisoning* its training data with carefully designed attacks. On the other hand, it has been demonstrated that bagging can reduce the influence of outliers in training data, especially if the most outlying observations are resampled with a lower probability. So, in [18] the authors argue that poisoning attacks can be viewed as a particular category of outliers, and, thus, bagging ensembles may be effectively exploited against them. Their preliminary results suggest that bagging ensembles can be a very promising defense strategy against poisoning attacks.

Another way of ensuring network security by any modern network management platform is to perform network traffic classification i.e., to properly associate network traffic flows to the network applications (e.g. FTP, HTTP, BitTorrent, etc.) that generate them. In several network scenarios, however, it is quite unrealistic to assume that all the classes an IP flow can belong to are *a priori* known. In these cases, in fact, some network protocols may be known, but novel protocols can appear so giving rise to *unknown* classes. In [34] the authors propose to face the problem of classifying IP flows by means of a multiple classifier approach based on the BKS combiner. This has been explicitly devised in order to effectively address the problem of the unknown traffic, too.

Finally, it is worth noting that MCSs have been also applied for building a host-based intrusion detection system. In [69] a new ensemble-based technique, called incremental Boolean combination (*incrBC*), is proposed in order to make adaptive anomaly detection from system call sequences. The proposed approach is based on an incremental learning of new training data according to a learn-and-combine approach: when a new block of training data becomes available, it is used to generate a new pool of classifiers by varying training hyperparameters and random initializations. The responses from the newly-trained classifiers are then combined to those of the previously-trained classifiers by applying Boolean combination in the ROC space. Since the pool size grows indefinitely over time, *incrBC* integrates model management strategies to limit the pool size without significantly degrading performance.

2.10   Miscellanea

In this section we will briefly review a number of applications which were so peculiar, and so unique, they did not fit in any of the categories exposed above. The aim of this section is to prove how versatile and flexible multiple classification techniques can be, and how easily they can be adaptable to manifold scenarios.

Food related processes sometimes involve the employment of decision support systems. The authors of [90] propose boosting of Multi-Layer Perceptrons for the discrimination of different types of coffee. Input data is produced by an Electronic Nose, and experiments are carried out using two groups of coffees, namely blends and monovarieties, consisting of seven classes each. In [104] five wheat varieties characterized are described by means of nine geometric and three color descriptive features. Pair-wise SLP or SV classifiers are used as base classifiers for multiple classifier systems performing wheat type recognition. Human communication related issues are dealt with in [135], where both independent component analysis and principle component analysis are used to derive an independent set of gestural primitives for visual sign-language, employing existing sign linguistics as a reference point in the feature reduction. In [67] the authors work on sentence recognition, using base classifiers, each proposing a set of N ordered candidate sentences. Sentences are constructed, by using such atomic elements, according to the Borda rule [35] or based on high similarity between base classifiers. Musical instruments are classified in [45] by using fuzzy inference systems operating on different features subsets. A two stage strategy is used, consisting in a fuzzy clustering followed by a supervised learning phase. Two combiners are used, one combining outputs from five sub-networks processing the "low" part of the feature space, and five other processing the "high" part, where low and high are referred to the analyzed part of the spectrum. The problem of life insurance applications underwriting is tackled in [19] where the authors discuss the issue of monitoring decisions quality of an on-line classifier, using a range of five classification methods and seven different fusion strategies. The final architecture comprises three off-line classifiers and a fusion module. Question answering systems are deal with in [28], where the use of diverse data fusion methods is investigated, with the aim to improve the performance of the passage retrieval component. The authors of [141] face the problem of small sample size by evaluating the performance of Reusing, Validation and Stacking; the authors of [2] investigate the reason of the good performance of Random Linear Oracle (RLO) ensembles of Naive Bayes classifiers, compared to Random Subclasses (RS). The comparative study suggests that RS is similar to RLO method, whereas RS is statistically better than or similar to Bagging and AdaBoost.M1 for most of the datasets. The similar performance of RLO and RS suggest that the creation of local structures (subclasses) is the main reason for the success of RLO. The another conclusion of this study is that RLO is more useful for classifiers (linear classifiers etc.) that have limited flexibility in their class boundaries. These classifiers can not learn complex class boundaries. Creating subclasses makes new, easier to learn, class boundaries. Boosting, instead, has proven to improve the predictive performance of unstable learners, but not of stable learners such as SVM [126]. The authors propose a method for improving performance of stable learners via boosting as well. Automated fault detection is an increasingly important problem in aircraft maintenance and operation. In [96] a method is proposed, where the mismatch between the actual flight maneuver being performed and the maneuver predicted by a classifier is a strong indicator that a fault is present. Web page classification is performed, in [3], by means of a generic multiple classifier system based solely on pairwise classifiers. This task can be profitably performed, for example, in enhancing the accuracy of search engines and in summarizing web content for small-screen handheld devices. Finally, in [32] a sequential classifier combination approach for pattern recognition in wireless sensor networks is proposed. Pattern recognition tasks are common in wireless sensor networks. Typically, a wireless sensor network consists of a base station and a (large) number of sensors. Due to restricted battery capacity, minimizing the energy consumption is a main concern. Assuming that each feature needed for classification is acquired by a sensor, a sequential classifier combination can minimize the number of features used for classification while maintaining a given correct classification rate.

3 Tools

In this section we will describe different tools currently available for implementing a Multiple Classifier System.

We will focus on which should be the features to be considered in order to select a tool and will give some hints for choosing a specific tool, given the user's requirements.

Since it is out of our scope to describe all the existing tools, we will present only the most significant ones, that can be considered as the most representative of their own category, according to the categorization we will illustrate in the next subsection.

3.1 Tool Categorization

A first difference among tools is if they have been explicitly designed and developed for pattern recognition/data mining purposes or if they are general purpose tools for mathematical calculus with one or more packages devoted to pattern recognition.

For this motivation hereinafter we will define *Native Tools* the first category, and *Adapted Tools* the second one (see Table 4).

Table 4 The two main categories

Category	Considered Tools
Native	Weka, KNIME
Adapted	PRTools (Matlab)

Native Tools are focused on data-mining problems and give to the user a more friendly interface.

In order to make a correct choice we have to ask ourselves what level of integration with the mathematical framework we want. In other words, it must be taken into account if we have the need to work at a *low level*, by considering some mathematical details and developing new strategies into the classification system, or we can work at an *high level*, that is, we do not have the need of modifying the existing algorithms but we are mainly focused on the data analysis and processing.

If we work at a *low level* the correct choice are the *Adapted tools*, whereas, if we are working at an *higher level*, we probably have to choice a *Native tool*.

From the above considerations, it is also possible to distinguish users into:

- **data user:** The user focused on the data analysis.
- **algorithmic user:** The user interested in the classification process optimization.

After this first categorization, we have to consider other important features for selecting the best tool. The most significant ones are:

- **Input/Output Data Format:** In some scenarios it is important to have the opportunity to use data from different sources and/or from remote sockets/databases.
- **GUI vs CLI:** Graphic User Interface (GUI) is of course more user friendly than a Command Line Interface (CLI), but in terms of allocated resources and for batch executions it is more efficient to have a CLI instead of a GUI.
- **Repository, Package Manager:** The presence of a package repository to extend the tools and the amount of classifiers and features implemented sometimes makes the difference among tools.
- **Documentation:** A complete and well-written documentation makes a tool usable.
- **Community:** The presence/absence of a community it is often the main reason to select a tool, in particular when we are trying to use a tool for the first time. The community could be seen as a sort of *dynamic* documentation.
- **Extensibility:** In some scenarios, it is important to create or to extend some classifiers or *nodes* to reach the problem solution.
- **External Tools Integration:** The most important tools are linked each other, giving to the users the opportunity to use the same classifiers/classification system within each one of them.
- **Efficiency and Developing Environment:** When running time is important, and/or for properly managing large datasets, some tools are not applicable and the user is obliged to use a tool instead of another one.
- **License:** In some cases, the best tool is not the cheaper one, whereas sometimes it is useless to pay to complete a task which could be solvable with an *open source* approach.

Table 5 Tools comparison

Features	Tools		
	Native Tools		Adapted Tools
	KNIME	WEKA	PRTools (Matlab)
I/O Data Format	Arff, cvs, database	Arff, cvs, Xrff, database	cvs
GUI/CLI	GUI	GUI/CLI	CLI
Repository	yes	yes	yes
Community	yes	yes	no
Extensibility	yes	yes	no
External Tools	yes	yes	no
Efficiency	Java	Java	C
License	GPLv3	GPLv3	Non-profit/Commercial

After presenting these features, it is now important to point out which could be the best tool for the Multiple Classifier System we want to design or to develop.

Sometimes, the same tool that appears to be simple and efficient for a general purpose use, becomes very difficult to use when we need to integrate different classifiers in an MCS.

So, in order to make a good choice, the questions that we have to answer are:

- Is this tool designed to use an MCS approach?
- Has this tool just implemented one or more Multiple Classifier strategies?
- Is it possible to implement a user-defined combination strategy?
- Is it provided with feedbacks about the iteration-status involved during the MCS steps?
- Which is the amount of work that has to be done in order to obtain a result?

Obviously the perfect tool has been not implemented yet. In the following we will describe the tools under exam in more details, with reference to the introduced features. A synthetic comparison is reported in Table 5.

3.2 Weka

Weka [60, 137] is a collection of machine learning algorithms for pattern recognition and data mining tasks. The algorithms can either be applied directly to a dataset or called from your own Java code. Weka contains tools for data pre-processing, classification, regression, clustering, association rules, and visualization. It is also well-suited for developing new machine learning schemes. Weka is an open source software issued under the GNU General Public License.

It provides an easy GUI Interface named *Weka Knowledge Explorer* that harnesses the power of the Weka software. Each of the major Weka packages (Filters, Classifiers, Clusterers, Associations, and Attribute Selection) is represented in the Explorer along with a Visualization tool which allows datasets and the predictions of Classifiers and Clusterers to be visualized in two dimensions.

The **preprocess panel** (Fig. 6) is the start point for knowledge exploration. From this panel the user can load datasets, browse the characteristics of attributes and apply any combination of Weka's unsupervised filters to the data.

The **classifier panel** (Fig. 7) allows the user to configure and execute any of the Weka classifiers on the current dataset. It is possible choose to perform a cross validation or test on a separate dataset. Classification errors can be visualized in a pop-up data visualization tool. If the classifier produces a decision tree it can be displayed graphically in a pop-up tree visualizer.

From the **cluster panel** (Fig. 8) it is possible to configure and execute any of the Weka clusters on the current dataset. Clusters can be visualized in a pop-up data visualization tool.

From the **associate panel** (Fig. 9) it is possible to mine the current dataset for association rules using the Weka associators.

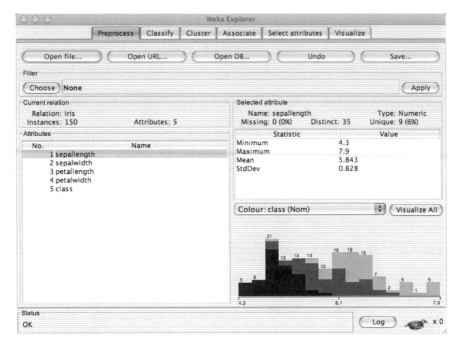

Fig. 6 The Weka Preprocess Panel

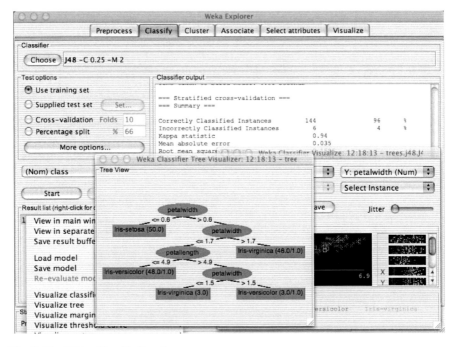

Fig. 7 The Weka Classify Panel

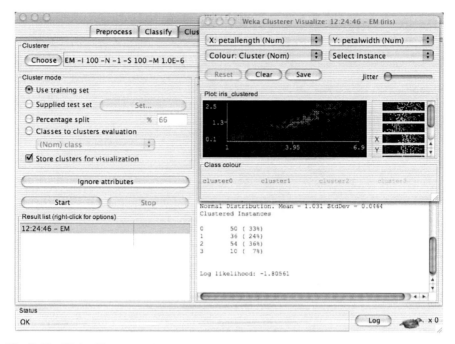

Fig. 8 The Weka Cluster Panel

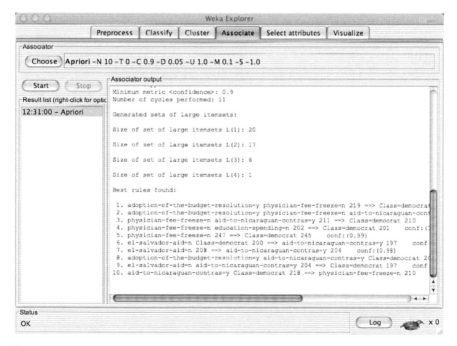

Fig. 9 The Weka Associate Panel

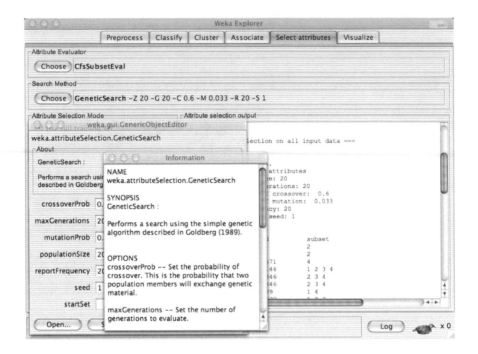

Fig. 10 The Weka Select Attributes Panel

From the **select attributes panel** (Fig. 10) allows you to configure and apply any combination of Weka attribute evaluator and search method to select the most pertinent attributes in the dataset. If an attribute selection scheme transforms the data, then the transformed data can be visualized in a pop-up data visualization tool.

Other panels are also available and it is possible to go in more details by looking at the online documentation.

As regards the MCS point of view, in Weka a certain number of Multiple Classifier Systems are implemented within the section *META*. Multiple Classifier Systems are considered as *meta*-classifiers. In particular, we can find implemented:

- **AdaBoostM1:** This Meta-Classifier implements the Freund & Schapire's Adaboost.M1 method [48].
- **Bagging:** This is the bagging approach as described in Section 1.3.
- **Decorate:** DECORATE [91] is a meta-learner for building diverse ensembles of classifiers by using specially constructed artificial training examples. Comprehensive experiments have demonstrated that this technique is consistently more accurate than the base classifier, Bagging and Random Forests. DECORATE also obtains higher accuracy than Boosting on small training sets, and achieves comparable performance on larger training sets.
- **MultiBoostAB:** MultiBoosting [132] is an extension to the highly successful AdaBoost technique for forming decision committees. MultiBoosting can be viewed as combining AdaBoost with wagging. It is able to harness both AdaBoost's high

bias and variance reduction with wagging's superior variance reduction. Using C4.5 as the base learning algorithm, Multi-boosting is demonstrated to produce decision committees with lower error than either AdaBoost or wagging signif-icantly more often than the reverse over a large representative cross-section of UCI data sets. It offers the further advantage over AdaBoost of suiting parallel execution.

- **Stacking:** As described in Section 1.3.
- **StackingC:** A more efficient version of Stacking [116].
- **RandomSubSpace:** As described in Section 1.3.
- **RandomForest:** This MCS approach [20] is not included within the META sec-tion, but it is located in the classifier section.

3.3 KNIME

KNIME (Konstanz Information Miner) [15, 16] is a user-friendly and comprehen-sive open-source data integration, processing, analysis, and exploration platform. From day one, KNIME has been developed using rigorous software engineering practices and is currently being used actively by over 6,000 professionals all over the world, in both industry and academia. KNIME.com provides support and main-tenance subscriptions for the open-source platform as well as enterprise extensions and services for the deployment of KNIME in a corporate environment.

KNIME, is a modular data exploration platform that enables the user to visually create data flows (often referred to as pipelines), selectively execute some or all analysis steps, and later investigate the results through interactive views on data and models.

KNIME was developed (and will continue to be expanded) by the Chair for Bioinformatics and Information Mining at the University of Konstanz, Germany. The group headed by Michael Berthold also uses KNIME for teaching and research at the University. Quite a number of new data analysis methods developed at the chair are integrated in KNIME.

The KNIME base version already incorporates over 100 processing nodes for data I/O, preprocessing and cleansing, modeling, analysis and data mining as well as various interactive views, such as scatter plots, parallel coordinates and others. It integrates all analysis modules of the well known Weka data mining environment and additional plugins allow R-scripts to be run, offering access to a vast library of statistical routines.

KNIME is based on the Eclipse platform and, through its modular API, easily extensible. When desired, custom nodes and types can be implemented in KNIME within hours thus extending KNIME to comprehend and provide first-tier support for highly domain-specific data. This modularity and extensibility permits KNIME to be employed in commercial production environments as well as teaching and research prototyping settings.

KNIME is released under a dual licensing scheme. The open source license (GPL) allows KNIME to be downloaded, distributed, and used freely.

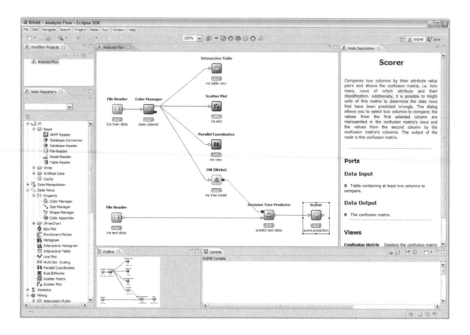

Fig. 11 An Example of the KNIME Data Analysis Workflow

The figure 11 shows KNIME in action and are intended to illustrate how KNIME nodes interact with one another and with other KNIME features.

Workflows are developed interactively in KNIME and are generally built incrementally rather than all at once. Prior to submitting data to the *J48* and *Decision Tree Prediction* nodes, the author of this workflow probably has first chosen to visually inspect the training data with the *Interactive Table,Scatter Plot*, and *Parallel Coordinates* nodes. So, the modeling and the prediction part of this workflow was probably added after the visual inspection encouraged the author to believe that such a choice could bear fruit.

Considering the MCS aspects, in KNIME, starting from the 2.4.0 version, the *Bagging* and *Boosting* approaches are natively implemented. The implementation consists of two *meta-nodes* that contains a collection of *simple-nodes* properly linked each other. All this *simple-nodes* are contained into a sub-folder *Utility-Nodes* within the main folder *Ensemble Learning* in the *Node Repository*. This technical choice give to the users the possibility to create other ensembles strategies starting from them.

In the figure 12 the content of the *Bagging Meta-Node* is shown. It is worth noting that the system gives the opportunity to change the base classifier by simply changing the relative nodes (the Decision Tree - learner and predictor - in this example).

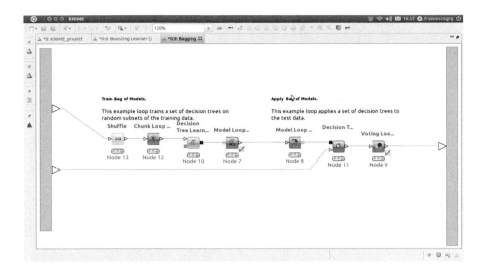

Fig. 12 The KNIME Bagging Meta-Node (expanded)

It is also possible to use all the WEKA Multiple Classifier approaches wrapped in KNIME nodes (*AdaBoostM1, Bagging, DECORATE, LogitBoost, MultiBoostAB, RandomSubSpace, Stacking, StackingC, Random Forest*). They has been already described in Section 3.2.

3.4 PRTools

PRTools is a toolbox for pattern recognition implemented in Matlab. It is developed by the Pattern Recognition Group of the Delft University in the Netherlands. It is very well documented, and it is probably the best toolbox for pattern recognition within Matlab. It can be freely used for academic research. One of the possible reference to learn more about this tool is the manual by Duin et al. [42].

The main difference between PRTools and the two previously described tools is that this is not a native tool but a toolbox within a more complex mathematical software. On the other hand, in PRTools there are natively implemented MCS capabilities.

The main PRTools functions for implementing an MCS are:

- **averagec:** Combining linear classifiers by averaging coefficients
- **baggingc:** Bootstrapping and aggregation of classifiers
- **votec:** Voting combining classifier
- **maxc:** Maximum combining classifier
- **minc:** Minimum combining classifier
- **meanc:** Averaging combining classifier
- **medianc:** Median combining classifier

- **prodc:** Product combining classifier
- **traincc:** Train combining classifier
- **parsc:** Parse classifier or map
- **parallel:** Parallel combining of classifiers
- **stacked:** Stacked combining of classifiers
- **sequential:** Sequential combining of classifiers

4 Conclusions

In this Chapter, after a survey of the main ideas underlying the multiple classifier approach, we have presented a review of more than 100 papers that have applied a Multiple Classifier System (MCS) to a specific domain, highlighting the relationships between the application domain and specific problem on one side, and the adopted multiple classifier approach on the other side. A description of some of the tools currently available for implementing a multiple classifier system completed our work.

All together, these parts provide useful information to practitioners and applied researchers for deciding which MCS approach best fits their needs.

References

[1] International workshop on multiple classifier systems. Web Page,
    http://www.diee.unica.it/mcs/
[2] Ahmad, A., Brown, G.: A study of random linear oracle ensembles. In: Benediktsson, et al. [12], pp. 488–497
[3] Alam, H., Rahman, A.F.R., Tarnikova, Y.: Solving problems two at a time: Classification of web pages using a generic pair-wise multiple classifier system. In: Windeatt, Roli [134], pp. 385–394
[4] Alpaydin, E.: Introduction To Machine Learning. MIT Press (2004)
[5] Ariu, D., Giacinto, G.: A modular architecture for the analysis of http payloads based on multiple classifiers. In: Sansone, et al. [112], pp. 330–339
[6] Asdornwised, W., Jitapunkul, S.: Automatic target recognition using multiple description coding models for multiple classifier systems. In: Windeatt, Roli [134], pp. 336–345
[7] Azizi, N., Farah, N., Sellami, M., Ennaji, A.: Using diversity in classifier set selection for arabic handwritten recognition. In: Gayar, et al. [52], pp. 235–244
[8] Azmy, W.M., Atiya, A.F., El-Shishiny, H.: Forecast combination strategies for handling structural breaks for time series forecasting. In: Gayar, et al. [52], pp. 245–253.
[9] Azmy, W.M., El Gayar, N., Atiya, A.F., El-Shishiny, H.: Mlp, gaussian processes and negative correlation learning for time series prediction. In: Benediktsson, et al. [12], pp. 428–437
[10] Batista, L., Granger, E., Sabourin, R.: A multi-classifier system for off-line signature verification based on dissimilarity representation. In: Gayar, et al. [52], pp. 264–273
[11] Batista, L., Granger, E., Sabourin, R.: Dynamic ensemble selection for off-line signature verification. In: Sansone, et al. [112], pp. 157–166
[12] Benediktsson, J.A., Kittler, J., Roli, F. (eds.): MCS 2009. LNCS, vol. 5519. Springer, Heidelberg (2009)

[13] Benediktsson, J.A., Sveinsson, J.R.: Consensus based classification of multisource remote sensing data. In: Kittler, Roli [73], pp. 280–289

[14] Benfenati, E., Mazzatorta, P., Neagu, D., Gini, G.C.: Combining classifiers of pesticides toxicity through a neuro-fuzzy approach. In: Roli, Kittler [107], pp. 293–303

[15] Berthold, M.R., Cebron, N., Dill, F., Gabriel, T.R., Kötter, T., Meinl, T., Ohl, P., Sieb, C., Thiel, K., Wiswedel, B.: KNIME: The konstanz information miner. In: Preisach, C., Burkhardt, H., Schmidt-Thieme, L., Decker, R. (eds.) GfKl. Studies in Classification, Data Analysis, and Knowledge Organization, pp. 319–326. Springer (2007)

[16] Berthold, M.R., Cebron, N., Dill, F., Gabriel, T.R., Kötter, T., Meinl, T., Ohl, P., Thiel, K., Wiswedel, B.: KNIME - the konstanz information miner: version 2.0 and beyond. SIGKDD Explorations 11(1), 26–31 (2009)

[17] Bertolami, R., Bunke, H.: Multiple classifier methods for offline handwritten text line recognition. In: Haindl, et al. [59], pp. 72–81

[18] Biggio, B., Corona, I., Fumera, G., Giacinto, G., Roli, F.: Bagging classifiers for fighting poisoning attacks in adversarial classification tasks. In: Sansone, et al. [112], pp. 350–369

[19] Bonissone, P.P., Eklund, N., Goebel, K.: Using an ensemble of classifiers to audit a production classifier. In: Oza, et al. [95], pp. 376–386

[20] Breiman, L.: Random forests. Machine Learning 45(1), 5–32 (2001)

[21] Breiman, L., Breiman, L.: Bagging predictors. Machine Learning, 123–140 (1996)

[22] Briem, G.J., Benediktsson, J.A., Sveinsson, J.R.: Boosting, bagging, and consensus based classification of multisource remote sensing data. In: Kittler, Roli [74], pp. 279–288

[23] Bruzzone, L., Cossu, R.: A robust multiple classifier system for a partially unsupervised updating of land-cover maps. In: Kittler, Roli [74], pp. 259–268

[24] Bruzzone, L., Cossu, R., Prieto, D.F.: Combining parametric and nonparametric classifiers for an unsupervised updating of land-cover maps. In: Kittler, Roli [73], pp. 290–299

[25] Cappelli, R., Maio, D., Maltoni, D.: Combining fingerprint classifiers. In: Kittler, Roli [73], pp. 351–361

[26] Chawla, N.V., Bowyer, K.W.: Designing multiple classifier systems for face recognition. In: Oza, et al. [95], pp. 407–416

[27] Chindaro, S., Sirlantzis, K., Fairhurst, M.C.: Analysis and modelling of diversity contribution to ensemble-based texture recognition performance. In: Oza, et al. [95], pp. 387–396

[28] Christensen, H.U., Arroyo, D.O.: Applying data fusion methods to passage retrieval in qas. In: Haindl, et al. [59], pp. 82–92

[29] Cordella, L.P., Foggia, P., Sansone, C., Tortorella, F., Vento, M.: A cascaded multiple expert system for verification. In: Kittler, Roli [73], pp. 330–339

[30] Cordella, L.P., Limongiello, A., Sansone, C.: Network intrusion detection by a multi-stage classification system. In: Roli, et al. [109], pp. 324–333

[31] Cordella, L.P., De Santo, M., Percannella, G., Sansone, C., Vento, M.: A multi-expert system for movie segmentation. In: Roli, Kittler [107], pp. 304–313

[32] Csirik, J., Bertholet, P., Bunke, H.: Sequential classifier combination for pattern recognition in wireless sensor networks. In: Sansone, et al. [112], pp. 187–196

[33] Dahmen, J., Keysers, D., Ney, H.: Combined classification of handwritten digits using the 'virtual test sample method'. In: Kittler, Roli [74], pp. 109–118

[34] Dainotti, A., Pescapè, A., Sansone, C., Quintavalle, A.: Using a behaviour knowledge space approach for detecting unknown ip traffic flows. In: Sansone, et al. [112], pp. 360–369

[35] de Borda, J.-C.: Memoire sur les elections au scrutin. Memoires de l'Academie Royale des Sciences, 657–664 (1781)

[36] Degtyarev, N., Seredin, O.: A geometric approach to face detector combining. In: Sansone, et al. [112], pp. 299–308

[37] Dietterich, T.G.: Ensemble methods in machine learning. In: Kittler, Roli [73], pp. 1–15

[38] Dietterich, T.G., Bakiri, G.: Solving multiclass learning problems via error-correcting output codes. Journal of Artificial Intelligence Research 2, 263–286 (1995)

[39] Dolenko, S.A., Orlov, Y.V., Persiantsev, I.G., Shugai, J.S., Dmitriev, A.V., Suvorova, A.V., Veselovsky, I.S.: Solar wind data analysis using self-organizing hierarchical neural network classifiers. In: Kittler, Roli [74], pp. 289–298

[40] Du, P., Li, G., Zhang, W., Wang, X., Sun, H.: Consistency measure of multiple classifiers for land cover classification by remote sensing image. In: Benediktsson, et al. [12], pp. 398–407

[41] Du, P., Sun, H., Zhang, W.: Target identification from high resolution remote sensing image by combining multiple classifiers. In: Benediktsson, et al. [12], pp. 408–417

[42] Duin, R.P.W., Juszczak, P., de Ridder, D., Paclík, P., Pekalska, E., Tax, D.M.J.: PR-Tools 4.0, a Matlab toolbox for pattern recognition (2004),
http://www.prtools.org

[43] Ebrahimpour, R., Kabir, E., Yousefi, M.R.: View-based eigenspaces with mixture of experts for view-independent face recognition. In: Haindl, et al. [59], pp. 131–140

[44] Erdogan, H., Erçil, A., Ekenel, H.K., Bilgin, S.Y., Eden, I., Kirisçi, M., Abut, H.: Multi-modal person recognition for vehicular applications. In: Oza, et al. [95], pp. 366–375

[45] Fanelli, A.M., Castellano, G., Buscicchio, C.A.: A modular neuro-fuzzy network for musical instruments classification. In: Kittler, Roli [73], pp. 372–382

[46] Fiérrez-Aguilar, J., Garcia-Romero, D., Ortega-Garcia, J., Gonzalez-Rodriguez, J.: Speaker verification using adapted user-dependent multilevel fusion. In: Oza, et al. [95], pp. 356–365

[47] Foggia, P., Sansone, C., Tortorella, F., Vento, M.: Automatic classification of clustered microcalcifications by a multiple classifier system. In: Kittler, Roli [74], pp. 208–217

[48] Freund, Y., Schapire, R.E.: Experiments with a new boosting algorithm. In: Proc. 13th International Conference on Machine Learning, pp. 148–156. Morgan Kaufmann (1996)

[49] Frinken, V., Fischer, A., Bunke, H.: Combining neural networks to improve performance of handwritten keyword spotting. In: Gayar, et al. [52], pp. 215–224

[50] Fröba, B., Rothe, C., Küblbeck, C.: Statistical sensor calibration for fusion of different classifiers in a biometric person recognition framework. In: Kittler, Roli [73], pp. 362–371

[51] Fröba, B., Zink, W.: On the combination of different template matching strategies for fast face detection. In: Kittler, Roli [74], pp. 418–428

[52] El Gayar, N., Kittler, J., Roli, F. (eds.): MCS 2010. LNCS, vol. 5997. Springer, Heidelberg (2010)

[53] Giacinto, G., Roli, F., Didaci, L.: A modular multiple classifier system for the detection of intrusions in computer networks. In: Windeatt, Roli [134], pp. 346–355

[54] Gini, G.C., Lorenzini, M., Benfenati, E., Brambilla, R., Malvé, L.: Mixing a symbolic and a subsymbolic expert to improve carcinogenicity prediction of aromatic compounds. In: Kittler, Roli [74], pp. 126–135

[55] Gordon, J., Shortliffe, E.H.: The dempster-shafer theory of evidence. In: Buchanan, B.G., Shortliffe, E.H. (eds.) Rule-Based Expert Systems, pp. 272–292. Addison Wesley Publishing Company, Reading (1984)

[56] Günter, S., Bunke, H.: New boosting algorithms for classification problems with large number of classes applied to a handwritten word recognition task. In: Windeatt, Roli [134], pp. 326–335

[57] Günter, S., Bunke, H.: Ensembles of classifiers derived from multiple prototypes and their application to handwriting recognition. In: Roli, et al. [109], pp. 314–323

[58] Hady, M.F.A., Schwenker, F.: Combining committee-based semi-supervised and active learning and its application to handwritten digits recognition. In: Gayar, et al. [52], pp. 225–234

[59] Haindl, M., Kittler, J., Roli, F. (eds.): MCS 2007. LNCS, vol. 4472. Springer, Heidelberg (2007)

[60] Hall, M., Frank, E., Holmes, G., Pfahringer, B., Reutemann, P., Witten, I.H.: The WEKA data mining software: an update. SIGKDD Explorations 11(1), 10–18 (2009)

[61] Higgins, J.E., Dodd, T.J., Damper, R.I.: Application of multiple classifier techniques to subband speaker identification with an hmm/ann system. In: Kittler, Roli [74], pp. 369–377

[62] Ho, T.K.: The random subspace method for constructing decision forests. IEEE Transactions on Pattern Analysis and Machine Intelligence 20(8), 832–844 (1998)

[63] Huang, Y.S., Suen, C.Y.: A method of combining multiple experts for the recognition of unconstrained handwritten numerals. IEEE Trans. Pattern Anal. Mach. Intell. 17(1), 90–94 (1995)

[64] Ianakiev, K.G., Govindaraju, V.: Architecture for classifier combination using entropy measures. In: Kittler, Roli [73], pp. 340–350

[65] Jain, A.K., Duin, R.P.W., Mao, J.: Statistical pattern recognition: A review. IEEE Trans. Pattern Anal. Mach. Intell. 22(1), 4–37 (2000)

[66] Jaser, E., Kittler, J., Christmas, W.J.: Building classifier ensembles for automatic sports classification. In: Windeatt, Roli [134], pp. 366–374

[67] Jiang, X., Yu, K., Bunke, H.: Classifier combination for grammar-guided sentence recognition. In: Kittler, Roli [73], pp. 383–392

[68] Khademi, M., Shalmani, M.T.M., Kiapour, M.H., Kiaei, A.A.: Recognizing combinations of facial action units with different intensity using a mixture of hidden markov models and neural network. In: Gayar, et al. [52], pp. 304–313

[69] Khreich, W., Granger, E., Miri, A., Sabourin, R.: Incremental boolean combination of classifiers. In: Sansone, et al. [112], pp. 340–349

[70] Kittler, J., Ballette, M., Czyz, J., Roli, F., Vandendorpe, L.: Decision level fusion of intramodal personal identity verification experts. In: Roli, Kittler [107], pp. 314–324

[71] Kittler, J., Hatef, M., Duin, R.P.W., Matas, J.: On combining classifiers. IEEE Transactions on Pattern Analysis and Machine Intelligence 20, 226–239 (1998)

[72] Kittler, J., Poh, N., Merati, A.: Cohort based approach to multiexpert class verification. In: Sansone, et al. [112], pp. 319–329

[73] Kittler, J., Roli, F. (eds.): MCS 2000. LNCS, vol. 1857. Springer, Heidelberg (2000)

[74] Kittler, J., Roli, F. (eds.): MCS 2001. LNCS, vol. 2096. Springer, Heidelberg (2001)

[75] Kittler, J., Sadeghi, M.: Physics-based decorrelation of image data for decision level fusion in face verification. In: Roli, et al. [109], pp. 354–363

[76] Ko, A.H.-R., Sabourin, R., de Souza Britto Jr., A.: A new hmm-based ensemble generation method for numeral recognition. In: Haindl, et al. [59], pp. 52–61

[77] Kumar, S., Ghosh, J., Crawford, M.M.: A hierarchical multiclassifier system for hyperspectral data analysis. In: Kittler, Roli [73], pp. 270–279

[78] Kumazawa, I.: Shape matching and extraction by an array of figure-and-ground classifiers. In: Kittler, Roli [73], pp. 393–402

[79] Kuncheva, L.I.: Combining Pattern Classifiers: Methods and Algorithms. Wiley Interscience (2004)

[80] Lam, L.: Classifier combinations: Implementations and theoretical issues. In: Kittler, Roli [73], pp. 77–86

[81] Di Lecce, V., Dimauro, G., Guerriero, A., Impedovo, S., Pirlo, G., Salzo, A.: A multiexpert system for dynamic signature verification. In: Kittler, Roli [73], pp. 320–329

[82] Li, P., Chan, K.L., Fu, S., Krishnan, S.M.: An abnormal ecg beat detection approach for long-term monitoring of heart patients based on hybrid kernel machine ensemble. In: Oza, et al. [95], pp. 346–355

[83] Lienemann, K., Plötz, T., Fink, G.A.: On the application of svm-ensembles based on adapted random subspace sampling for automatic classification of nmr data. In: Haindl, et al. [59], pp. 42–51

[84] Lienemann, K., Plötz, T., Fink, G.A.: Stacking for ensembles of local experts in metabonomic applications. In: Benediktsson, et al. [12], pp. 498–508

[85] Loog, M., Li, Y., Tax, D.M.J.: Maximum membership scale selection. In: Benediktsson, et al. [12], pp. 468–477

[86] Lu, Y.: Knowledge integration in a multiple classifier system. Appl. Intell. 6(2), 75–86 (1996)

[87] Marasco, E., Johnson, P., Sansone, C., Schuckers, S.: Increase the security of multibiometric systems by incorporating a spoofing detection algorithm in the fusion mechanism. In: Sansone, et al. [112], pp. 309–318

[88] Marcialis, G.L., Roli, F.: High security fingerprint verification by perceptron-based fusion of multiple matchers. In: Roli, et al. [109], pp. 364–373

[89] Marcialis, G.L., Roli, F.: Serial fusion of fingerprint and face matchers. In: Haindl, et al. [59], pp. 151–160

[90] Masulli, F., Pardo, M., Sberveglieri, G., Valentini, G.: Boosting and classification of electronic nose data. In: Roli, Kittler [107], pp. 262–271

[91] Melville, P., Mooney, R.J.: Diverse ensembles for active learning. In: Brodley, C.E. (ed.) ICML. ACM International Conference Proceeding Series, vol. 69. ACM (2004)

[92] Merler, S., Furlanello, C., Larcher, B., Sboner, A.: Tuning cost-sensitive boosting and its application to melanoma diagnosis. In: Kittler, Roli [74], pp. 32–42

[93] Minguillón, J., Tate, A.R., Arús, C., Griffiths, J.R.: Classifier combination for in vivo magnetic resonance spectra of brain tumours. In: Roli, Kittler [107], pp. 282–292

[94] Mohamed, T.A., El Gayar, N., Atiya, A.F.: A co-training approach for time series prediction with missing data. In: Haindl et al. [59], pp. 93–102

[95] Oza, N.C., Polikar, R., Kittler, J., Roli, F. (eds.): MCS 2005. LNCS, vol. 3541. Springer, Heidelberg (2005)

[96] Oza, N.C., Tumer, K., Tumer, I.Y., Huff, E.M.: Classification of aircraft maneuvers for fault detection. In: Windeatt, Roli [134], pp. 375–384

[97] Powalka, R.K., Sherkat, N., Whitrow, R.J.: Multiple recognizer combination topologies. In: Simner, M.L., Leedham, C.G., Thomassen, A.J.W.M. (eds.) Handwriting and Drawing Research: Basic and Applied Issues. IOS Press (1995)

[98] Prabhakar, S., Jain, A.K.: Decision-level fusion in fingerprint verification. In: Kittler, Roli [74], pp. 88–98

[99] Pranckeviciene, E., Baumgartner, R., Somorjai, R.L.: Using domain knowledge for in the random subspace method: Application: Application to the classification of biomedical spectra. In: Oza, et al. [95], pp. 336–345

[100] Procopio, M.J., Kegelmeyer, W.P., Grudic, G.Z., Mulligan, J.: Terrain segmentation with on-line mixtures of experts for autonomous robot navigation. In: Benediktsson, et al. [12], pp. 385–397

[101] Rahman, F., Tarnikova, Y., Kumar, A., Alam, H.: Second guessing a commercial 'black box' classifier by an 'in house' classifier: Serial classifier combination in a speech recognition application. In: Roli, et al. [109], pp. 374–383

[102] Rajan, S., Ghosh, J.: An empirical comparison of hierarchical vs. two-level approaches to multiclass problems. In: Roli, et al. [109], pp. 283–292

[103] Rajan, S., Ghosh, J.: Exploiting class hierarchies for knowledge transfer in hyperspectral data. In: Oza, et al. [95], pp. 417–427

[104] Raudys, S., Baykan, Ö.K., Babalik, A., Denisov, V., Bielskis, A.A.: Classifiers fusion in recognition of wheat varieties. In: Haindl, et al. [59], pp. 62–71

[105] Re, M., Valentini, G.: Ensemble based data fusion for gene function prediction. In: Benediktsson, et al. [12], pp. 448–457

[106] Re, M., Valentini, G.: An experimental comparison of hierarchical bayes and true path rule ensembles for protein function prediction. In: Gayar, et al. [52], pp. 294–303

[107] Roli, F., Kittler, J. (eds.): MCS 2002. LNCS, vol. 2364. Springer, Heidelberg (2002)

[108] Roli, F., Kittler, J., Fumera, G., Muntoni, D.: An experimental comparison of classifier fusion rules for multimodal personal identity verification systems. In: Roli, Kittler [107], pp. 325–336

[109] Roli, F., Kittler, J., Windeatt, T. (eds.): MCS 2004. LNCS, vol. 3077. Springer, Heidelberg (2004)

[110] Sadeghi, M., Khoshrou, S., Kittler, J.: Confidence based gating of colour features for face authentication. In: Haindl, et al. [59], pp. 121–130

[111] Samadzadegan, F., Bigdeli, B., Ramzi, P.: A multiple classifier system for classification of lidar remote sensing data using multi-class svm. In: Gayar, et al. [52], pp. 254–263

[112] Sansone, C., Kittler, J., Roli, F. (eds.): MCS 2011. LNCS, vol. 6713. Springer, Heidelberg (2011)

[113] Sansone, C., Paduano, V., Ceccarelli, M.: Combining 2d and 3d features to classify protein mutants in hela cells. In: Gayar, et al. [52], pp. 284–293

[114] De Santo, M., Percannella, G., Sansone, C., Vento, M.: Combining audio-based and video-based shot classification systems for news videos segmentation. In: Oza, et al. [95], pp. 397–406

[115] Schettini, R., Brambilla, C., Cusano, C.: Content-based classification of digital photos. In: Roli, Kittler [107], pp. 272–281

[116] Seewald, A.K.: How to make stacking better and faster while also taking care of an unknown weakness. In: Sammut, C., Hoffmann, A.G. (eds.) Machine Learning, Proceedings of the Nineteenth International Conference (ICML 2002), University of New South Wales, Sydney, Australia, July 8-12, pp. 554–561. Morgan Kaufmann (2002)

[117] Serrano, Á., de Diego, I.M., Conde, C., Cabello, E., Bai, L., Shen, L.: Fusion of support vector classifiers for parallel gabor methods applied to face verification. In: Haindl, et al. [59], pp. 141–150

[118] Sirlantzis, K., Fairhurst, M.C., Hoque, S.: Genetic algorithms for multi-classifier system configuration: A case study in character recognition. In: Kittler, Roli [74], pp. 99–108

[119] Sirlantzis, K., Hoque, S., Fairhurst, M.C.: Input space transformations for multi-classifier systems based on n-tuple classifiers with application to handwriting recognition. In: Windeatt, Roli [134], pp. 356–365

[120] Slavík, P., Govindaraju, V.: Use of lexicon density in evaluating word recognizers. In: Kittler, Roli [73], pp. 310–319

[121] Smits, P.C.: Combining supervised remote sensing image classifiers based on individual class performances. In: Kittler, Roli [74], pp. 269–278

[122] Suen, C.Y., Lam, L.: Multiple classifier combination methodologies for different output levels. In: Kittler, Roli [73], pp. 52–66

[123] Sun, S.: Ensemble learning methods for classifying eeg signals. In: Haindl, et al. [59], pp. 113–120

[124] Sun, S.: An improved random subspace method and its application to eeg signal classification. In: Haindl, et al. [59], pp. 103–112

[125] Svetnik, V., Liaw, A., Tong, C., Wang, T.: Application of breiman's random forest to modeling structure-activity relationships of pharmaceutical molecules. In: Roli, et al. [109], pp. 334–343

[126] Ting, K.M., Zhu, L.: Boosting support vector machines successfully. In: Benediktsson, et al. [12], pp. 509–518

[127] Tulyakov, S., Govindaraju, V.: Neural network optimization for combinations in identification systems. In: Benediktsson, et al. [12], pp. 418–427

[128] Visentini, I., Kittler, J., Foresti, G.L.: Diversity-based classifier selection for adaptive object tracking. In: Benediktsson, et al. [12], pp. 438–447

[129] Wan, W., Fraser, D.: A multiple self-organizing map scheme for remote sensing classification. In: Kittler, Roli [73], pp. 300–309

[130] Wang, X., Tang, X.: Experimental study on multiple lda classifier combination for high dimensional data classification. In: Roli, et al. [109], pp. 344–353

[131] Waske, B., Benediktsson, J.A., Sveinsson, J.R.: Classifying remote sensing data with support vector machines and imbalanced training data. In: Benediktsson, et al. [12], pp. 375–384

[132] Webb, G.I.: Multiboosting: A technique for combining boosting and wagging. Machine Learning 40(2), 159–196 (2000)

[133] Wilczok, E., Lellmann, W.: Design and evaluation of an adaptive combination framework for ocr result strings. In: Windeatt, Roli [134], pp. 395–404

[134] Windeatt, T., Roli, F. (eds.): MCS 2003. LNCS, vol. 2709. Springer, Heidelberg (2003)

[135] Windridge, D., Bowden, R.: Induced decision fusion in automated sign language interpretation: Using ica to isolate the underlying components of sign. In: Roli, et al. [109], pp. 303–313

[136] Witten, I.H., Frank, E.: Data Mining: Practical Machine Learning Tools and Techniques with Java Implementations. Morgan Kaufmann, San Francisco (2000)

[137] Witten, I.H., Frank, E., Hal, M.A.: Data Mining: Practical Machine Learning Tools and Techniques, 3rd edn. Morgan Kaufmann, Burlington (2011)

[138] Wolpert, D.H.: Stacked generalization. Neural Networks 5, 241–259 (1992)

[139] Xu, J.-W., Singh, V., Govindaraju, V., Neogi, D.: A cascade multiple classifier system for document categorization. In: Benediktsson, et al. [12], pp. 458–467

[140] Yousri, N.A.: A multi-objective sequential ensemble for cluster structure analysis and visualization and application to gene expression. In: Gayar, et al. [52], pp. 274–283

[141] Zhang, C.-X., Duin, R.P.W.: An empirical study of a linear regression combiner on multi-class data sets. In: Benediktsson, et al. [12], pp. 478–487

Chapter 11
Self Organisation and Modal Learning: Algorithms and Applications

Dominic Palmer-Brown and Chrisina Jayne

Abstract. Modal learning in neural computing [33] refers to the strategic combination of modes of adaptation and learning within a single artificial neural network structure. Modes, in this context, are learning methods that are transferable from one learning architecture to another, such as weight update equations. In modal learning two or more modes may proceed in parallel in different parts of the neural computing structure (layers and neurons), or they occupy the same part of the structure, and there is a mechanism for allowing the neural network to switch between modes.

From a theoretical perspective any individual mode has inherent limitations because it is trying to optimise a particular objective function. Since we cannot in general know a priori the most effective learning method or combination of methods for solving a given problem, we should equip the system (the neural network) with the means to discover the optimal combination of learning modes during the learning process. There is potential to furnish a neural system with numerous modes. Most of the work conducted so far concentrates on the effectiveness of two to four modes. The modal learning approach applies equally to supervised and unsupervised (including self organisational) methods. In this chapter, we focus on modal self organisation.

Examples of modal learning methods include the Snap-Drift Neural Network (SDNN) [5, 25, 28, 33, 32] which toggles its learning between two modes, an adaptive function neural network, in which adaptation applies simultaneously to both the weights and to the shape of the individual neuron activation functions, and the combination of four learning modes, in the form of Snap-drift ADaptive FUnction Neural Network [17, 18, 33]. In this chapter, after reviewing modal learning in general, we present some examples methods of modal self organisation. Self organisation is taken in the broadest context to include unsupervised methods. We review the simple unsupervised modal method called snap-drift [5, 25, 28, 32], which combines Learning Vector Quantization [21, 22, 23, 37] with a 'Min' or Fuzzy AND

Dominic Palmer-Brown · Chrisina Jayne
London Metropolitan University, 166-220 Holloway Road, London, N7 8DB, UK
e-mail: {d.palmer-brown,c.jayne}@londonmet.ac.uk

M. Bianchini et al. (Eds.): *Handbook on Neural Information Processing*, ISRL 49, pp. 379–400.
DOI: 10.1007/978-3-642-36657-4_11　　　　　© Springer-Verlag Berlin Heidelberg 2013

method. Snap-drift is then applied to the Self-Organising Map [34]. The methods are utilised in numerous real-world problems such as grouping learners' responses to multiple choice questions, natural language phrase recognition and pattern classification on well known datasets. Algorithms, dataset descriptions, pseudocode and Matlab code are presented.

1 Introduction

Modal learning is a new approach to neural computing and it refers to the combination of more than one mode of learning within a single neural network. It contrasts with hybrid [42] or multi-modal approaches [6] where the modes of learning occur in different modules and/or at strictly separated times. There are several reasons to explore modal learning. One motivation is to overcome the inherent limitations of any given mode (for example some modes memorise specific features, others average across features, and both approaches may be relevant according to the circumstances); another motivation is from neuroscience, cognitive science and human learning, where multiple modes are required to explain behaviour; and a third reason is non-stationary input data, or time-variant learning objectives, where the required mode is a function of time. Combining modes in one network also presents some efficiency gains, and the possibility of maximising the flexibility (through the parallel application of many forms of plasticity) of the neural network as a learning agent.

Twenty years ago there were already several forms of artificial neural network, each utilising a different form of learning. Along the way, many forms of learning have been hailed as superior forms. However, decades after the introduction of Kohonen learning, SOMs [20], and Backpropagation [38] they are still being used alongside more recent methods such as Bayesian [30] and SVMs [41]. No single method or mode prevails. A wide range of methods are still in use, simply because there are significant problems and datasets for which each method is suitable and effective. In this context, Modal Learning arises from the desire to equip a single neural network or module with the power of several modes of learning, to achieve learning results that a single mode could not achieve, through exploitation of the complementary nature of each mode. A mode in this context is defined as an adaptation method for learning that could be applied in more than one type of architecture or network. It is analogous to a human mode of learning, such as learning by analogy or category learning. Modes of learning map onto learning objectives. Well known modes therefore include the Delta Rule [43], Backpropagation (BP), Learning Vector quantization (LVQ) [22], and Hebbian Learning [14]. In contrast, the Adaptive Resonance Theory (ART) [3] or Bayesian neural networks [30] define architectures and approaches to learning, within which particular modes are used.

In general, the objective of learning may be unknown, changing, or difficult to quantify. Even when the objective of learning is transparent, the optimal mode is not a given. Therefore it is not possible to know a-priori which is the most suitable learning mode. For example, Backpropagation is good for approximation and

transformation, but if the features within the data need to be assimilated (or mem-
orised) directly, then learning vector quantization or SVM [41] would be more
appropriate. The learning agent should be able to establish the most effective mode
or combination of modes during, and as part of, the learning process.

Each mode is inherently limited because it is tied to a particular objective function. A simple illustration of the potential benefits of a modal learning approach
can be seen in the following sequence of class boundaries in a $2D$, 2 class problem (Fig. 1). The example illustrates how it is necessary to increase complexity
to an extent that is not well supported by the data in order to find a single mode
solution. In contrast a solution with more than one mode allows problem decomposition to occur, and each mode can be relatively simple in structure. A solution
can be achieved by combining a straight line (perceptron), a simple curve (multi-
layer perceptron) and a cluster. This requires 3 modes of learning (see Fig. 2 left).
Alternatively combining a curve (multilayer perceptron) and a cluster requires only
2 modes of learning (Fig. 2 right).

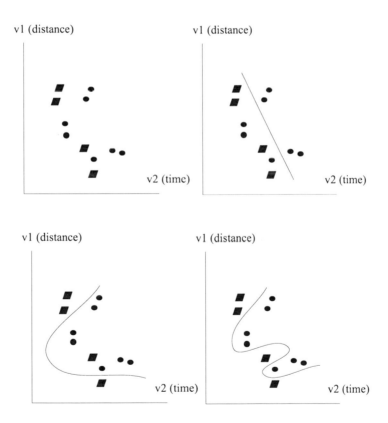

Fig. 1 Increasingly complex solutions to a 2-class problem

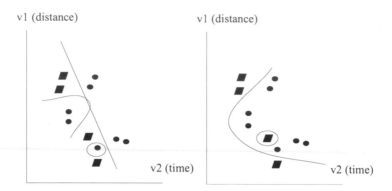

Fig. 2   3 mode solution and 2 mode solution

Rather than trying to solve the whole problem with a single mode, a simpler learnt solution is achievable by combining modes of learning. When we look at human and machine learning in a wider context, there are many reasons and motivations to consider modal learning, as it allows for a range of learning methods to be taken into account, along the spectrum from memorisation to generalisation.

Modal learning in general is achieved by optimising the objective functions for each mode. This involves minimising $\sum_i O_i$, where O_i is the objective function for mode i. For example, if we are combining the Delta Rule and Learning Vector Quantisation, we must minimise $[\sum(squared\ errors) + \sum(Input - Weight)]$.

Snap-Drift Neural Network (SDNN), introduced in   [5, 25, 28, 33, 32], is an example of a modal learning method which toggles its weight update equation between two modes: 'Min'('Fuzzy AND') and Learning Vector Quantization [21, 22, 23, 37]. Another example is the Adaptive Function Neural network [17], in which adaptation applies simultaneously to both the weights and to the shape of the individual neuron activation functions [17, 18]. The Snap-drift ADaptive FUnction Neural Network [17, 18, 33] combines four learning modes: two modes for the weights in the snap-drift layer and two modes in the adaptive function layer; one mode adapts weights and the other adapts the function.

In this chapter the SDNN algorithm and its application to self-organisation [34] are reviewed. Section 2 gives a detailed description of the SDNN algorithm, its architecture and pseudocode. In Section 3 the Snap-Drift SOM (SDSOM) which adopts the Kohonen SOM architecture [23] is presented. Section 4 illustrates the applications of SDNN and SDSOM to publically available data. The effect of applying snap-drift is evaluated in respect to classification performance and data visualisation (the shape of the resultant maps). Section 4 shows also the application of SDNN to a real-world problem related to e-learning and more specifically grouping learners responses to multiple choice questions. The Appendix at the end of the chapter includes Matlab code for the SDNN with detailed comments showing how the method can be used in practice.

2 Snap-Drift Neural Network

2.1 Description

The Snap-Drift Neural Network (SDNN) is an unsupervised algorithm able to adapt rapidly, for example in non-stationary environments where new patterns are introduced over time. The standard snap-drift neural network (SDNN) algorithm has been successfully applied for continuous learning in many diverse applications [8, 25, 26, 27, 28]. An example application of the unsupervised snap-drift algorithm is the analysis and interpretation of data representing interactions between trainee computer network managers and a simulated network management system [25], where it helped to identify patterns of the user behaviour. Another application is feature discovery and clustering of speech waveforms recorded from non-stammering and stammering speakers [28]. Phonetic properties of non-stammering and stammering speech were discovered, and rapid automatic classification into stammering and non stammering speech was found to be possible. Most recently, snap-drift has been successfully applied to categorising student responses to multiple choice questions in a virtual learning context [9, 32].

It is essentially a simple modal learning method, which swaps periodically between the two learning modes (snap and drift). Snap is based on the fuzzy AND (Min) of input and weight; and drift is based on learning vector quantization (LVQ) [24]. Snap-Drift harnesses the complementary strengths of the two modes of learning which are dynamically combined in a rapid form of adaptation.

The learning process is unsupervised and unlike error minimisation and maximum likelihood methods in MLPs [38]. Those methods perform optimisation for classification by for example pushing features in the direction that minimizes classification error. In such methods there is no requirement for the learned weight vector to be a significant feature within the input data. In contrast, SDNN swaps its learning mode to find, in an unsupervised fashion, to find a rich set of features in the data and uses them to group the data into categories. The effect of the learning process for a single neuron is illustrated in Fig. 3, for one cluster of data in two dimensions. The weight vectors are normalised and so they are maintained at unit length. The weight vector for the neuron will settle to an angle somewhere between the snap and drift angles. So, each weight vector is bounded by snap and drift: snap gives the angle of the minimum values (on all dimensions) and drifting gives the average angle of the patterns grouped under the neuron. Hence, snap essentially provides an anchor vector pointing at the 'bottom left hand corner' of the pattern group for which the neuron wins. This represents a feature that is common to all the patterns in the group and gives a high probability of rapid (in terms of epochs) convergence (both snap and drift are convergent, but snap is faster). Drift tilts the vector towards the centroid angle of the group and ensures that an average, generalised feature is included in the final vector. The angular range of the pattern-group membership depends on the proximity of neighbouring groups (competition), but can also be controlled by adjusting a threshold on the weighted sum of inputs to the neurons.

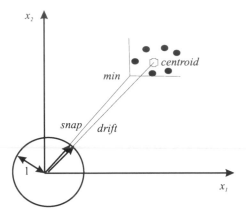

Fig. 3 Snap and drift weight vectors

2.2 Architecture

The architecture of the Snap-Drift Neural network is shown in (Fig. 4). It consists of an input layer, a distributed d layer for feature extraction and a selection s layer for feature classification [25]. The distributed d layer groups the input patterns according to their features using snap-drift. The D most activated(winning) d nodes whose weight vectors best match the current input pattern are used as the input data to the selection, s layer, for the purposes of feature classification. In the d layer, the output nodes with the highest net input are accepted as winners.

In the s layer, a quality assurance threshold is applied. If the net input of the most active s node is above the threshold, that s node is accepted as the winner, and defines the category of the input pattern; otherwise a new uncommitted output node is recruited as the winner and its weights initialized with the current d layer pattern

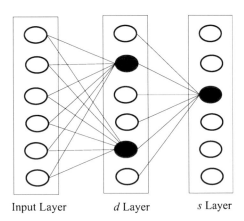

Fig. 4 Snap-drift architecture

(the D winning nodes). The threshold influences the granularity of the categories. If the threshold is zero, relatively few categories will be formed, based on very different combinations of d nodes; but if the threshold is set high, even a small proportion of the winning d nodes not matching the current weights of the winning s node leads to the recruitment a new s node, and thus finer grained categories are formed.

2.3 Algorithm

In essence the snap-drift weight update algorithm can be stated as:

$$Snap - drift = \alpha(Snap) + (1 - \alpha)(drift) \tag{1}$$

Snap-drift learning uses a combination of fuzzy AND (or MIN) learning (snap), and Learning Vector Quantisation (drift) [24].

The learning of both of the layers in the neural system follows:

$$w_{ji}^{(new)} = \alpha(I \cap w_{ji}^{(old)}) + (1 - \alpha)(w_{ji}^{(old)} + \beta(I - w_{ji}^{(old)})) , \tag{2}$$

where $w_{ji} = weights\ vectors$; $I = input\ vectors$, and $\beta = the\ drift\ learning\ rate$. The β learning rate may have different values for the d and s layers. When $\alpha = 1$, fast, minimalist (snap) learning is invoked:

$$w_{ji}^{(new)} = I \cap w_{ji}^{(old)} \tag{3}$$

This works for binary data, otherwise equation (3) becomes the fuzzy AND of the weight with the data, $Min(I, w_{ji}^{(old)})$. Consequently, snap encodes, within the weights, the common elements of all patterns for which the neuron wins. In contrast, when $\alpha = 0$, (2) simplifies to:

$$w_{ji}^{(new)} = w_{ji}^{(old)} + \beta(I - w_{ji}^{(old)}) \tag{4}$$

which implements a simple form of clustering (drift towards the centroid of the pattern group) or LVQ, at a speed determined by β. Finally, after either snap or drift, weights are normalized:

$$w_{ji}^{(new)} = \frac{w_{ji}^{(new)}}{\left| w_{ji}^{(new)} \right|} \tag{5}$$

The snap and drift modes provide complementary features; snap capturing the common elements of the group of patterns as represented by the minimum values on each input dimension, whereas drift captures the average values of the group of patterns. Snap also has the effect of contributing to rapid convergence. The two modes provide different features that are overlaid within a single network and a shared set of weight vectors.

This is a summary of the SDNN algorithm:

1. Initialise parameters:

   n_d number of nodes in the d layer

   n_s number of nodes in the s layer

   D number of winning nodes in the d layer

   t quality assurance threshold

   λ learning rate d layer

   μ learning rate s layer

   Initialise weights $\mathbf{W_d}$ between input and d layer with randmoly selected patterns and normilise these

   Initialise and normlaise weights $\mathbf{W_s}$ between d layer and s (output) layer to small numbers between 0 and 1

   Intialise the output vector $\mathbf{s_0}$ from s layer to 0

   Initialize $\alpha = 1$ or 0, (equation (1))

2. For each epoch (t) and for each input pattern $\mathbf{x}$

   a. Find the D winning nodes at d layer with the largest net input, where $netinput = \mathbf{x}.\mathbf{W_d}$

   b. Adapt the weights $\mathbf{W_d}$ according to the snap or drift learning procedure snap-drift: (equation (2))

   c. Set the the output of d layer $\mathbf{d_0}$ as follows: the elements corresponding to the D winning nodes to 1 and the rest to 0

   d. Calculate the input vector $\mathbf{s_i}$ in the s layer as the product of $\mathbf{d_0}$ and the weights $\mathbf{W_s}$, $\mathbf{s_i} = \mathbf{d_0}.\mathbf{W_s}$

   e. Set $maxVal =$ the activation value of node in s layer with the largest input component from the vector $\mathbf{s_i}$

   f. Test the threshold condition:

   IF $(maxVal > t)$

   THEN

   Weights $\mathbf{W_s}$ are adapted according to the snap or drift learning procedure in equation (2))

   ELSE

   Set the value of the first element in the vector $\mathbf{s_0}$ that equals 0 to 1. (This recruits the first uncomitted node.)

   Adapt weights $\mathbf{W_s}$ according to the equation (2)

3   Snap-Drift Self-Organising Map

3.1   *Description*

The self-organising feature map algorithm (SOM) developed by Kohonen [23] has been used widely in clustering analysis and visualization of high-dimensional data

[37]. The SOMs can also be used for pattern classification by applying fine tuning of the map with LVQ learning algorithms [21, 23, 24]. The Kohonen feature map was inspired by the idea that self-organising maps resemble the topologically organised maps found in the cortices of the brain [20]. The SOM algorithm is based on unsupervised learning realised by finding the best matching node (the winner) on the map to the input vector and adapting the weights of the winner and the topological neighbourhood nodes. After the training each node on the map identifies particular input vectors and the organisation of the map reflects the organisation of the input data. In this work, SDNN is deployed in a self-organising map, to ascertain whether the advantages of snap-drift over LVQ alone (drift, without snap) transfer into the formation of topological maps. We are interested in classification performance and data visualisation (the shape of the resultant maps).

3.2   Architecture

The SDSOM has the same architecture Fig. 5 as a standard SOM, with a layer of input nodes connecting to the self organising map layer. A shrinking neighbourhood is used during training, as in SOM, with the weight vector of each neighbour of the winning node being adapted according to the input pattern. The difference in SD-SOM is the weight update, which consists of either snap (*min* of input and weight) or drift (LVQ, as in SOM) (equation (2)).

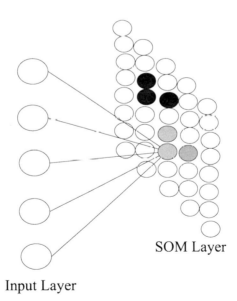

SOM Layer

Input Layer

Fig. 5 SOM architecture

3.3   Algorithm

This is a summary of the SDSOM algorithm:

1. Initialize parameters: $\alpha = 1$ or 0, (equation (1))
   Set size of the SOM layer map
   Initialize neighborhood size
   Initialize weights w between input and SOM layer with the values of randomly selected input patterns
   Normalize weights
   Initialize learning rate for drift mode in the range (0,1)
2. For each epoch, swap α between 0 or 1

   a. For each input pattern **x**
      i. Find the winning node in SOM with the largest net input which is the product of the pattern and the weights **x**.**W**
      ii. Update weights of the winning node and its neighbour nodes according to the current learning mode (equation (2))
      iii. Normalize weights (equation (5))
   b. Decrease the neighborhood size by 1

3. Label som layer nodes

The shaded nodes in Fig. 5 represent different classes or labels. Nodes receive the class label of the majority of the patterns for which they win. There is generally a tendency for neighbouring nodes to have the same class, given the nature of SOMs, but this is not forced by the labelling algorithm.

4   Applications

4.1   Applications of SDNN and SDSOM to Publicly Available Data

A range of data sets are chosen to represent a variety of learning challenges. They vary in terms of the number of input variables, the number of classes, and the level of separability of the classes. Since they are all known and freely available they provide useful benchmark comparisons with a number of neural computing and other machine learning techniques.

4.1.1   Description of Data

Animal Data. The Animal data presents a simple classification problem. It is artificial data and consists of 16 animals described by 13 attributes such as size, number of legs etc. [37]. The 16 animals are grouped into three classes (the first one represents bird, the second represents carnivore and the third represents herbivore).

Iris Data. The Iris data set has three classes setosa, versicolor and virginica [10, 7]. The iris data has 150 patterns, each with 4 attributes. The class distribution is 33.3%

for each of 3 classes. One of the classes is linearly separable from the other two, and the two are linearly inseparable from each other.

Wine Data. The Wine data set is the result of a chemical analysis of wines grown in the same region in Italy but derived from three different cultivars [11]. The analysis determines the quantities of 13 constituents (input variables) found in each of the three types of wines. There are 178 patterns with the following distribution: class 1:59, class 2:71, class 3:48.

Optical and Pen-Based Recognition of Handwritten Digits (OCR) Data. The OCR data set [1, 19] consists of 3823 training and 1797 testing patterns. Each pattern has 64 attributes which are integer numbers between 0 and 16. There are 10 classes corresponding to the digits 0 to 9. The 64 attributes are extracted from normalised bitmaps of handwritten digits by 43 people.

The experiments with the OCR data set use the already existing division of training/testing patterns 3828 and 1797 respectively, as originally proposed by Kaynak [19]. This facilitates direct performance comparisons between SDSOM and alternative algorithms that have been applied to the same data.

Natural Language Processing Data (NLP). The Lancaster Parsed Corpus [LPC] is a corpus of English sentences excerpted from printed publications of the year 1961, and is a subset of the Lancaster-Oslo/Bergen Corpus [12]. Words are tagged with their syntactic categories and each sentence in the LPC has undergone syntactic analysis. Phrase recognition is a well defined and well known application and a benchmark for testing the performance of neural networks in the field of Natural Language Processing (NLP) [29, 39]. The individual input patterns are encoded in binary according to the structure of the pre-tagged corpus [12]. This is achieved by separating the input layer into several regions for each tag, where each region corresponds to a different symbol type. A total of 45 bits are needed to encode all symbol types. In syntactic terms, there is a variety of terminal and non-terminal symbols tags. The terminal symbol groups are: punctuation (Pu), conjunction (Co), nouns (NP), verbs (VP) and prepositions (PP). The non-terminal symbol groups are Sentences (S), Finite clauses (F), Non-finite clauses (T), major phrase types (V) and minor phrase types (M). Together with a maximum of 4 Look Back symbols and 1 Look Ahead symbol [40], this makes a total of 15 input fields. By using linear binary coding for each symbol type within each input field, the size of an input pattern is 45 x 15 = 675 bits when using a standard time delay neural network input arrangement such as in [27]. By sampling pre-tagged sentences from LPC, we generate 254 input patterns, from all stages of parsing, typically involving mixtures of terminal and non-terminal symbols. Table 1 shows the number of input patterns for each symbol type.

For this problem half of the input set (127 patterns) is used for training. There are 2 types of test data: Natural Test data (ND) and Pure test data (PD) [40]. The ND consists of the remaining 127 patterns, which contains a mixture of some patterns also present in the training set because they happen to occur more than once (naturally occurring syntactic repetition), and new input patterns that have never encountered before. The PD consists entirely of input patterns that have never been

Table 1 Symbol types and number of input sequences

Symbol Type	Symbol and Description	Number of sequences.
Minor Phrase	E = Label used for existential 'there'	2
Major Phrase	J = An adjective phrase	4
	= A noun phrase	83
	Na = A noun phrase marked as subject of the verb	14
	P = A prepositional phrase	16
	Po = A prepositional phrase beginning with preposition 'of'	8
	R = An adverb phrase	12
	Rq = an adverb phrase beginning with a wh- word, e.g. 'How do you feel?' or 'How long'	1
	V = A finite 'verb phrase' i.e. one that exclude objects, complements	49
	Vi = Non-finite verb phrase	5
Sentence	S = Sentence	50
	S& = Compound sentence	2
	S+ = Compound sentence	2
Non-finite and Verbless Clause	Ti = Infinitive clause	4

encountered before (natural repetition removed). For each run the training patterns are selected at random from the entire data set. The average number of patterns for the PD is 102 (st. dev 4).

4.1.2 Experiments and Results

Experiments are carried out with SDNN, SDSOM and SOM. SDNN and SDSOM are typically trained between 200 and 250 epochs while SOM is trained for 500 epochs. This is long enough for the SDNN groups and the SOM maps to be stable in all cases. Summary of the results are presented in Table 2. The SDNN results for the OCR data in Table 2 are based on combining SDNN and the Adaptive Function Neural Network as previously published in [18], and the NLP results are from [27].

In order to perform a labelling of nodes for the purposes of classification the number of patterns for which the node wins is accumulated for each class and for each node. The majority class, with the highest number of patterns, becomes the class label of that node. The training classification score is the percentage of patterns categorised by nodes of the correct class. The training class labels are retained for use in testing. Nodes in the s (SDNN) layer or SOM map that by the end of training have no associated patterns for which they win are not labelled. During testing,

Table 2 Mean % correct classification for training and test sets based on 10 runs. Standard deviation given in the brackets.

Method/Data set	Animal	Iris	Wine	OCR	NLP - ND	NLP - PD
SDNN train	100(0)	98(0.82)	95.8(0.8)	99.53	93.47	93.47
SDNN test	100(0)	100(0)	95.6(0.9)	94.99	89.65	86.98
SDSOM train	100(0)	100(0)	100(0)	99.7(0)	100(0)	100(0)
SDSOM test	100(0)	98.8(1.5)	91.7(2.2)	97.3(0)	81.8(1.7)	77.6(2.4)
SOM train	100(0)	100 (0)	100 (0)	99.6(0)	100(0)	100(0)
SOM test	100(0)	95.7 (3.1)	85(3.8)	97.3(0)	79.8(3.1)	75.9(5.1)

if a winning node is unlabelled (which is rare) then the most active labelled node provides the class (correct or incorrect).

The Animal data presents a relatively easy classification task because each pattern differs quite significantly, therefore it is a simple challenge for any method to separate or classify them individually without the need for generalised rules. All methods perform well. There is however an important qualitative difference between the SDSOM and SOM results. SDSOM has projected the classes onto the map in a linearly separable fashion; two straight lines can separate the three animal classes on the map. This is not possible in the SOM, which mixes the herbivores and carnivores to a greater extent. The snap mode finds some common elements that are specific to herbivores that are not based on the overall similarity of herbivores across all dimensions, which is the limitation of LVQ, or any method that calculates overall similarity. This is a characteristic of modal learning. By superposition in the weight vectors of features from both modes, the network assimilates a combination of overall similarity of pattern groups and specific within-pattern features.

The Iris maps differ substantially between SOM and SDSOM, see Fig. 6. The SOM map presents a widely dispersed set of points (Fig. 6 right). They are nonetheless in clear regions associated with the three classes. However, the lines between classes in the map are curved with several changes of direction and there is no margin between the classes, even in the case of the linearly separable classes. In the SDSOM map, the margin between setosa and the other two classes is significant, and the linearly inseparable virginica is more tightly grouped than in SOM (Fig. 6 left). These factors give a classification advantage to SDSOM of 98.8% as opposed to 95.7%, and the Students t test indicates a 99.5% probability of the higher SDSOM rate being statistically significant. The SDNN (without a map but with d and s layers) performs well on this data set giving 100% accuracy on the test set.

The average separation on the Wine data map of the classes is larger in SDSOM (see Fig. 7), and the classification is 91.7% as opposed to 85% with SOM. This increase is 99.99% likely to be statistically significant. The additional layer (d layer) of SDNN allows it to perform very well on this data.

The OCR data maps for SDSOM and SOM are shown in Fig. 8 left and right respectively. The accuracy of classification for both methods for the test set is 97.3%. In common with the Iris data set the SOM map (Fig. 8 right) presents clearly the different classes but fills the entire space with no margin between the classes, while in the SDSOM map (Fig. 8 left) the classes are more tightly grouped and with larger margins in between them.

The results for the NLP data show that the average correct classifications for the ND and PD using SDSOM are slightly higher than the ones obtained with SOM, although these differences are not statistically significant. The lower performance of both SDSOM and SOM is due to the number of input patterns for each of the symbols, which varies greatly and is insufficient for effective training in some cases. For example, symbols E, S& and S+ only have 2 input patterns corresponding to them. Chance dictates that all of the input patterns may happen to be used exclusively as either testing or training data. When these inputs are selected for testing, e.g. S+ and S&, unsurprisingly they tend to be recognized as S. The SDNN performs better on this task [27]. It uses the performance guided version of snap-drift, whereby the mode is swapped only when performance on an s node declines.

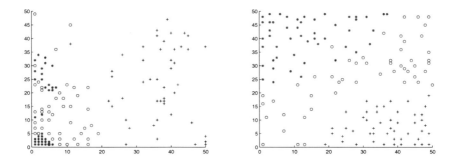

Fig. 6 SDSOM and SOM 50x50 applied to Iris Data set o (verisicolor) + (setosa) * (virginica)

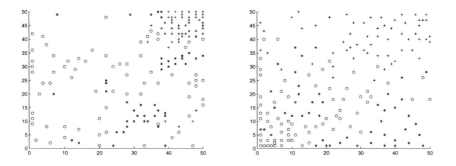

Fig. 7 SDSOM and SOM 50x50 applied to Wine Data set + (class 1) o (class 2) * (class 3)

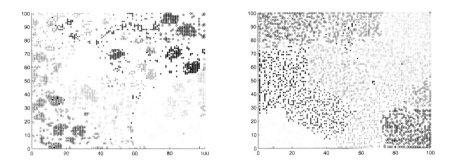

Fig. 8 SDSOM and SOM 100x100 applied to OCR Data set. The different colours represent the 10 different classes (digits 0 to 9)

4.1.3   Application of SDNN to E-Learning

Snap-Drift has been applied to online formative assessments of multiple choice questions with the purpose of providing feedback based on the common features in groups of answers [32], [9]. When students attempt on-line formative assessments they generate data that is invaluable for understanding their learning. That data is generally lost. However it can be captured, analysed and used as the basis for providing immediate feedback to the students as well as providing lecturers and tutors with a detailed picture of the learning of their students.

There are many studies investigating the role of different types of feedback in web-based assessments that report positive results from the use of Multiple Choice Questions (MCQs) in online tests for formative assessments (e.g. [4], [15], [36]). In these studies it is assumed that all the possible errors for a question can be predicted and a generic and focused feedback can be written for that question. However, this kind of feedback relates to a specific question rather than a combination of questions.

The diagnostic feedback developed here differs in that it does not reveal which questions were wrong; instead, the students are encouraged by the feedback to reflect on misunderstood concepts (that relate to their combination of errors on all the questions), and then to attempt the test again. Predicting all possible mistakes and writing generic and focused feedback for a combination of questions would be a daunting task and would not be feasible for large test banks (2 questions with 5 possible answers creates 25 possible answer combinations; 5 questions creates 3125 combinations, and so on). As the question bank grows the number of possible answer combinations increases exponentially, so that automation is essential for at least part of the process.

The neural network and in particular the Snap-Drift approach can address these problems by providing an efficient means of discovering a relatively small numbers of groups of similar answers so that responses can be targeted to the answers given by a very wide range of students with different states of knowledge. First the Snap-Drift neural network (SDNN) is trained in order to learn the different categories of student answers. The data used for training can be collected from students' responses for questions on a particular topic in a subject from the previous cohorts of students. In order to pre-process the data each response from the students is encoded into binary form, in preparation to be presented as input patterns for the SDNN. Here is an example of a possible format for 5 questions

$$A = 00001; \ B = 00010; \ C = 00100; \ D = 01000; \ E = 10000$$

and, a response such as $[D, D, C, B, A]$ will be encoded as

$$[0, 1, 0, 0, 0, 0, 1, 0, 0, 0, 0, 0, 1, 0, 0, 0, 0, 0, 1, 0, 0, 0, 0, 0, 1].$$

During training, on presentation of each input pattern, the SDNN will learn to group the input patterns according to their general features. The groups are recorded, and represent different states of knowledge in relation to a given topic, in the sense that responses within that group share a particular combination of answers to certain questions. In other words, they contain the same mistake(s) and/or answer the same question(s) correctly. For example, for one group every response might have in common answer B to question 3 and answer A to question 4. The other answers to the other questions will vary within the group, but the group is formed by the neural network based on the commonality in some question answers (two of them in that case). From one group to another the precise number of common responses varies between 2 and the number of questions.

The training relies upon having representative training data. The number of responses required to train the system so that it can generate reasonable groups, varies from one domain to another. When new responses are still creating new groups, more training data is required. Once new responses are not creating new groups, it is because those new responses are similar to previous responses, and enough responses to train the system reliably are already available.

Snap-drift is suitable because it is an unsupervised, easy-to-apply, quick and effective means of discovering groupings, and is capable of discovering both clearly separable clusters (drift) and groups that are characterized by precise features that may represent only a fraction of the structure of patterns (snap). After training the SDNN provides grouped of answers and their common features. Suitable feedback is written for each group of answers. A particular feedback derives from the responses to several questions and is therefore not tied to any particular question, and so the learner is encouraged to retake the same test, receiving different feedbacks depending on their evolving state of knowledge.

SDNN is integrated with an on-line system of multiple choice questions tests. Students login into the system with their student id numbers. The student responses, time and student id are recorded in the database after each student's submission of answers. The students are prompted to select a module and a topic and this leads to the screen with a specific set of multiple choice questions. On submission of the answers the system converts these into a binary vector which is fed into the SDNN. The SDNN produces a group number and the system retrieves the corresponding feedback for this group from the feedback file and sends it to the student's browser.

They are then prompted to go back and try the same questions again or to select a different topic. Responses, recorded in the database, can also be used for monitoring student progress and for identifying misunderstood concepts that can be addressed in subsequent face-to-face sessions. The collected data facilitates analysis of how the feedback influences the learning of individual students and it can be used for retraining the neural network. Subsequently the content of the feedback can be improved. Once designed, MCQs and feedbacks can be reused for subsequent cohorts of students.

5 Conclusions and Future Work

Combining two modes of unsupervised learning and self-organisation in one neural network produces results that are not achievable with single modes. Modal learning with up to four modes, combining snap-drift, the delta rule, and an adaptive function method has demonstrated the potential for combining several modes [18]. Future work needs to explore mechanisms for controlling and optimising the selection of modes in real time. Performance guided mode switching has proved effective [25], [35] where a periodic or occasional system performance measure is available. Toggling between modes every epoch works well for two unsupervised modes as illustrated in this chapter. However, with multiple unsupervised modes toggling would not be an option. Borrowing an idea from real-time systems, a queue of modes could be maintained with an abandoned mode passing to the back of the queue. Alternatively, performance measures associated with each mode could be used as the priority values to create a queue with the most recently effective modes at the front of the queue.

Appendix

Matlab Code Snap-Drift

```
%%%%%%%%%%%%%%%%%%%%%%%%%%%%%%%%%%%%%%%%%%%%%%%%%%%%%%%%%%%%%%%%
function sd =
    SnapDrift(input,dlayer,slayer,dnode,threshold,learning_rateD,learning_rateS)
%SnapDrift class constructor
%training patterns matrix
sd.input = input;
%the number of nodes in the d layer
sd.dlayer = dlayer;
%number of nodes in the s layer
sd.slayer = slayer;
%dnode is D, the number of winning nodes in the d layer
sd.dnode = dnode;
%threshold is the quality assurance threshold
sd.threshold = threshold;
sd.learning_rateD = learning_rateD;
sd.learning_rateS = learning_rateS;
[sd.no_pattern, sd.ilayer] = size(sd.input);
```

```
%Initialise and normalise weights between input
%and d layer.
for j = 1:sd.dlayer
    r = floor(rand(1)*sd.no_pattern)+1;
    %initialise weights with randmoly selected patterns
    sd.weights_ilayer_dlayer(:,j) = sd.input(r,:);
end
for j = 1:sd.dlayer
    norm_weight = norm(sd.weights_ilayer_dlayer(:,j));
    sd.weights_ilayer_dlayer(:,j) =
      sd.weights_ilayer_dlayer(:,j)/norm_weight;
end

%Initialise and normlaise weights between d layer
%and s  (output) layer
sd.weights_dlayer_slayer = ones(sd.dlayer,sd.slayer);
for j = 1:sd.slayer
    norm_weight = norm(sd.weights_dlayer_slayer(:,j));
    sd.weights_dlayer_slayer(:,j) =
      sd.weights_dlayer_slayer(:,j)/norm_weight;
end
%Initilalise output layer s (for large D, initialise some weights
%to 1's, with probability D/d)
sd.output = zeros(sd.slayer,1);
end %end function sd

%%%%%%%%%%%%%%%%%%%%%%%%%%%%%%%%%%%%%%%%%%%%%%%%%%%%%%%%%%%%%
function sd = train(sd,epoch)
for q = 1:epoch
   for counter = 1:sd.no_pattern
      dlayer_input = sd.input(counter,:)*sd.weights_ilayer_dlayer;
      %input from dlayer to slayer
      sInput = zeros(sd.dlayer,1);
      %learning between d layer and input layer on D most active nodes
      for s = 1:sd.dnode
          [maxVal, ind] = max(dlayer_input);
          dInput = sd.input(counter,:)';
      sd.weights_ilayer_dlayer =
          snap_drift_learn(sd.learning_rateD,sd.weights_ilayer_dlayer,q,
          ind,dInput);
          dlayer_input(ind) = 0;
          sInput(ind) = 1;
      end %for sd.dnode
      %learning between d layer and s layer
      slayer_input = (sInput)'*sd.weights_dlayer_slayer;
      [maxVal, ind] = max(slayer_input);
      if (maxVal > sd.threshold)
          sd.weights_dlayer_slayer =
             snap_drift_learn(sd.learning_rateS,sd.weights_dlayer_slayer,
             q,ind,sInput);
          sd.output(ind) = 1;
      else
          j = 1; %recruit a new, as yet untrained, s node
          while (j<=sd.slayer & sd.output(j)>0)
             j = j + 1;
          end %end while
          if (j<=sd.slayer)
              sd = snap_drift_ds_layer(sd,q,j,sInput);
              sd.output(j) = 1;
          end
      end %if threshold
   end %for counter
end %for epoch
end %train
```

```
%%%%%%%%%%%%%%%%%%%%%%%%%%%%%%%%%%%%%%%%%%%%%%%%%%%%%%%%%%%%
function weights = snap_drift_learn(lr,weights,q,ind,input)
if (mod(q,2) == 0)
    %adapt weights of the winning node using snap
    weights(:,ind) = min(weights(:,ind),input);
else
    %adapt weights of the winning node using drift
    weights(:,ind) = weights(:,ind) + lr*(input - weights(:,ind));
end %end if
norm_weight = norm(weights(:,ind));
if (norm_weight)>0
    weights(:,ind) = weights(:,ind)/norm_weight;
end
end %snap_drift_learn

%%%%%%%%%%%%%%%%%%%%%%%%%%%%%%%%%%%%%%%%%%%%%%%%%%%%%%%%%%%%
function label(sd,train_class,input_test,test_class,clNum)
%label s nodes based on response to training data data
%Set class frequency matrix of s nodes to 0
committed_class_counter = zeros(sd.slayer,clNum);
%For each pattern increment its class counter for the winning s node
for counter = 1:sd.no_pattern
    dlayer_input_train = sd.input(counter,:)*sd.weights_ilayer_dlayer;
    %input from dlayer to slayer
    sInput = zeros(sd.dlayer,1);
    for s = 1:sd.dnode
        [maxVal, ind] = max(dlayer_input_train);
        dlayer input_train(ind) = 0;
        sInput(ind) = 1;
    end %for sd.dnode
    slayer_input = (sInput)'*sd.weights_dlayer_slayer;
    [maxVal, ind] = max(slayer_input);
    committed_class_counter(ind,train_class(counter)+1) =
            committed_class_counter(ind,train_class(counter)+1)+1;
end %for counter
error_train = 0;
output_map = zeros(sd.slayer,1);
%Find the maximum class counter for each s node and assign that as its
%class in the output_map array
for s = 1:sd.slayer
    [maxVal,ind] = max(committed_class_counter(s,:));
    if (committed_class_counter(s,ind)>0)
        output_map(s) = ind;
    end
    for j=1:clNum
        if (j ~= ind)
            error_train = error_train + committed_class_counter(s,j);
        end
    end
end
% % train error
error_train = error_train/sd.no_pattern;

%label test data
[m,n] = size(input_test);
test_error = 0;
output_test_map = zeros(sd.slayer,1);
committed_class_counter = zeros(sd.slayer,clNum);
%For each pattern in the test set
%find the most active s node with a class identity in the output_map
for counter = 1:m
    dlayer_input_test = input_test(counter,:)*sd.weights_ilayer_dlayer;
    %input from dlayer to slayer
    sInput = zeros(sd.dlayer,1);
    for s = 1:sd.dnode
        [maxVal, ind] = max(dlayer_input_test);
        dlayer_input_test(ind) = 0;
```

```
            sInput(ind) = 1;
        end %for sd.dnode
        slayer_input = (sInput)'*sd.weights_dlayer_slayer;
        [maxVal, ind] = max(slayer_input);
        while (output_map(ind) == 0)
            slayer_input(ind) = 0;
            [maxVal, ind] = max(slayer_input);
        end
        if (output_map(ind) - (test_class(counter)+1))>0
            test_error = test_error + 1;
        end
        output_test_map(ind) = output_map(ind);
    end %for counter
    % % test error
    test_error = test_error/m;
    end %label function


    %%%%%%%%%%%%%%%%%%%%%%%%%%%%%%%%%%%%%%%%%%%%%%%%%%%
    %Sample use of the Snap Drift Matlab code
    %Save all functions in separate files named as the function names
    %
    %load the data (e.g. iris data)
    input_train = load('irisTrain.txt');
    input_test = load('irisTest.txt');
    train_class = load('irisTrainClass.txt');
    test_class = load('irisTrestClass.txt');

    %Instantiate the SnapDrift object with suitable parameter values
    %Set learning constants in the range 0,1
    %D (dnode) is the number of winning dlayer nodes, and therefore
    %the number of features required for categorisation or
    %classification in the s layer
    %The size of the d (dlayer) is normally much greater than D
    %Threshold: because weights are normalised, thresholds of greater than 1
    %tend to force additional s categories.
    %It does not matter if the max number of s nodes is greater
    %than required, snap-drift recruits new s nodes only when required.

    sd = SnapDrift(input_train,50,10,10,2.5,0.1,0.2);

    %call the train method on the sd object
    sd = train(sd,50);
    %call the label method if classification required
    %and classes of training data are available
    classNum = 3; % 3 classes in this case
    label(sd,train_class,input_test,test_class,classNum);
```

References

1. Alpaydin, E., Kaynak, C.: Cascading Classifiers. Kybernetika 34, 369–374 (1998)
2. Burge, P., Shawe-Taylor, J.: An Unsupervised Neural Network Approach to Profiling the Behavior of Mobile Phone Users for Use in Fraud Detection. Journal of Parallel and Distributed Computing 61(7), 915–925 (2001)
3. Carpenter, G.A., Grossberg, S.: Adaptive resonance theory. In: Arbib, M.A. (ed.) The Handbook of Brain Theory and Neural Networks, Second Edition, 2nd edn., pp. 87–90. MIT Press, Cambridge (2003)
4. Dafoulas, G.A.: The role of feedback in online learning communities. In: Fifth IEEE International Conference on Advanced Learning Technologies, pp. 827–831 (2005)

5. Donelan, H., Pattinson, C., Palmer-Brown, D.: The Analysis of User Behaviour of a Network Management Training Tool using a Neural Network. Journal of Systemics, Cybernetics and Informatics 3(5) (2006)
6. Dotan, Y., Intrator, N.: Multimodality exploration by an unsupervised projection pursuit neural network. IEEE Transactions on Neural Networks 9(3), 464–472 (1998)
7. Duda, R.O., Hart, P.E.: Pattern Classification and Scene Analysis, p. 218. John Wiley & Sons, New York (1973)
8. Ekpenyong, F., Palmer-Brown, D., Brimicombe, A.: Extracting road information from recorded GPS data using snap-drift neural network. Neurocomputing 73, 24–36 (2009)
9. Fernandez Aleman, J.L., Palmer-Brown, D., Jayne, C.: Effects of Response Driven Feedback in Computer Science Learning. IEEE Transactions on Education 99 (2010), doi:10.1109/TE.2010.2087761
10. Fisher, R.A.: The use of multiple measurements in taxonomic problems. Annual Eugenics 7, Part II, 179–188 (1936); also in Contributions to Mathematical Statistics. John Wiley, NY (1950)
11. Forina, M., Lanteri, S., Armanino, C., et al.: PARVUS–an extendible package for data exploration, classification and correlation. Institute of Pharmaceutical and Food Analysis and Technologies, Via Brigata Salerno, 16147 Genoa, Italy (1991)
12. Garside, R., Leech, G., Varadi, T.: Manual of Information to Accompany the Lancaster Parsed Corpus. University of Oslo (1987)
13. Gupta, L., McAvoy, M.: Investigating the prediction capabilities of the simple recurrent neural network on real temporal sequences. Pattern Recognition 33(i12), 2075–2081 (2000)
14. Hebb, D.O.: The organization of behavior. Wiley & Sons, New York (1949)
15. Higgins, E., Tatham, L.: Exploring the potential of Multiple Choice Questions in Assessment. Learning & Teaching in Action 2(1) (2003)
16. Horton, P., Nakai, K.: A Probablistic Classification System for Predicting the Cellular Localization Sites of Proteins. Intelligent Systems in Molecular Biology, 109–115 (1996)
17. Kang, M., Palmer-Brown, D.: A Multilayer ADaptive FUnction Neural Network (MADFUNN) for Analytical Function Recognition. IJCNN (2006); part of the IEEE World Congress on Computational Intelligence, WCCI 2006, Vancouver, BC, Canada, pp. 1784–1789 (2006)
18. Kang, M., Palmer-Brown, D.: A Modal Learning Adaptive Function Neural Network Applied to Handwritten Digit Recognition. Information Sciences 178(20), 3802–3812 (2008)
19. Kaynak, C.: Methods of Combining Multiple Classifiers and Their Applications to Handwritten Digit Recognition. MSc Thesis, Institute of Graduate Studies in Science and Engineering, Bogazici University (1995)
20. Kohonen, T.: Self-organised formation of topologically correct feature maps. Biological Cybernetics 43 (1982)
21. Kohonen, T.: Learning Vector Quantisation. Helsinki University of Technology, Laboratory of Computer and Information Science, Report TKK-F-A-601 (1986)
22. Kohonen, T.: Learning Vector Quantisation. Neural Networks 1, 303 (1988)
23. Kohonen, T.: Self-Organisation and Asssociative Memory, 3rd edn. Springer, Heilderberg (1989)
24. Kohonen, T.: Improved Versions of Learning Vector Quantization. In: Proc. of IJCNN 1990, Washington, DC, vol. 1, pp. 545–550 (1990)
25. Lee, S.W., Palmer-Brown, D., Roadknight, C.M.: Performance guided Neural Network for Rapidly Self Organising Active Network Management. Neurocomputing 61C, 5–20 (2004a)

26. Lee, S.W., Palmer-Brown, D., Roadknight, C.M.: Reinforced Snap Drift Learning for Proxylet Selection in Active Computer Networks. In: Proc. of IJCNN 2004, Budapest, Hungary, vol. 2, pp. 1545–1550 (2004b)

27. Lee, S.W., Palmer-Brown, D.: Snap-drift learning for phrase recognition. In: Proc. IEEE IJCNN 2005, Montréal, Québec, Canada, vol. 1, pp. 588–592 (2005)

28. Lee, S.W., Palmer-Brown, D.: Phonetic Feature Discovery in Speech Using *Snap-Drift* Learning. In: Kollias, S.D., Stafylopatis, A., Duch, W., Oja, E. (eds.) ICANN 2006, Part II. LNCS, vol. 4132, pp. 952–962. Springer, Heidelberg (2006)

29. Mayberry III, M.R., Miikkulainen, R.: SARDSRN: A Neural Network ShiftReduce Parser. In: Proc. of the 16th IJCAI, Stockholm, Sweden, pp. 820–825 (1999)

30. MacKay, D.J.C.: Bayesian methods for supervised neural networks. In: Arbib, M.A. (ed.) The Handbook of Brain Theory and Neural Networks, pp. 144–149. MIT Press, Cambridge (1998)

31. Palmer-Brown, D., Tepper, J., Powell, H.: Connectionist Natural Language Parsing. Trends in Cognitive Sciences 6(10), 437–442 (2002)

32. Palmer-Brown, D., Draganova, C., Lee, S.W.: Snap-Drift Neural Network for Selecting Student Feedback. In: Proc. IJCNN 2009, Atlanta, USA, pp. 391–398 (2009)

33. Palmer-Brown, D., Lee, S.W., Draganova, C., Kang, M.: Modal Learning Neural Networks. WSEAS Transactions on Computers 8(2), 222–236 (2009)

34. Palmer-Brown, D., Draganova, C.: Snap-Drift Self Organising Map. In: Diamantaras, K., Duch, W., Iliadis, L.S. (eds.) ICANN 2010, Part II. LNCS, vol. 6353, pp. 402–409. Springer, Heidelberg (2010)

35. Palmer-Brown, D., Draganova, C.: Recurrent Snap Drift Neural Network for Phrase Recognition. In: WCCI 2010 IEEE World Congress on Computational Intelligence, IJCNN 2010, Barcelona, Spain, pp. 3445–3449 (2010)

36. Payne, A., Brinkman, W.-P., Wilson, F.: Towards Effective Feedback in e-Learning Packages: The Design of a Package to Support Literature Searching, Referencing and Avoiding Plagiarism. In: Proceedings of HCI 2007 Workshop: Design, Use and Experience of e-Learning Systems, pp. 71–75 (2007)

37. Ritter, H., Kohonen, T.: Self-Organizing Semantic Maps. Biological Cybernetics 61, 241–254 (1989)

38. Rumelhart, D.E., Hinton, G.E., Williams, R.J.: Learning representations by back-propagation errors. Nature 323, 533–536 (1986)

39. Rushton, J.N.: Natural Language Parsing using Simple Neural Networks. In: Proc. of MLMTA, Las Vegas, Nevada, pp. 137–141 (2003)

40. Tepper, J., Powell, H.M., Palmer-Brown, D.: A Corpus based Connectionist Architecture for Large scale Natural Language Parsing. Connection Science 14(2), 93–114 (2002)

41. Vapnik, V.: The Nature of Statistical Learning Theory. Springer, New York (1995)

42. Webb, A.R., Lowe, D.: A hybrid optimisation strategy for adaptive feed-forward layered newtorks. RSRE Memorandum 4193, Royal Signals and Radar Establishemnt, St Andrews Road, Malvern, UK (1988)

43. Widrow, B., Hoff, M.E.: Institute of Radio Engineers WESCON Convention Record, Adaptive switching circuits, pp. 96–104. Institute of Radio Engineers, New York (1960)

Chapter 12
Bayesian Networks, Introduction and Practical Applications

Wim Wiegerinck, Willem Burgers, and Bert Kappen

Abstract. In this chapter, we will discuss Bayesian networks, a currently widely accepted modeling class for reasoning with uncertainty. We will take a practical point of view, putting emphasis on modeling and practical applications rather than on mathematical formalities and the advanced algorithms that are used for computation. In general, Bayesian network modeling can be data driven. In this chapter, however, we restrict ourselves to modeling based on domain knowledge only. We will start with a short theoretical introduction to Bayesian networks models and inference. We will describe some of the typical usages of Bayesian network models, e.g. for reasoning and diagnostics; furthermore, we will describe some typical network behaviors such as the explaining away phenomenon, and we will briefly discuss the common approach to network model design by causal modeling. We will illustrate these matters by a detailed modeling and application of a toy model for medical diagnosis. Next, we will discuss two real-world applications. In particular we will discuss the modeling process in some details. With these examples we also aim to illustrate that the modeling power of Bayesian networks goes further than suggested by the common textbook toy applications. The first application that we will discuss is for victim identification by kinship analysis based on DNA profiles. The distinguishing feature in this application is that Bayesian networks are generated and computed on-the-fly, based on case information. The second one is an application for petrophysical decision support to determine the mineral content of a well based on borehole measurements. This model illustrates the possibility to model with continuous variables and nonlinear relations.

1 Introduction

In modeling intelligent systems for real world applications, one inevitably has to deal with uncertainty. This uncertainty is due to the impossibility to model all the

Wim Wiegerinck · Willem Burgers · Bert Kappen
SNN Adaptive Intelligence, Donders Institute for Brain, Cognition and Behaviour,
Radboud University Nijmegen, Geert Grooteplein 21, 6525 EZ Nijmegen, The Netherlands
e-mail: {w.wiegerinck,w.burgers,b.kappen}@science.ru.nl

M. Bianchini et al. (Eds.): *Handbook on Neural Information Processing*, ISRL 49, pp. 401–431.
DOI: 10.1007/978-3-642-36657-4_12   © Springer-Verlag Berlin Heidelberg 2013

different conditions and exceptions that can underlie a finite set of observations. Probability theory provides the mathematically consistent framework to quantify and to compute with uncertainty. In principle, a probabilistic model assigns a probability to each of its possible states. In models for real world applications, the number of states is so large that a sparse model representation is inevitable. A general class with a representation that allows modeling with many variables are the Bayesian networks [26, 18, 7].

Bayesian networks are nowadays well established as a modeling tool for expert systems in domains with uncertainty [27]. Reasons are their powerful but conceptually transparent representation for probabilistic models in terms of a network. Their graphical representation, showing the conditional independencies between variables, is easy to understand for humans. On the other hand, since a Bayesian network uniquely defines a joint probability model, inference — drawing conclusions based on observations — is based on the solid rules of probability calculus. This implies that the mathematical consistency and correctness of inference are guaranteed. In other words, all assumptions in the method are contained in model, i.e., the definition of variables, the graphical structure, and the parameters. The method has no hidden assumptions in the inference rules. This is unlike other types of reasoning systems such as e.g., Certainty Factors (CFs) that were used in e.g., MYCIN — a medical expert system developed in the early 1970s [29]. In the CF framework, the model is specified in terms of a number of if-then-else rules with certainty factors. The CF framework provides prescriptions how to invert and/or combine these if-then-else rules to do inference. These prescriptions contain implicit conditional independence assumptions which are not immediately clear from the model specification and have consequences in their application [15].

Probabilistic inference is the problem of computing the posterior probabilities of unobserved model variables given the observations of other model variables. For instance in a model for medical diagnoses, given that the patient has complaints x and y, what is the probability that he/she has disease z? Inference in a probabilistic model involves summations or integrals over possible states in the model. In a realistic application the number of states to sum over can be very large. In the medical example, the sum is typically over all combinations of unobserved factors that could influence the disease probability, such as different patient conditions, risk factors, but also alternative explanations for the complaints, etc. In general these computations are intractable. Fortunately, in Bayesian networks with a sparse graphical structure and with variables that can assume a small number of states, efficient inference algorithms exists such as the junction tree algorithm [18, 7].

The specification of a Bayesian network can be described in two parts, a qualitative and a quantitative part. The qualitative part is the graph structure of the network. The quantitative part consists of specification of the conditional probability tables or distributions. Ideally both specifications are inferred from data [19]. In practice, however, data is often insufficient even for the quantitative part of the specification. The alternative is to do the specification of both parts by hand, in collaboration with domain experts. Many Bayesian networks are created in this way. Furthermore, Bayesian networks are often developed with the use of software packages such as Hugin (www.hugin.com), Netica (www.norsys.com) or Bayes-Builder (www.snn.ru.nl). These packages typically contain a graphical user interface (GUI) for modeling and an inference engine based on the junction tree algorithm for computation.

We will discuss in some detail a toy application for respiratory medicine that is modeled and inferred in this way. The main functionality of the application is to list the most probable diseases given the patient-findings (symptoms, patient background revealing risk factors) that are entered. The system is modeled on the basis of hypothetical domain knowledge. Then, it is applied to hypothetical cases illustrating the typical reasoning behavior of Bayesian networks.

Although the networks created in this way can be quite complex, the scope of these software packages obviously has its limitations. In this chapter we discuss two real-world applications in which the standard approach to Bayesian modeling as outlined above was infeasible for different reasons: the need to create models on-the-fly for the data at hand in the first application and the need to model continuous-valued variables in the second one.

The first application is a system to support victim identification by kinship analysis based on DNA profiles (Bonaparte, in collaboration with NFI). Victims should be matched with missing persons in a pedigree of family members. In this application, the model follows from Mendelian laws of genetic inheritance and from principles in DNA profiling. Inference needs some preprocessing but is otherwise reasonably straightforward. The graphical model structure, however, depends on the family structure of the missing person. This structure will differ from case to case and a standard approach with a static network is obviously insufficient. In this application, modeling is implemented in the engine. The application generates Bayesian networks on-the-fly based on case information. Next, it does the required inferences for the matches.

The second model has been developed for an application for petrophysical decision support (in collaboration with SHELL E&P). The main function of this application is to provide a probability distribution of the mineral composition of a potential reservoir based on remote borehole measurements. In this model, variables are continuous valued. One of them represents the volume fractions of 13 minerals, and is therefore a 13-D continuous variable. Any sensible discretization in a standard Bayesian network approach would lead to a blow up of the state space. Due to non-linearities and constraints, a Bayesian network with linear-Gaussian distributions [3] is also not a solution.

The chapter is organized as follows. First, we will provide a short introduction to Bayesian networks in section 2. In the next section we will discuss in detail modeling basics and the typical application of probabilistic reasoning in the medical toy model. Next, in sections 4 and 5 we will discuss the two real-world applications. In these chapters, we focus on the underlying Bayesian network models and the modeling approaches. We will only briefly discuss the inference methods that were applied whenever they deviate from the standard junction tree approach. In section 6, we will end with discussion and conclusion.

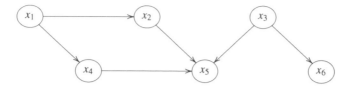

Fig. 1 DAG representing a Bayesian network $P(x_1)P(x_2|x_1)P(x_3)P(x_4|x_1)P(x_5|x_2,x_3,x_4)$
$P(x_6|x_3)$

2   Bayesian Networks

In this section, we first give a short and rather informal review of the theory of
Bayesian networks (subsection 2.1). Furthermore in subsection 2.2, we briefly discuss Bayesian networks modeling techniques, and in particular the typical practical
approach that is taken in many Bayesian network applications.

2.1   Bayesian Network Theory

To introduce notation, we start by considering a joint probability distribution, or
probabilistic model, $P(X_1,\ldots,X_n)$ of n stochastic variables $X_1,\ldots,X_n$. Variables X_j
can be in state x_j. A state, or value, is a realization of a variable. We use shorthand
notation

$$P(x_1,\ldots,x_n) = P(X_1 = x_1,\ldots,X_n = x_n) \tag{1}$$

to denote the probability (in continuous domains: the probability density) of variables X_1 in state x_1, variable X_2 in state x_2 etc.

A Bayesian network is a probabilistic model P on a finite directed acyclic graph
(DAG). For each node i in the graph, there is a random variable X_i together with a
conditional probability distribution $P(x_i|x_{\pi(i)})$, where $\pi(i)$ are the parents of i in the
DAG, see figure 1. The joint probability distribution of the Bayesian network is the
product of the conditional probability distributions

$$P(x_1,\ldots,x_n) = \prod_{i=1}^{n} P(x_i|x_{\pi(i)}) \,. \tag{2}$$

Since any DAG can be ordered such that $\pi(i) \subseteq 1,\ldots i-1$ and any joint distribution
can be written as

$$P(x_1,\ldots,x_n) = \prod_{i=1}^{n} P(x_i|x_{i-1},\ldots,x_1) \,, \tag{3}$$

it can be concluded that a Bayesian network assumes

$$P(x_i|x_{i-1},\ldots,x_1) = P(x_i|x_{\pi(i)}) \,. \tag{4}$$

In other words, the model assumes: given the values of the direct parents of a variable X_i, this variable X_i is independent of all its other preceding variables in the graph.

Since a Bayesian network is a probabilistic model, one can compute marginal distributions and conditional distributions by applying the standard rules of probability calculus. For instance, in a model with discrete variables, the marginal distribution of variable X_i is given by

$$P(x_i) = \sum_{x_1} \cdots \sum_{x_{i-1}} \sum_{x_{i+1}} \cdots \sum_{x_N} P(x_1, \ldots, x_N) . \tag{5}$$

Conditional distributions such as $P(x_i|x_j)$ are obtained by the division of two marginal distributions

$$P(x_i|x_j) = \frac{P(x_i, x_j)}{P(x_j)} . \tag{6}$$

The bottleneck in the computation is the sum over combinations of states in (5). The number of combinations is exponential in the number of variables. A straightforward computation of the sum is therefore only feasible in models with a small number of variables. In sparse Bayesian networks with discrete variables, efficient algorithms that exploit the graphical structure, such as the junction tree algorithm [21, 18, 7] can be applied to compute marginal and conditional distributions. In more general models, exact inference is infeasible and approximate methods such as sampling have to be applied [22, 3].

2.2 Bayesian Network Modeling

The construction of a Bayesian network consists of deciding about the domain, what are the variables that are to be modeled, and what are the state spaces of each of the variables. Then the relations between the variables have to be modeled. If these are to be determined by hand (rather than by data), it is a good rule of thumb to construct a Bayesian network from cause to effect. Start with nodes that represent independent root causes, then model the nodes which they influence, and so on until we end at the leaves, i.e., the nodes that have no direct influence on other nodes. Such a procedure often results in sparse network structures that are understandable for humans [27].

Sometimes this procedure fails, because it is unclear what is cause and what is effect. Is someone's behavior an effect of his environment, or is the environment a reaction on his behavior? In such a case, just avoid the philosophical dispute, and return to the basics of Bayesian networks: a Bayesian network is not a model for causal relations, but a joint probability model. The structure of the network represents the conditional independence assumptions in the model and nothing else.

A related issue is the decision whether two nodes are really (conditionally) independent. Usually, this is a matter of simplifying model assumptions. In the true world, all nodes should be connected. In practice, reasonable (approximate) assumptions are needed to make the model simple enough to handle, but still powerful enough for practical usage.

When the variables, states, and graphical structure is defined, the next step is to determine the conditional probabilities. This means that for each variable x_i, the conditional probabilities $P(x_i|x_{\pi(i)})$ in eqn. (4) have to be determined. In case of a finite number of states per variable, this can be considered as a table of $(|x_i| - 1) \times |x_{\pi(i)}|$ entries between 0 and 1, where $|x_i|$ is the number of states of variable x_i and $|x_{\pi(i)}| = \prod_{j \in \pi(i)} |x_j|$ the number of joint states of the parents. The -1 term in the $(|x_i| - 1)$ factor is due to the normalization constraint $\sum_{x_i} P(x_i|x_{\pi(i)}) = 1$ for each parent state. Since the number of entries is linear in the number of states of the variables and exponential in the number of parent variables, a too large state space as well as a too large number of parents in the graph makes modeling practically infeasible.

The entries are often just the result of educated guesses. They may be inferred from data, e.g. by counting frequencies of joint occurrences of variables in state x_i and parents in states $x_{\pi(i)}$. For reliable estimates, however, one should have sufficiently many data for each joint state $(x_i, x_{\pi(i)})$. So in this approach one should again be careful not to take state space and/or number of parents too large. A compromise is to assume a parametrized tables. A popular choice for binary variables is the noisy-OR table [26]. The table parametrization can be considered as an educated guess. The parameters may then be estimated from data.

Often, models are constructed using Bayesian network software such as the earlier mentioned software packages. With the use of a graphical user interface (GUI), nodes can be created. Typically, nodes can assume only values from a finite set. When a node is created, it can be linked to other nodes, under the constraint that there are no directed loops in the network. Finally — or during this process — the table of conditional probabilities are defined, manually or from data as mentioned above. Many Bayesian networks that are found in literature fall into this class, see e.g., www.norsys.com/netlibrary/. In figure 2, a part of the ALARM network as represented in BayesBuilder (www.snn.ru.nl/) is plotted. The ALARM network was originally designed as a network for monitoring patients in intensive care [2]. It consists of 37 variables, each with 2, 3, or 4 states. It can be considered as a relatively large member of this class of models. An advantage of the GUI based approach is that a small or medium sized Bayesian network, i.e., with up to a few dozen of variables, where each variable can assume a few states, can be developed quickly, without the need of expertise on Bayesian networks modeling or inference algorithms.

3   An Example Application: Medical Diagnosis

In this section we will consider the a Bayesian network for medical diagnosis of the respiratory system. This is model is inspired on the famous 'ASIA network' described in [21].

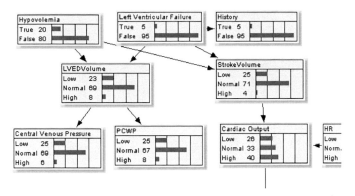

Fig. 2 Screen shot of part of the 'Alarm network' in the BayesBuilder GUI

3.1 Modeling

We start by considering the the following piece of qualitative 'knowledge':

> The symptom *dyspnoea* (shortness of breath) may be due to the diseases *pneumonia*, *lung cancer*, and/or *bronchitis*. Patients with *pneumonia*, and/or *bronchitis* often have a very nasty *wet coughing*. Pneumonia, and/or *lung cancer* are often accompanied by a heavy *chest pain*. Pneumonia is often causing a severe *fever*, but this may also be caused by a *common cold*. However, a *common cold* is often recognized by a *runny nose*. Sometimes, *wet coughing*, *chest pain*, and/or *dyspnoea* occurs unexplained, or are due to another cause, without any of these diseases being present. Sometimes diseases co-occur. A *weakened immune-system* (for instance, homeless people, or HIV infected) increases the probability of getting an *pneumonia*. Also, *lung cancer* increases this probability. *Smoking* is a serious risk factor for *bronchitis* and for *lung cancer*.

Now to build a model, we first have to find out which are the variables. In the text above, these are the ones printed in *italics*. In a realistic medical application, one may want to model multi-state variables. For simplicity, however, we take in this example all variables binary (true/false). Note that by modeling diseases as separate variables rather than by mutually exclusive states in a single disease variable, the model allows diseases to co-occur.

The next step is to figure out a sensible graphical structure. In the graphical representation of the model, i.e., in the DAG, all these variables are represented by nodes. The question now is which arrows to draw between the nodes. For this, we will use the principle of causal modeling. We derive these from the 'qualitative knowledge' and some common sense. The general causal modeling assumption in this medical domain is that risk factors 'cause' the diseases, so risk factors will point to diseases, and diseases 'cause' symptoms, so diseases will point to symptoms.

We start by modeling risk factors and diseases. Risk factors are *weakened immune-system* (for *pneumonia*), *smoking* (for *bronchitis* and for *lung cancer*), and *lung cancer* (also for *pneumonia*). The nodes for *weakened immune-system* and

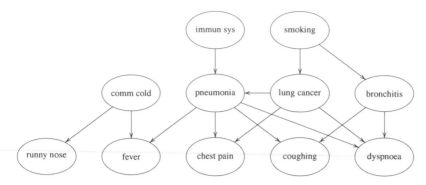

Fig. 3 DAG for the respiratory medicine toy model. See text for details.

smoking have no incoming arrows, since there are no explicit causes for these variables in the model. We draw arrows from these node to the diseases for which they are risk factors. Furthermore, we have a node for the disease *common cold*. This node has no incoming arrow, since no risk factor for this variable is modeled.

Next we model the symptoms. The symptom *dyspnoea* may be due to the diseases *pneumonia*, *lung cancer*, and/or *bronchitis*, so we draw an arrow from all these diseases to *dyspnoea*. In a similar way, we can draw arrows from *pneumonia*, and *bronchitis* to *wet coughing*; arrows from *pneumonia*, and *lung cancer* to *chest pain*; arrows from *pneumonia* and *common cold* to *fever*; and an arrow from *common cold* to *runny nose*. This completes the DAG, which can be found in figure 3. (In the figures and in some of the text in the remainder of the section we abbreviated some of the variable names, e.g. we used *immun sys* instrad of *weakened immune-system*, etc.)

The next step is the quantitative part, i.e., the determination of the conditional probability tables. The numbers that we enter are rather arbitrary guesses and we do not pretend them to be anyhow realistic. In determining the conditional probabilities, we used some modeling assumptions such as that the probability of a symptom in the presence of an additional causing diseases is at least as high as the probability of that symptom in the absence of that disease. The tables as presented in figure 4. In these tables, the left column represents the probability values in the true state, $P(variablename) \equiv P(variablename = true)$, so $P(variablename = false) = 1 - P(variablename)$. The other columns indicate the joint states of the parent variables.

3.2 Reasoning

Now that the model is defined, we can use it for reasoning, e.g. by entering observational evidence into the system and doing inference, i.e. computing conditional probabilities given this evidence. To do the computation, we have modeled the system in BayesBuilder. In figure 5 we show a screen shot of the Bayesian network as modeled in BayesBuilder. The program uses the junction tree inference algorithm to

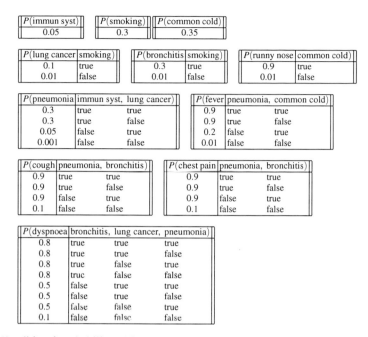

P(immun syst)		P(smoking)		P(common cold)
0.05		0.3		0.35

P(lung cancer	smoking)		P(bronchitis	smoking)		P(runny nose	common cold)
0.1	true		0.3	true		0.9	true
0.01	false		0.01	false		0.01	false

P(pneumonia	immun syst,	lung cancer)		P(fever	pneumonia,	common cold)
0.3	true	true		0.9	true	true
0.3	true	false		0.9	true	false
0.05	false	true		0.2	false	true
0.001	false	false		0.01	false	false

P(cough	pneumonia,	bronchitis)		P(chest pain	pneumonia,	bronchitis)
0.9	true	true		0.9	true	true
0.9	true	false		0.9	true	false
0.9	false	true		0.9	false	true
0.1	false	false		0.1	false	false

P(dyspnoea	bronchitis,	lung cancer,	pneumonia)
0.8	true	true	true
0.8	true	true	false
0.8	true	false	true
0.8	truc	false	false
0.5	false	true	true
0.5	false	true	false
0.5	false	false	true
0.1	false	false	false

Fig. 4 Conditional probability tables parametrizing the respiratory medicine toy model. The numbers in the nodes represent the marginal probabilities of the variables in state 'true'. See text for details.

compute the marginal node probabilities and displays them on screen. The marginal node probabilities are the probability distributions of each of the variables in the absence of any additional evidence. In the program, evidence can be entered by clicking on a state of the variable. This procedure is sometimes called 'clamping'. The node probabilities will then be conditioned on the clamped evidence. With this, we can easily explore how the models reasons.

3.2.1 Knowledge Representation

Bayesian networks may serve as a rich knowledge base. This is illustrated by considering a number of hypothetical medical guidelines and comparing these with Bayesian network inference results. These results will also serve to comment on some of the typical behavior in Bayesian networks.

1. In case of high *fever* in absence of a *runny nose*, one should consider *pneumonia*.

   Inference.   We clamp *fever* = *true* and *runny nose* = *false* and look at the conditional probabilities of the four diseases. We see that in particular the probability of pneumonia is increased from about 2% to 45%. See figure 6.

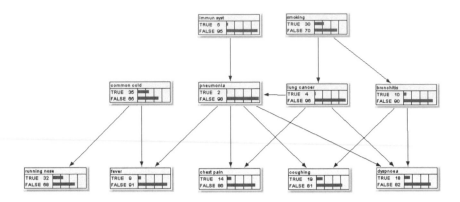

Fig. 5 Screen shot of the respiratory medicine toy model in the BayesBuilder GUI. Red bars present marginal node probabilities.

>   Comment.   There are two causes in the model for *fever*, namely has parents *pneumonia* and *common cold*. However, the absence of a *common cold* makes *common cold* less likely. This makes the other explaining cause *pneumonia* more likely.

2. *Lung cancer* is often found in patients with *chest pain*, *dyspnoea*, no *fever*, and usually no *wet coughing*.

>   Inference.   We clamp *chest pain* = *true*, *dyspnoea* = *true*, *fever* = *false*, and *coughing* = *false* We see that probability of *lung cancer* is raised 0.57. Even if we set *coughing* = *true*, the probability is still as high as 0.47.
>   Comment.   *Chest pain* and *dyspnoea* can both be caused by *lung cancer*. However, *chest pain* for example, can also be caused by *pneumonia*. The absence of in particular *fever* makes *pneumonia* less likely and therefore *lung cancer* more likely. To a lesser extend this holds for absence of *coughing* and *bronchitis*.

3. *Bronchitis* and *lung cancer* are often accompanied, e.g patients with *bronchitis* often develop a *lung cancer* or vice versa. However, these diseases have no known causal relation, i.e., *bronchitis* is not a cause of *lung cancer*, and *lung cancer* is not a cause of *bronchitis*.

>   Inference.   According to the model, $P(lung\ cancer|bronchitis = true) = 0.09$ and $P(bronchitis|lung\ cancer = true) = 0.25$. Both probabilities are more than twice the marginal probabilities (see figure 3).
>   Comment.   Both diseases have the same common cause: *smoking*. If the state of smoking is observed, the correlation is broken.

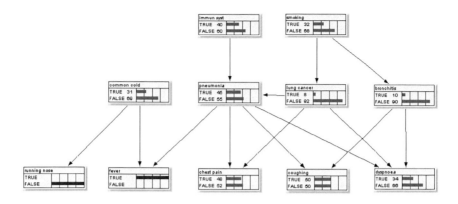

Fig. 6 Model in the state representing medical guideline 1, see main text. Red bars present conditional node probabilities, conditioned on the evidence (blue bars).

3.2.2 Diagnostic Reasoning

We can apply the system for diagnosis. the idea is to enter the patient observations, i.e. symptoms and risk factors into the system. Then diagnosis (i.e. finding the cause(s) of the symptoms) is done by Bayesian inference. In the following, we present some hypothetical cases, present the inference results and comment on the reasoning by the network.

1. Mr. Appelflap calls. He lives with his wife and two children in a nice little house in the suburb. You know him well and you have good reasons to assume that he has no risk of a weakened immune system. Mr. Appelflap complains about high fever and a nasty wet cough (although he is a non-smoker). In addition, he sounds rather nasal. What is the diagnosis?

   Inference.   We clamp the risk factors *immun sys = false, smoking = false* and the symptoms *fever = true, runny nose = true*. We find all disease probabilities very small, except *common cold*, which is almost certainly true.

   Comment.   Due to the absence of risk factors, the prior probabilities of the other diseases that could explain the symptoms is very small compared to the prior probability of *common cold*. Since *common cold* also explains all the symptoms, that disease takes all the probability of the other causes. This phenomenon is called 'explaining away': pattern of reasoning in which the confirmation of one cause (*common cold*, with a high prior probability and confirmed by *runny nose*) of an observed event (*fever*) reduces the need to invoke alternative causes (*pneumonia* as an explanation of *fever*).

2. The salvation army calls. An unknown person (looking not very well) has arrived in their shelter for homeless people. This person has high fever, a nasty wet cough (and a runny nose.) What is the diagnosis?

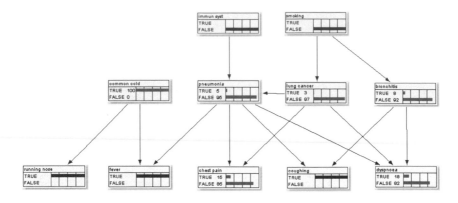

Fig. 7 Diagnosing mr. Appelflap. Primary diagnosis: *common cold*. See main text.

Inference.   We suspect a weakened immune system, so the system we clamp
the risk factor *immun sys* = *true*. As in the previous case, the symptoms are
fever = *true*, *runny nose* = *true*. However, now we not only find *common cold*
with a high probability ($P = 0.98$), but also *pneumonia* ($P = 0.91$).

Comment.   Due to the fact that with a *weakened immune system*, the prior probability of *pneumonia* is almost as high as the prior probability of *common cold*.
Therefore the conclusion is very different from the previous cas. Note that for
this diagnosis, it is important that diseases can co-occur in the model.

3. A patient suffers from a recurrent pneumonia. This patient is a heavy smoker but
otherwise leads a 'normal', healthy live, so you may assume there is no risk of a
weakened immune system. What is your advice?

Inference.   We clamp *immun sys* = *false*, *smoking* = *true*, and *pneumonia* =
true. As a result, we see that there is a high probability of *lung cancer*.

Comment.   The reason is that due to *smoking*, the prior of disease is increased.
More importantly, however, is that *weakened immune system* is excluded as
cause of the *pneumonia*, so that *lung cancer* remains as the most likely explanation of the cause of the recurrent *pneumonia*.

3.3   Discussion

With the toy model, we aimed to illustrate the basic principles of Bayesian network modeling. With the inference examples, we have aimed to demonstrate some
of typical reasoning capabilities of Bayesian networks. One features of Bayesian
networks that distinguish them from e.g. conventional feedforward neural networks
is that reasoning is in arbitrary direction, and with arbitrary evidence. Missing data

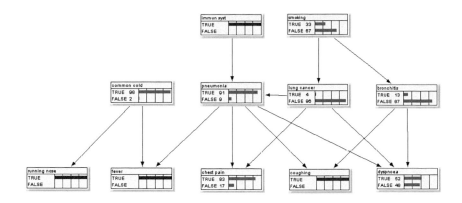

Fig. 8 Salvation army case. Primary diagnosis: *pneumonia*. See main text.

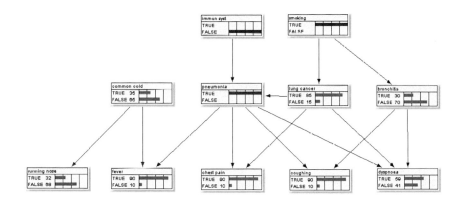

Fig. 9 Recurrent pneumonia case. Primary diagnosis: *lung cancer*. See main text.

or observations are dealt with in a natural way by probabilistic inference. In many applications, as well as in the examples in this section, the inference question is to compute conditional node probabilities. These are not the only quantities that one could compute in a Bayesian networks. Other examples are are correlations between variables, the probability of the joint state of the the nodes, or the entropy of a conditional distribution. Applications of the latter two will be discussed in the next sections.

In the next sections we will discuss two Bayesian networks for real world applications. The modeling principles are basically the same as in the toy model described in this section. There are some differences, however. In the first model, the network consists of a few types of nodes that have simple and well defined relations among each other. However, for each different case in the application, a different network has to be generated. It does not make sense for this application to try to build these networks beforehand in a GUI. In the second one the complexity is more in the variables themselves than in the network structure. Dedicated software has been written for both modeling and inference.

4 Bonaparte: A Bayesian Network for Disaster Victim Identification

Society is increasingly aware of the possibility of a mass disaster. Recent examples are the WTC attacks, the tsunami, and various airplane crashes. In such an event, the recovery and identification of the remains of the victims is of great importance, both for humanitarian as well as legal reasons. Disaster victim identification (DVI), i.e., the identification of victims of a mass disaster, is greatly facilitated by the advent of modern DNA technology. In forensic laboratories, DNA profiles can be recorded from small samples of body remains which may otherwise be unidentifiable. The identification task is the match of the unidentified victim with a reported missing person. This is often complicated by the fact that the match has to be made in an indirect way. This is the case when there is no reliable reference material of the missing person. In such a case, DNA profiles can be taken from relatives. Since their profiles are statistically related to the profile of the missing person (first degree family members share about 50% of their DNA) an indirect match can be made.

In cases with one victim, identification is a reasonable straightforward task for forensic researchers. In the case of a few victims, the puzzle to match the victims and the missing persons is often still doable by hand, using a spread sheet, or with software tools available on the internet [10]. However, large scale DVI is infeasible in this way and an automated routine is almost indispensable for forensic institutes that need to be prepared for DVI.

Bayesian networks are very well suited to model the statistical relations of genetic material of relatives in a pedigree [12]. They can directly be applied in kinship analysis with any type of pedigree of relatives of the missing persons. An additional advantage of a Bayesian network approach is that it makes the analysis tool more transparent and flexible, allowing to incorporate other factors that play a role — such as measurement error probability, missing data, statistics of more advanced genetic markers etc.

Recently, we have developed software for DVI, called Bonaparte. This development is in collaboration with NFI (Netherlands Forensic Institute). The computational engine of Bonaparte uses automatically generated Bayesian networks and Bayesian inference methods, enabling to correctly do kinship analysis on the basis of DNA profiles combined with pedigree information. It is designed to handle large scale events, with hundreds of victims and missing persons. In addition, it has graphical user interface, including a pedigree editor, for forensic analysts. Data-interfaces to other laboratory systems (e.g., for the DNA-data input) will also be implemented.

In the remainder of this section we will describe the Bayesian model approach that has been taken in the development of the application. We formulate the computational task, which is the computation of the likelihood ratio of two hypotheses.

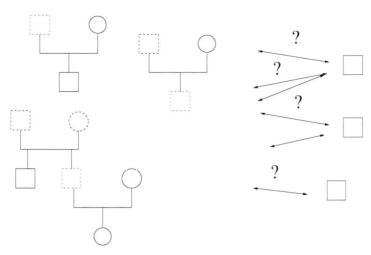

Fig. 10 The matching problem. Match the unidentified victims (blue, right) with reported missing persons (red, left) based on DNA profiles of victims and relatives of missing persons. DNA profiles are available from individuals represented by solid squares (males) and circles (females).

The main ingredient is a probabilistic model P of DNA profiles. Before discussing the model, we will first provide a brief introduction to DNA profiles. In the last part of the section we describe how P is modeled as a Bayesian network, and how the likelihood ratio is computed.

4.1 Likelihood Ratio of Two Hypotheses

Assume we have a pedigree with an individual MP who is missing (the Missing Person). In this pedigree, there are some family members that have provided DNA material, yielding the profiles. Furthermore there is an Unidentified Individual UI, whose DNA is also profiled. The question is, is $UI - MP$? To proceed, we assume that we have a probabilistic model P for DNA evidence of family members in a pedigree. To compute the probability of this event, we need hypotheses to compare. The common choice is to formulate two hypotheses. The first is the hypothesis H_1 that indeed $UI = MP$. The alternative hypothesis H_0 is that UI is an unrelated person U. In both hypotheses we have two pedigrees: the first pedigree has MP and family members FAM as members. The second one has only U as member. To compare the hypotheses, we compute the likelihoods of the evidence from the DNA profiles under the two hypotheses,

- Under H_p, we assume that $MP = UI$. In this case, MP is observed and U is unobserved. The evidence is $E = \{DNA_{MP} + DNA_{FAM}\}$.
- Under H_d, we assume that $U = UI$. In this case, U is observed and MP is unobserved. The evidence is $E = \{DNA_U + DNA_{FAM}\}$.

Under the model P, the likelihood ratio of the two hypotheses is

$$LR = \frac{P(E|H_p)}{P(E|H_d)} .$$
(7)

If in addition a prior odds $P(H_p)/P(H_d)$ is given, the posterior odds $P(H_p|E)/P(H_d|E)$ follows directly from multiplication of the prior odds and likelihood ratio,

$$\frac{P(H_p|E)}{P(H_d|E)} = \frac{P(E|H_p)P(H_p)}{P(E|H_d)P(H_d)} .$$
(8)

4.2 DNA Profiles

In this subsection we provide a brief introduction on DNA profiles for kinship analysis. A comprehensive treatise can be found in e.g. [6]. In humans, DNA found in the nucleus of the cell is packed on chromosomes. A normal human cell has 46 chromosomes, which can be organized in 23 pairs. From each pair of chromosomes, one copy is inherited from father and the other copy is inherited from mother. In 22 pairs, chromosomes are homologous, i.e., they have practically the same length and contain in general the same genes (functional elements of DNA). These are called the autosomal chromosomes. The remaining chromosome is the sex-chromosome. Males have an X and a Y chromosome. Females have two X chromosomes.

More than 99% of the DNA of any two humans of the general population is identical. Most DNA is therefore not useful for identification. However, there are well specified locations on chromosomes where there is variation in DNA among individuals. Such a variation is called a genetic marker. In genetics, the specified locations are called loci. A single location is a locus.

In forensic research, the short tandem repeat (STR) markers are currently most used. The reason is that they can be reliable determined from small amounts of body tissue. Another advantage is that they have a low mutation rate, which is important for kinship analysis. STR markers is a class of variations that occur when a pattern of two or more nucleotides is repeated. For example,

$$(CATG)_3 = CATGCATGCATG .$$
(9)

The number of repeats x (which is 3 in the example) is the variation among the population. Sometimes, there is a fractional repeat, e.g. $CATGCATGCATGCA$, this would be encoded with repeat number $x = 3.2$, since there are three repeats and two additional nucleotides. The possible values of x and their frequencies are well documented for the loci used in forensic research. These ranges and frequencies vary between loci. To some extend they vary among subpopulations of humans. The STR loci are standardized. The NFI uses CODIS (Combined DNA Index System) standard with 13 specific core STR loci, each on different autosomal chromosomes.

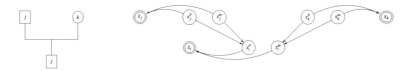

Fig. 11 A basic pedigree with father, mother, and child. Squares represent males, circles represent females. Right: corresponding Bayesian network. Double ringed nodes are observables. x_j^p and x_j^m represents paternal and maternal allele of individual j. See text.

The collection of markers yields the DNA profile. Since chromosomes exist in pairs, a profile will consist of pairs of markers. For example in the CODIS standard, a full DNA profile will consist of 13 pairs (the following notation is not common standard)

$$\bar{\mathbf{x}} = (^1x^1, {}^1x^2), (^2x^1, {}^2x^2), \ldots, (^{13}x^1, {}^{13}x^2),\qquad(10)$$

in which each $^\mu x^s$ is a number of repeats at a well defined locus μ. However, since chromosomes exists in pairs, there will be two alleles $^\mu x^1$ and $^\mu x^2$ for each location, one paternal — on the chromosome inherited from father — and one maternal. Unfortunately, current DNA analysis methods cannot identify the phase of the alleles, i.e., whether an allele is paternal or maternal. This means that $(^\mu x^1, {}^\mu x^2)$ cannot be distinguished from $(^\mu x^2, {}^\mu x^1)$. In order to make the notation unique, we order the observed alleles of a locus such that $^\mu x^1 \leq {}^\mu x^2$.

Chromosomes are inherited from parents. Each parent passes one copy of each pair of chromosomes to the child. For autosomal chromosomes there is no (known) preference which one is transmitted to the child. There is also no (known) correlation between the transmission of chromosomes from different pairs. Since chromosomes are inherited from parents, alleles are inherited from parents as well. However, there is a small probability that an allele is changed or mutated. This mutation probability is about 0.1%.

Finally in the DNA analysis, sometimes failures occur in the DNA analysis method and an allele at a certain locus drops out. In such a case the observation is $(^\mu x^1, F)$, in which "F" is a wild card.

4.3 A Bayesian Network for Kinship Analysis

In this subsection we will describe the building blocks of a Bayesian network to model probabilities of DNA profiles of individuals in a pedigree. First we observe that inheritance and observation of alleles at different loci are independent. So for each locus we can make an independent model P_μ. In the model description below, we will consider a model for a single locus, and we will suppress the μ dependency for notational convenience.

4.3.1 Allele Probabilities

We will consider pedigrees with individuals i. In a pedigree, each individual i has two parents, a father $f(i)$ and a mother $m(i)$. An exception is when a individual is a founder. In that case it has no parents in the pedigree.

Statistical relations between DNA profiles and alleles of family members can be constructed from the pedigree, combined with models for allele transmission. On the given locus, each individual i has a paternal allele x_i^f and an maternal allele x_i^m. f and m stands for 'father' and 'mother'. The pair of alleles is denoted as $x_i = (x_i^f, x_i^m)$. Sometimes we use superscript s which can have values $\{f, m\}$. So each allele in the pedigree is indexed by (i, s), where i runs over individuals and s over phases (f, m). The alleles can assume N values, where N as well as the allele values depend on the locus.

An allele from a founder is called 'founder allele'. So a founder in the pedigree has two founder alleles. The simplest model for founder alleles is to assume that they are independent, and each follow a distribution $P(a)$ of population frequencies. This distribution is assumed to be given. In general $P(a)$ will depend on the locus. More advanced models have been proposed in which founder alleles are correlated. For instance, one could assume that founders in a pedigree come from a single but unknown subpopulation [1]. This model assumption yields corrections to the outcomes in models without correlations between founders. A drawback is that these models may lead to a severe increase in required memory and computation time. In this chapter we will restrict ourself to models with independent founder alleles.

If an individual i has its parents in the pedigree the allele distribution of an individual given the alleles of its parents are as follows,

$$P(x_i | x_{f(i)}, x_{m(i)}) = P(x_i^f | x_{f(i)}) P(x_i^m | x_{m(i)}) , \tag{11}$$

where

$$P(x_i^f | x_{f(i)}) = \frac{1}{2} \sum_{s=f,m} P(x_i^f | x_{f(i)}^s) , \tag{12}$$

$$P(x_i^m | x_{m(i)}) = \frac{1}{2} \sum_{s=f,m} P(x_i^m | x_{m(i)}^s) . \tag{13}$$

To explain (12) in words: individual i obtains its paternal allele x_i^f from its father $f(i)$. However, there is a 50% chance that this allele is the *paternal* allele $x_{f(i)}^f$ of father $f(i)$ and a 50% chance that it is his *maternal* allele $x_{f(i)}^m$. A similar explanation applies to (13).

The probabilities $P(x_i^f | x_{f(i)}^s)$ and $P(x_i^m | x_{m(i)}^s)$ are given by a mutation model $P(a|b)$, which encodes the probability that allele of the child is a while the allele on the parental chromosome that is transmitted is b. The precise mutation mechanisms for the different STR markers are not known. There is evidence that mutations from father to child are in general about 10 times as probable as mutations from mother to child. Gender of each individual is assumed to be known, but for notational convenience we suppress dependency of parent gender. In general, mutation tends to decrease with the difference in repeat numbers $|a-b|$. Mutation is also locus dependent [4].

Several mutation models have been proposed, see e.g. [8]. As we will see later, however, the inclusion of a detailed mutation model may lead to a severe increase in required memory and computation time. Since mutations are very rare, one could ask if there is any practical relevance in a detailed mutation model. The simplest mutation model is of course to assume the absence of mutations, $P(a|b) = \delta_{a,b}$. Such model enhances efficient inference. However, any mutation in any single locus would lead to a 100% rejection of the match, even if there is a 100% match in the remaining markers. Mutation models are important to get some model tolerance against such case. The simplest non-trivial mutation model is a uniform mutation model with mutation rate μ (not to be confused with the locus index μ),

$$P(a|a) = 1 - \mu, \tag{14}$$

$$P(a|b) = \mu/(N-1) \quad \text{if } a \neq b. \tag{15}$$

Mutation rate may depend on locus and gender.

An advantage of this model is that the required memory and computation time increases only slightly compared to the mutation free model. Note that the population frequency is in general not invariant under this model: the mutation makes the frequency more flat. One could argue that this is a realistic property that introduces diversity in the population. In practical applications in the model, however, the same population frequency is assumed to apply to founders in different generations in a pedigree. This implies that if more unobserved references are included in the pedigree to model ancestors of an individual, the likelihood ratio will (slightly) change. In other words, formally equivalent pedigrees will give (slightly) different likelihood ratios.

4.3.2 Observations

Observations are denoted as $\bar{x}_i$, or $\bar{x}$ if we do not refer to an individual. The parental origin of an allele can not be observed, so alleles $x^f = a, x^m = b$ yields the same observation as $x^f = b, x^m = a$. We adopt the convention to write the smallest allele first in the observation: $\bar{x} = (a,b) \Leftrightarrow a \leq b$. In the case of an allele loss, we write $\bar{x} = (x, F)$ where F stands for a wild card. We assume that the event of an allele loss can be observed (e.g. via the peak hight [6]). This event is modeled by L. With $L = 1$ there is allele loss, and there will be a wild card F. A full observation is coded as $L = 0$. The case of loss of two alleles is not modeled, since in that case we simply have no observation.

The observation model is now straightforwardly written down. Without allele loss ($L = 0$), alleles y result in an observation $\bar{y}$. This is modeled by the deterministic table

$$P(\bar{x}|y, L = 0) = \begin{cases} 1 & \text{if } \bar{x} = \bar{y}, \\ 0 & \text{otherwise.} \end{cases} \tag{16}$$

Note that for a given y there is only one $\bar{x}$ with $\bar{x} = \bar{y}$.

With allele loss ($L = 1$), we have

$$P(\bar{x} = (a,F)|(a,b),L=1) = 1/2 \quad \text{if } a \neq b \tag{17}$$

$$P(\bar{x} = (b,F)|(a,b),L=1) = 1/2 \quad \text{if } a \neq b \tag{18}$$

$$P(\bar{x} = (a,F)|(a,a),L=1) = 1 . \tag{19}$$

I.e., if one allele is lost, the alleles (a,b) leads to an observation a (then b is lost), or to an observation b (then a is lost). Both events have 50% probability. If both alleles are the same, so the pair is (a,a), then of course a is observed with 100% probability.

4.4 Inference

By multiplying all allele priors, transmission probabilities and observation models, a Bayesian network of alleles x and DNA profiles of individuals $\bar{x}$ in a given pedigree is obtained. Assume that the pedigree consists of a set of individuals $\mathcal{I} = 1,\ldots,K$ with a subset of founders $\mathcal{F}$, and assume that allele losses L_j are given, then this probability reads

$$P(\{\bar{x},x\}_\mathcal{I}) = \prod_j P(\bar{x}_j|x_j,L_j) \prod_{i \in \mathcal{I} \backslash \mathcal{F}} P(x_i|x_{f(i)},x_{m(i)}) \prod_{i \in \mathcal{F}} P(x_i) . \tag{20}$$

Under this model the likelihood of a given set DNA profiles can now be computed.

If we have observations $\bar{x}_j$ from a subset of individuals $j \in \mathcal{O}$, the likelihood of the observations in this pedigree is the marginal probability $P(\{x\}_\mathcal{O})$, which is

$$P(\{\bar{x}\}_\mathcal{O}) = \sum_{x_1} \cdots \sum_{x_K} \prod_{j \in \mathcal{O}} P(\bar{x}_j|x_j,L_j) \prod_{i \in \mathcal{I} \backslash \mathcal{F}} P(x_i|x_{f(i)},x_{m(i)}) \prod_{i \in \mathcal{F}} P(x_i) . \tag{21}$$

This computation involves the sum over all states of allele pairs x_i of all individuals.

In general, the allele-state space can be prohibitively large. This would make even the junction tree algorithm infeasible if it would straightforwardly be applied. Fortunately, a significant reduction in memory requirement can be achieved by "value abstraction": if the observed alleles in the pedigree are all in a subset A of M different allele values, we can abstract from all unobserved allele values and consider them as a single state z. If an allele is z, it means that it has a value that is not in the set of observed values A. We now have a system in which states can assume only $M+1$ values which is generally a lot smaller than N, the number of a priori possible allele values. This procedure is called value abstraction [14]. The procedure is applicable if for any $a \in A$, $L \in \{0,1\}$, and $b_1,b_2,b_3,b_4 \notin A$, the following equalities hold

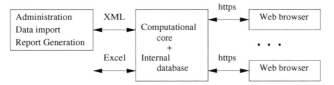

Fig. 12 Bonaparte's basic architecture

$$P(a|b_1) = P(a|b_2) \tag{22}$$

$$P(\bar{x}|a,b_1,L) = P(\bar{x}|a,b_2,L) \tag{23}$$

$$P(\bar{x}|b_1,a,L) = P(\bar{x}|b_2,a,L) \tag{24}$$

$$P(\bar{x}|b_1,b_2,L) = P(\bar{x}|b_3,b_4,L) \tag{25}$$

If these equalities hold, then we can replace $P(a|b)$ by $P(a|z)$ and $P(\bar{x}|a,b)$ by $P(\bar{x}|a,z)$ etc. in the abstracted state representation. The conditional probability of z then follows from

$$P(z|x) = 1 - \sum_{a \in A} P(a|x) \tag{26}$$

for all x in $A \cup z$. One can also easily check that the observation probabilities satisfy the condition. The uniform mutation model satisfies condition (22) since $P(a|b) = \mu/(N-1)$ for any $a \in A$ and any $b \notin A$. Note that condition (22) does not necessarily holds for a general mutation model, so value abstraction could then not be applied.

Using value abstraction as a preprocessing step, a junction tree-based algorithm can straightforwardly applied to compute the desired likelihood. In this way, likelihoods and likelihood ratios are computed for all loci, and reported to the user.

4.5 The Application

Bonaparte has been designed to facilitate large scale matching. The application has a multi-user client-server based architecture, see fig. 12. Its computational core and the internal database runs on a server. All match results are stored in internal database. Rewind to any point in back in time is possible. Via an XML and secure https interfaces, the server connects to other systems. Users can login via a web-browser so that no additional software is needed on the clients. A live demo version for professional users is available on www.bonaparte-dvi.com.

The application is currently being deployed by the Netherlands Forensic Institute NFI. On 12 May 2010, Afriqiyah Airways Flight 8U771 crashed on landing near Tripoli International Airport. There were 103 victims. One child survived the crash. A large number of victims were blood-relatives. The Bonaparte program has been successfully used for the matching analysis to identify the victims. The program has two advantages compared to NFI's previous approach. Firstly, due to fully automated processing, the identification process has been significantly accelerated. Secondly, unlike the previous approach, the program does not need reference samples from first degree relatives since it processes whole pedigree information. For this accident, this was important since in some cases parents with children crashed together and for some individuals, no reference samples from living first degree relatives were available. Bonaparte could do the identification well with samples from relatives of higher degree.

4.6 Summary

Bonaparte is an application of Bayesian networks for victim identification by kinship analysis based on DNA profiles. The Bayesian networks are used to model statistical relations between DNA profiles of different individuals in a pedigree. By Bayesian inference, likelihood ratios and posterior odds of hypotheses are computed, which are the quantities of interest for the forensic researcher. The probabilistic relations between variables are based on first principles of genetics. A feature of this application is the automatic, on-the-fly derivation of models from data, i.e., the pedigree structure of a family of a missing person. The approach is related to the idea of modeling with templates, which is discussed in e.g. [20].

5   A Petrophysical Decision Support System

Oil and gas reservoirs are located in the earth's crust at depths of several kilometers, and when located offshore, in water depths of a few meters to a few kilometers. Consequently, the gathering of critical information such as the presence and type of hydrocarbons, size of the reservoir and the physical properties of the reservoir such as the porosity of the rock and the permeability is a key activity in the oil and gas industry.

Pre-development methods to gather information on the nature of the reservoirs range from gravimetric, 2D and 3D seismic to the drilling of exploration and appraisal boreholes. Additional information is obtained while a field is developed through data acquisition in new development wells drilled to produce hydrocarbons, time-lapse seismic surveys and in-well monitoring of how the actual production of hydrocarbons affects physical properties such as the pressure and temperature. The purpose of information gathering is to decide which reservoirs can be developed economically, and how to adapt the means of development best to the particular nature of a reservoir.

The early measurements acquired in exploration, appraisal and development boreholes are a crucial component of the information gathering process. These measurements are typically obtained from tools on the end of a wireline that are lowered into the borehole to measure the rock and fluid properties of the formation. Their is a vast range of possible measurement tools [28]. Some options are very expensive and may even risk other data acquisition options. In general acquiring all possible data imposes too great an economic burden on the exploration, appraisal and development. Hence data acquisition options must be exercised carefully bearing in mind the learnings of already acquired data and general hydrocarbon field knowledge.

Also important is a clear understanding of what data can and cannot be acquired later and the consequences of having an incorrect understanding of the nature of a reservoir on the effectiveness of its development.

Making the right data acquisition decisions, as well as the best interpretation of information obtained in boreholes forms one of the principle tasks of petrophysicists. The efficiency of a petrophysicist executing her/his task is substantially influenced by the ability to gauge her/his experience to the issues at hand. Efficiency is hampered when a petrophysicists experience level is not yet fully sufficient and by the rather common circumstance that decisions to acquire particular types of information or not must be made in a rush, at high costs and shortly after receiving other information that impact on that very same decision. Mistakes are not entirely uncommon and almost always painful. In some cases, non essential data is obtained at the expense of extremely high cost, or essential data is not obtained at all; causing development mistakes that can jeopardize the amount of hydrocarbon recoverable from a reservoir and induce significant cost increases.

The overall effectiveness of petrophysicists is expected to improve using a decision support system (DSS). In practice a DSS can increase the petrophysicists' awareness of low probability but high impact cases and alleviate some of the operational decision pressure.

In cooperation with Shell E&P, SNN has developed a DSS tool based on a Bayesian network and an efficient sampler for inference. The main tasks of the application is the estimation of compositional volume fractions in a reservoir on the basis of measurement data. In addition it provides insight in the effect of additional measurements. Besides an implementation of the model and the inference, the tool contains graphical user interface in which the user can take different views on the sampled probability distribution and on the effect of additional measurements.

In the remainder of this section, we will describe the Bayesian network approach for the DSS tool. We focus on our modeling and inference approach. More details are described elsewhere [5].

5.1 Probabilistic Modeling

The primary aim of the model is to estimate the compositional volume fractions of a reservoir on the basis of borehole measurements. Due to incomplete knowledge, limited amount of measurements, and noise in the measurements, there will be uncertainty in the volume fractions. We will use Bayesian inference to deal with this uncertainty.

The starting point is a model for the probability distribution $P(\mathbf{v}, \mathbf{m})$ of the compositional volume fractions $\mathbf{v}$ and borehole measurements $\mathbf{m}$. A causal argument *"The composition is given by the (unknown) volume fractions, and the volume fractions determine the distribution measurement outcomes of each of the tools"* leads us to a Bayesian network formulation of the probabilistic model,

$$P(\mathbf{v}, \mathbf{m}) = \prod_{i=1}^{Z} P(m_i|\mathbf{v})P(\mathbf{v}) . \tag{27}$$

In this model, $P(\mathbf{v})$ is the so-called *prior*, the prior probability distribution of volume fractions before having seen any data. In principle, the prior encodes the generic geological and petrophysical knowledge and beliefs [30]. The factor $\prod_{i=1}^{Z} P(m_i|\mathbf{v})$ is the *observation model*. The observation model relates volume fractions $\mathbf{v}$ to measurement outcomes m_i of each of the Z tools i. The observation model assumes that *given* the underlying volume fractions, measurement outcomes of the different tools are independent. Each term in the observation model gives the probability density of observing outcome m_i for tool i given that the composition is $\mathbf{v}$. Now given a set of measurement outcomes $\mathbf{m}^o$ of a subset *Obs* of tools, the probability distribution of the volume fractions can be updated in a principled way by applying *Bayes' rule*,

$$P(\mathbf{v}|\mathbf{m}^o) = \frac{\prod_{i \in Obs} P(m_i^o|\mathbf{v})P(\mathbf{v})}{P(\mathbf{m}^o)} . \tag{28}$$

The updated distribution is called the *posterior* distribution. The constant in the denominator $P(\mathbf{m}^o) = \int_{\mathbf{v}} \prod_{i \in Obs} P(m_i^o|\mathbf{v})P(\mathbf{v}) d\mathbf{v}$ is called the *evidence*.

In our model, $\mathbf{v}$ is a 13 dimensional vector. Each component represents the volume fraction of one of 13 most common minerals and fluids (water, calcite, quartz, oil, etc.). So each component is bounded between zero and one. The components sum up to one. In other words, the volume fractions are confined to a simplex $\mathbb{S}^K = \{\mathbf{v}|0 \leq v_j \leq 1, \sum_k v_k = 1\}$. There are some additional physical constraints on the distribution of $\mathbf{v}$, for instance that the total amount of fluids should not exceed 40% of the total formation. The presence of more fluids would cause a collapse of the formation.

Each tool measurement gives a one-dimensional continuous value. The relation between composition and measurement outcome is well understood. Based on the physics of the tools, petrophysicists have expressed these relations in terms of deterministic functions $f_j(\mathbf{v})$ that provide the idealized noiseless measurement outcomes of tool j given the composition $\mathbf{v}$ [30]. In general, the functions f_j are nonlinear. For most tools, the noise process is also reasonably well understood — and can be described by either a Gaussian (additive noise) or a log-Gaussian (multiplicative noise) distribution.

A straightforward approach to model a Bayesian network would be to discretize the variables and create conditional probability tables for priors and conditional distributions. However, due to the dimensionality of the volume fraction vector, any reasonable discretization would result in an infeasible large state space of this variable. We therefore decided to remain in the continuous domain.

The remainder of this section describes the prior and observation model, as well as the approximate inference method to obtain the posterior.

5.2 The Prior and the Observation Model

The model has two ingredients: the prior of the volume fractions $P(\mathbf{v})$ and the observation model $P(m_j|\mathbf{v})$.

There is not much detailed domain knowledge available about the prior distribution. Therefore we decided to model the prior using conveniently parametrized family of distributions. In our case, $\mathbf{v} \in \mathbb{S}^K$, this lead to the Dirichlet distribution [22, 3]

$$Dir(\mathbf{v}|\alpha,\mu) \propto \prod_{j=1}^{K} v_j^{\alpha\mu_j-1} \delta\left(1 - \sum_{i=1}^{K} v_i\right). \tag{29}$$

The two parameters $\alpha \in \mathbb{R}_+$ (precision) and $\mu \in \mathbb{S}^K$ (vector of means) can be used to fine-tune the prior to our liking. The delta function — which ensures that the simplex constraint holds — is put here for clarity, but is in fact redundant if the model is constraint to $\mathbf{v} \in \mathbb{S}^K$. Additional information, e.g. the fact that the amount of fluids may not exceed 40% of the volume fraction can be incorporated by multiplying the prior by a likelihood term $\Phi(\mathbf{v})$ expressing this fact. The resulting prior is of the form

$$P(\mathbf{v}) \propto \Phi(\mathbf{v})Dir(\mathbf{v}|\alpha,\mu). \tag{30}$$

The other ingredients in the Bayesian network are the observation models. For most tools, the noise process is reasonably well understood and can be reasonably well described by either a Gaussian (additive noise) or a log-Gaussian (multiplicative noise) distribution. In the model, measurements are modeled as a deterministic tool function plus noise,

$$m_j = f_j(\mathbf{v}) + \xi_j, \tag{31}$$

in which the functions f_j are the deterministic tool functions provided by domain experts. For tools where the noise is multiplicative, a log transform is applied to the tool functions f_j and the measurement outcomes m_j. A detailed description of these functions is beyond the scope of this paper. The noises ξ_j are Gaussian and have a tool specific variance σ_j^2. These variances have been provided by domain experts. So, the observational probability models can be written as

$$P(m_i|\mathbf{v}) \propto \exp\left(-\frac{(m_j - f_j(\mathbf{v}))^2}{2\sigma_j^2}\right). \tag{32}$$

5.3 Bayesian Inference

The next step is given a set of observations $\{m_i^o\}$, $i \in$ Obs, to compute the posterior distribution. If we were able to find an expression for the evidence term, i.e., for the marginal distribution of the observations $P(\mathbf{m}^o) = \int_\mathbf{v} \prod_{i \in \text{Obs}} P(m_i^o|\mathbf{v})P(\mathbf{v})d\mathbf{v}$ then the posterior distribution (28) could be written in closed form and readily evaluated. Unfortunately $P(\mathbf{m}^o)$ is intractable and a closed-form expression does not exist. In order to obtain the desired compositional estimates we therefore have to resort to approximate inference methods. Pilot studies indicated that sampling methods gave the best performance.

The goal of any sampling procedure is to obtain a set of N samples $\{x_i\}$ that come from a given (but maybe intractable) distribution π. Using these samples we can approximate expectation values $\langle A \rangle$ of a function $A(x)$ according to

$$\langle A \rangle = \int_x A(x)\pi(x)\mathrm{d}x \approx \frac{1}{N}\sum_{i=1}^{N} A(x_i) . \tag{33}$$

For instance, if we take $A(x) = x$, the approximation of the mean $\langle x \rangle$ is the sample mean $\frac{1}{N}\sum_{i=1}^{N} x_i$.

An important class of sampling methods are the so-called Markov Chain Monte Carlo (MCMC) methods [22, 3]. In MCMC sampling a Markov chain is defined that has an equilibrium distribution π, in such a way that (33) gives a good approximation when applied to a sufficiently long chain $x_1, x_2, \ldots, x_N$. To make the chain independent of the initial state x_0, a burn-in period is often taken into account. This means that one ignores the first $M \ll N$ samples that come from intermediate distributions and begins storing the samples once the system has reached the equilibrium distribution π.

In our application we use the hybrid Monte Carlo (HMC) sampling algorithm [11, 22]. HMC is a powerful class of MCMC methods that are designed for problems with continuous state spaces, such as we consider in this section. HMC can in principle be applied to any noise model with a continuous probability density, so there is no restriction to Gaussian noise models. HMC uses Hamiltonian dynamics in combination with a Metropolis [23] acceptance procedure to find regions of higher probability. This leads to a more efficient sampler than a sampler that relies on random walk for phase space exploration. HMC also tends to mix more rapidly than the standard Metropolis Hastings algorithm. For details of the algorithm we refer to the literature [11, 22].

In our case, $\pi(\mathbf{v})$ is the posterior distribution $p(\mathbf{v}|m_i^o)$ in (28). The HMC sampler generates samples $\mathbf{v}_1, \mathbf{v}_2, \ldots, \mathbf{v}_N$ from this posterior distribution. Each of the N samples is a full K-dimensional vector of volume fractions constraint on $\mathbb{S}^K$. The number of samples is of the order of $N = 10^5$, which takes a few seconds on a standard PC. Figure 13 shows an example of a chain of 10 000 states generated by the sampler. For visual clarity, only two components of the vectors are plotted (quartz and dolomite). The plot illustrates the multivariate character of the method: for example, the traces shows that the volume fractions of the two minerals tend to be mutually exclusive: either 20% quartz, or 20% dolomite but generally not both. From the traces, all kind of statistics can be derived. As an example, the resulting one dimensional marginal distributions of the mineral volume fractions are plotted.

The performance of the method relies heavily on the quality of the sampler. Therefore we looked at the ability of the system to estimate the composition of a (synthetic) reservoir and the ability to reproduce the results. For this purpose, we set the composition to a certain value $\mathbf{v}^*$. We apply the observation model to generate measurements $\mathbf{m}^o$. Then we run HMC to obtain samples from the posterior

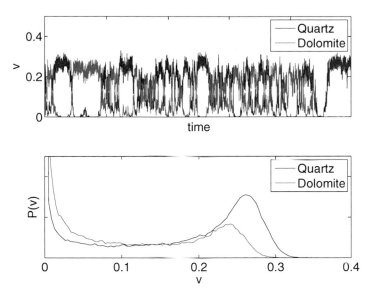

Fig. 13 Diagrams for quartz and dolomite. Top: time traces (10 000 time steps) of the volume fractions of quartz and dolomite. Bottom: Resulting marginal probability distributions of both fractions.

$P(\mathbf{v}|\mathbf{m}^o)$. Consistency is assessed by comparing results of different runs to each other and by comparing them with the "ground truth" $\mathbf{v}^*$. Results of simulations confirm that the sampler generates reproducible results, consistent with the underlying compositional vector [5]. In these simulations, we took the observation model to generate measurement data (the generating model) equal to the observation model that is used to compute the posterior (the inference model). We also performed simulations where they are different, in particular in their assumed variance. We found that the sampler is robust to cases where the variance of the generating model is smaller than the variance of the inference model. In the cases where the variance of the generating model is bigger, we found that the method is robust up to differences of a factor 10. After that we found that the sampler suffered severely from local minima, leading to irreproducible results.

5.4   *Decision Support*

Suppose that we have obtained a subset of measurement outcomes $\mathbf{m}^o$, yielding a distribution $P(\mathbf{v}|\mathbf{m}^o)$. One may subsequently ask the question which tool t should be deployed next in order to gain as much information as possible?

When asking this question, one is often interested in a specific subset of minerals and fluids. Here we assume this interest is actually in one specific component u. The question then reduces to selecting the most informative tool(s) t for a given mineral u.

We define the informativeness of a tool as the expected decrease of uncertainty in the distribution of v_u after obtaining a measurement with that tool. Usually, entropy is taken as a measure for uncertainty [22], so a measure of informativeness is the expected entropy of the distribution of v_u after measurement with tool t,

$$
\langle H_{u,t}|\mathbf{m}^o\rangle \equiv -\int P(m_t|\mathbf{m}^o)\int P(v_u|m_t,\mathbf{m}^o)
$$
$$
\times \log\left(P(v_u|m_t,\mathbf{m}^o)\right) dv_u dm_t .
$$

(34)

Note that the information of a tool depends on the earlier measurement results since the probabilities in (34) are conditioned on $\mathbf{m}^o$.

The most informative tool for mineral u is now identified as that tool t^* which yields in expectation the lowest entropy in the posterior distribution of v_u:

$$
t^*_{u|\mathbf{m}^o} = \underset{t}{\operatorname{argmin}} \langle H_{u,t}|\mathbf{m}^o\rangle
$$

In order to compute the expected conditional entropy using HMC sampling methods, we first rewrite the expected conditional entropy (34) in terms of quantities that are conditioned only on the measurement outcomes $\mathbf{m}^o$,

$$
\langle H_{u,t}|\mathbf{m}^o\rangle = -\int\int P(v_u,m_t|\mathbf{m}^o)
$$
$$
\times \log\left(P(v_u,m_t|\mathbf{m}^o)\right) dv_u dm_t
$$
$$
+ \int P(m_t|\mathbf{m}^o)\int \log\left(P(m_t|\mathbf{m}^o)\right) dm_t .
$$

(35)

Now the HMC run yields a set $V = \{v_1^j, v_2^j, \ldots, v_K^j\}$ of compositional samples (conditioned on $\mathbf{m}^o$). We augment these by a set $M = \{m_1^j = f_1(\mathbf{v}^j) + \xi_1^j, \ldots, m_Z^j = f_Z(\mathbf{v}^j) + \xi_Z^j\}$ of synthetic tool values generated from these samples (which are indexed by j) by applying equation (31). Subsequently, discretized joint probabilities $P(v_u, m_t|\mathbf{m}^o)$ are obtained via a two-dimensional binning procedure over v_u and m_t for each of the potential tools t. The binned versions of $P(v_u, m_t|\mathbf{m}^o)$ (and $P(m_t|\mathbf{m}^o)$) can be directly used to approximate the expected conditional entropy using a discretized version of equation (35).

The outcome of our implementation of the decision support tool is a ranking of tools according to the expected entropies of their posterior distributions. In this way, the user can select a tool based on a trade-off between expected information and other factors, such as deployment costs and feasibility.

5.5   The Application

The application is implemented in C++ as a stand alone version with a graphical user interface running on a Windows PC. The application has been validated by petrophysical domain experts from Shell E&P. The further use by Shell of this application is beyond the scope of this chapter.

5.6 Summary

This chapter described a Bayesian network application for petrophysical decision support. The observation models are based on the physics of the measurement tools. The physical variables in this application are continuous-valued. A naive Bayesian network approach with discretized values would fail. We remained in the continuous domain and used the hybrid Monte Carlo algorithm for inference.

6 Discussion

Human decision makers are often confronted with highly complex domains. They have to deal with various sources of information and various sources of uncertainty. The quality of the decision is strongly influenced by the decision makers experience to correctly interpret the data at hand. Computerized decision support can help to improve the effectiveness of the decision maker by enhancing awareness and alerting the user to uncommon situations that may have high impact. Rationalizing the decision process may alleviate some of the decision pressure.

Bayesian networks are widely accepted as a principled methodology for modeling complex domains with uncertainty, in which different sources of information are to be combined, as needed in intelligent decision support systems. We have discussed in detail three applications of Bayesian networks. With these applications, we aimed to illustrate the modeling power of the Bayesian networks and to demonstrate that Bayesian networks can be applied in a wide variety of domains with different types of domain requirements. The medical model is a toy application illustrating the basic modeling approach and the typical reasoning behavior. The forensic and petrophysical models are real world applications, and show that Bayesian network technology can be applied beyond the basic modeling approach.

The chapter should be read as an introduction to Bayesian network modeling. There has been carried out much work in the field of Bayesian networks that is not covered in this chapter, e.g. the work on Bayesian learning [16], dynamical Bayesian networks [24], approximate inference in large, densely connected models [9, 25], templates and structure learning [20], nonparametric approaches [17, 13], etc.

Finally, we would like to stress that the Bayesian network technology is only one side of the model. The other side is the domain knowledge, which is maybe even more important for the model. Therefore Bayesian network modeling always requires a close collaboration with domain experts. And even then, the model is of course only one of many ingredients of an application, such as user-interface, data-management, user-acceptance etc. which are all essential to make the application a success.

Acknowledgements. The presented work was partly carried out with support from the Intelligent Collaborative Information Systems (ICIS) project, supported by the Dutch Ministry of Economic Affairs, grant BSIK03024. We thank Ender Akay, Kees Albers and Martijn Leisink (SNN), Mirano Spalburg (Shell E & P), Carla van Dongen, Klaas Slooten and Martin Slagter (NFI) for their collaboration.

References

1. Balding, D., Nichols, R.: DNA profile match probability calculation: how to allow for population stratification, relatedness, database selection and single bands. Forensic Science International 64(2-3), 125–140 (1994)
2. Beinlich, I., Suermondt, H., Chavez, R., Cooper, G., et al.: The ALARM monitoring system: A case study with two probabilistic inference techniques for belief networks. In: Proceedings of the Second European Conference on Artificial Intelligence in Medicine, vol. 256. Springer, Berlin (1989)
3. Bishop, C.: Pattern Recognition and Machine Learning. Springer (2006)
4. Brinkmann, B., Klintschar, M., Neuhuber, F., Hühne, J., Rolf, B.: Mutation rate in human microsatellites: influence of the structure and length of the tandem repeat. The American Journal of Human Genetics 62(6), 1408–1415 (1998)
5. Burgers, W., Wiegerinck, W., Kappen, B., Spalburg, M.: A Bayesian petrophysical decision support system for estimation of reservoir compositions. Expert Systems with Applications 37(12), 7526–7532 (2010)
6. Butler, J.: Forensic DNA Typing: Biology, Technology, and Genetics of STR Markers. Academic Press (2005)
7. Castillo, E., Gutierrez, J.M., Hadi, A.S.: Expert Systems and Probabilistic Network Models. Springer (1997)
8. Dawid, A., Mortera, J., Pascali, V.: Non-fatherhood or mutation? A probabilistic approach to parental exclusion in paternity testing. Forensic Science International 124(1), 55–61 (2001)
9. Doucet, A., Freitas, N.D., Gordon, N. (eds.): Sequential Monte Carlo Methods in Practice. Springer, New York (2001)
10. Drábek, J.: Validation of software for calculating the likelihood ratio for parentage and kinship. Forensic Science International: Genetics 3(2), 112–118 (2009)
11. Duane, S., Kennedy, A., Pendleton, B., Roweth, D.: Hybrid Monte Carlo Algorithm. Phys. Lett. B 195, 216 (1987)
12. Fishelson, M., Geiger, D.: Exact genetic linkage computations for general pedigrees. Bioinformatics 18(suppl. 1), S189–S198 (2002)
13. Freno, A., Trentin, E., Gori, M.: Kernel-based hybrid random fields for nonparametric density estimation. In: European Conference on Artificial Intelligence (ECAI), vol. 19, pp. 427–432 (2010)
14. Friedman, N., Geiger, D., Lotner, N.: Likelihood computations using value abstraction. In: Proceedings of the Sixteenth Conference on Uncertainty in Artificial Intelligence, pp. 192–200. Morgan Kaufmann Publishers (2000)
15. Heckerman, D.: Probabilistic interpretations for mycin's certainty factors. In: Kanal, L., Lemmer, J. (eds.) Uncertainty in Artificial Intelligence, pp. 167–196. North-Holland (1986)
16. Heckerman, D.: A tutorial on learning with Bayesian networks. In: Innovations in Bayesian Networks. SCI, vol. 156, pp. 33–82. Springer, Heidelberg (2008)
17. Hofmann, R., Tresp, V.: Discovering structure in continuous variables using Bayesian networks. In: Advances in Neural Information Processing Systems (NIPS), vol. 8, pp. 500–506 (1995)
18. Jensen, F.: An Introduction to Bayesian Networks. UCL Press (1996)
19. Jordan, M.: Learning in Graphical Models. Kluwer Academic Publishers (1998)
20. Koller, D., Friedman, N.: Probabilistic Graphical Models: Principles and Techniques. The MIT Press (2009)

21. Lauritzen, S., Spiegelhalter, D.: Local computations with probabilities on graphical structures and their application to expert systems. Journal of the Royal Statistical Society. Series B (Methodological), 157–224 (1988)
22. MacKay, D.: Information Theory, Inference and Learning Algorithms. Cambridge University Press (2003)
23. Metropolis, N., Rosenbluth, A., Rosenbluth, M., Teller, A., Teller, E.: Equation of state calculations by fast computing machines. The Journal of Chemical Physics 21(6), 1087 (1953)
24. Murphy, K.: Dynamic Bayesian Networks: Representation, Inference and Learning. Ph.D. thesis, UC Berkeley (2002)
25. Murphy, K.P., Weiss, Y., Jordan, M.I.: Loopy belief propagation for approximate inference: An empirical study. In: Proceedings of Uncertainty in AI, pp. 467–475 (1999)
26. Pearl, J.: Probabilistic Reasoning in Intelligent systems: Networks of Plausible Inference. Morgan Kaufmann Publishers, Inc. (1988)
27. Russell, S., Norvig, P., Canny, J., Malik, J., Edwards, D.: Artificial Intelligence: A Modern Approach. Prentice Hall (2003)
28. Schlumberger: Log Interpretation Principles/Applications. Schlumberger Limited (1991)
29. Shortliffe, E., Buchanan, B.: A model of inexact reasoning in medicine. Mathematical Biosciences 23(3-4), 351–379 (1975)
30. Spalburg, M.: Bayesian uncertainty reduction for log evaluation. SPE International SPE88685 (2004)

Chapter 13
Relevance Feedback in Content-Based Image Retrieval: A Survey

Jing Li and Nigel M. Allinson

In content-based image retrieval, relevance feedback is an interactive process, which builds a bridge to connect users with a search engine. It leads to much improved retrieval performance by updating a query and similarity measures according to a user's preference; and recently techniques have matured to some extent. Most previous relevance feedback approaches exploit short-term learning (intra-query learning) that deals with the current feedback session but ignoring historical data from other users, which potentially results in a great loss of useful information. In the last few years, long-term learning (inter-query learning), by recording and collecting feedback knowledge from different users over a variety of query sessions has played an increasingly important role in multimedia information searching. It can further improve the retrieval performance in terms of effectiveness and efficiency. In the published literature, no comprehensive survey of both short-term learning and long-term learning RF techniques has been conducted. To this end, the goal of this chapter is to address this omission and offer suggestions for future work.

1   Introduction

Due to the dramatically increasing popularity of digital imaging devices - cameras, scanners, webcams, etc., millions of new images are created every day. For example, the largest internet photo sharing site, *Flickr* (www.flickr.com), grows by some 4,000 images per minute. How to search such large-scale image databases effectively and efficiently is becoming urgent and challenging. Image retrieval,

Jing Li
School of Information Engineering, Nanchang University, PRC
e-mail: jing.li.2003@gmail.com

Nigel M. Allinson
School of Computer Science, University of Lincoln, UK
e-mail: nallinson@lincoln.ac.uk

M. Bianchini et al. (Eds.): *Handbook on Neural Information Processing*, ISRL 49, pp. 433–469.
DOI: 10.1007/ 978-3-642-36657-4_13   © Springer-Verlag Berlin Heidelberg 2013

searching similar images to a query in a large-scale image database, is a vital way to manage this goal. Historically, most image retrieval techniques were keywords-based and involved manual annotation of database images. However, such an approach is impractical in real-world applications mainly because of: i) near impossibility to annotate each image with meaningful keywords; ii) users having a different understanding of the keyword dictionary and this may change over time; and iii) different users possess contrasting opinions of an identical image. Consequently, keywords may not be specific enough to represent an image.

Content-based image retrieval (CBIR) aims to avoid the above-mentioned problems by considering an image through its contents, i.e., low-level visual features such as colour, texture, shape, layout, etc. After feature extraction, the most similar images are returned to the user according to some similarity metric. However, psychological research (Tversky 1977) suggests that humans perceive images by their concepts, which are referred to high-level semantic contents. Therefore, there is a semantic gap (Rui et al. 1998) between the low-level visual features and high-level image concepts. As shown in Fig. 1, these two images have quite similar visual features. However, their semantic contents are totally different: the left image is with a chameleon; the right one is with a frog. That is to say retrieval performance based only on low-level visual features is unlikely to be sufficient; and it is most important to take into account the effective interaction between users and search engines. Hence, relevance feedback (RF), originating from document retrieval (Rocchio 1966), has been introduced to bridge the gap. Traditional RF methods in CBIR include the following two basic steps (Rui et al. 1998): i) when retrieved images are returned to the user, some relevant and irrelevant images are labelled as positive and negative samples, respectively; and ii) the retrieval system refines the retrieved results based on these labelled feedback samples. These two steps are conducted iteratively until the user is satisfied with the presented results.

Fig. 1 Semantic gap: these two images have similar visual features but they are completely different in semantic contents

Over the last decade, RF techniques have been developed based on diverse machine learning techniques: feature selection (Zhou and Huang 2001; Tao et al. 2006), semi-supervised learning (Li et al. 2006), query modification (Kushki et al. 2004), random sampling of both feedback samples and image features (Tao et al. 2006), density estimation of positive samples (Chen et al. 2001), negative samples analysis (Tao et al. 2007), samples imbalance problem (Tao et al. 2007, 2006), and distance metric learning (Giacinto and Roli 2004). In this chapter, we divide RF approaches into two categories: i) short-term learning (intra-query learning); and ii) long-term learning (inter-query learning). Short-term learning considers the current feedback information for future retrieval; while long-term learning utilizes feedback knowledge recorded and collected from multiple users over a variety of query sessions. Fruitful progress has been made on short-term learning and there are some literature reviews (Smeulders et al. 2000; Vasconcelos and Kunt 2001; Zhou and Huang 2003). However, there is a common drawback in short-term learning techniques: the feedback knowledge from different users during previous feedback process is discarded. In other words, the RF procedure will commence with a *tabula rasa* for each session. To this end, long-term learning has gained increasing attention in the last few years since it can further improve the retrieval performance in terms of both effectiveness and efficiency by utilizing multiple user log information.

Most of long-term learning algorithms are combined with short-term learning methods and there is some overlap between them. Nevertheless, we categorize them from different aspects. RF approaches based on short-term learning can be separated according to the way labelled samples are treated: i) one-class: for positive samples only; ii) two-class: one class for positive samples and the other for negative samples; and iii) multi-class: several classes for positive samples or negative samples. RF techniques using long-term learning are categorized into the following types: i) latent semantic indexing-based techniques; ii) correlation-based approaches; iii) clustering-based algorithms; iv) feature representation-based methods; v) similarity measure modification-based approaches; and vi) other techniques.

This chapter is organized as follows. In Section 2, the basic elements of CBIR are briefly reviewed. In Section 3, we introduce some representative short-term learning methods. A variety of long-term learning RF techniques are reviewed in Section 4. Section 5 summarizes with suggestions for future work. This chapter should form a unique reference source for this exciting and timely topic.

2   Content-Based Image Retrieval

The retrieval process in CBIR can be detailed as follows: firstly, low-level visual features of each image in the database, e.g., colour, shape, texture, layout, etc., are extracted and stored. The features are the "content" in CBIR. When a query is input, its features are extracted and a similarity (or dissimilarity) metric is utilized to compare them with those previously encoded. The larger (or smaller) the score is, the more similar the image is to the query, and then images can be ranked according to the score. Finally, the most similar images are returned to the user. Several CBIR systems have been designed to access databases conveniently for both research and commercial purposes; where significant milestones include QBIC (Niblack et al. 1993), MARS (Ortega et al. 1997; Rui et al. 1998), PicHunter (Cox et al. 2000), PicToSeek (Gevers and Smeulders 2000), PicSOM (Laaksonen et al. 2000), and SIMPLIcity (Wang et al. 2001). In the following sub-sections, we will introduce the basic elements in CBIR, which are i) low-level feature extraction; ii) similarity measure; iii) classification methods; and iv) current databases.

2.1 Low-Level Feature Extraction

In CBIR (Long et al. 2003), low-level visual contents, e.g., colour, texture, shape, spatial layout, etc., are extracted at each pixel of an image. This approach has been widely applied in a variety of computer vision applications, e.g., content-based image retrieval (CBIR), scene classification, robot localization, and biometrics.

Because of the robustness to background clutter, rotation, and scaling, colour (Jain and Vailaya 1996; Manjunath et al. 2001; Niblack et al. 1993) is the most extensively used feature for global-level image representation. Colour information can be extracted in various colour spaces, where RGB, CIELAB, and HSV are the most widely applied in CBIR and will be introduced as follows. In such a colour space, each pixel in an image can be represented by a point in a 3D coordinate. RGB colour space has been extensively utilized in computer graphics. Its three primary colour components, i.e., red, green, and blue, match the input channels of human eyes and a colour can be formed by adding these components together with different amounts. The perceptually uniform CIE L*a*b* space includes all colours that are visible to humans. This colour space is based on the concept that colour is the pairwise combination of red and yellow, red and blue, green and yellow, green and blue. Herein, 'L' is the lightness coordinate, 'a*' is the position on the green-red axis, and 'b*' is the blue-yellow coordinate. It can model the human visual perception of colour differences by calculating the distance between two colour pairs in the colour space since it is directly proportional to that perceived by human eyes. The HSV perceptual colour space models the way as humans perceive colours, where hue ('H') defines the colour type, saturation ('S') represents the purity of the colour, and value ('V') is the brightness. The RGB, CIE L*a*b*, and HSV colour models are illustrated in Fig. 2, where the differences among these three colour spaces can be discovered. Humans are less sensitive to the chromatic information that is invariant to illumination changes. For this reason, the opponent colour representation (R-G, 2B-R-G, R+G+B) can down-sample the first two chromatic axes by separating the brightness information onto the third axis. The widely used colour features include mean colour, colour histogram (Swain and Ballard 1991), etc. Mean colour is the simplest colour feature that computes the averaged intensity for each colour component. The colour histogram is a histogram of pixel intensities by calculating the number of pixels in each quantized colour bin. It is easy to compute and robust to rotation, translation, and small viewpoint changes. However, it does not take into account the spatial organization of colours. After discretising the colour space into a predefined number of colours, colour coherent vector (Pass et al. 1996) divides a colour histogram into two parts: coherent and incoherent. For each colour, it records the number of coherent pixels and incoherent pixels where a pixel is defined as coherent if it is in a large similarly-coloured region and vice versa. However, colour is sensitive to varying lighting conditions – especially with shadow and reflection, and therefore only utilizing colour information is not sufficient to represent an image.

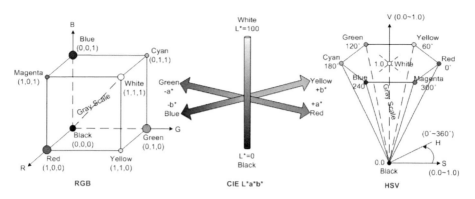

Fig. 2 The RGB, CIE L*a*b*, and HSV colour models

Texture (Chang and Kuo 1993; Manjunath and Ma 1996; Manjunath et al. 2001) is another importance feature that has been widely used for analysis, i.e., texture classification, texture synthesis, texture segmentation, and image compression. Texture is composed of repetitive small patterns (also called texels) in an image, such as fabrics, grass, bricks, hair, etc., and depends on scales. One of the simplest texture features is the edgeness, which can be calculated from the edge pixel rate in a region. Except for it, texture representation techniques can be divided into two categories: structural-based and statistical-based methods. Structural features, e.g., adjacency graphs, are effective for regular texture patterns since a hierarchy of spatial arrangement of them can depict a texture; statistical features, including co-occurrence matrices, Tamura features, wavelet transforms,, etc., consider the spatial arrangement of intensities in an image and are usually adopted for global texture representation. Gray level co-occurrence matrix measures spatial relations between two pixels by an angle and distance. It is sensitive to rotation and representatively large and sparse. Gabor filtering convolves an image with a bank of Gabor filters (Daugman 1980; Daugman 1985) with different scales and orientations. For global texture feature representation, mean and standard deviation of the filtered image can be utilized. Discrete cosine transform (DCT) (Jain 1989) converts an image from the spatial domain to the frequency domain and represents the image data as a sum of basic cosine functions at different frequencies. In the frequency domain, most data energy is concentrated in the lower spatial frequencies, and therefore most of the visually significant information about the image can be described by a few DCT coefficients. Tamura features consist of coarseness, contrast, directionality, line-likeness, regularity, and roughness, among which the first three features are more related to human perception and have been widely utilized. However, they are performed only at a single scale and cannot deal with scale changes. Based on multiresolution analysis, wavelet transform (Daubechies 1990) is implemented by successively decomposing a signal into different levels (different frequencies with different resolutions) with a series of basis functions derived from translations and dilations of a mother wavelet – a prototype for generating other functions in the transform process. At each level, four frequency sub-bands are computed, which are low-low, low-high, high-low, and high-high. After decomposition, mean and standard deviation are calculated for each sub-band at each decomposition level to form a feature vector. Basically, texture features are not suitable for the non-textured images which cannot be recognized by shapes, e.g., natural scene images. Nevertheless, they actually work well for those images containing clustered objects, e.g., man-made objects like shoes and buildings.

Compared with colour and texture features, shape features (Jain and Vailaya 1996; Niblack et al. 1993) are not so widely applied because they are usually related to object contours which require accurate segmentation. Nevertheless, automatic object segmentation still remains unsolved not only because it is time consuming but also its performance cannot be guaranteed. The edge direction histogram (Manjunath et al. 2001), one of the most effective global shape features, groups edge pixels into discrete direction bins by first detecting edge points by an edge detector and then computing the gradient direction for each of them. Afterwards, the number of edge points falling into each direction bin will be counted and histogram matching can be implemented by either bin-by-bin distance (e.g., histogram intersection) or cross-bin distance (e.g., earth mover's distance).

However, the primitive features mentioned above do not take into account the spatial information of an image and may give rise to many false positives. For instance, given a pair of scenes — one with blue sky and the other with sea, the classification performance based only on colour histograms may not be satisfactory since they may be similar and cannot be distinguished from each other. Spatial layout, e.g., left/right and upper/lower visual fields, reveals not only the physical locations of objects but also spatial relationships among them. Therefore, the performance can be largely improved by embedding the spatial information into the colour histogram, given the truth that the sky should be located at the upper position of an image while the sea should be at the bottom of an image. The simplest way to consider spatial layout is to divide an image into un-overlapping blocks and then discover their correlations. However, this method does not always reflect the true spatial relations of the image. Region-based techniques consider the spatial information by segmenting an image into subparts and representing an image by their spatial relationships. These techniques have more discriminative power but require segmentation whose performance cannot be guaranteed.

The above-mentioned low-level features mainly belong to global feature representation, i.e., they are extracted from all of the pixels in a whole image, e.g., colour histogram. More recently, local features (local descriptors) (Li and Allinson 2008; Mikolajczyk et al. 2005; Mikolajczyk and Schmid 2005), which describe local information in a small region or an object of interest, have shown their superiority in various kinds of applications in computer vision research, such as robust matching (Tuytelaars and Gool 2004), image retrieval (Schmid and Mohr 1997), etc. In CBIR, the scale invariant feature transform (SIFT) (Lowe 2004) is one of the most widely used local features. SIFT consists of a detector and a descriptor: the SIFT detector finds the local maximums of a series of difference of Gaussian (DoG) images; the SIFT descriptor is a 3D histogram for gradient magnitudes and orientations representation. This feature is invariant to changes in partial illumination, background clutter, occlusion, and transformations in terms of rotation and scaling. Jain et al. (2009) applied SIFT to matching tattoo images on human bodies for targeting on crime issues, where tattoo locations and tattoo class labels are utilized to improve the retrieval performance while reducing the retrieval time.

2.2   Similarity Measure

After low-level visual feature extraction, similarity measures are conducted to compare the query image with each image in the database and then the most similar results will be returned to the user. Here, we introduce the frequently used distance measures, including the Minkowski-form distance, the Kullback-Leibler Divergence, and the Earth mover's distance.

Minkowski-Form Distance

The Minkowski-form distance between the query image $\vec{I}$ and $\vec{J}$ is defined as

$$D(\vec{I},\vec{J}) = \left(\sum_{k} \left| p_k(\vec{I}) - p_k(\vec{J}) \right|^q \right)^{1/q} . \tag{1}$$

where p_k is the number of pixels in the k-th bin. When $q=1$, $q=2$, and $q=\infty$, it becomes the L_1 distance (Manhattan distance), L_2 distance (Euclidean distance), and L_∞ distance, respectively. It is the most widely used distance metric in content-based image retrieval (Carson et al. 2002).

Kullback-Leibler Divergence

The Kullback-Leibler (KL) Divergence (Minka and Picard 1997) is a distance measure related to the relative entropy between two probability distributions. Given the query image $\vec{I}$, an image $\vec{J}$ in the database, and p_k is the number of pixels in the k-th bin, the KL divergence between $p_k(\vec{I})$ and $p_k(\vec{J})$ is denoted as

$$D(\vec{I}\|\vec{J}) = \sum_{i} p_k(\vec{I}) \log \frac{p_k(\vec{I})}{p_k(\vec{J})} . \tag{2}$$

The properties of the KL divergence are as follows: 1) $D(\vec{I}\|\vec{J}) \geq 0$; 2) $D(\vec{I}\|\vec{J}) \neq D(\vec{J}\|\vec{I})$; and 3) $D(\vec{I}\|\vec{J}) = 0 \Leftrightarrow \vec{I} = \vec{J}$. Because KL divergence is generally asymmetrical, it is not a metric and the symmetric KL-divergence is denoted as following:

$$D(\vec{I}\|\vec{J}) = \int_{-\infty}^{+\infty} p_k(\vec{I}) \log \frac{p_k(\vec{I})}{p_k(\vec{J})} dx + \int_{-\infty}^{+\infty} p_k(\vec{J}) \log \frac{p_k(\vec{J})}{p_k(\vec{I})} dx \tag{3}$$

Earth Mover's Distance

Applied to colour and texture in image databases, Earth mover's distance (EMD) (Rubner et al. 2000) measures the distance between two distributions by revealing the least cost to transform one distribution to the other.

By introducing a "signature" for each distribution, EMD allows for partial matching between two distributions with variable-size structures. A signature $\vec{s}_j = (\vec{m}_j, w_j)$ represents a set of dominant clusters extracted from an original distribution, where $\vec{m}_j$ is the d-dimensional mean of the j-th cluster and w_j is the number of pixels belonging to the j-th cluster. Generally speaking, simple images correspond to short signatures and complex images have long signatures.

The computation of EMD can be considered as a linear programming problem. Given the query image $\vec{I}$ and an image $\vec{J}$ in the database, the overall cost of matching $\vec{I}$ to $\vec{J}$ is

$$\Upsilon = \sum_{i \in \vec{I}} \sum_{j \in \vec{J}} c_{ij} f_{ij} \ . \tag{4}$$

where the cost c_{ij} is the ground distance between element i in the signature of $\vec{I}$ and element j in the signature of $\vec{J}$. To minimize the overall cost, a set of flows f_{ij} should be discoverd subject to the following constraints:

$$\begin{aligned} f_{ij} &\geq 0 \qquad i \in \vec{I}, j \in \vec{J} \\ \sum_{i \in \vec{I}} f_{ij} &= y_j \quad j \in \vec{J} \\ \sum_{j \in \vec{J}} f_{ij} &\leq x_i \quad i \in \vec{I} \end{aligned} \tag{5}$$

where x_i and y_j are the demanding amount to fill the two signatures to equal length, subject to

$$\sum_{j \in \vec{J}} y_j \leq \sum_{i \in \vec{I}} x_i \ . \tag{6}$$

Finally, the EMD is defined as

$$\begin{aligned} \text{EMD}(\vec{I}, \vec{J}) &= \sum_{i \in \vec{I}} \sum_{j \in \vec{J}} c_{ij} f_{ij} \Big/ \sum_{i \in \vec{I}} \sum_{j \in \vec{J}} f_{ij} \\ &= \sum_{i \in \vec{I}} \sum_{j \in \vec{J}} c_{ij} f_{ij} \Big/ \sum_{j \in \vec{J}} y_j \end{aligned} \tag{7}$$

2.3   Classification Methods

A number of classification methods have been applied to modelling the relevance feedback as a classification problem. Here, we introduce the most widely adopted methods in CBIR, which are artificial neural networks (ANNs), support vector machines (SVMs), and ensemble learning.

2.3.1 Artificial Neural Networks

Inspired by the parallel structure of biological neural systems, artificial neural networks (ANNs) are composed of layers of simple processing units, i.e., artificial neurons, as shown in Fig. 3. Generally, a multi-layer neural network involves an input layer, a hidden layer, and an output layer, where the neurons reflect the relationships between the input $\{\vec{x}_i\}_{i=1}^n$ and the output y and they can generate nonlinear output to learn the nonlineaity from the training data. As seen in Fig. 4, there is an example neural network consisting of an input layer with five neurons, a hidden layer with three neurons, and an output layer with two neurons, where each input $\vec{x}_i$ is weighted by w_i, and then mapped and summed via a function into the output $y = F(z)$ with $z = \sum_{i=1}^n w_i \vec{x}_i$. Herein, F can be a sigmoid function, a radial basis function (RBF), or etc.

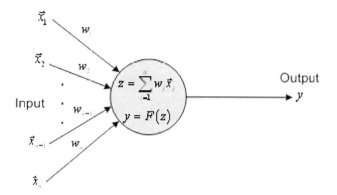

Fig. 3 An artificial neuron

For a classification problem, there are usually m output corresponding to m categories. In the learning process of an ANN, series of training examples are presented to the network and the weight defined as the strength connecting between an input and a neuron is automatically adjusted to minimize the error – the difference between the output and the target value. The complexity of the architecture of an ANN depends on some factors, e.g., the number of training examples, the number of hidden layers, and the number of neurons in each hidden layer. Typically, a single hidden layer is enough to model complex data, that is, a three-layer network with sufficient number of neurons in the hidden layer can deal with arbitrary high-dimensional data. More hidden layers rarely bring an improvement in performance while the model gets slower and more complicated., Using more hidden layers may cause a problem of converging to a local minimum and it requires the use of the random initialization methods to achieve better global optimization. Nevertheless, its advantages on performance have been shown in some certain applications, such as cascade correlation (Fahlman and Lebiere 1990) and ZIP code recognition (LeCun et al. 1989). Although ANNs require long training times, the evaluation stage is very fast, which is very important for on-line applications. Moreover, ANNs are quite robust to noisy training examples (training data with errors) or complex sensor data from cameras because of their good generalization ability on unseen data after the learning stage. ANNs have been utilized in various applications, such as robot control, face recognition, etc. If long training time is acceptable, it would be a good choice to apply them in a classification task.

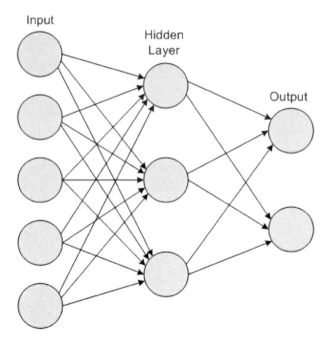

Fig. 4 An example of ANNs

2.3.2   SVM

The support vector machine (SVM) (Vapnik 1995; Burges 1998) is a binary classifier, which maximizes the margin between positive examples and negative examples, as shown in Fig. 5. Because of its good generalization ability and no requirement for prior knowledge about the data, it has been utilized as one of the most popular classifiers in CBIR.

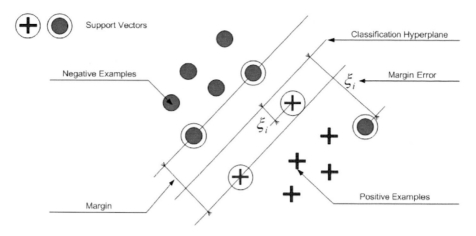

Fig. 5 SVM maximizes the margin between positive examples and negative examples

Considering a problem of classifying a set of linearly separable training examples $\{(\vec{x}_i, y_i)\}_{i=1}^{N}$ with $\vec{x}_i \in \mathfrak{R}^L$ and their associated class labels $y_i \in \{+1, -1\}$, SVM separates these two classes by a hyperplane

$$\vec{w}^T \cdot \vec{x} + b = 0, \tag{8}$$

where $\vec{x}$ is an input vector, $\vec{w}$ is an adaptive weight vector, and b is a bias. The optimal hyperplane, which maximizes the geometric margin $2/\|\vec{w}\|$ between two classes, can be obtained by

$$\min_{\vec{w}, b, \vec{\xi}} \|\vec{w}\|^2 / 2 + C \sum_{i=1}^{N} \xi_i$$
$$\text{s.t.} \quad \begin{aligned} y_i (\vec{w}^T \cdot \vec{x}_i + b) \geq 1 - \xi_i, & \quad 1 \leq i \leq N \\ \vec{\xi} \geq 0 \end{aligned} \tag{9}$$

where $\vec{\xi} = \left[\xi_1, \xi_2, ..., \xi_N\right]^T$ is the vector of all slack variables to deal with the nonlinear separable problem. For linearly separable training examples, we can set $\vec{\xi} = 0$. By introducing a Lagrange multiplier α_i, the Lagrangian is

$$L(\vec{w}, b, \vec{\xi}, \vec{\alpha}, \vec{\kappa}) = \frac{1}{2}\|\vec{w}\|^2 + C\sum_{i=1}^{N}\xi_i - \sum_{i=1}^{N}\alpha_i \left(y_i\left(\vec{w}^T \cdot \vec{x}_i + b\right) - 1 + \xi_i\right) - \sum_{i=1}^{N}\kappa_i\xi_i, \tag{10}$$

and the solution is determined by

$$\max_{\vec{\alpha}, \vec{\kappa}} \min_{\vec{w}, b, \vec{\xi}} L(\vec{w}, b, \alpha), \tag{11}$$

which can be achieved by the Karush-Kuhn-Tucker (KKT) conditions

$$\frac{\partial L}{\partial \vec{w}} = 0 \Rightarrow \vec{w} = \sum_{i=1}^{N} \alpha_i y_i \vec{x}_i$$

$$\frac{\partial L}{\partial b} = 0 \Rightarrow \vec{\alpha}^T \vec{y} = 0 \tag{12}$$

$$\frac{\partial L}{\partial \xi} = 0 \Rightarrow C - \vec{\alpha} - \vec{\kappa} = 0$$

Therefore, the parameters $\vec{w}$ and b can be obtained using the Wolfe dual problem

$$\max_{\vec{\alpha}} \quad Q(\alpha) = \sum_{i=1}^{N} \alpha_i - \frac{1}{2} \sum_{i,j=1}^{N} \alpha_i \alpha_j y_i y_j \left(\vec{x}_i^T \cdot \vec{x}_j \right)$$

$$\text{s.t.} \quad \begin{aligned} & 0 \le \alpha_i \le C \\ & \vec{\alpha}^T \vec{y} = 0 \end{aligned} \tag{13}$$

Most of α_i are zeros, and $\vec{x}_i$ corresponding to $\alpha_i > 0$ are referred to as the support vectors.

In the dual format, data points only appear in the inner product. To solve the nonlinearly separable problem, the data points from the low-dimensional input space L are mapped onto a higher dimensional feature space H (the Hilbert Inner Product Space) by the replacement

$$\vec{x}_i \cdot \vec{x}_j \rightarrow \phi(\vec{x}_i) \cdot \phi(\vec{x}_j) = K\left(\vec{x}_i, \vec{x}_j\right), \tag{14}$$

where $K\left(\vec{x}_i, \vec{x}_j\right)$ is a kernel function with entries $\phi(\vec{x}_i)$, $\phi(\vec{x}_j) \in \Re^H$. A lot of standard kernel functions can be embedded in SVM, such as linear kernels $K\left(\vec{x}_i, \vec{x}_j\right) = \vec{x}_i^T \cdot \vec{x}_j$, polynomial kernels $K\left(\vec{x}_i, \vec{x}_j\right) = \left(\vec{x}_i \cdot \vec{x}_j + c\right)^d$, Gaussian radial basis function (RBF) $K\left(\vec{x}_i, \vec{x}_j\right) = \exp\{-\|\vec{x}_i - \vec{x}_j\|^2 / 2\sigma^2\}$, and so forth. Then, the kernel version of the Wolfe dual problem is

$$Q(\alpha) = \sum_{i=1}^{N} \alpha_i - \frac{1}{2} \sum_{i,j=1}^{N} \alpha_i \alpha_j y_i y_j K\left(\vec{x}_i \cdot \vec{x}_j\right). \tag{15}$$

Finally, for a given kernel function, the SVM classifier is given as

$$F\left(\vec{x}\right) = \text{sgn}\left(f(\vec{x})\right). \tag{16}$$

where $f(\vec{x}) = \sum_{i=1}^{l} \alpha_i y_i K\left(\vec{x}_i, \vec{x}_j\right) + b$ is the output hyperplane decision function of SVM.

In traditional SVM-based RF algorithms, $f(\vec{x})$ is used for measuring the dissimilarity between the query image and an example image. For a given example, a high $f(\bar{x})$ indicates it is far away from the decision boundary and thus has high prediction confidence; while a low $f(\bar{x})$ shows it is close to the boundary and its corresponding prediction confidence is low.

2.3.3   Ensemble Learning

Weak classifiers, which only perform slightly better than random guessing, may not be able to provide satisfactory classification results. Moreover, overfitting occurs when the constructed classifier is more complex than the structure of the data, i.e., the classifier can achieve zero training error but generalizes poorly on unseen data. By integrating a series of component classifiers (usually weak classifiers) into a strong classifier through a voting rule, ensemble learning can avoid overfitting and significantly improve the generalization ability of each single weak classifier, e.g., a SVM. The basic idea is that examples that are misclassified by one weak classifier may be correctly classified by another, as diagrammatically illustrated in Fig. 6. Bagging (Breiman 1995) and boosting (Schapire 1990) are the most influential algorithms in ensemble learning, where bagging constructs a set of weak classifiers in parallel and combines them by a majority voting rule; boosting sequentially builds a weak classifier based on previously misclassified training examples and makes the final decision by a weighted voting rule. In the following sections, we introduce bagging, boosting, and Adaboost (Freund and Schapire 1996) – a representative algorithm of boosting.

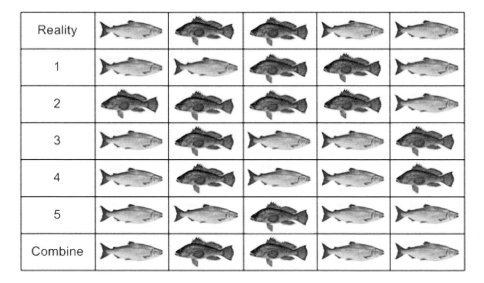

Fig. 6 Ensemble learning example with five component classifiers to differentiate between Salmon (white) and Sea bass (black)

2.3.3.1 Bagging

Bagging (Breiman 1995) is short for bootstrap aggregating. Given a dataset D with N training examples, bootstrap is to independently select N' $(N' \leq N)$ points from N with replacement to create different training sets. After m bootstraps, m newly constructed training sets are obtained, each of which is used to train an individual classifier and all of them are later aggregated by a majority voting rule. One thing is worth noting: with replacement, repetitive examples or missing examples may occur. As shown in Fig. 7, in the training data Tn, (G,H,J) has been selected twice while some other examples have never been selected.

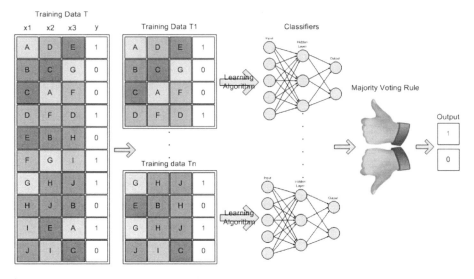

Fig. 7 Bagging procedure with $N = 10$ and $N' = 4$

2.3.3.2 Boosting

To enhance the classification performance of weak classifiers, boosting (Schapire 1990) builds a strong classifier by iteratively adding a weak classifier with more attention to the most informative data points, i.e., misclassified examples at the previous learning step. The main idea is to re-weight the training data each time adding a new classifier, i.e., misclassified examples will be assigned more weight while correctly classified examples will be given less weight. After the learning stage, the output of all weak classifiers is integrated by a weighted voting.

Adaboost

AdaBoost (Freund and Schapire 1996), an abbreviation for adaptive boosting, is an important boosting algorithm that has been universally applied in a variety of computer vision applications, such as CBIR and biometrics. It focuses on difficult training examples and sequentially adapts the weak classifier by adjusting the weights of training examples at each round. Starting with equal weights for all training examples, at the next iteration, the weights of misclassified examples will be increased while the weights of correctly classified ones will be reduced. Finally, a strong classifier is constructed by combining these weak classifiers with the weighted majority voting, i.e., better weak classifiers get more weight and *vice versa*. AdaBoost is particularly suitable for applications with many features. Considering a binary classification problem, the procedure of Adaboost (Duda 1999; Zhou 2009) is given in Table 1.

Table 1 Adaboost Algorithm

Input: The data set D : $\{(\vec{x}_i, y_i)\}_{i=1}^{N}$ with $y_i \in \{+1,-1\}$, Learning algorithm ℓ, Number of learning rounds t.
Output: A series of weak classifiers h_t and a strong classifier $H(X) = \text{sign}(f(X)) = \text{sign}\left(\sum_{t=1}^{T} \alpha_t h_t(X)\right)$ with $X = [\vec{x}_1, \vec{x}_2, ..., \vec{x}_N]$.

1.	Weight initialization: $D_1(i) = 1/N$.
2.	*Conduct the following steps iteratively.*
3.	For $t = 1$ to T
4.	Train h_t from D using distribution D_t: $h_t = \ell(D, D_t)$;
5.	Error estimation of h_t: $\varepsilon_t = \text{Pr}_{i \sim D_i}\left[h_t(\vec{x}_i) \neq y_i\right]$;
6.	If $\varepsilon_t \leq 1/2$
7.	Weight assignment for h_t: $\alpha_t = (1/2)\ln\left((1-\varepsilon_t)/\varepsilon_t\right)$;
8.	Update the distribution: $$D_{t+1}(i) = (D_t(i)/Z_t) \times \begin{cases} \exp(-\alpha_t) \text{ if } h_t(\vec{x}_i) = y_i \\ \exp(\alpha_t) \text{ if } h_t(\vec{x}_i) \neq y_i \end{cases}$$ $$= D_t(i)\exp(-\alpha_t y_i h_t(\vec{x}_i))/Z_t$$ where Z_t is a normalization factor to make sure D_{t+1} is a distribution;
9.	End
10.	End

2.4   Current Databases

Most previous CBIR algorithms were tested on a subset of the Corel image gallery. Recently, two other databases, i.e., TinyImage (Torralba et al. 2008) and ImageNet (Deng et al 2009), have shown their advantages in memory cost and
retrieval precision, respectively, and have much potential for future use.

2.4.1   Corel Image Gallery

Corel image gallery (Wang et al. 2001), a real-world image database, is the most
popularly used database in CBIR. Based on the semantic concepts, images in the
database are separated into different folders (categories), e.g., dog, flower, etc.,
each of which contains 100 images. Since images in a folder not always correspond to its concept, they can be manually labelled with different numbers of concepts (Guo et al. 2002; Li et al. 2006; Bian and Tao 2010).

2.4.2   TinyImage

TinyImage (Torralba et al. 2008) contains 79 million tiny images (low-resolution
images with the size of 32×32 pixels) which densely cover all visual object
classes. It was generated by utilizing 7 different search engines, namely Altavista,
Ask, Flickr, Cydral, Google, Picsearch and Webshots, where each image in the database is labelled with one of the non-abstract nouns of the 75,062 English words
in WordNet. After removing duplicate and uniform images, 79,302,017 images of
760 GB were selected and stored in a single hard-disk over an 8-month period. Although there are only a few hundred bits for each image, the images include the
most salient information for classification. Moreover, with reduced memory requirement, TinyImage enables fast indexing and ease for storing and management.

2.4.3   ImageNet

In contrast to TinyImage, ImageNet (Deng et al. 2009) is constructed around the
semantic hierarchy of WordNet. It contains 3.2 million of clean (full-resolution)
images belonging to 12 sub-trees with 5,247 synsets (categories), constituting 10%
WordNet sysnets. A snap shot from mammal to giant panda is given in Fig. 8. On
average, there are 600 images in each category and the average size of an image is
around 400×350 pixels.

   Image data were collected using the following scheme: 1) search candidate images from the internet without duplicates, where queries were translated to several
languages to obtain a larger range of diverse candidate images; and 2) select clean
candidate images through the service of the Amazon Mechanical Turk, i.e., each
candidate image is judged by multiple users.

   ImageNet (www.image-net.org) is the largest image dataset that is available in
computer vision research, which is in terms of the total number of images, the
number of categories, and the number of images for each category. It is a high-
resolution set of images and includes various kinds of changes on appearance,
positions, viewpoints, background clutter, and occlusions from objects.

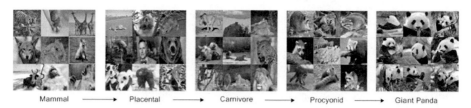

Mammal ⟶ Placental ⟶ Carnivore ⟶ Procyonid ⟶ Giant Panda

Fig. 8 A snap shot of the mammal subtree from ImageNet with nine sample images for each synset

3   Short-Term Learning RF

Short-term learning (intra-query learning) RF is defined as being restricted to one user for the current feedback session only and the learning flowchart is given in Fig. 9. These learning algorithms will be divided into one-class, two-class, and multi-class modelling algorithms based on the way they treat the positive samples and negative samples. In either short-term or long-term RF learning techniques, SVM-based RF achieves the most attention and plays an important role in CBIR because of its good generalization ability and computation efficiency. The original SVM-based RF (Zhang et al. 2001) maximizes the margin between positive samples and negative samples. At each feedback round, only the support vectors are used for training. However, when labelled feedback samples are limited, the retrieval/classification performance may be unsatisfactory. This is caused by: 1) the SVM is unstable for a small-sized training set since its optimal hyperplane, which is determined by support vectors, can be very sensitive to the training examples; 2) the SVM's optimal hyperplane may be unbalanced when the number of negative feedback samples is far greater than that of positive feedback samples; and 3) overfitting may occur when the number of training examples is much less than the dimension of the feature vector, resulting in zero training errors but poor generalization for unseen data points. To this end, enhanced SVM-based RF techniques were proposed to alleviate the aforementioned problems, which will be shown in the following sub-sections.

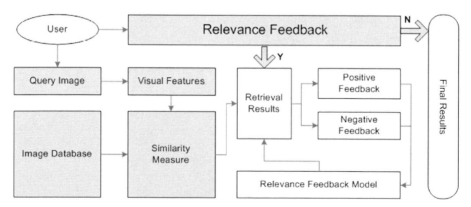

Fig. 9 The flowchart of the short-term learning RF in the image retrieval system, which comes from (Li et al. 2006)

3.1   One-Class

To solve the unbalance problem in SVM-based RF, Hoi et al. (2004) proposed the biased SVM to include the most positive samples by describing the data with a pair of sphere hyperplanes, where the inner one embraces most of the positive samples while the outer one pushes negative samples away.

Su et al. (2001) applied principal component analysis (PCA) (Duda et al. 2001) to decrease the noise in image features and reduce the dimension of the feature space. It models all positive examples with a Gaussian distribution and updates its parameters based on the positive examples in a query session. This method reduces the number of training examples that are needed for parameter calculation and therefore it improves the retrieval speed and reduces the memory requirement. However, it is more reasonable to model positive examples with a Gaussian mixture model (GMM) (Duda et al. 2001) since this can reduce the unimodal problem (Torre and Kanade 2005) caused by approximating each class by a single Gaussian.

3.2   Two-Class

Active learning SVM (Tong and Chang 2001) combines active learning with SVMs. At each feedback session, it selects points that are nearest to the decision boundaries, which are deemed as the most informative images; meanwhile, it pushed the most positive sample away from the boundary on the positive side. By selecting the most uncertain query points, active learning SVM reduces the computation cost.

Based on user-labelled positive and negative feedbacks, constrained similarity measure-based SVM (Guo et al. 2002) employed both SVM and Adaboost to learn an irregular nonsphere boundary to separates all the images in the database into two groups, where images inside the boundary are ranked by their Euclidean distances to the query and images outside the boundary are ranked by their distance to the boundary. By considering the perceptual similarity between images, the retrieval performance can be improved and it is found SVM-based method outperforms Adaboost-based algorithm in learning the boundary.

Making use of unlabelled samples, Li et al. (2006) put forward the multitraining support vector machine (MTSVM) to improve the performance of CBIR. MTSVM combines the co-training technique (Blum and Mitchell 1998) with the random subspace method (RSM) (Ho 1998) and so inherits the merits of both of them. Co-training can augment the number of labelled examples with unlabelled examples. First, it individually trains a pair of sub-classifiers on different aspects of a small number of labelled samples, and then each sub-classifier independently labels several unlabelled samples at each time and enlarges the training set by including the newly labelled samples. After several iterations, the sub-classifiers are combined to generate a stronger classifier. RSM, which incorporates the benefits of bootstrap and aggregation, is an example of a random sampling method. By bootstrapping on the feature set, multiple classifiers can be generated through training on the multiple sampled features. Afterwards, aggregation of the generated classifiers is conducted by a multiple classifier combination rule. However, co-training should meet the requirement that component classifiers are unrelated or weakly related while RSM requires the weak classifiers are adequately trained. To this end, the combination of co-training and RSM is carried out by bootstrapping in the feature space through the RSM followed by labelling a small number of positive and negative feedbacks from the unlabelled data set for co-training. Finally, each data point is determined to be relevant or irrelevant by the aggregation of all weak classifiers through the majority voting rule. In this way, both the requirements are satisfied, and the overfitting and unstable problems can be alleviated.

Tao et al. (2006) proposed the asymmetric bagging and random subspace SVM (ABRS-SVM), which is a combination of asymmetric bagging-based SVM (AB-SVM) and random subspace SVM (RS-SVM) and thus inherits the merits of bagging (Breiman 1995) and random subspace method (RSM) (Ho 1998). In AB-SVM, bagging is implemented by bootstrapping on negative feedback samples and performing aggregation by the majority voting rule. It can solve the unstable and unbalance problems; while in RS-SVM, RSM executes the bootstrap in the feature space to deal with the overfitting problem.

Bian and Tao (2010) explored the intrinsic low-dimensional manifold $\Re^L$ of low-level visual features embedded in a high-dimensional feature space $\Re^H$ by the biased discriminant Euclidean embedding (BDEE) – a mapping consisting of two parts: discrimination preservation (inter-class) and local geometry preservation (intra-class). In $\Re^L$, discrimination preservation aims to maximize distances between positive examples and negative examples and keep distances between positive examples as close as possible; in $\Re^H$, local geometry preservation keeps linear reconstruction coefficients obtained in $\Re^H$ and use them to preserve the local geometry of positive examples. Also, BDEE were extended into semi-BDEE, a semi-supervised version of BDEE by taking into account the unlabelled samples.

3.3   Multi-class

Based on the assumption that users are only interested in the positive class, relevance feedback was regarded as a biased classification problem by Zhou and Huang (2001), where the biased discriminant transform (BDT) were proposed by considering the discriminant information from negative samples of x classes and the classification problem became a $(1 + x)$-class problem.

Instead of considering the learning problem as a binary classification problem, Hoi and Lyu (2004) proposed a group-based RF scheme by dividing positive examples into two groups, each of which is used for learning by an ensemble of SVMs.

However, all the short-term learning RF algorithms lack feedback information from different users at different sessions. To this end, long-term learning RF is becoming more popular and will be reviewed in Section 4.

4   Long-Term Learning RF

Basically, long-term learning (Li and Allinson 2008) consolidates user log information during each feedback session. Before we introduce long-term learning RF techniques in CBIR, we will first introduce some related work in other research areas. Text retrieval exploits some of the long-term learning methods used in CBIR. Experiments were undertaken on two groups by Anick (2003) to examine whether log information helps improve the precision of web search. One group involved the AlTaVista search engine without terminological feedback, whereas the other was provided with feedback. It is demonstrated that users who make use of the log information achieves better search precision. Cui et al. (2003) estimated the correlations between a query term and document terms to automatically expand queries by updating weighted links between query sessions and the query space (i.e., all query terms) and between query sessions and document space (i.e., all document terms). User log files serve as a bridge to link query space with document space. Based on a series of linear transformations, Tai et al. (2002) combined user feedback information of relevant documents with the similarity information of the original document to build a model, which improves the retrieval performance of the vector space model, a traditional information retrieval model where queries and documents are expressed as vectors in a high-dimensional space. In this way, previous feedback information can be used in future feedback sessions. Here, the simple principal component analysis is applied to reduce vector space to a low-dimensional semantic space, which is constructed by concepts. Apart from text retrieval, long-term learning has also been applied in video retrieval (Sav et al. 2005), where the semantic concept of the query was modelled by a Gaussian mixture model that was adjusted by accumulated user feedback information.

We divide RF techniques using long-term learning into the following categories: i) latent semantic indexing-based techniques (Deerwester et al. 1990; Heisterkamp 2002; Koskela and Laaksonen 2003; Chen et al. 2005;); ii) correlation-based approaches (Yang et al. 2002; Zhou et al. 2002; Zhuang et al. 2002; Jiang et al. 2005; Urban and Jose 2006); iii) clustering-based algorithms (Han et al. 2005; Yoshizawa and Schweitzer 2004); iv) feature representation-based methods (Cord and Gosselin 2006; Jing et al. 2004); v) similarity measure modification-based approaches (Chang and Yeung 2005; Fournier and Cord 2002; He et al. 2004; Hoi et al. 2006); and vi) other techniques (Gosselin and Cord 2005; He 2003; Nedovic and Marques 2005; Wacht et al. 2006; Yin et al. 2005). The generic structure of long-term learning RF techniques is represented in Fig. 10.

4.1   *Latent Semantic Indexing-Based Techniques*

Typically, the latent semantic indexing (LSI) technique (Deerwester et al. 1990) is utilized to build a latent semantic space and the essence of LSI is the singular value decomposition (SVD) of the term-by-document matrix. Given a query document where the terms are known, a set of similar documents can be obtained.

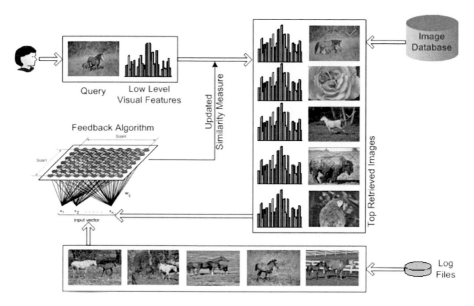

Fig. 10 The long-term learning algorithm

In CBIR, the "term" is referred to as a query; while the "document" refers to an image in the database. Obviously, the first step is to construct the matrix $\mathbf{M}_{m \times n}$, where m represents the number of terms and n denotes the number of documents. After decomposition, $\mathbf{M}$ is approximated by $\tilde{\mathbf{M}}$, which is composed of the top k largest singular values associated with their singular vectors. The decomposition and approximation processes are denoted by $\mathbf{M}_{m \times n} = \mathbf{U}_{m \times r} \mathbf{S}_{r \times r} \mathbf{V}_{n \times r}^{T} \approx \tilde{\mathbf{M}}_{m \times n} = \tilde{\mathbf{U}}_{m \times k} \tilde{\mathbf{S}}_{k \times k} \tilde{\mathbf{V}}_{n \times k}^{T}$, where $\mathbf{U}$ and $\mathbf{V}$ are orthogonal and $\mathbf{S}$ is a diagonal matrix, $r \leq \min(m,n)$ is the rank of $\mathbf{M}$ and $k \leq r$. The SVD process of $\mathbf{M}_{m \times n}$ is described in Fig. 11; while the approximation step is given in Fig. 12.

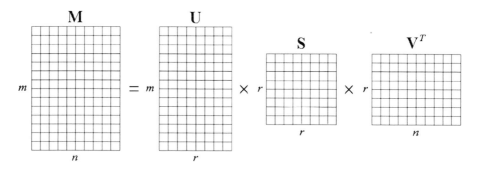

Fig. 11 The SVD process of the term-by-document matrix

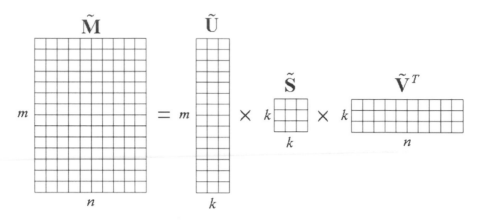

Fig. 12 The approximation step of the term-by-document matrix

Heisterkamp (2002) constructed **M** from the accumulated user feedback information. Differing from the typical definition, "document" is referred to the user labelled feedbacks and "words of vocabulary" refers to images in the database. Moreover, the goal here is to discover similar terms when provided with an unknown term. Terms are compared by $\tilde{\mathbf{M}}\tilde{\mathbf{M}}^T = \tilde{\mathbf{U}}\tilde{\mathbf{S}}^2\tilde{\mathbf{U}}^T$; while documents are compared by $\tilde{\mathbf{M}}^T\tilde{\mathbf{M}} = \tilde{\mathbf{V}}\tilde{\mathbf{S}}^2\tilde{\mathbf{V}}^T$. To deal with a previously unknown query document, a vector of its component terms is constructed as a pseudo-document $\mathbf{T}_q$ and then mapped into the latent semantic space by $\mathbf{F}_q = \tilde{\mathbf{U}}^T\mathbf{T}_q$. The distance between the query and each labelled feedback is then defined as the distance between $\mathbf{F}_q$ and its corresponding column of $\tilde{\mathbf{S}}\tilde{\mathbf{V}}^T$. By utilizing the LSI, log file information from various feedback sessions is accumulated. It was found that the LSI is robust even with poor log file input. Nevertheless, the approach achieves strong robustness only when a large amount of relevance feedback information is available.

Koskela and Laaksonen (2003) investigated long-term learning within the Pic-SOM system (Laaksonen et al. 2000), where relevant images are labelled as positive and all others are considered as negative. The definitions of "document" and "words of vocabulary" are defined as the same as by Heisterkamp (2002). The LSI technique is utilized to map the user labelled samples into the latent semantic space, which is mainly for dimensionality reduction. In this way, the number of labelled samples is reduced to the number of queries in the training data. Since each image may only be present once in each feedback session, frequency weighting of the labelled samples is utilized. That is, elements of **M** are weighted by the number of labelled images with co-occurring terms. Therefore, images with similar semantic contents will be clustered together in the latent semantic space. Besides low-level features, a user interaction feature is also applied. It is defined by the rows of $\mathbf{W} = \tilde{\mathbf{U}}\tilde{\mathbf{S}}$, each of which corresponds to an image. By utilizing this user interaction feature, the retrieval precision is greatly improved while no extra human intervention is needed.

Chen et al. (2005) proposed a region-based long-term RF approach according to the feedback knowledge extracted by the LSI technique, where "documents" are equated to images and "words of vocabulary" refers to regions. The user-provided feedbacks are accumulated in **M** by adding 1 to the component if a returned image is labelled as positive. To extract the region of interest (named the semantic region) for each image, a multiple instance learning algorithm is applied based on the feedback information for the whole image, where a one-class support vector machine (SVM) is used to simulate the regions' distribution as a hyper-sphere and classify positive regions from negative ones. The semantic region, which is in a positive labelled sample and with minimum distance to the query region, will be labelled as positive. With accumulated user feedback information, the number of positive regions will be augmented for training the one-class SVM classifier. The proposed framework achieves high retrieval accuracy by combining short-term learning and long-term learning in region-based image retrieval. However, it should be noted that this method may encounter the "over-segment" problem in image segmentation by utilizing Blob-world (Carson et al. 2002), a technique that segments an image into regions through combining and grouping extracted image features and then subsequent process will be conducted based on the region features. Because the semantic region may be segmented into several sub-regions in some images, it may result in more than one positive region in a positive sample. Moreover, information about negative samples is ignored.

4.2   Correlation-Based Approaches

Correlation-based long-term learning approaches aim to discover the semantic correlations between images, which are usually represented by a matrix or a two/multi-layer graph. By exploring semantic similarities between images based on accumulated log files, the performance of CBIR can be improved in terms of effectiveness or efficiency.

Peer indexing, an image indexing method based on user feedback information, was presented by Yang et al. (2002). The peer index of an image refers to a set of images which are semantically correlated to it. Given an image I_i, its peer index is defined as $P(I_i) = \left\{ \left\langle x_{i\_1}, w_{i\_1} \right\rangle ; \left\langle x_{i\_2}, w_{i\_2} \right\rangle ; \cdots ; \left\langle x_{i\_k}, w_{i\_k} \right\rangle ; \cdots ; \left\langle x_{i\_n}, w_{i\_n} \right\rangle \right\}$, where x_{i_k} denotes its correlated image I_k and w_{i_k} indicates the associated weight. Given a query, some images are labelled as relevant or irrelevant. The query and each of its relevant images will be put into each other's peer index; whereas the query and each irrelevant image will be removed from each other's peer index. A schematic of the approach is given in Fig. 13. The peer indices are updated using the accumulated log file information. The semantic correlations among images are obtained through a cooperative framework, which combines peer indexing with low-level features. This method can greatly improve retrieval accuracy but the computational efficiency with respect to the storage and peer index filtering can be poor.

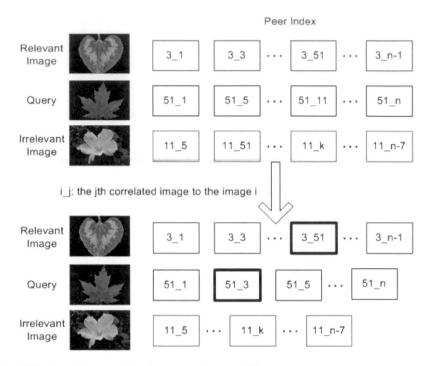

Fig. 13 The learning algorithm for peer index acquisition

The number of feedback samples is augmented by stored feedback information (Zhou et al. 2002). Positive samples labelled by users are collected into the feedback database, which is described by $R_{m \times n}$, where m is the number of feedback iterations and n denotes the number of images in the image gallery. That is, the component r_{ij} in the matrix represents the relevance score of image I_j at i-th feedback iteration F_i. An example feedback database is shown in Fig. 14. Subsequently, the correlations between images in the database and the current feedback are measured by collaborative filtering. By analyzing the log file information, this method speeds up the feedback process. However, it is heuristic-based and lacks empirical studies from real world users. What is more, negative samples are not considered.

Zhuang et al. (2002) developed a two-layer graph model for long-term learning, as shown in Fig. 15. It consists of a semantic layer and a visual layer. The semantic layer refers to correlations between images at the semantic level, which is represented by semantic links; the visual layer refers to visual similarity between images at the feature level, which is described by visual links. Each link is associated with a weight, indicating the extent at which two images are correlated. A learning strategy is utilized to adjust the semantic layer based on the relevance information from previous feedbacks. That is, after the user labelling some relevant and

Images

	I_1	I_2	I_3	I_4	I_5	I_6	$\cdots$	I_n
							$\cdots$	
F_1	1							1
F_2	1				1			
F_3	1	1			1			1
$\vdots$								
F_m	1	1			1			1

Feedback Iterations

Fig. 14 The feedback database

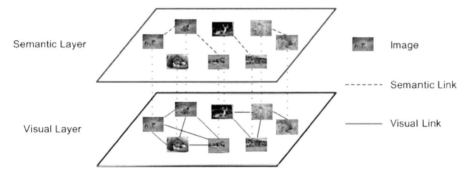

Fig. 15 The two-layer graph model

irrelevant images with respect to a query, a semantic link between each relevant image and the query is generated; while the current semantic link between an irrelevant image and the query is discarded. After this step, similarities between the query and its candidates will be recomputed by a similarity propagation process in order to generalize with either semantic links or visual links by adapting the weights of the semantic links. This method reduces the search space from the whole database to a more limited graph model. However, it requires large storage and the computation complexity is high for updating the model.

Jiang et al. (2005) combined long-term RF with hidden annotation to improve the effectiveness and efficiency in CBIR. This framework consists of two parts: the semantic concept learning and hidden annotation. Based on feedback knowledge, semantic concepts of images in the database are automatically learned by a multi-layer semantic representation (MSR). MSR includes many semantic layers, each of which represents a hard partition in the semantic space and contains a set of concepts, as shown in Fig. 16. Here, N should be as small as possible; while $m, n, p, ..., q$ should be as large as possible. Subsequently, the hidden annotation is iteratively conducted as follows: 1) a small subset of images belonging to each learned concept are predicted by a semi-supervised learning algorithm and annotated by an annotator; and 2) the annotated images are put into the semantic knowledge of the image database. The MSR reflects the correlations among images and attempts to discover the hidden concepts among images, and hence the semantic gap between low-level features and high-level concepts is reduced. Moreover, the proposed method is claimed to be more accurate than simple keyword-based representation.

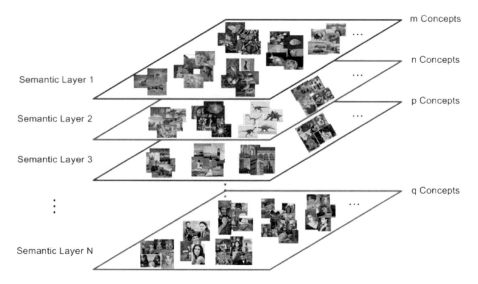

Fig. 16 The Multi-layer semantic representation

Based on the assumption that images in the same group may have the same common semantic concept, Urban and Jose (2006) reflected semantic correlations between images within a group in a single graph, which they called the image-context graph (ICG). It is composed of images, terms and low-level features and represents correlations between them. An illustrative graph with three images, four terms, and two kinds of low-level visual features is shown in Fig. 17. The links in ICG reflect: i) correlations between images and their low-level features; ii) low-level feature similarities; and iii) semantic correlations. The link weights are adapted by combining long-term learning and short-term learning. During the long-term learning, for each positive feedback, if it has never been labelled before, a new link will be constructed between the feedback and each of all other labelled positive samples; otherwise, its link weight will be added by 1. For each negative feedback sample, the weight will be decremented by 1. In short-term learning, weights of low-level features can be regarded as being proportional to the sum of weighted distance between each positive sample and the query. Finally, link weights are adjusted by corresponding feature weights. The ICG model is good at combining different features and achieves better performance than traditional retrieval methods that firstly treat each kind of feature separately and then integrate them into a vector. However, how to automatically update the link weights is not discussed.

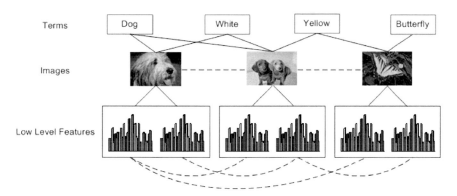

Fig. 17 The image-context graph

4.3   Clustering-Based Algorithms

Employing accumulated information from each feedback session, images in the database are clustered into groups with different semantic concepts. In this way, clustering-based techniques achieve refined retrieval results of CBIR by utilizing the information from each cluster or group.

Han et al. (2005) developed a semantic-based memory learning model to deal with the sparsity of user log information. Based on the accumulated feedback knowledge, semantic correlations between images are updated by the ratio of the co-positive-feedback frequency and the co-feedback frequency. The former frequency indicates the co-occurrence that two images are both labelled as positive samples to the same query; while the latter represents the occurrence that they are both labelled as feedback samples and at least one of them is labelled as positive. After that, a learning strategy is utilized to analyze the semantic correlations to predict future relevant images. It has the following main steps: i) cluster images in the database into a few semantic-correlated clusters by k-means clustering, according to the obtained semantic correlations; 2) assume that each cluster is related to a specific semantic subject, based on the ranking of images; 3) discover hidden semantic correlations between images based on the ranking, which represents how likely an image contains the corresponding concept in its cluster; and iv) estimate the semantic relevance between each image and each labelled feedback sample by a probabilistic model. The similarity measure is also adapted according to feedback knowledge by combining low-level feature-based short-term learning and the semantic-based memory learning, which supplements each other.

This method is easy to implement and annotate images; and so be able to general-ize keyword annotation from labelled feedbacks to unlabelled images. However, it is worth noting that correlations of negative samples are not considered and the theoretical verification of this learning algorithm is not provided.

A semantic grouping method, which also accumulates user feedback informa-tion, was proposed by Yoshizawa and Schweitzer (2004). The key point is that for current feedback session, either the set of negative samples or the set of positive samples that have been retrieved before and previously labelled as positive is firstly classified into a class, respectively. Hence, these two classes are integrated into one group. In this way, this semantic group is updated by some rules after each query session, which can provide useful information for future feedbacks and gradually enhance the retrieval performance.

4.4   Feature Representation-Based Methods

By adjusting the relative weighting of feature vectors based on accumulated feed-back knowledge, feature representation-based approaches can provide improved retrieval accuracy in CBIR.

Feature representation of images is refined by modifying feature vectors ac-cording to a subset of labels accumulated during feedback sessions (Cord and Gosselin 2006). Vectors belonging to the same concept are accumulated around the corresponding centre; while those intermediate among different concepts are distributed between concept centres. A diagrammatic representation with six con-cepts is shown in Fig. 18. The representation of images is allowed to be improved even if the prior information of the concepts is unknown, such as their sizes, their numbers, and their structures. This technique is able to handle mixed concepts. However, it is under a constraint that the dimensionality of feature vectors is higher than the number of concepts, which is not always applicable to real-world applications.

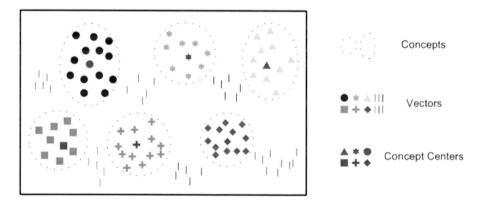

Fig. 18 Vector-concept representation

Jing et al. (2004) proposed a region-based representation method and combined it with RF, where each region is represented by the low-level features extracted from it and a corresponding weight. On the basis of the region-based representation, a query point movement (QPM)-based RF algorithm is developed. All segmented regions in positive samples are consolidated to form a new region as an optimal query. The weights of these regions are adjusted according to the accumulated user's feedback information, which is weighted on the latest positive samples and thus make the optimal query as a pseudo image. The weighting is accumulated by $RI_i(l) = \left((l-1)*RI_i(l-1) + LRI_i\right)/l$, where $l > 0$ and LRI_i denotes the up-to-date region weight of a given region R_i in a positive sample at the end of a feedback session. By this means, the accumulated region's importance can be used for subsequent query sessions, which result in more accurate estimation of averaged region importance for all positive samples. This method inherits merits from both RF and region-based representations, it also improves the retrieval accuracy for future feedback sessions. However, spatial relationships between the segmented regions are not taken into account.

4.5   Similarity Measure Modification-Based Approaches

There are some long-term learning approaches based on similarity measure modification. They adapt similarity scores at the end of each feedback session.

Chang and Yeung (2005) proposed a stepwise adaptation algorithm to modify the similarity measure based on the accumulated feedback knowledge. For each feedback session, the algorithm is integrated with pairwise constraints, which are separated into a family of small subsets, each of a fixed size ω. When the number of newly collected pairwise constraints reaches ω, a new distance metric will be obtained using a semi-supervised metric adaptation method. The small subsets of constraints will then be removed. Therefore, only temporary storage is required for log file information. Moreover, this method improves both the retrieval effectiveness and efficiency of CBIR.

Based on accumulated relevance information in preceding feedback sessions, a semantic similarity measure is developed and refined by user annotations (Fournier and Cord 2002). At the end of each feedback session, the long-term measure between a query R and a target image T is updated by $S_{LT}(R,T) = S_{LT}(R,T) + \mu\left(S_D^T - S_{LT}(R,T)\right)$, where S_D^T is the desired similarity according to user annotations; for a number of queries $\mathbf{R} = \{R_i, 1 \le i \le l\}$ and target images $\mathbf{T}$, the measure is $S_{LT}(\mathbf{R},\mathbf{T}) = \frac{1}{l}\sum_{i=1}^{l} S_{LT}(R_i,T)$. While computing the long-term similarity, visual similarity $S_V(\mathbf{R},\mathbf{T})$ is measured simultaneously. The final similarity score is then obtained by $S(\mathbf{R},\mathbf{T}) = S_V(\mathbf{R},\mathbf{T}) \cdot S_{LT}(\mathbf{R},\mathbf{T})$. This approach has shown its effectiveness. However, it has not been verified in large- scale databases or by various users.

He et al. (2004) assumed that images lie in a low-dimensional Riemannian manifold. Based on the accumulated log file information, a radial basis function network is trained to optimally project the low-level feature space into the hidden semantic space. After this, a distance matrix is constructed to measure distances between images in the semantic space. The distances between positive samples are reduced; whereas those between negative ones are increased. As RF progresses, the matrix will reveal the distances between images in the semantic space. From this assumption, RF only needs to be conducted in the sub-manifold in question rather than the whole ambient space. However, no full justification is provided to verify this assumption is reliable. In addition, whether the projection is one-to-one or many-to-one is ambiguous.

Hoi et al. (2006) computed relevance information between a query and images in the database by a relevance score $f_q(I_i) = \frac{1}{2}(f_{LG}(I_i) + f_{LL}(I_i))$, where q is a query, I_i is an image in the database, $f_{LG}(I_i)$ is the relevance function based on user log data information, and $f_{LL}(I_i)$ is the relevance function for low-level features. Both $f_{LG}(I_i)$ and $f_{LL}(I_i)$ are normalized to $[0,1]$. For each image, $f_{LG}(I_i)$ is computed by the difference between the similarity to positive samples and the similarity to negative samples. Afterwards, relevance information is integrated with traditional RF to construct a unified log-based framework. Because the log data inevitably contain errors, a new SVM classifier is developed based on the regularization theory to deal with this effective source of noise. This framework shows an enhanced effectiveness in CBIR based on some empirical studies. Unfortunately, its computation efficiency in training the novel SVM and computing the relevance information of log files is poor.

Leung and Auer (2008) considered the user-uncertainty in distinguishing relevance images that look similar and proposed a probabilistic model to estimate the probability of an image to be selected as relevant if it is equally similar to the other. It divides the search space into m regions by a weighted k-means, where the size of a region is determined by the sum of weights in the whole region and each centre image in a region will be returned to the user. In the current feedback session, the most relevant image to the user will get more weight in the next feedback iteration.

4.6   Others

A kernel matrix (Gosselin and Cord 2005), which aims to reflect the nonlinear property of semantic knowledge, is adapted by $K_{t+1} = (1-\rho)K_t + \rho K_{y_i}$, where t is the current feedback session, $\rho \in [0,1]$ is the weighting parameter and $y_i \in \{-1,0,1\}$ are the labels of the provided samples at the t-th session. This approach was developed to model mixed categories, and it is also, as claimed, able to learn any kind of semantic links through a similarity matrix framework.

He et al. (2003) developed a RF framework by combining both long-term learning and short-term learning. The long-term learning is based on constructing a semantic space from the labelled positive feedbacks after several iterations. The semantic space is represented by a vector space model and is updated according to further accumulated feedbacks. The short-term learning aims to model traditional RF methods with query refinement based on low-level features. Considering the storage limitation, dimensionality of the semantic space is reduced through the use of SVD. Retrieved results are improved by taking into account the semantic space as well as low-level visual features. However, negative samples are not considered during all feedback sessions.

A novel user interface utilizing long-term learning approach was proposed by Nedovic and Marques (2005). It is based on the assumption that the judgment of humans always outperforms any hypotheses based on extracted low-level features. Once two images are determined as similar by a user, they are firmly believed to be similar to each other. Colour and shape are extracted as low-level features and their weights are updated according to user labelled positive feedbacks. With this interface, retrieved images can be labelled as relevant or irrelevant in the form of semantics; meanwhile, users can view semantically similar images that are previously labelled by different users at various feedback sessions. This user interface has the advantage that no additional human effort is required in accumulating user feedback information than in a traditional RF process. However, it does not consider the negative feedbacks and has not been fully verified because of an absence of adequate empirical studies.

The optimal semantic space (Wacht et al. 2006), which is composed of high-level semantic information of each image in the database, was obtained by both positive and negative feedback information accumulated from all feedback sessions. A query is semantically refined by integrating the long-term learning with the short-term learning. The long-term learning is used for constructing the semantic space; while the short-term leaning is for updating the query weight vector, which includes the following steps: 1) initial retrieval results are obtained according to low-level feature; 2) the query is represented by high-level features based on the user-labelled feedbacks, each of which is represented by a semantic vector; 3) images are sorted according to high-level features; 4) the query weight vector is updated by the semantic space and user-labelled feedbacks; and 5) steps 3 and 4 are repeated if the user is not satisfied with retrieved results. The semantic space can better represent each image in the database, and thus high retrieval accuracy is always achieved for databases with similar semantic categories. Moreover, this method is not sensitive to the sizes of image databases.

An image relevance reinforcement model (Yin et al. 2005) is constructed to combine different RF approaches in CBIR. Given a query, the optimal RF technique is automatically selected for each feedback session. Moreover, a shared long-term memory is maintained for storing the relevance information over various feedback iterations from different users. The relevance information includes the latest query formulation, feature importance, and *a priori* probabilities of relevant and irrelevant images for each previously used query image in the database. Therefore, subsequent feedback iterations can be conducted based on the previous relevance information. In this way, the learning process is speeded up and it is suggested that multiple RF techniques perform better than a single RF algorithm. Furthermore, a concept digesting strategy was proposed to avoid large memory requirements.

5   Summary

Relevance feedback techniques in content-based image retrieval are comprehensively reviewed and categorized into short-term learning and long-term learning in this chapter. Short-term learning is regarded as a learning process restricted to one user for the current feedback session and divided into one-class, two-class, and multi-class approaches; long-term learning is based on accumulated user feedback information from different users in various feedback sessions and categorized into the following types: i) latent semantic indexing-based techniques; ii) correlation-based approaches; iii) clustering-based algorithms; iv) feature representation-based methods; v) similarity measure modification-based approaches; and vi) other techniques. Most long-term learning algorithms are combined with short-term learning methods and improved retrieval performance has been reported in terms of both effectiveness and efficiency. Although some techniques are sensitive to the quality of log files and require high storage space, they point in the direction of utilizing log file information during feedback sessions. Future work could focus on further improving the retrieval performance and reducing the storage requirement, which may be achieved through preprocessing on accumulated user log files, e.g., discard a subset of log files with useless information from *non-serious* users. Statistically analyzing log files may also obtain significant results to accelerate the retrieval process as well as improving the retrieval effectiveness. Additionally, the methods of text analysis can be adopted to deal with user log files, e.g., to use vector approximation to speed up the retrieval process. Furthermore, machine learning models, such as the prevailing manifold learning, could be applied to clustering the items of log files in order to enhance the retrieval performance.

References

[1] Anick, P.: Using Terminological Feedback for Web Search Refinement - A Log-based Study. In: Proc. Int'l ACM SIGIR Conf. Research and Development in Information Retrieval, pp. 88–95 (2003)
[2] Bian, W., Tao, D.: Biased Discriminant Euclidean Embedding for Content-Based Image Retrieval. IEEE Transactions on Image Processing 19(2), 545–554 (2010)
[3] Blum, A., Mitchell, T.: Combining Labeled and Unlabeled Data with Cotraining. In: COLT: Proc. Workshop on Computational Learning Theory (1998)
[4] Breiman, L.: Bagging Predictors. Machine Learning 24(2), 123–140 (1995)
[5] Burges, J.C.: A Tutorial on Support Vector Machines for Pattern Recognition. Data Mining and Knowledge Discovery 2(2), 121–167 (1998)

[6]   Carson, C., Belongie, S., Greenspan, H., Malik, J.: Blobworld: Image Segmentation Using Expectation-Maximization and Its Application to Image Querying. IEEE Trans. Pattern Analysis and Machine Intelligence 24(8), 1026–1038 (2002)

[7]   Chang, T., Kuo, C.: Texture Analysis and Classification with Tree-Structured Wavelet Transform. IEEE Trans. Image Processing 2(4), 429–441 (1993)

[8]   Chang, H., Yeung, D.Y.: Stepwise Metric Adaptation Based on Semi-Supervised Learning for Boosting Image Retrieval Performance. In: British Machine Vision Conference (2005)

[9]   Chen, X., Zhang, C., Chen, S.C., Chen, M.: A Latent Semantic Indexing Based Method for Solving Multiple Instance Learning Problem in Region-Based Image Retrieval. In: Proc. IEEE Int'l Symposium Multimedia, pp. 37–45 (2005)

[10] Chen, Y., Zhou, X., Huang, T.S.: One-class SVM for Learning in Image Retrieval. In: Proc. IEEE Int'l Conf. Image Processing, vol. 1, pp. 34–37 (2001)

[11] Cord, M., Gosselin, P.H.: Image Retrieval using Long-Term Semantic Learning. In: IEEE Int'l Conf. Image Processing, pp. 2,909–2,912 (2006)

[12] Cox, I.J., Miller, M.L., Minka, T.P., Papathomas, T.V., Yianilos, P.N.: The Bayesian Image Retrieval System, PicHunter: Theory, Implementation and Psychophysical Experiments. IEEE Trans. Image Processing 9(1), 20–37 (2000)

[13] Cui, H., Wen, J., Nie, J., Ma, W.: Query Expansion by Mining User Logs. IEEE Trans. Knowledge and Data Engineering 15(4), 829–939 (2003)

[14] Daubechies, I.: The Wavelet Transform, Time-frequency Localization and Signal Analysis. IEEE Trans. Information Theory 36(5), 961–1005 (1990)

[15] Daugman, J.G.: Two-Dimensional Spectral Analysis of Cortical Receptive Field Profile. Vision Research 20, 847–856 (1980)

[16] Daugman, J.G.: Uncertainty Relation for Resolution in Space, Spatial Frequency, and Orientation Optimized by Two–Dimensional Visual Cortical Filters. J. Optical Society of America 2(7), 1,160–1,169 (1985)

[17] Deerwester, S., Dumais, S.T., Furnas, G.W., Landauer, T.K., Harshman, R.: Indexing by latent semantic analysis. J. American Society for Information Science 41, 391–407 (1990)

[18] Deng, J., Dong, W., Socher, R., Li, L.J., Li, K., Li, F.F.: ImageNet: A Large-scale Hierarchical Image Database. In: IEEE Conf. Computer Vision and Pattern Recognition (2009)

[19] Duda, R.O., Hart, P.E., Stork, D.G.: Pattern Classification. Wiley, John & Sons, Incorporated (1999)

[20] Fahlman, S.E., Lebiere, C.: The Cascade-Correlation Learning Architecture. In: Advances in Neural Information Processing Systems 2, pp. 524–532 (1990)

[21] Freund, Y., Schapire, R.E.: Experiments with a New Boosting Algorithm. In: Proc. Int'l Conf. Machine Learning, pp. 148–156 (1996)

[22] Fournier, J., Cord, M.: Long-term Similarity Learning in Content-based Image Retrieval. In: Proc. IEEE Int'l Conf. Image Processing, vol. 1, pp. 1,441–1,444 (2002)

[23] Gevers, T., Smeulders, A.: Pictoseek: Combining color and shape invariant features for image retrieval. IEEE Trans. Image Processing 9(1), 102–119 (2000)

[24] Giacinto, G., Roli, F.: Instance-Based Relevance Feedback for Image Retrieval. In: Advances Neural Information Processing Systems, pp. 489–496 (2004)

[25] Gosselin, P.H., Cord, M.: Semantic Kernel Learning for Interactive Image Retrieval. In: IEEE Int'l Conf. Image Processing, vol. 1, 1,177–1,180 (2005)

[26] Guo, G., Jain, A.K., Ma, W., Zhang, H.: Learning Similarity Measure for Natural Image Retrieval with Relevance Feedback. IEEE Trans. Neural Networks 12(4), 811–820 (2002)

[27] Han, J., Ngan, K.N., Li, M., Zhang, H.J.: A Memory Learning Framework for Effective Image Retrieval. IEEE Trans. Image Processing 14(4), 511–524 (2005)

[28] He, X., King, O., Ma, W., Li, M., Zhang, H.: Learning a Semantic Space from User's Relevance Feedback for Image Retrieval. IEEE Trans. Circuits and Systems for Video Technology 13(1), 39–48 (2003)

[29] He, X., Ma, W.Y., Zhang, H.J.: Learning an Image Manifold for Retrieval. In: Proc. ACM Int'l Conf. Multimedia, pp. 17–23 (2004)

[30] Heisterkamp, D.R.: Building a Latent Semantic Index of an Image Database from Patterns of Relevance Feedback. In: Proc. Int'l Conf. Pattern Recognition, vol. 4, pp. 134–137 (2002)

[31] Ho, T.K.: The Random Subspace Method for Constructing Decision Forests. IEEE Trans. Pattern Analysis and Machine Intelligence 20(8), 832–844 (1998)

[32] Hoi, C.H., Chan, C.H., Huang, K., Lyu, M.R., King, I.: Biased Support Vector Machine for Relevance Feedback in Image Retrieval. In: Proc. Int'l Joint Conf. Neural Networks, pp. 3189–3194 (2004)

[33] Hoi, C.H., Lyu, M.R.: Group-based Relevance Feedback with Support Vector Machine Ensembles. In: Int'l Conf. Pattern Recognition, vol. 3, pp. 874–877 (2004)

[34] Hoi, S.C.H., Lyu, M.R., Jin, R.: A Unified Log-based Relevance Feedback Scheme for Image Retrieval. IEEE Trans. Knowledge and Data Engineering 18(4), 509–524 (2006)

[35] Hsu, C.T., Li, C.Y.: Relevance Feedback Using Generalized Bayesian Framework With Region-Based Optimization Learning. IEEE Trans. Image Processing 14(10), 1,617–1,631 (2005)

[36] Jain, A., Vailaya, A.: Image Retrieval Using Color and Shape. Pattern Recognition 29(8), 1233–1244 (1996)

[37] Jain, A.K.: Fundamentals of Digital Image Processing. Prentice Hall Inc., New Jersey (1989)

[38] Jain, A.K., Lee, J.E., Jin, R., Gregg, N.: Content-based Image Retrieval: An Application to Tattoo Images. In: IEEE Int'l Conf. Image Processing, vol. 2, pp. 745–742 (2009)

[39] Jiang, W., Er, G., Dai, Q., Gu, J.: Hidden Annotation for Image Retrieval with Long-Term Relevance Feedback Learning. Pattern Recognition 38(11), 2,007–2,021 (2005)

[40] Jing, F., Li, M., Zhang, H., Zhang, B.: Relevance Feedback in Region-Based Image Retrieval. IEEE Trans. Circuits and Systems for Video Technology 14(5), 672–681 (2004)

[41] Koskela, M., Laaksonen, J.: Using Long-Term Learning to Improve Efficiency of Content-Based Image Retrieval. In: Proc. Int'l Workshop on Pattern Recognition in Information Systems, pp. 72–79 (2003)

[42] Kushki, A., Androutsos, P., Plataniotis, K.N., Venetsanopoulos, A.N.: Query Feedback for Interactive Image Retrieval. IEEE Trans. Circuits and Systems for Video Technology 14, 644–655 (2004)

[43] Laaksonen, J., Koskela, M., Laakso, S., Oja, E.: PicSOM—Content-based Image Retrieval with Self-Organizing Maps. Pattern Recognition Letters 21(13-14), 1,199–1,207 (2000)

[44] LeCun, Y., Boser, B., Denker, J.S., Henderson, D., Howard, R.E., Hubbard, W., Jackel, L.D.: Backpropagation Applied to Handwritten Zip Code Recognition. Neural Computation 1(4), 541–551 (1989)

[45] Leung, A.P., Auer, P.: An Efficient Search Algorithm for Content-Based Image Retrieval with User Feedback. In: IEEE Int'l Conf. Data Mining Workshops, pp. 884–890 (2008)

[46] Li, J., Allinson, N.M.: A Comprehensive Review of Current Local Features for Computer Vision. Neurocomputing 71(10-12), 1771–1787 (2008)

[47] Li, J., Allinson, N.M.: Long-term Learning in Content-based Image Retrieval. Int'l J. Imaging Systems and Technology 18(2-3), 160–169 (2008)

[48] Li, J., Allinson, N., Tao, D., Li, X.: Multitraining Support Vector Machine for Image Retrieval. IEEE Trans. Image Processing 15(11), 3,597–3,601 (2006)

[49] Long, F., Zhang, H., Feng, D.: Fundamentals of Content-based Image Retrieval. In: Multimedia Information Retrieval and Management, pp. 1–12. Springer (2003)

[50] Lowe, D.G.: Distinctive Image Features from Scale-Invariant Keypoints. Int'l J. Computer Vision 60(2), 91–110 (2004)

[51] Lu, Y., Zhang, H., Liu, W., Hu, C.: Joint Semantics and Feature Based Image Retrieval Using Relevance Feedback. IEEE Trans. Multimedia 5(3), 339–347 (2003)

[52] Manjunath, B., Ma, W.: Texture Features for Browsing and Retrieval of Image Data. IEEE Trans. Pattern Analysis and Machine Intelligence 18(8), 837–842 (1996)

[53] Manjunath, B., Ohm, J., Vasudevan, V., Yamada, A.: Color and Texture Descriptors. IEEE Trans. Circuits and Systems for Video Technology 11(6), 703–715 (2001)

[54] Mikolajczyk, K., Schmid, C.: A Performance Evaluation of Local Descriptors. IEEE Trans. Pattern Analysis and Machine Intelligence 27(10), 1,615–1,630 (2005)

[55] Mikolajczyk, K., Tuytelaars, T., Schmid, C., Zisserman, A., Matas, J., Schaffalitzky, F., Kadir, T., Van Gool, L.: A Comparison of Affine Region Detectors. Int'l J. Computer Vision 65(1/2), 43–72 (2005)

[56] Minka, T.P., Picard, R.W.: Interactive Learning with a "Society of Models". Pattern Recognition 30(4), 565–581 (1997)

[57] Müller, H., Müller, W., Squire, D.M., Marchand-Maillet, S., Pun, T.: Long-Term Learning from User Behavior in Content-Based Image Retrieval. Technical report, University of Geneva (2000)

[58] Nakazato, M., Dagli, C., Huang, T.S.: Evaluating Group-based Relevance Feedback for Content-based Image Retrieval. In: Proc. IEEE Int'l. Conf. on Image Processing, pp. 599–602 (2003)

[59] Nedovic, V., Marques, O.: A Collaborative, Long-term Learning Approach to Using Relevance Feedback in Content-based Image Retrieval Systems. In: Int'l Symposium ELMAR, pp. 143–146 (2005)

[60] Niblack, W., Barber, R., Equitz, W., Flickner, M., Glasman, E., Petkovic, D., Yanker, P., Faloutsos, C., Taubin, G.: The QBIC project: Quering Images by Content using Color, Texture, and Shape. In: Proc. SPIE Storage and Retrieval for Image and Video Databases, pp. 173–187 (1993)

[61] Ortega, M., Rui, Y., Chakrabarti, K., Mehrotra, S., Huang, T.S.: Supporting similarity queries in MARS. In: Proc. ACM Int'l Conf. Multimedia, pp. 403–413 (1997)

[62] Pass, G., Zabih, R., Miller, J.: Comparing Images Using Color Coherence Vectors. In: Proc. ACM Int'l Conf. Multimedia, pp. 65–73 (1996)

[63] Rocchio, J.J.: Document Retrieval System: Optimization and Evaluation. PhD dissertation, Harvard Computational Lab, Harvard University, Cambridge, MA (1996)

[64] Rubner, Y., Tomasi, C., Guibas, L.J.: The Earth Mover's Distance as a Metric for Image Retrieval. Int'l J. Computer Vision 40(2), 99–121 (2000)

[65] Rui, Y., Huang, T.S., Ortega, M., Mehrotra, S.: Relevance Feedback: A Power Tool in Interactive Content-based Image Retrieval. IEEE Trans. Circuits and Systems for Video Technology 8(5), 644–655 (1998)

[66] Sav, S., O'Connor, N., Smeaton, A., Murphy, N.: Associating Low-level Features with Semantic Concepts using Video Objects and Relevance Feedback. In: Int'l Workshop on Image Analysis for Multimedia Interactive Services (2005)

[67] Schapire, R.E.: The Strength of Weak Learnability. Machine Learning 5(2), 197–227 (1990)

[68] Schmid, C., Mohr, R.: Local Grayvalue Invariants for Image Retrieval. IEEE Trans. Pattern Analysis and Machine Intelligence 19(5), 530–534 (1997)

[69] Smeulders, A.W.M., Worring, M., Santini, S., Gupta, A., Jain, R.: Content-based image retrieval at the end of the early years. IEEE Trans. Pattern Analysis and Machine Intelligence 22(12), 1,349–1,380 (2000)

[70] Su, Z., Li, S., Zhang, H.: Extraction of Feature Subspaces for Content-Based Retrieval Using Relevance Feedback. In: Proc. of ACM Multimedia, pp. 98–106 (2001)

[71] Su, Z., Zhang, H., Li, S., Ma, S.: Relevance Feedback in Content-Based Image Retrieval: Bayesian Framework, Feature Subspaces, and Progressive Learning. IEEE Trans. Image Processing 12(8), 924–937 (2003)

[72] Swain, M., Ballard, D.: Color Indexing. Int'l J. Computer Vision 7(1), 11–32 (1991)

[73] Tai, X., Ren, F., Kita, K.: Long-Term Relevance Feedback Using Simple PCA and Linear Transformation. In: Proc. Int'l Workshop on Database and Expert Systems Applications, pp. 261–268 (2002)

[74] Tamura, H., Mori, S., Yamawaki, T.: Texture Features Corresponding to Visual Perception. IEEE Trans. Systems, Man, and Cybernetics 8(6), 460–473 (1978)

[75] Tao, D., Tang, X., Li, X., Wu, X.: Asymmetric Bagging and Random Subspace for Support Vector Machines-based Relevance Feedback in Image Retrieval. IEEE Trans. Pattern Analysis and Machine Intelligence 28(7), 1,088–1,099 (2006)

[76] Tao, D., Tang, X., Li, X., Rui, Y.: Kernel Direct Biased Discriminant Analysis: A New Content-based Image Retrieval Relevance Feedback Algorithm. IEEE Trans. Multimedia 8(4), 716–727 (2006)

[77] Tao, D., Li, X., Maybank, S.J.: Negative Samples Analysis in Relevance Feedback. IEEE Trans. Knowledge and Data Engineering 19(4), 568–580 (2007)

[78] Tong, S., Chang, E.: Support Vector Machine Active Learning for Image Retrieval. In: Proc. ACM Int'l Conf. Multimedia, pp. 107–118 (2001)

[79] Torralba, A., Fergus, R., Freeman, W.T.: 80 Million Tiny Images: A Large Dataset for Non-parametric Object and Scene Recognition. IEEE Trans. Pattern Analysis and Machine Intelligence 30(11), 1958–1970 (2008)

[80] Torre, F.D.L., Kanade, T.: Multimodal Oriented Discriminant Analysis. In: Int'l Conf. Machine Learning (2005)

[81] Tversky, A.: Features of Similarity. Psychological Review 84(4), 327–352 (1977)

[82] Urban, J., Jose, J.M.: Adaptive Image Retrieval using a Graph Model for Semantic Feature Integration. In: Proc. ACM Int'l Workshop on Multimedia Information Retrieval, pp. 117–126 (2006)

[83] Vapnik, V.: The Nature of Statistical Learning Theory. Springer (1995)

[84] Vasconcelos, N., Kunt, M.: Content-based Retrieval from Image Databases: Current Solutions and Future Directions. In: Proc. IEEE Int'l Conf. Image Processing, vol. 3, pp. 6–9 (2001)

[85] Wacht, M., Shan, J., Qi, X.: A Short-Term and Long-Term Learning Approach for Content-Based Image Retrieval. In: Proc. IEEE Int'l Conf. Acoustics, Speech, and Signal Processing, vol. 2, pp. 389–392 (2006)

[86] Wang, J.Z., Li, J., Wiederhold, G.: SIMPLIcity: Semantics-Sensitive Integrated Matching for Picture Libraries. IEEE Trans. Pattern Analysis and Machine Intelligence 23(9), 947–963 (2001)

[87] Yang, J., Li, Q., Zhuang, Y.: Image Retrieval and Relevance Feedback using Peer Indexing. In: Proc. IEEE Int'l Conf. Multimedia and Expo, vol. 2, pp. 409–412 (2002)

[88] Yin, P.Y., Bhanu, B., Chang, K.C., Dong, A.: Integrating Relevance Feedback Techniques for Image Retrieval Using Reinforcement Learning. IEEE Trans. Pattern Analysis and Machine Intelligence 27(10), 1,536–1,551 (2005)

[89] Yoshizawa, T., Schweitzer, H.: Long-term Learning of Semantic Grouping from Relevance-feedback. In: Proc. ACM SIGMM Int'l Workshop on Multimedia Information Retrieval, pp. 165–172 (2004)

[90] Zhang, L., Lin, F., Zhang, B.: Support Vector Machine Learning for Image Retrieval. In: Proc. IEEE Int. Conf. Image Processing, vol. 2, pp. 21–724 (2001)

[91] Zhou, X.S., Huang, T.S.: Comparing Discriminating Transformations and SVM for Learning during Multimedia Retrieval. In: Proc. ACM International Conference on Multimedia (2001)

[92] Zhou, X.S., Huang, T.S.: Relevance feedback in image retrieval: A comprehensive review. Multimedia Systems 8(6), 536–544 (2003)

[93] Zhou, X.S., Huang, T.S.: Small Sample Learning during Multimedia Retrieval using Biasmap. In: Proc. IEEE Int'l Conf. Computer Vision and Pattern Recognition, vol. 1, pp. 11–17 (2001)

[94] Zhou, X.D., Zhang, L., Liu, L., Zhang, Q., Shi, B.: A Relevance Feedback Method in Image Retrieval by Analyzing Feedback Log File. In: Proc. Int'l Conf. Machine Learning and Cybernetics, vol. 3, pp. 1,641–1,646 (2002)

[95] Zhou, Z.H.: Ensemble Learning. In: Encyclopedia of Biometrics, pp. 270–273 (2009)

[96] Zhuang, Y., Yang, J., Li, Q., Pan, Y.: A Graphic-theoretic Model for Incremental Relevance Feedback in Image Retrieval. In: Proc. Int'l Conf. Image Processing, vol. 1, pp. 413–416 (2002)

Chapter 14
Learning Structural Representations of Text Documents in Large Document Collections

Ah Chung Tsoi, Markus Hagenbuchner, Milly Kc, and ShuJia Zhang

Abstract. The main aim of this chapter is to study the effects of structural representation of text documents when applying a connectionist approach to modelling the domain. While text documents are often processed un-structured, we will show in this chapter that the performance and problem solving capability of machine learning methods can be enhanced through the use of suitable structural representations of text documents. It will be shown that the extraction of structure from text documents does not require a knowledge of the underlying semantic relationships among words used in the text. This chapter describes an extension of the bag of words approach. By incorporating the "relatedness" of word tokens as they are used in the context of a document, this results in a structural representation of text documents which is richer in information than the bag of words approach alone. An application to very large datasets for a classification and a regression problem will show that our approach scales very well. The classification problem will be tackled by the latest in a series of techniques which applied the idea of self organizing map to graph domains. It is shown that with the incorporation of the relatedness information as expressed using the Concept Link Graph, the resulting clusters are tighter when compared them with those obtained using a self organizing map alone using a bag of words representation. The regression problem is to rank a text corpus. In this case, the idea is to include content information in the ranking of documents and compare them with those obtained using PageRank. In this case, the results are inconclusive due possibly to the truncation of the representation of the Concept Link Graph representations. It is conjectured that the ranking of documents will be sped up if we include the Concept Link Graph representation of all documents together with their hyperlinked structure. The methods described in this chapter are capable of solving real world and data mining problems.

Ah Chung Tsoi
Macau University of Science and Technology, Macau SAR, China
e-mail: actsoi@must.edu.mo

Markus Hagenbuchner · Milly Kc · ShuJia Zhang
University of Wollongong, Australia
e-mail: {markus,millykc,shujia}@uow.edu.au

M. Bianchini et al. (Eds.): *Handbook on Neural Information Processing*, ISRL 49, pp. 471–503.
DOI: 10.1007/978-3-642-36657-4_14   © Springer-Verlag Berlin Heidelberg 2013

1 Introduction

The number of documents being created, by enterprises, as part of the business functions, the government, or individuals, continue to increase at a rapid rate. Most of these documents are "born digital", that is, they are created in the digital format or they exist on the Internet. These documents can be created by word processors, formatted outputs generated by computers, or generated automatically by the computers. Increasingly it is difficult for enterprises to keep track of the various documents generated, and to know their contents. With the popularity of the XML (Extended Markup Language) as a formatting language for documents, especially its adoption by Microsoft Word as the default output format, more and more documents are created in the XML format. This document formatting language allows a user to denote a paragraph, a sentence or a word. There exist query languages for XML (such as XQuery) which treat XML documents as a source of information, and allows for an interaction with XML documents in a similar manner to interaction with databases. However, such query languages do not scale well with large document collections, requiring that all documents are formatted in XML, and have limitations to applications such as document categorization, document ranking, and many others.

Machine learning offers a more generic approach to solving practical problems [1]. In fact, it has been proven that some machine learning methods offer universal approximation properties [2, 3], and hence, can solve any problem that can be described by a nonlinear function with mild restrictions on its continuity properties, the nonlinear function can be pointwise dis-continuous. Machine learning methods require that data is presented in a suitable form. First, the data needs to be represented in a form that can be understood by the machine learning method [1]. For example, many machine learning methods require that information is presented in vectorial form. Second, machine learning methods require in general that the data is represented in a form that can support a given learning task. For example, a machine learning method would be unable to solve a document categorization problem if the input data does not contain any information that is relevant for the task.

A traditional and still very common method to represent a text document is to use what is known as a "bag of words" (BoW) approach [4]. In this approach, the words in a document are first de–stemmed so that they only retain their word stems, and these stemmed words modulo a set of commonly occurring words, are taken as the elements of an n–vector $\mathbf{w}$. A document D can then be represented as $\mathbf{d}^T\mathbf{w}$, where $\mathbf{d} \in \mathcal{R}^n$ and its elements represent the number of times of occurrence of the corresponding word–stem in the document and the superscript T denotes the transpose of the vector. In this representation of a document, the context among words is stripped. Each word is assumed to occur in isolation from the other words, and each word exists independently of one another. This representation of the document is popular with most users since the algorithm is relatively simple and fast. Given its simplicity, and its "agnostic" nature about the context in which a word occurs, its successes in various applications concerning textual analysis is quite surprising.

On the other end of the spectrum, there are a number of methods which strive to build a semantic understanding of a document [5, 6]. In such an approach [6], each word occurring in the document is tagged. Its part of speech, e.g., if it is a verb, a noun, a preposition, is provided, and its meaning in the context is known. Thus, in this model, the full context in which a word occurs is known, and its relationships to its neighbours are known. This method while is more accurate than the "bag of words" representation it is time–consuming, and expensive to build.

So far a user needs to choose: if one goes for simplicity, then the "bag of words" is a good choice, as the model is fast to build; on the other hand, if one goes for comprehensiveness, then the semantic representation is a good choice, as it provides the meaning (after suitable disambiguation of the word which might have a number of meanings in different contexts has been conducted) and the full context under which each word in the document occurs.

In this chapter, we wish to propose a balanced approach: we wish to build a representation model of a document using an approach similar to that used in the "bag of words" approach, but in our model we will provide a way in which the context under which a word occurs is encoded [7]. In our approach, we do not require to know the meaning of the words. The context under which the word is used will be obtained from the way in which the word is used within the text corpus. This approach thus frees one from the tedious task of providing words occurring in the text with meanings[1]. The resulting representation of the document will be in terms of a graph, in which the nodes represent words, and the links connecting the nodes represent the relationships among the words; the stronger the link, the more closely the two words are linked with one another, and the weaker the link, the two words are more loosely connected. We call such an approach the Concept Link Graph (CLG) approach [7].

Once a document is represented in terms of a graph, one can proceed to analyse the documents. Here in this chapter we are concerned with two particular such analyses of documents contained in a corpus, viz., how the documents can be clustered together, and how the importance of the documents are ranked relative to one another. Both problems are important questions to ask concerning a given corpus. In the first problem, one asks the question: if we are given a text corpus, how could the documents in this corpus be grouped together. Thus the assumption is that the text corpus can be grouped into different groups, with respect to a "similarity" criterion or measure. In other words, given a similarity measure, how can we group the documents into different groups, where the intra-group similarity is larger than the inter-group similarity. The grouping of the documents into groups will depend on the "similarity" criterion. Indeed, one of the questions which one could ask is: what is the similarity criterion used? Different similarity criteria will yield different sets of clusters. Thus, for example, if one wishes to group the documents together according to a "distance" criterion, then one can make use of the spectral clustering

---

[1] With the availability of the WordNet, this task is made much easier. However, it is still quite a tedious task of disambiguating words from the various possible meanings that it can take and then to provide it with the exact meaning within a context. Our proposed approach is totally data–driven. It does not require such devices. The obvious disadvantage is that the words are not endowed with meanings, and thus the analysis performed on texts is only good for machine representations, rather than for humans to understand.

technique [8]; if one wishes to group the documents together according to how they correlate with one another, then one could use a principal component analysis [9] or a canonical correlation analysis [10]. In our case, we wish to use the following criterion: if the "features" of two documents are close to one another in the feature space, then they should remain close in a lower dimensional display space. The display space is often assumed to be the two dimensional plane (this is done for convenience of showing the clusters for visualization purposes). In this case, we can use the so–called self organizing map (SOM) algorithm [11].

For the second problem, given a text corpus, the question to ask is: which document is more important than the others. Here again one needs to ask the question: what is the criterion of "importance" that one would use to rank the documents. One possibility is to define the importance of a document by a degree of popularity of the document. One such scheme is "popularity voting" where a document is considered more important if many other documents link to it [12]. This is shown by the number of in–links or references that the particular document receives. It helps if the documents that "voted" for the importance of the particular document are themselves "important" documents, that is they are voted for by others as important. The PageRank algorithm is such an algorithm which determines the rank of a document based on the link topology of the network alone [12]. The content of the documents are not considered explicitly, though it is considered implicitly through the linking of one document to another: two documents are linked and hence there is a link between them. In other words, they must be related somehow for such a link to exist. If two documents are not related, there does not exist a link between them. This chapter addresses an extended question: can the ranking of documents be improved by considering document content as well as the inter-document link structure? However, this is an ill-posed question, as we do not know *a priori* how to measure the "importance" of the documents. Thus, even if we can rank the documents according to their links and contents, we do not know if the ranked order of documents is correct or not. Sure, we can measure how fast the computed rank using both content and link structure approaches the PageRank, to demonstrate that the inclusion of content assists the computation of the PageRank, however, this is begging the question: why do we need to use the content information in the first instance, since the link structure information alone can compute the PageRank, albeit possibly at a slower rate.

Such an argument is overlooking a particular aspect of documents that exist in the Internet, viz., the Internet is an open collection in that documents are added to it every second of the day, every day of the year. The size of the Internet is huge, some claiming that it contains multiple billions of web pages [13, 14]. It is claimed that every day there are millions of new documents being created on the Internet [15, 16]. Even with the resources available to Google [17], it is not possible to crawl the Internet and retrieve all the documents in real time, and, hence, to obtain a complete view of the Internet at any moment of time. This causes the link structure of the documents on the Internet to be always open with respect to the view of any search engine. The PageRank computed is thus a relative measure only as it is based on incomplete information. The "true" PageRank, assuming that it is possible to have the entire structure of the Internet available, is never going to be achievable. Thus it makes sense to investigate if the inclusion of content of the web pages will accelerate the computation of the PageRank, or to accelerate the process of ranking of the web pages.

Intuitively speaking, one feels that the inclusion of the content information should help, though this is by no means demonstrated so far. This question is not as academic as it appears. There is one situation in which such an investigation would be helpful, viz., in the case of a distributed crawler. In a distributed crawler implementation, there are a number of crawlers distributed in various parts of the Internet. Each crawler crawls a limited portion of the Internet, and retrieves the information. There are at least two options in which the distributed crawler can do concerning the computation of the PageRank of the documents retrieved: (1) to pool all the documents together in a central place, and then compute the PageRank in the usual manner, (2) to compute the PageRank *in situ* in each distributed crawler and then combine them in a central place to form the composite PageRank. The pooling together of the retrieved documents from each distributed crawler poses considerable communication problems, especially if the number of retrieved documents is large. Hence, it might be better if the PageRank can be computed locally first and then the processed results are combined centrally to provide an overall ranking of all the retrieved documents. But the computation of locally retrieved web pages from the link structure alone would pose some issues, as there will be "edge" effects in that there may be many "missing" links in which the computation of the PageRank is required to estimate, if we were to compute the PageRank from the link structure alone. This is exactly the problem which we wish to investigate: would the provision of the content and link structure assist in the accelerated computation of the PageRank. If this is feasible, then, this will provide a good method for the ranking of locally retrieved web pages.

Now let us return to the issue of representation of documents. As indicated earlier, in this chapter we will discuss an approach to represent a document in terms of a graph. This appears to be not a common approach, favoured by other researchers. The main reason is that most of the machine learning techniques, process only vectorial inputs. If the data is presented in terms of a graph, then the graph needs to be "flattened" into a vectorial form first by removing all the contextual information. Then, the vectorial inputs at each node are concatenated together to form a long vector, padding with zeros where necessary. Thus, there is little incentive with traditional machine learning methods in representing a document as a graph, because one knows eventually the contextual information will be ignored in the processing of the documents.

In this chapter, our intuition is that the contextual information (expressed in topological terms) will assist in the task at hand, whether it is to cluster the corpus of documents into groups, or to rank them in terms of importance. Thus, when we are provided with graph representation of the documents, we wish to use algorithms which can process these graphs while taking into account of their contextual relationships. Towards that end, we have extended the classic self organizing map (SOM) [1] so that it can handle graph inputs [18]. This extension is applied to clustering the corpus into clusters. For the computation of the PageRank, we have extended the classic multilayer perceptron (MLP) [1] such that it can take graph inputs. This is called a graph neural network [19]. Thus, in this chapter, we will retain the topological relationships among the nodes in a graph as far as possible in the computational process.

The structure of this chapter will be as follows: Section 2 describes our approach to representing document content in terms of a graph. In Section 3, we will present a general framework in which a graph can be processed. Section 4 presents ways in which graphs can be trained in an unsupervised fashion, while in Section 5, we will present the parameter estimation of a graph neural network. The need to describe these two methods separately is due to the fact that in the unsupervised learning paradigm, the parameter estimation is closely coupled with the self organizing map model, while in the supervised learning paradigm, at least as it is formulated in terms of a graph neural network, the model can be separated from the parameter estimation step. Hence we prefer to consider them separately, even though they could be considered under the same general framework. Section 6 provides the application of the self organizing map idea presented in Section 4 to the problem of classifying documents in the Wikipedia dataset while in Section 7, the graph neural network idea is applied to the problem of importance ranking of documents obtained from the Internet. In Section 9, some conclusions are drawn, and some future directions of research will be indicated.

2 Representation of Unstructured or Semi-structured Text Documents

Before we will consider the Concept Link Graph approach [7], we will briefly consider the "bag of words" (BoW) approach [4] as a background. Consider that we are given a text document, with W' words. The words are de-stemmed to obtain just the stem of the word. Secondly, a set of commonly occurring words, like "the", "a", "of", etc. are deleted from the list of words. Assuming that we are left with W words, we can arrange these W words in a W dimensional vector $\mathbf{w}$, called a "vocabulary" vector, for obvious reasons, with each word being an element of the vector. Then, the document can be represented as $\mathbf{a}^T \mathbf{w}$, where $\mathbf{a}$ is a W dimensional vector, whose i-th element, is an integer, corresponding to the number of occurrence of the word w_i in the i-th position of the vector $\mathbf{w}$. Thus, a text corpus T consisting of D documents can be represented by a $W \times D$ matrix T, where W is the dimension of the vector formed by the union of all the vocabulary vectors of the individual documents in the corpus $\mathcal{T}$. Note that the BoW approach does not take into account of the context upon which the word occurs, as each word in the composite vocabulary $\mathbf{w}$ vector is independent. That such a simple method works and have been working well in most text processing tasks is quite surprising.

The Concept Link Graph (CLG) approach [7] is a modification of this BoW approach by extending it to accept the correlations among the words. In the CLG approach, instead of considering all words, we will only consider the set of nouns occurring in the text document. As in the BoW approach, we will consider only the de–stemmed version of the nouns, and we will take away a set of commonly occurring nouns. To avoid the need to compile an exhaustive list of nouns, the extraction of nouns from a text document is carried out based on grammatical rules in English to work out the probability of a word being a noun[2]. For example, if an untagged word follows an article (a relatively short list), and precedes a verb (the suffix provides clues, and a list of common verbs is also kept), then there is a very high probability that it is a noun. Or, if an untagged word follows an adjective (the suffix provides clues), then it is likely to be a noun as well.

For the set of nouns extracted from the set of D documents in the corpus, we will be able to form a term–document matrix as follows: form a W dimensional "vocabulary" vector $\mathbf{w}$ consisting of all the de-stemmed nouns. Then, the nouns of each document can be modelled as $\mathbf{a}^T \mathbf{w}$, where $\mathbf{a}$ is a W dimensional vector, whose elements correspond to the number of occurrence of the de-stemmed nouns in the document. Concatenate all the $\mathbf{a}$ vectors together to form a $W \times D$ term–document matrix, $D = \begin{bmatrix} \mathbf{a}_1 \ \mathbf{a}_2 \ \dots \ \mathbf{a}_D \end{bmatrix}$. The $W \times W$ term–term association matrix can then be formed as DD^T.

Given the term-term association matrix, we cluster related nouns together in groups by feeding the term-term association matrix to a self-organizing map (SOM) method [1]. The self organizing map is a popular and convenient method for clustering high dimensional vectors. A main advantage of the SOM is that the clustering involves a projection to a low dimensional display space where feature vectors from the high dimensional feature space will remain close in the low dimensional display space [1]. In our case, the number of nouns in the document set is large. But what we wish to do is to cluster them into groups which are related to one another according to the ways in which such nouns occurred in the documents. We will describe the SOM in Section 4, and hence here it suffices to indicate that the outcome of this step is that the nouns are clustered into groups, which are related to one another according to the ways in which the nouns occurred in the document set. We denote each cluster as a "concept" as the words in the same cluster are related to one another.

Once equipped with the set of clustered nouns, or concepts, we can then proceed to build a graph representation of each document. We will start by decomposing a document into paragraphs[3]. Each word occurring in a paragraph is replaced by the cluster in which the word falls in, or the concept in which the word belongs. The frequency of occurrence of the concept in the paragraph is computed. Then, a concept–paragraph matrix can be formed for each paragraph of the document. Assuming that there are P paragraphs in a document and there are a total of C concepts in the document set, then each paragraph can be represented by $\mathbf{p}^T \mathbf{c}$, where the C dimensional vector $\mathbf{c}$ represents the concepts, while the i-th elements of the C dimensional vector $\mathbf{p}$ represent the number of occurrence of the i-th concept in the paragraph.

---

[2] The way that nouns are extracted as indicated in this chapter is different to those indicated in [7].

[3] This is just a convenient way to start. If necessary, it is possible to start with any division of a document into segments, and a term–segment matrix can be formed, similar to the ways in which a term–paragraph matrix can be formed as indicated in this section.

The $C \times P$ concept–paragraph matrix for a particular document can be formed by concatenating all the paragraphs of a document together $P = \begin{bmatrix} \mathbf{p}_1 \, \mathbf{p}_2 \cdots \mathbf{p}_P \end{bmatrix}$. A singular value decomposition can be performed on the $C \times C$ concept–concept matrix: $PP^T = U\Sigma V^T$, where U and V are unitary matrices, and Σ is a diagonal matrix, with the diagonal elements arranged in descending order of magnitude. It is possible to interpret the diagonals of the matrix Σ as "themes", with the strength of the theme given by the value of the diagonal element. Normally, this concept–concept matrix is fully populated, i.e., all its elements are non-zero. However, it is possible to force some of the elements in the concept–concept matrix to be zero by thresholding their values. Thus, when an element is below a certain threshold it is forced to be zero[4]. The thresholds can be used to remove the relationship between weakly related concepts, and can be effective in the removal of features which are of least value. A thresholded concept–concept matrix can be interpreted as a graph, in which the nodes of the graph represent the concepts, and the link between two nodes is the corresponding element in the PP^T matrix.

3   General Framework for Processing Graph Structured Data

In this section, we will provide a general framework for modelling graphs [3]. Consider a node n in the graph. This node will be linked to other nodes through links. Assume for simplicity sake that the graph contains only un-directed links[5]. Let $\mathbf{x}_n(t)$ be the *state* vector at time t of the node n. Then, the state equation is given in the following parametric form:

$$\mathbf{x}_n(t) = \mathcal{F}_{\mathbf{w}}\left(\mathbf{x}_n(t-1), \mathbf{u}_n(t), \mathbf{x}_{[ne]}(t-1), \mathbf{u}_{[ne]}(t-1), t\right) \qquad (1)$$

where $\mathbf{x}$ is an s_n dimensional vector, denoting the state of the node n. $\mathbf{u}_n(t)$ denotes the vector containing the link and node labels of the node n, $\mathbf{x}_{[ne]}(t)$ denotes the state of nodes in the topological neighborhood of the node n, and $\mathbf{u}_{[ne]}(t)$ is the vector containing the link and node labels of the nodes which are in the topological neighborhood of the node n. $\mathcal{F}_{\mathbf{w}}$ is an s_n dimensional nonlinear function. The parameters in the model are represented by the vector $\mathbf{w}$.

For some of the nodes there will be an associated output label. Thus, in these cases, we can model the outputs as follows:

$$\mathbf{y}_n(t) = \mathcal{G}_{\mathbf{w}}\left(\mathbf{x}_n(t), \mathbf{u}_n(t), t\right) \qquad (2)$$

---

[4] This heuristic is a convenient method for forcing some of the elements of the concept–concept matrix to be zero, as long as the threshold value is small and the values used for this chapter are all small values. A more vigorous method will need to be implemented along the lines of obtaining a sparse representation of the inverse covariance matrix [20].

[5] It will be slightly more complex in terms of notations if we consider directed graphs, as one would need to consider the ancestors and the descendants of the current node n.

where $\mathbf{y}_n(t)$ denotes the set of outputs of the n-th node at time t. $\mathcal{G}_{\mathbf{w}}$ is a nonlinear function vector, with an unknown parameter set $\mathbf{w}^6$.

The unknown parameters $\mathbf{w}$ can be learned from a given set of training examples [3]. For the graph which represents the Internet, only some of the nodes are labelled. This is often called a semi–supervised learning problem, as some of the nodes are labelled (have an output) and the rest of the nodes are unlabelled (do not have an output).

The unknown parameters in the model Eq.(1) and Eq.(2) can be estimated, from the set of labelled nodes, and then the outputs of the unlabelled nodes can be predicted [3].

4   Self Organizing Maps for Structures

In this section we will consider the issue of grouping graph inputs according to some similarity criterion. In our case, the similarity criterion used is: two graph inputs which are close in the high dimensional feature space, remain to be close in the lower dimensional display space [11]. The display space in our case is often two-dimensional. This is the case for self organizing map (SOM) applications.

In the following we will first consider the classic SOM situation [11], and then we will see how we can modify the SOM situation to accept graph inputs. For the classic SOM, the problem can be stated as follows [11, 1]: given a training data set, $\mathcal{T} = \{\mathbf{u}_i, i = 1, 2, \ldots, N_T\}$, where $\mathbf{u}_i$ is the i-th training sample of dimension N_s. Our aim is to group the sample vectors, $\mathbf{u}_i$ which are close in the high dimensional feature space, N_s, so that they remain close to one another in the lower dimensional display space. The classic SOM first establishes a $N \times M$ grid in the display space; at the intersection of the grid points, there is a N_s dimensional vector $\mathbf{x}_i$. Thus, the SOM has a total of $N \times M \times N_s$ unknown parameters. The vectors $\mathbf{x}_i$ are initialized randomly. The classic SOM training algorithm consists of two steps:

Competition step: Randomly draw a sample $\mathbf{u}_i$ from the training dataset $\mathcal{T}$, compute the following:

$$j = \arg \min_i \|\mathbf{u}_i - \mathbf{x}_i\|^2 \tag{3}$$

where $\mathbf{x}_i$ is a N_s dimensional vector.

Parameter update step: The vectors $\mathbf{x}_i$ in a neighborhood of the winning vector j, $ne[j]$, are updated as follows:

$$\mathbf{x}_i \leftarrow \mathbf{x}_i - \alpha(t)d(i,j)(\mathbf{u}_i - \mathbf{x}_i) \tag{4}$$

where $\alpha(t)$ is a variable which will decrease from its initial value $\alpha(0)$ to 0. A linear decay rate for $alpha$ by $\frac{1}{t}$ is most common [11] though it is possible to consider non-linear decay rates. The distance function d is defined as follows:

---

[6] Here we abuse the notations slightly by using the same $\mathbf{w}$ to denote the unknown parameters in Eq(1) and Eq (2) respectively. Normally these two unknown parameter sets will be different.

$d(i,j) = \exp\left\{-\frac{\|\mathbf{c}_i - \mathbf{c}_j\|^2}{2\sigma(t)^2}\right\}$, $\mathbf{c}_j$ is the coordinate of the winning neuron in the $N \times M$ map, $\mathbf{c}_i$ is the coordinate of the neuron in the neighborhood j of the winning neuron i. $\sigma(t)$ is a monotonically decreasing function of t to ≈ 1.

These two steps are run, say, t_{max} times where t_{max} is a preset integer representing the number of training iterations. There are two mechanisms for which the algorithm will stop updating: (1) when $\alpha(t) = 0$ and (2) when the updating neighborhood is shrunk to 1. The convergence of the SOM has not been proven. However, it is one of the most popular clustering algorithms used by users, as it allows visualization of the clusters on the two dimensional display space [11]. Moreover, the linear computational time complexity of the SOM algorithm renders it useful for large scale clustering problems.

To extend this algorithm to graph inputs, we will modify the competition step. The updating step remains the same as shown in Eq(4). To extend the SOM to accept graph inputs, the key problem is how one node connects into another node. In this case, we assume that each node of the graph is modelled by a SOM with a two dimensional display map with a $N \times M$ grid. In addition, we assume that each neuron in the map is modelled by a $N_s + 2 \times N \times M$ vector $\mathbf{w}_k$. The reason why we need to augment the dimension of the neurons in the map will become clear below.

A simple way to proceed would be to consider the mapping of the Q nodes which are linked to a particular node n [18]. Consider one of the parent nodes, m which has a link with the current node n. Assume that we present node m to the SOM, and found through the application of Eq.(3) that the winning neuron has coordinates m_x, m_y on the display map. One simple way to incorporate the effect of this node m with that of the current node n would be to assume that in the location of the coordinate of m_x, m_y of the display map for node n there is an additional input, the strength of which is from node m. The additional input affects the current node through its distance from the center of the current node n. In other words, in the $N_s + 2 \times N \times M$ dimensional vector denoting the current node n, we will decompose it as follows [18]:

$$\mathbf{w} = \begin{bmatrix} \mathbf{x}_i & \mathbf{c}_{q_1} & \mathbf{c}_{q_2} & \cdots & \mathbf{c}_{q_Q} \end{bmatrix},$$

where $\mathbf{x}_i$ is N_s dimensional vector (the same dimension as the input vector), $\mathbf{c}_j, j = 1, 2, \ldots, Q$ are two dimensional vectors, denoting the centers of the best matching vectors in the Q nodes which are connected with node n. So, for node n, we can modify the competitive step as follows:

$$j = \arg\min_i \|\mathbf{u}_i - \mathbf{x}_i\|^2 + \beta \sum_{n_{[ne]}} \|\mathbf{c}_i - \mathbf{c}_{q_k}\|^2, \tag{5}$$

where β is used to weigh the influence of the neighbors $n_{[ne]}$ of the node n. The vector $\mathbf{c}_{q_k}$, where $q_k, k = 1, 2, \ldots, Q$, are the coordinates of the winning neurons of the neighbours of the current node n. This formulation can handle the graph inputs.

It is noted that normally in a graph not every node is connected with every other nodes. But by attaching a $2 \times N \times M$ vector to each node we provide for the occurrence of such unlikely event. This vector will be quite sparse, as not all the nodes are connected together. Hence one would wish to see if there is any way in which this state vector dimension can be reduced. One way in which this can be handled is by considering a state for each neuron [21]. In this case we will augment the vector describing each neuron $\mathbf{x}_i$ by a $N \times M$ vector $\mathbf{v}_i$ (instead of $2 \times N \times M$ vector as in the previous case. $\mathbf{v}_i$ will denote the number of times that the i-th neuron is a winner for the nodes contained in the neighborhood of the current node n. This "hard" coding of the results of the competition (as each neighbour has only one output in the whole map: the winning neuron) can be tempered by using a "soft" coding as follows:

$$j = \arg\min_i \left\{ \mu \|\mathbf{u}_i - \mathbf{w}_i\|^2 + (1 - \mu) \sum_{\ell=1}^{N \times M} \sum_{k \in ne[\ell]} \frac{1}{\sqrt{2\pi}\sigma(t)} \exp\left\{ -\frac{\|\mathbf{c}_\ell - \mathbf{c}_k\|^2}{2\sigma(t)^2} \right\} \right\}$$

(6)

where $\mathbf{c}_\ell$ denotes the coordinates of the neuron in the node which is connected to the current node n. $\sigma(t)$ is a monotonically decreasing function in t. The constant μ is designed to moderate the effect of the influence of the current node or those in the neighborhood of the current node n. This is called the PMGraphSOM (probabilistic measure graph self organizing map) [21].

It turns out that for processing small graphs, even the $N \times M$ dimensional state vector is too large, as the sparse graphs will result in only a few of the elements to be non-zeros. Thus, to further modify the competitive step, we could sort the values of strengths of the state. In other words, in Eq(6), we have:

$$j = \arg\min_i \left\{ \mu \|\mathbf{u}_i - \mathbf{w}_i\|^2 + (1 - \mu) \sum_{\ell-1}^{N \times M} \sum_{k \in ne[\ell]} \kappa_k \right\}$$

(7)

where κ_k is the sorted version of the following sequence in decreasing order with elements which are below a threshold τ will be considered to be 0:

$$\kappa_k = \frac{1}{\sqrt{2\pi}\sigma(t)} \exp\left\{ -\frac{\|\mathbf{c}_\ell - \mathbf{c}_k\|^2}{2\sigma(t)^2} \right\}$$

(8)

This competitive operation is found to be able to overcome the issue of sparseness in the $N \times M$ state vector. This is called a compact SOM (compact self organizing map) [22].

5 Graph Neural Networks

In this section we will briefly describe the graph neural network (GNN) [3]. The general formulation of the GNN is already provided in Section 3. Here what we like to do is to implement the encoding network $\mathcal{F}_w$ by a multilayer perceptron (MLP) with one hidden layer. This is a particular case of the more general framework provided in Section 3. In this case, the GNN will consist of nodes, each one of which is modelled using an MLP without the output layer. In other words, in each node, we have

$$\mathbf{x}_i = \mathbf{F_w}(B\mathbf{u}_i + B_i\mathbf{x}_{ne[i]}) \tag{9}$$

where $\mathbf{x}_i$ denotes the state of the node i, $\mathbf{u}_i$ is the input to the node i, and $\mathbf{x}_{ne[i]}$ is the states of the nodes which are in the neighborhood of node i. The dimension of $\mathbf{x}_{ne[i]}$ will depend on the in-degree of node i. In order to reduce the number of parameters, we will assume that each node is modelled by the same MLP architecture [3]. In other words, each node is modelled by a state vector $\mathbf{x}_i$ of dimension n_i. If the in-degree of node i is Q, then the dimension of $\mathbf{x}_{ne[i]}$ will be $Q \times n_i$. Matrices B and B_i will be of appropriate dimensions. $\mathbf{F_w}$ is a vector nonlinearity. The subscript $\mathbf{w}$ denotes the parameter set used in the MLP (the input to hidden weights together with the biases). For nodes which accept outputs, then the outputs at these nodes will be modelled as follows:

$$\mathbf{y}_i = \mathbf{G_w}(C\mathbf{x}_i) \tag{10}$$

where $\mathbf{G_w}$ is a vector nonlinearity. The subscript $\mathbf{w}$ denotes the parameter set of the hidden layer neuron to output weights. In some situations it may be possible to use a linear output function $\mathbf{G_w}$ [3].

Given a semi-supervised learning approach, where in a graph, some nodes are provided with known outputs, while the other nodes are without any outputs, then it is possible to train the parameters in Eqs(9) and (10) such that the sum of square errors of nodes with outputs is a minimum [3]. This can be achieved using a gradient descent method in the form of back propagation of errors. Thus, the weights $\mathbf{w}$ in $\mathbf{G_w}$ and $\mathbf{F_w}$ are updated using:

$$\Delta\mathbf{w} = \alpha * \frac{\partial\mathbf{G_w}}{\partial\mathbf{w}} \text{ for the MLP realizing function } \mathbf{G}, \text{ and} \tag{11}$$

$$\Delta\mathbf{w} = \alpha * \frac{\partial\mathbf{F_w}}{\partial\mathbf{w}} \text{ for the MLP realizing function } \mathbf{F}. \tag{12}$$

These derivatives are obtained by using the standard chain rule of differentiation and is analogous to the derivation of the standard error back propagation algorithm in MLPs [1]. Hence, we will omit the details of their derivation here.

6   Clustering of the Wikipedia Dataset

Wikipedia is a "community" encyclopedia, in that it is a collaborative effort by the worldwide community to contribute articles to the encyclopedia. This has been very successful, in that, currently it has over 3 million articles, covering information on almost any subject in the world, and that often this is the first place where someone would look for the information on the Internet. The articles are continuously updated by volunteers around the globe and hence the information is up–to–date.

In the creation of an article on a particular topic, the organizers of Wikipedia allow the creator of the document to label it in categories. There is no pre-defined set of categories for the author of the article to choose from, and, as a result, currently there are over 80,000 categories in which the Wikipedia articles can be assigned, and new categories can be introduced at any time. The consequence is that an author of an article is left with a virtually impossible task of identifying one or more suitable categories from the list of existing categories. The result is that authors introduce new categories in order to avoid to have to look up existing categories causing the introduction of a large number of categories whose names are rather un-informative. There are numerous categories in Wikipedia which are named *category1*, *category2* and so on, or are named according to the title of an article. Moreover, a quick inspection of some of the categories would convince the user that there might be only one or two articles in some of the categories. However, this defies the purpose of categorizing documents. Hence, the question is: what would be the practical number of categories in the Wikipedia dataset[7]. Or the question could be: Given a new document, what would be the most appropriate category of this document? These questions are not just academic questions. If we can group the articles into categories, then this will facilitate the easy retrieval of a particular article in the Wikipedia dataset, or any other dataset consisting of text documents.

One way to tackle this issue is to represent each document using the BoW approach. Thus, each document is represented by a vector, the elements of which represent the frequency of occurrence of the de-stemmed word in the document. Then one may use a self organizing map to cluster the documents in the two dimensional display space. This will form the baseline result in this study. The SOM approach ignores the hyperlinks which link the documents to one another, as each document is considered an independent document. The hyperlinks to other documents are ignored when processing the corpus.

Another approach would be to recognize the underlying structure of the interconnected nature of the documents in the dataset, and use an approach like the SOM for structures, e.g., PMGraphSOM, to process the data, as the SOM for structures takes into consideration of the interconnected nature of the documents. Thus, here the procedure will be to process the documents using Concept Link Graph approach. This will provide a graph representation of each document. Then, the SOM for structures is used to cluster these documents into groups. Now because of the possible explosion due to the number of hyperlinked documents in the dataset, and these hyperlinked documents are in turn containing hyperlinks to other documents, we will need to consider the "level of depth" in which the hyperlinks are considered. What this means is that for level 0, each document will be considered by itself as a standalone document. All the hyperlinks contained in the document will be ignored. For level 1, each document will include links to other documents; but it will not consider the documents which are linked to the linked documents. For level 2, the hyperlinks

---

[7] The organizers of the INEX competition used the Wikipedia dataset as a text corpus in one of their annual competitions. In 2008, the organizers grouped the Wikipedia dataset into 39 categories without indicating how such number was obtained. The question is: is 39 the actual number of groups in which the Wikipedia dataset can be grouped?

in the linked documents will be considered, but the documents which are linked to these linked documents will not be considered, and so on.

Wikipedia does not provide a reduced target set on which we could evaluate the clustering results. Instead, we will use a dataset made available in 2008 by the Initiative for the Evaluation of XML Retrieval (INEX). INEX provides a subset of the Wikipedia database consisting of $54,889$ XML documents for which a reduced target set of 39 categories is proposed. The following presents some experimental results based on this dataset. The experiments aim to:

- Cluster documents according to a similarity measure.
- Discover correlations between the given categories and the clusters produced by PMGraphSOM.
- Visualize the differences between the clusters formed by using SOM and those formed by using PMGraphSOM when trained on a collection of documents.

The INEX dataset provides 39 categories as target values. Since each document can belong to one or more categories, this means that there is a 39-dimensional binary target vector available for each document where a non-zero value in the n-th position in the target vector indicates that the associated document belongs to the n-th category. It is important to note that the SOM and PMGraphSOM are trained unsupervised. This means that the target information is not used during the training process. We will use the target information solely for the evaluation purposes, and in order to visualize the behaviours of the SOM and PMGraphSOM respectively.

We created two sets of data. First, a CLG representation is produced for each of the documents by following the procedures described in Section 2, where the nodes in the CLG are labelled by a 100-dimensional label which uniquely identifies the concept that is represented by the node. As was described in Section 2, the concepts are generated on the basis of the BoW and a consolidation procedure which clusters the vectors produced by the BoW approach. This means that the combined set of labels attached to all the nodes in the same CLG are a refined representation of the BoW vector for a given document. For this reason, and in order to allow for a direct comparison between the SOM and the PMGraphSOM, we generated a second dataset that represents each document by a vector. The vector is generated by element wise summing all the labels of the CLG associated with the same document. Hence, this second dataset consists of $54,889$ vectors each of which is of dimension 100.

We first demonstrate the clustering of the documents when training a SOM on the feature vectors. This approach does not take any structural information of a document or document content into account.

A variety of SOMs were trained. Figure 1 presents the mapping produced after training four SOMs which differed in the size of the feature map used in each case. The size of a SOM can influence the quality of the clusters formed. The larger a SOM is, i.e., the larger the feature map size is, the more freedom exists by which the vectors can be mapped. However, the purpose of clustering is to group data vectors which are similar according to some similarity criterion together, and thus, it is effectively used to produce a level of compression on the data vectors. A SOM

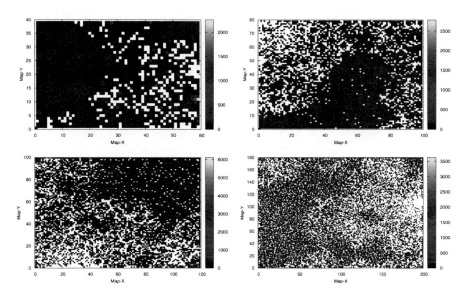

Fig. 1 Trained SOM on the feature vectors. Shown are the results of four SOMs which differ in terms of the size of the feature map used in each case, as indicated.

that is too large can defy such purposes. On the other hand, the smaller the SOM is, the more restrictions exist by which the vectors can be mapped. It then becomes possible that very dense clusters are formed which do not allow for the formation of smaller sub-clusters. The clustering results depicted in Figure 1 show that the quality of the clustering remains similar across a large range of feature map sizes: there is an area of relatively dense mappings and an area of relatively sparse mappings in all SOMs. Moreover, the mappings are color coded in order to indicate the frequency of activations at the same x-y coordinates in the cluster. The color chart next to each illustration provides an indication of the number of mappings at any one location, where the white color refers to locations at which there was no mapping at all. Visually, the training of the SOM did not seem to have resulted in the formation of 39 clusters as provided by the target values. We then evaluated the clustering performance as follows:

1. Select neuron locations for which there were at least 20 activations.
2. Count the number of documents from a particular category mapped on each selected neuron location and assign the majority class information to the neuron.
3. For each document, given the mapping location of such document, find out the nearest location from the neuron locations selected above.
4. Classify the document into the category assigned to the winning neuron in Step 2.
5. For each category, compute the classification accuracy by using the number of correct classifications divided by the total number of classifications.
6. Compute both the micro and the macro classification accuracy measures.

Table 1 INEX2009 SOM training results

TrainConfig	Micro Acc	Macro Acc
Map size=60x40 $\sigma(0) = 40, \alpha = 0.9, t = 100$	0.1177	0.1743
Map size=100x80 $\sigma(0) = 40, \alpha = 0.9, t = 100$	0.1129	0.161
Map size=120x100 $\sigma(0) = 40, \alpha = 0.9, t = 100$	0.0973	0.1432
Map size=200x180 $\sigma(0) = 40, \alpha = 0.9, t = 100$	0.0766	0.1327

The results of training the PMGraphSOM and the SOM are shown in Table 1. It is observed that the results appear to show that the clustering performance improves with smaller networks; an observation which will be re–assessed later in this chapter. Moreover, Table 1 shows that even the best SOM performs at well below 20% on any of the two performance measures. One could argue that this relatively poor performance is due to the missing of some of the information that is crucial for clustering the documents. The structural relationship between the various concepts was missing in the dataset used for the training of the SOM. Hence, in the following, we will consider the experimental results obtained when training a variety of PMGraph-SOMs on the CLG representation of the documents. The CLGs generated features a total of $436, 883$ nodes and $3, 904, 208$ links. The algorithm detected 100 unique concepts. We labeled each node by a 100-dimensional binary label that features a single non-zero element at the n-th position if the associated node represented the n-th concept. The largest CLG featured 60 nodes, the smallest CLG featured just a single node. The training of a PMGraphSOM on graphs results in the mapping of all nodes from a given dataset. Since each document is represented as a CLG, and since each CLG may feature a number of nodes, and hence, the processing of one document would result in the mapping of all nodes in the corresponding CLG. This would make it almost impossible to visually distinguish the mapping of nodes from different Concept Link Graphs in a dataset consisting $54, 889$ graphs. Thus, in order to improve the visualization of the mappings, we introduce a super node to each of the CLGs. The super node is labelled by a zero vector, and it connects to all the other nodes in the same CLG. Thus, the mapping of a super node can be interpreted as the mapping of the associated document.

An experiment was conducted with the following training configuration: map size $= 100 \times 80, \sigma(0) = 40, \mu = 0.01, t = 50$, and $\alpha = 0.9$. The final mappings produced are visualized in Figure 2. It is observed that the trained map can produce a number of dense clusters in the display space, and that the cluster centers are activated much more frequently as is shown by the color scale. Thus, the visual inspection of this very first experiment has already shown that the documents are clustered in a much more pronounced fashion than those obtained using the SOM. Moreover, it is much clearer to observe the various clusters that have formed.

We could have applied the K-means clustering algorithm in order to compute a prototype representation for each cluster. Instead, we used a much simpler thresholding method. Neurons which have been activated by at least 50 supernodes are then considered to be the center of a cluster. We used 50 as the threshold value

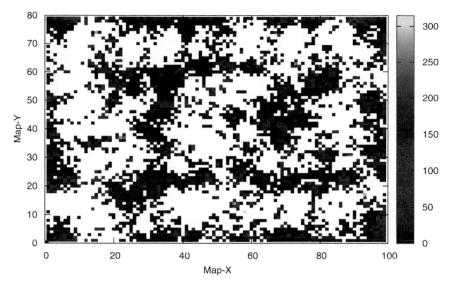

Fig. 2 Mapping of supernodes on a fully trained PMGraphSOM. The color scale indicates the frequency of mappings at a location. No mappings occurred in areas marked white.

since this resulted in a number of cluster centers which covered the entire display area. A smaller threshold value would result in a number of cluster centers which is much larger than the actual number of clusters, while a larger threshold value would cause an increase of overlapping clusters. We trained a number of PMGraphSOMs by varying the μ value and feature map sizes, and computed both the macro accuracy and the micro accuracy measures analogous to those computed for the SOM situation. The performances are summarized in Table 2.

Table 2 INEX2009 PMGraphSOM training results

TrainConfig		Micro Acc	Macro Acc
map=100x80, $\sigma(0) = 40$, $\mu = 0.1$, $t = 100$, $\alpha = 0.9$		0.0398	0.0996
map=100x80, $\sigma(0) = 40$, $\mu = 0.2$, $t = 100$, $\alpha = 0.9$		0.0402	0.1008
map=100x80, $\sigma(0) = 40$, $\mu = 0.01$, $t = 50$, $\alpha = 0.9$		0.0396	0.0991
map=140x100, $\sigma(0) = 40$, $\mu = 0.3$, $t = 100$, $\alpha = 0.9$		0.0402	0.0977
map=200x150, $\sigma(0) = 40$, $\mu = 0.1$, $t = 100$, $\alpha = 0.9$		0.0391	0.0949

Table 2 shows that the classification performance of the PMGraphSOM, when compared to those obtained using the SOM, is at least 50% worse, and that neither the network size nor the choice of μ had any significant effect on the performance figures. We analysed this result further by computing a confusion matrix. This is shown in Figure 3. Figure 3 plots the number of correspondences between the 39 target classes, and the 236 clusters obtained by the PMGraphSOM. The confusion matrix confirms that none of the clusters corresponds in any significant way to the

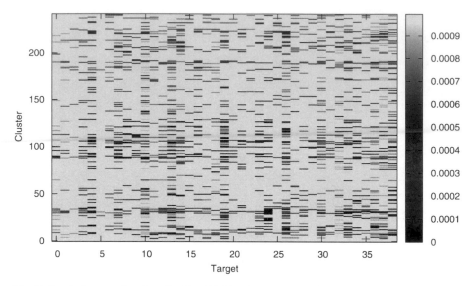

Fig. 3 Confusion matrix of the PMGraphSOM that was shown in Figure 2

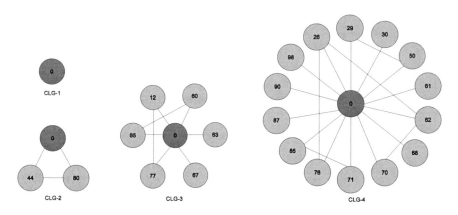

Fig. 4 Visualization of the CLGs mapped at distant locations

target classes. This appears as a surprise given that the PMGraphSOM produced much better well formed clusters when compared to those obtained using the SOM.

We investigated further by looking at individual CLGs that were mapped at the same location, within the same cluster, or at distant locations. A selection of such CLGs are visualized in Figure 4 and in Figure 5 respectively.

The figures show CLGs with their super node. The super node is marked by a zero. All other nodes are marked by an integer value which indicates the concept that is represented by the node. Figure 4 presents a selection of CLGs that were mapped at distant locations. For example, all empty Concept Link Graphs are mapped near the bottom-center at the coordinates $(58, 11)$, Concept Link Graphs

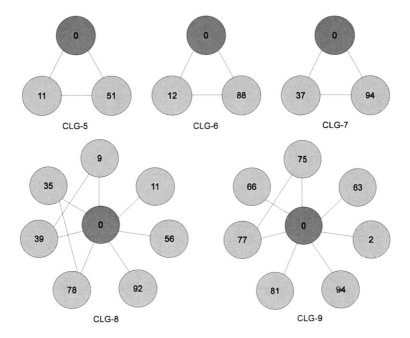

Fig. 5 Visualization of the CLGs mapped at the same location (top), or that were mapped in nearby locations (the two graphs below)

that contain one single node (CLG-1) are mapped at the bottom-right hand corner at the coordinates $(93, 17)$, whereas CLGs having three nodes (CLG-2) are clustered around the coordinates $(11, 21)$, some of the CLGs that contain around seven nodes (CLG-3) are located at the bottom-left hand corner around the coordinates $(8, 0)$, CLGs contain more concepts that have more than ten nodes (CLG-4) are mapped at the coordinates $(99, 7)$. In other words, CLGs that differ significantly are observed to be in different clusters. It can be clearly observed that these CLGs are very different in terms of structure, size, and concepts covered. Moreover, CLGs that featured the same number of nodes but differed significantly in topology were found to be mapped within the same region (such as those found in the bottom-left hand corner) but within different clusters. Thus, to this extent, the PMGraphSOM performed a mapping that resulted in distinct graphs to be mapped at locations at a distance from each other. In comparison, Figure 5 presents a selection of CLGs that were mapped at the same or nearby locations within the same cluster. For example, CLG-5 and CLG-6 are mapped at the same coordinates $(11, 21)$, and CLG-7 is mapped as a direct neighbor at $(12, 21)$. These three CLGs featured the same number of nodes, and the same topology, but differed in concepts covered by these CLGs. As another example, CLG-8 was mapped at the coordinates $(8, 0)$ whereas CLG-9 is mapped as a direct neighbor at coordinates $(8, 1)$. Both graphs feature an identical number of nodes and a very similar topology. The topology of CLG-9 is identical to the

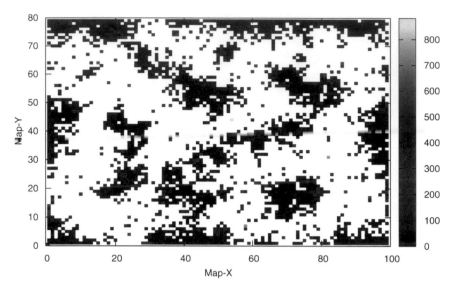

Fig. 6 Trained INEX2009 CLGs on PMGraphSOM using $\mu = 0.2$

topology of CLG-8 with just one link added. But again, these CLGs did differ in the concepts covered. This confirms that the PMGraphSOM did cluster the CLGs such that CLGs that are similar in size and topology are mapped at nearby locations. However, the PMGraphSOM did not cluster the CLGs according to the concepts covered by these CLGs. Hence, the PMGraphSOM clustered the documents based on the content structure rather than based on the actual contents. The PMGraphSOM used in this investigation was trained by using $\mu = 0.01$; placing a significant focus on encoding structural information. the PMGraphSOM did so successfully, and this explains the relatively poor classification performance on a task that requires the grouping based on document contents as well.

The following experiment trains a PMGraphSOM by shifting the focus away from structural information. This was performed by setting $\mu = 0.2$. This effectively doubles the impact of the concepts on the mapping of the nodes. The mapping produced by a map fully trained using these parameters is shown in Figure 6. When compared to the clustering shown in Figure 2, it can be observed that the number of clusters is significantly reduced. Moreover, as was shown in Table 2, the classification performance of this PMGraphSOM remains largely unchanged. The reason for this observation can be found in the similarity measure used when training the PMGraphSOM, and in the labelling scheme of the nodes. The PMGraphSOM uses the Euclidean Distance as a similarity measure, and the nodes are labelled by a 100-dimensional binary vector featuring one single non-zero value. This means that any two different data labels are orthogonal in the Euclidean space. The labels are either identical (distance is zero) or orthogonal (distance is 2). If we had trained a PMGraphSOM using $\mu = 1$, then this would have caused the mapping of all CLGs onto exactly 100 different locations as there are 100 different labels. Since all labels are equidistant, and hence, these mappings would occur at the same distance to each other. No clustering at all would occur. This means that any further increase of the μ parameter would not be helpful.

6.1 Discussion of Results

This set of experiments has evaluated an unsupervised training scheme on a dataset which provided target values. The training is entirely driven by the training data provided, and the training parameters provided. The experiments have shown that the PMGraphSOM does cluster the documents according to a criterion provided with the data and the parameters. We have observed that the documents were clustered successfully by content structure rather than by content itself. This is an ability that the standard SOM does not possess. However, it was also shown that the clustering produced may not necessarily be according to some expected pre-conceived clustering results. This was shown by assuming that the expected clustering is content based. If we had presented a CLG by labelling the nodes in a way that maintains similarity of concepts in the Euclidean space then this would have resulted in a clustering that would more closely resemble the task given by INEX. However, it needs to be stressed that the PMGraphSOM is trained unsupervised. If target information is available, then a supervised training scheme should be used. In this section, we made use of a supervised learning problem to visualize the abilities and limitations of the PMGraphSOM, and to highlight the importance of the features that are provided by the data to the PMGraphSOM's training algorithm. The algorithm will successfully cluster the data based on the features provided but may not cluster according to any target value that may be available. In the following section, we will present a number of experimental findings when using the Graph Neural Network approach; a supervised learning scheme for graph data structures.

7 Ranking of Documents

In this section we will consider the ranking of documents. As indicated in Section 1, PageRank ranks the documents using the link structure information only. PageRank does not use any content information. In this section we wish to explore the question: what happens if we include information on the content of the documents. Would this help in ranking the documents. As indicated in Section 1, this question is ill-posed, in that "we do not know the answer, even if we found it". However a meaningful question to ask: how fast would the PageRank algorithm converge if we include content information. In other words, if we include content information, would it assist us in ranking the documents faster than the PageRank operation. Intuitively, the PageRank algorithm uses only link information. The links are formed because there are relationships between the two documents (in however implicit manner). Thus by adding the content information, we are providing more information to the ranking operation, and hence intuitively the convergence should be faster than if we only use the link information, as in the PageRank algorithm.

As indicated in Section 1, such a situation could occur when we wish to compute the rank of documents based on a locally crawled corpus. In this case, we will assume to have a distributed crawler, in which there are a number of different crawlers distributed in various parts of the world. The intent is to rank the documents contained in the local store first, before passing on such information to a central location in which the ranking of the distributed crawlers will be combined into an aggregated one. Once the distributed crawlers have crawled a large number of sites, this will mimic the situation in which the PageRank is computed. This is due to the size of the Internet so that it is not possible for any one crawler to crawl all the documents and compute the PageRank. The PageRank is necessarily computed on a truncated portion of the Internet, and contains many "missing" links. Thus it is meaningful for us to investigate if the inclusion of the contents of pages would assist in the ranking of the documents.

There have been efforts in the application of the PageRank equation to rank documents on incomplete information [23]. It was found that the PageRank algorithm is particularly lacking when computing the rank of border pages. Border pages are web documents which receive one or more links from external domains. While the PageRank algorithm is unable to sufficiently incorporate the document content information[8] it was shown that a machine learning approach can simulate the effect of the PageRank equation [25] in addition to being expandable to incorporate document content information as well [26]. It is known that the GNN is capable of simulating the PageRank algorithm for a given set of Web documents [24]. Here we will present a machine learning approach whose task will be to encode limited local information for the purpose of computing the global PageRank of web pages.

Towards this end, we have used a crawled corpus of web pages on the Internet which we had conducted since 2007. This corpus currently contains over 26.6 million web documents from $4,814$ domains, which were hosted on $1,751$ unique web sites. The size of a web site varies widely. This is shown in Figure 7.

The figure shows that the large majority (86.6%) of web sites contain fewer than 10,000 web pages. In fact, as the zoomed area shows, nearly 30% of all web sites contained less than 10 web pages. This implies that there is a significant number of sites which contain only few intra-domain links. In other words, an attempt to compute a global document rank based on local information that is available to an individual site is a challenging task affecting most of the sites in the World Wide Web. We tackle this challenge by conducting a set of experiments as follows:

1. Apply the PageRank algorithm to the link structure of the entire dataset. This will compute the global rank value based on "complete" information (within a closed dataset) for all documents in the corpus; complete in the sense that this is all the information obtained from the World Wide Web. The computed values will be

---

[8] There exists research in personalization of the PageRank which enables the PageRank algorithm to incorporate limited information about document content [24]. However, the approach is limited to reduce the document content to a very small set of key words.

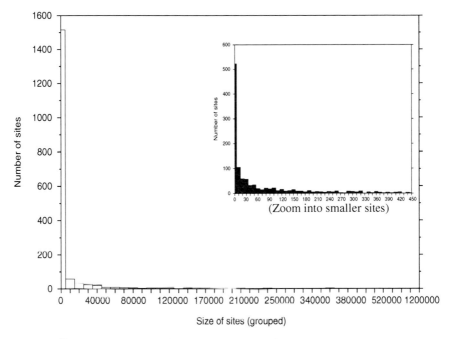

Fig. 7 Size distribution of sites in the dataset in groups of 10,000

used as the target values in the experiments, and the experimental results will be evaluated on how close they are to these target values. We will refer to these rank values as the *global PageRank* values.

2. Split the corpus into the various domains covered, then compute the PageRank of each domain. This will compute the rank of documents based on incomplete information since each subset does not contain web pages from other domains, and hence, the inter domain links are missing. This will provide the baseline information for this experiment. The work carried out in this step is similar to that attempted in [23]. We will refer to these values as the *local PageRank*, and we will refer to the intra-document structure used in this step as the *intra-domain graphs*.

3. Train a standard MLP so as to produce a mapping from the available, non-structured information to the global PageRank. This experiment will provide an insight into the limitations of traditional machine learning methods when applied to applications in a structured domain.

4. Train a GNN on the intra-domain graphs with the aim of producing a mapping to the global PageRank. This experiment will demonstrate the GNN's ability (and limitation) to approximate the global PageRank based on incomplete information.

5. Extend the training set used in the previous step by including the contents of the documents using a Concept Link Graph representation. This is to verify whether additional information available to a domain can support the learning task.

Table 3 Performance of the MLP obtained by using 1-dimensional and 2-dimensional inputs

Experiment	Number of iterations	MSE
MLP with 1-dimensional input	40	1.42731
MLP with 2-dimensional inputs	40	1.25886

Table 4 Performance by using GNN achievable for site 4

Experiment	MSE	WMSE
GNN on intra-document graph	0.01030	0.005500
GNN on intra-document graph labelled by CLG	0.00955	0.006025

We computed the global PageRank $X \in \Re^n$ by using Equation 13

$$X = dWX + (1 - d)E \tag{13}$$

where W is an $n \times n$ matrix with elements $w_{ij} = \frac{1}{h_j}$ if a hyperlink from node i to node j exists, h_j is the total number of outlinks of node j; otherwise $w_{ij} = 0$, and d is a parameter called the *damping factor*. We used $d = 0.7$ since this is often used as a default value in the literature [12]. As a result, the minimum rank value will be 0.3. For example, all pages which do not feature an inlink will have the rank value 0.3.

We then split the corpus into $4,814$ subsets according to the $4,814$ domains covered by the corpus, and computed the local PageRank on each of these subsets. We used Equation 13 and $d = 0.7$ as before.

A first set of experiments trains a standard MLP on the local PageRank scores (a 1-dimensional input), and the corresponding target is the global PageRank value of the same page. This was then extended by adding the total number of outlinks in the page to the input features. This created a 2-dimensional input vector. The average performance of a number of runs is shown in Table 3. We found that the MLP training algorithm converged after just 40 iterations, and that the mean squared error remained at a relatively high level when compared to the error of an untrained network. This experiment has provided a baseline result, and has shown that additional relevant input features can help the MLP to reduce the error. The main problem is that this given learning task requires the encoding of a (hyperlink-) structure whereas an MLP is not able to directly encode structures. This limitation causes the MLP to produce unsatisfactory results.

An incorporation of the hyperlink structure produces a dataset of graphs consisting of nodes that present the web documents in a corpus, and links that represent the hyperlinks. We created a dataset consisting of intra-domain graphs and labeled the nodes by a 1-dimensional vector containing the local page rank value of the associated document. A series of GNNs were then trained on this dataset. The average performance based on MSE is shown in the second row in Table 4.

We then created a CLG for each of the $4,814$ domains, and labeled the nodes in the intra-document graph by the associated CLG. A selection of CLGs is shown in

Figure 8 and Figure 9 respectively. The figures present the CLGs of relatively small documents since the larger documents resulted in much larger CLGs which could not be visualized in this chapter without compromising readability. Figure 8 presents the CLG of two documents from a domain with the IP address 158.104.100.60. It is a common observation that documents which were retrieved from the same domain (same IP address) exhibit related information. Hence, one can expect that the CLG graph representation of documents from one domain are more similar than the CLG representation of documents from a different domain. One purpose of Figure 8 and Figure 9 is to visualize some of the similarities and dissimilarities among CLGs from the same domain and from different domains. In Figure 8 and Figure 9 each CLG represents a document, each node represents a concept, and each link represents the relatedness between any pair of concepts. The number shown in bold font in each of the figures is a unique ID which we assigned to the concepts. It can be observed that the algorithm grouped related nouns together into the same concept. For example, concept 12 grouped the noun stems *success*, *profession*, *superb*, and *record* which makes much sense given that the nouns in other concepts are quite obviously unrelated. For example, as can be observed in Figure 8, the noun stems of concept 6 are very different to the noun stems of concept 12. In comparison, Figure 9 presents the CLGs of two documents from a domain with an IP address 12.110.113.146. It shows that some concepts can encompass a relatively large number of noun stems, and that the relatedness between concepts can be very significant (as is indicated by a value of 13.25 in Figure 9). CLGs are fully connected graphs. However, we have removed links whose weights are very small (less than 1.0). While this step was not strictly necessary, this was performed to reduce the turn around time of the experiments.

A series of GNNs were trained on the intra-document graphs now labeled by the CLG. The average result in MSE is summarized in row 3 of Table 4. The experiments have shown that the inclusion of document content allowed the system to improve the mean squared error by about 5%. This is an interesting finding since it shows that document content can be used as an additional feature in support of a learning task whose targets are purely based on the link structure of a corpus. Thus, when dealing with incomplete link information, it is possible to improve the document ranking by incorporating document content into the set of input features. The experiment has also demonstrated that the machine learning method described in Section 5 is indeed successful in encoding complex structures such as a graph-of-graphs which involves a graph, with nested graphs in each of the nodes as the situation shown in these examples. Each node of the top graph could in itself contain another document which can be described by another set of nodes, and these nodes can contain documents and can be represented by graphs as well.

We were interested in identifying whether the improvement affected certain categories of documents more than other categories. Given that the purpose of ranking is to create an order in a given set of documents, and given that most applications would be more interested in highly ranked pages than on lowly ranked pages, and hence, we investigated the GNN's performance on highly ranked documents when compared to the performance on lowly ranked documents. This was achieved by

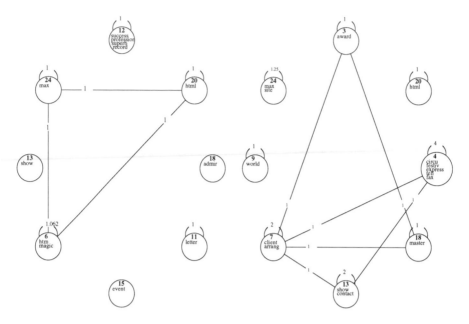

Fig. 8 CLG of document 5 (left) and document 17 (right) in site with an IP address 158.104.100.60

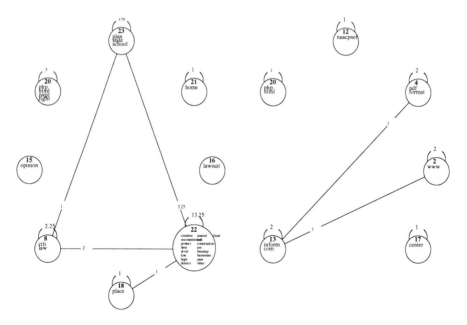

Fig. 9 CLG of document 16 (left) and document 26 (right) in site 12.110.113.146

weighting the error values according to the normalized target value associated with a node. In other words, the error of a highly ranked node is weighted much more strongly than the error of a lowly ranked node. This is shown in Table 4 in the column labeled by WMSE (weighted mean squared error). There are two important observations: First, the GNN generally performs better on highly ranked pages than on lowly ranked ones. Secondly, the introduction of the CLG did not improve the performance on the highly ranked documents. This means that the performance improvement on the MSE must have originated from improvements on the lowly ranked pages. This comes at no surprise since highly ranked pages generally exhibit numerous inlinks while lowly ranked pages feature very few inlinks. Thus, when dealing with incomplete link information, the impact of missing links is much stronger on lowly ranked pages than on highly ranked pages. For example, assume that there is a document featuring one inlink in the corpus. In the incomplete case, if this one link is missing then this results in the total loss of link information for this page. In contrast, a web page featuring many hundreds of inlinks will be impacted less when a number of links are missing in the incomplete case.

The MSE is an indicator on the accuracy of the mapping produced by the GNN. The PageRank algorithm is commonly used to create an ordering on documents. We analysed the outputs produced by the GNN by comparing the ordering of pages based on the output values generated by the GNN when compared to the order of the same documents when ordered by global PageRank values. We had found that the smaller domains, which are more easily affected by incomplete information, create the greatest challenge. Hence, we will investigate the ordering of pages within the smaller sites. Figure 10 shows the ordering of pages in a domain featuring 12 pages as produced by three experimental runs. For example, if a page is ranked 5-th by GNN, and ranked 5-th by the global PageRank then the point would be mapped on the diagonal of the graph. Hence, the closer the data points are to the diagonal, the better the results. We found that in general the trained GNN produces an ordering which more closely related to a PageRank ordering than it is to random ordering. Moreover, Figure 10 shows that the best performance was obtained when we labeled the nodes in the CLG graph by a 27-dimensional vector which indicated the concept the node represented. This produced a significantly better ordering when compared to labelling the nodes in the CLG by the ID value of the concepts, or when not labelling the nodes in the CLG at all.

Note that this Section makes no attempt to completely and comprehensively cover the task of ranking based on incomplete information. The purpose of this chapter is to show the advantages of machine learning approaches which are capable of encoding structural representation of large document collections. Such ability allowed us to encode the structured representation of documents within an interlinked domain. To the best of our knowledge, there is no other supervised machine learning method with such capabilities. Moreover, this section has demonstrated that the problem of ranking documents based on incomplete information is an application area which benefits from the described approach. Such an application is important for the realization of decentralized search engines.

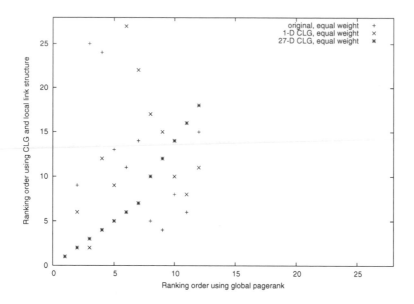

Fig. 10 Comparison of ranking order between global PageRank and estimated rank values

8 Related Work

A modern reasonably comprehensive survey of the types of issues facing text mining is contained in [27] where there are extensive discussions in various approaches in text mining. However, all the techniques used in [27] assume a vectorial input while in this chapter, we are more interested in graph inputs, i.e., in a structured domain. Hence, we will not repeat the types of discussions which are contained in [27]. We will only discuss work which pertain to using structured domain approaches.

The importance of the ability to learn structural representations of text documents is underlined by various other research approaches in the field [28]. Today, machine learning approaches capable of dealing with structured information are generating state-of-the-art performances on many benchmark problems [29, 30, 31, 32, 33], or provide the most generic framework capable of adapting to a variety of learning tasks [34]. Moreover, the provisioning of a document structure provides an additional feature that is often of significance to a given learning task [28].

Some of the earlier works on the categorization of text documents using document structure employed an xy-tree algorithm in order to represent the document structure [35]. The xy-tree algorithm is a recursive algorithm which segments a rendered text document into regular shaped geographical regions. The result is a tree representation, one for each given document, where each node corresponds to a segment (a section, column, a paragraph) of the documents. Such tree representation was then encoded by a Recursive Multi-layer Perceptron Network (RMLP) in order to solve document categorization problems [35].

An alternative approach found that some learning problems benefit from having the relatedness of words and sentences represented by a graph [28]. A graph representation of a document is obtained by representing sentences and words as nodes in a graph, and the context of sentences by a link in a graph. Two sentences or words that appear in succession are connected by a link in a graph [5]. The resulting graph is referred to as a *WordGraph*. WordGraphs have been encoded successfully by a Graph Neural Network for the purpose of text summarization [34].

The vast majority of text documents are described by markup languages such as XML, Postscript, HTML, LaTex, or RTF. Such text documents are rendered by parsers which parse the underlying markup language of a document. The parsing of any markup language creates a parsing tree. It is found that the nodes of a parsing tree almost always correspond to structural properties of a rendered document. The parsing tree can be used to represent a text document for some learning problems since the generation of parsing trees is computationally very time efficient. For example, the XML parsing tree has been used on large scale data mining projects for the purpose of clustering [29], and for solving the classification problem of large collections of text documents [26]. The performance produced by these approaches is unsurpassed [36].

There are numerous alternative approaches to the encoding of structural representation of text documents. For example, the Tensor Space Model approach has produced some good results [37]. The Tensor Space Model can capture both the structure and content of document, and was shown to scale linearly if the number of unique terms does not increase with the number of documents [37]. In the more common case in which the number of unique terms increases with the size of a document collection, the computational time complexity of the Tensor Space Model becomes quadratic, and hence, this limits its application to large document collections.

The Vector Space Model is popularly used for pattern mining, and has been considered for the modelling of semi-structured text documents [38]. Here the documents are represented as feature vectors. The feature vectors provide a description of document content as well as the local neighborhood of connected documents. The main issue here is the high dimensionality of the input matrix. The issue is addressed by exploiting the sparsity of the input matrix [39, 40]. However, the sparse representation introduces an overhead when the number of non-zero elements increases as the feature indices need to be stored as well. Thus, the Vector Space Model is unsuitable for dealing with large document collections which exhibit many features.

S-GRACE is a clustering approach specifically designed to work on parsing trees [41]. The method identifies the common set of nodes and edges. This results in a so-called s-graph for any pair of documents in a collection. The s-graphs are represented by bit strings which simplifies the computation of similarities between documents [42]. S-GRACE adopts a hierarchical method for the step-wise merging of similar pairs of clusters. Thus, S-GRACE can produce an arbitrary number of clusters from a document collection. The main problem with this approach is the s-graph generation which can be computationally very expensive for large document collections.

A number of clustering techniques for data trees engage the tree-edit distance measure [43, 44, 45]. The tree-edit distance can be engaged to cluster text documents which can be represented as a data tree (i.e. a parsing tree). While the tree-edit distance approaches are recognized for their ability to accurately compute the similarity between any pair of trees, the main problem remains the computational time complexity of these methods. Methods based on tree-edit distance are often limited to very small document collections [45].

Several methods have been proposed to determine the similarity between two structured objects by computing the similarity based on the paths within a data structure [39, 46]. These methods found their origin in data mining, and hence, a main advantage is that these approaches often scale very well. The drawback is that these methods only consider document structure. The text content of the documents is not considered.

Another interesting approach is a semantic clustering method called SemX-Clust [47]. The approach groups documents which share both structural similarities and content similarities. The approach requires substantial preprocessing since all documents are represented as a three tuple which describes the content and structure of a document, and an ontology knowledge base. Once pre-processing is completed, the approach scales well with large document collections.

9   Conclusions

In this chapter, we have considered two issues in text processing: clustering a corpus into various clusters, and the ranking of documents in a corpus. Our central question is: would the incorporation of content information help improve the operation at hand. We have introduced the idea of a Concept Link Graph as one way in which we can represent the content of an otherwise unstructured document. This Concept Link Graph allows us to capture some of the relatedness of the word tokens in the document relative to how they are used in the corpus, rather than resorting to semantic approaches. In the Concept Link Graph, the relatedness of the word tokens are inferred from the way in which the word tokens are used in the corpus. Then, we apply a PMGraphSOM to process the Concept Link Graph representation of the documents, and compare the results with those obtained from a SOM, which operated on a BoW representation of the documents, which ignores any relatedness of the word tokens used. It is shown that by incorporating the contents information, the clusters formed are "tighter" when inspected visually. We also tackled the issue of including content information in the computation of the ranks of the documents. Here the results are inconclusive in that we observe a speed up in some cases, but not a universal speedup, when compared with the PageRank computations. Here we surmise that the issue is because we only represent the content information partially, and have not been able to incorporate the entire content information when processing the ranking of the corpus. Thus, it is understandable that the results are inconclusive. However, it is conjectured that if we incorporate the full content information as represented using a Concept Link Graph structure, and overcome the long term dependency problem which plagued gradient descent methods, we will observe a universal speedup. This is left as a topic for future research.

Acknowledgements. The authors wish to acknowledge partial financial support by the Australian Research Council in the form of two Discovery Project grants: DP0774168 and DP0774617 respectively.

References

[1] Haykin, S.: Neural Networks, A Comprehensive Foundation. Prentice Hall (1998)

[2] Hornik, K.: Multilayer feedforward networks are universal approximators. Neural Networks 2(5), 359–366 (1989)

[3] Scarselli, F., Gori, M., Tsoi, A., Hagenbuchner, M., Monfardini, G.: Computational capabilities of graph neural networks. IEEE Transactions on Neural Networks 20, 81–102 (2009)

[4] McCallum, A.K.: Bow: A toolkit for statistical language modeling, text retrieval, classification and clustering (1996), http://www.cs.cmu.edu/~mccallum/bow

[5] Mihalcea, R., Tarau, P.: TextRank: Bringing order into texts. In: Proceedings of EMNLP, pp. 404–411. ACL, Barcelona (2004)

[6] Sowa, J.F.: Conceptual Structures: Information Processing in Mind and Machine. Addison-Wesley, Reading (1984)

[7] Chau, R., Tsoi, A.C., Hagenbuchner, M., Lee, V.: A conceptlink graph for text structure mining. In: Mans, B. (ed.) Thirty-Second Australasian Computer Science Conference (ACSC 2009), Wellington, New Zealand. CRPIT, vol. 91, pp. 129–137. ACS (2009)

[8] Belkin, M., Niyogi, P.: Laplacian eigenmaps for dimensionality reduction and data representation. Neural Computation 15(6), 1373–1396 (2003)

[9] Jolliffe, I.: Principal Component Analysis, 2nd edn. Springer-Verlag Inc., New York (2002)

[10] Hotelling, H.: Analysis of a complex of statistical variables into principal components. Journal of Educational Psychology 24, 417–441 (1933)

[11] Kohonen, T.: Self-Organisation and Associative Memory, 3rd edn. Springer (1990)

[12] Brin, S., Page, L.: The anatomy of a large-scale hypertextual web search engine. In: Proceedings of the 7th International Conference on World Wide Web (WWW), Brisbane, Australia, pp. 107–117 (1998)

[13] Chiang, W., Hagenbuchner, M., Tsoi, A.: The wt10g dataset and the evolution of the web. In: 14th International World Wide Web Conference, Alternate track papers and posters, Chiba city, Japan, pp. 938–939 (May 2005)

[14] Green, D.: The evolution of web searching. Online Information Review 24(2), 124–137 (2000)

[15] Despeyroux, T.: Practical semantic analysis of web sites and documents. In: WWW 2004: Proceedings of the 13th International Conference on World Wide Web, New York, USA, pp. 685–693 (May 2004)

[16] Netcraft, "Web server survey" (October 13 , 2005),
      http://news.netcraft.com/archives/web_server_survey.html

[17] The google platform,
      http://en.wikipedia.org/wiki/Google_platform
      (accessed July 07, 2011)

[18] Hagenbuchner, M., Sperduti, A., Tsoi, A.: A self-organizing map for adaptive process-ing of structured data. IEEE Transactions on Neural Networks 14, 491–505 (2003)

[19] Scarselli, F., Gori, M., Tsoi, A., Hagenbuchner, M., Monfardini, G.: The graph neural network model. IEEE Transactions on Neural Networks 20, 61–80 (2009)

[20] Yuan, M.: Efficient computation of the ll regularized solution path in gaussian graphical models. Journal of Computational and Graphical Statistics 17, 809–826 (2008)

[21] Zhang, S., Hagenbuchner, M., Tsoi, A.C., Sperduti, A.: Self Organizing Maps for the Clustering of Large Sets of Labeled Graphs. In: Geva, S., Kamps, J., Trotman, A. (eds.) INEX 2008. LNCS, vol. 5631, pp. 469–481. Springer, Heidelberg (2009)

[22] Hagenbuchner, M., Da San Martino, G., Tsoi, A.C., Spertudi, A.: Sparsity issues in self-organizing-maps for structures. In: Proceedings of European Symposium on Artificial Neural Networks, vol. ES2011-71 (2011)

[23] Chen, Y., Gan, Q., Suel, T.: Local methods for estimating pagerank values. In: Proceed-ings of the Thirteenth ACM International Conference on Information and Knowledge Management, CIKM 2004, pp. 381–389. ACM, New York (2004)

[24] Yong, S., Hagenbuchner, M., Tsoi, A.: Ranking web pages using machine learning ap-proaches. In: International Conference on Web Intelligence, Sydney, Australia, Decem-ber 9-12, vol. 3, pp. 677–680 (2008)

[25] Scarselli, F., Yong, S., Gori, M., Hagenbuchner, M., Tsoi, A., Maggini, M.: Graph neural networks for ranking web pages. In: Web Intelligence Conference, pp. 666–672 (2005)

[26] Zhang, S.J., Hagenbuchner, M., Scarselli, F., Tsoi, A.C.: Supervised Encoding of Graph-of-Graphs for Classification and Regression Problems. In: Geva, S., Kamps, J., Trotman, A. (eds.) INEX 2009. LNCS, vol. 6203, pp. 449–461. Springer, Heidelberg (2010)

[27] Feldman, R., Sanger, J.: The Text Mining Handbook. Cambridge University Press (2007)

[28] Tsoi, A.C., Hagenbuchner, M., Chau, R., Lee, V.: Unsupervised and supervised learning of graph domains. In: Bianchini, M., Maggini, M., Scarselli, F., Jain, L. (eds.) Innova-tions in Neural Information Paradigms and Applications, pp. 43–66. Springer, Heidel-berg (2009)

[29] Hagenbuchner, M., Sperduti, A., Tsoi, A.C., Trentini, F., Scarselli, F., Gori, M.: Cluster-ing XML Documents Using Self-organizing Maps for Structures. In: Fuhr, N., Lalmas, M., Malik, S., Kazai, G. (eds.) INEX 2005. LNCS, vol. 3977, pp. 481–496. Springer, Heidelberg (2006)

[30] Kc, M., Hagenbuchner, M., Tsoi, A.C., Scarselli, F., Sperduti, A., Gori, M.: XML Doc-ument Mining Using Contextual Self-organizing Maps for Structures. In: Fuhr, N., Lal-mas, M., Trotman, A. (eds.) INEX 2006. LNCS, vol. 4518, pp. 510–524. Springer, Heidelberg (2007)

[31] Yong, S.L., Hagenbuchner, M., Tsoi, A.C., Scarselli, F., Gori, M.: Document Mining Using Graph Neural Network. In: Fuhr, N., Lalmas, M., Trotman, A. (eds.) INEX 2006. LNCS, vol. 4518, pp. 458–472. Springer, Heidelberg (2007)

[32] Hagenbuchner, M., Tsoi, A., Sperduti, A., Kc, M.: Efficient clustering of structured documents using graph self-organizing maps. In: Comparative Evaluation of XML In-formation Retrieval Systems, pp. 207–221. Springer, Berlin (2008)

[33] Kc, M., Chau, R., Hagenbuchner, M., Tsoi, A.C., Lee, V.: A Machine Learning Ap-proach to Link Prediction for Interlinked Documents. In: Geva, S., Kamps, J., Trotman, A. (eds.) INEX 2009. LNCS, vol. 6203, pp. 342–354. Springer, Heidelberg (2010)

[34] Muratore, D., Hagenbuchner, M., Scarselli, F., Tsoi, A.C.: Sentence Extraction by Graph Neural Networks. In: Diamantaras, K., Duch, W., Iliadis, L.S. (eds.) ICANN 2010, Part III. LNCS, vol. 6354, pp. 237–246. Springer, Heidelberg (2010)

[35] de Mauro, C., Diligenti, M., Gori, M., Maggini, M.: Similarity learning for graph-based image representations. Pattern Recognition Letters 24, 1115–1122 (2003)

[36] Hagenbuchner, M., Kc, M., Tsoi, A.: XML Data Mining: Models, Methods, and Applications. In: Data Driven Encoding of Structures and Link Predictions in Large XML Document Collections. IGI Global (2010) (accepted for publication on May 30, 2010)

[37] Kutty, S., Nayak, R., Li, Y.: Xml documents clustering using tensor space model-a preliminary study. In: ICDM 2010 Workshop on Optimization Based Methods for Emerging Data Mining Problems, pp. 1167–1173 (December 13, 2010)

[38] Salton, G., McGill, M.: Introduction to modern information retrieval. McGraw-Hill, New York (1989)

[39] Leung, H., Chung, F., Chan, S., Luk, R.: Xml document clustering using common xpath. In: Proceedings of the International Workshop on Challenges in Web Information Retrieval and Integration, pp. 91–96. IEEE Computer Society, Washington, DC (2005)

[40] Vercoustre, A.-M., Fegas, M., Gul, S., Lechevallier, Y.: A Flexible Structured-Based Representation for XML Document Mining. In: Fuhr, N., Lalmas, M., Malik, S., Kazai, G. (eds.) INEX 2005. LNCS, vol. 3977, pp. 443–457. Springer, Heidelberg (2006)

[41] Wang, C., Hong, M., Pei, J., Zhou, H., Wang, W., Shi, B.-L.: Efficient Pattern-Growth Methods for Frequent Tree Pattern Mining. In: Dai, H., Srikant, R., Zhang, C. (eds.) PAKDD 2004. LNCS (LNAI), vol. 3056, pp. 441–451. Springer, Heidelberg (2004)

[42] Guha, S., Rastogi, R., Shim, K.: Rock: A robust clustering algorithm for categorical attributes. In: Proc.of the 15th Int. Conf. on Data Engineering (2000)

[43] Dalamagas, T., Cheng, T., Winkel, K., Sellis, T.: A methodology for clustering xml documents by structure. Information Systems 31(3), 187–228 (2006)

[44] Nierman, A., Jagadish, H.: Evaluating structural similarity in xml documents. In: Proceedings of International Workshop on Mining Graphs, Trees, and Sequences, pp. 61–66 (2002)

[45] Neuhaus, M., Bunke, H.: Self-organizing maps for learning the edit costs in graph matching. IEEE Transactions on Systems, Man, and Cybernetics, Part B 3(35), 503–514 (2005)

[46] Nayak, R., Tran, T.: A progressive clustering algorithm to group the xml data by structural and semantic similarity. IJPRAI 21(4), 723–743 (2007)

[47] Tagarelli, A., Greco, S.: Toward semantic xml clustering. In: Ghosh, J., Lambert, D., Skillicorn, D.B., Srivastava, J. (eds.) SDM, pp. 188–199. SIAM (2006)

Chapter 15
Neural Networks in Bioinformatics

Masood Zamani and Stefan C. Kremer

1 Introduction

Bioinformatics or computational biology is a multidisciplinary research area that combines molecular biology, computer science, and mathematics. Its aims are to organize, utilize and explore the vast amount of information obtained from biological experiments for understanding the relationships and useful patterns in data. Bioinformatics problems, such as protein structure prediction and sequence alignments, are commonly categorized as non-deterministic polynomial problems, and require sophisticated algorithms and powerful computational resources. Artificial Intelligence (AI) techniques have a proven track record in the development of many research areas in the applied sciences. Among the AI techniques, artificial neural networks (ANNs) and their variations have proven to be one of the more powerful tools in terms of their generalization and pattern recognition capabilities. In this chapter, we review a number of bioinformatics problems solved by different artificial neural network architectures.

In a field as young and diverse as bioinformatics, it is always a challenge to try to organize the scope of problems and their respective solutions in a sensible way. In text, this organization is further constrained to a mostly linear narrative. If we view biological systems as information processing devices, then we can trace a typical flow of information from DNA sequences, to RNA, and then to protein sequences (following the path of the "Central Dogma of Molecular Biology" [18]). From there, the information can be viewed to move on to protein structure, functionality and higher level of biological phenomena. We can use this flow to guide our narrative in this chapter.

Many problems in bioinformatics involve predicting later stages in the information flow from earlier ones. Bioinformatics methods capable of such predictions

Masood Zamani · Stefan C. Kremer
The School of Computer Science at the University of Guelph,
Guelph, Ontario
e-mail: {mzamani,skremer}@uoguelph.ca

M. Bianchini et al. (Eds.): *Handbook on Neural Information Processing*, ISRL 49, pp. 505–525.
DOI: 10.1007/978-3-642-36657-4_15   © Springer-Verlag Berlin Heidelberg 2013

can often eliminate costly, difficult, or time-consuming tasks in important biological research. For example, predicting protein structure and function based on amino acid sequence, is an essential component of modern drug design, and can replace expensive wet-lab work.

In our narrative we will situate problems, first, along their input data and secondly along their outputs as shown in Tables 1, 2 and 3. For each problem type we will proceed to describe, section by section, the nature of the data, its representation and any special considerations when using this data with artificial neural networks. We also consider the nature of the computational problem to be solved and discuss how to effectively apply a neurally inspired solution to it.

More specifically, this chapter is organized as follows. In Section 2 we discuss problems involving the analysis of DNA, including the detection of promoter regions, RNA coding regions, rare events, new motifs and DNA barcoding. Next, in Section 3 we turn our attention to peptide, or amino acid sequences. This section covers many problems related to the elucidation of structure and function of proteins, including: identifying secondary structure components, structural domains, disulphide bonds, contact points, solvent accessibility and protein binding sites, motifs, protein stability and protein interactions. Finally, in Section 4 we discuss the highest level of bioinformatic analysis including the diagnosis of cancers using spectrometry and microarray data.

This chapter is intended to provide an introduction to the predominant research areas and some of the approaches used within bioinformatics.

2   Analyzing DNA Sequences

In this section, we examine approaches that involve analyzing DNA sequences. DNA is a class of molecules that consist of a helical pair of polymers. The polymers are complementary and encode identical information. Each polymer is composed of many nucleotides that are joined in sequential fashion along a *backbone*. The information encoded in DNA can be viewed as a very long sequence of 4-base symbols since there are only four standard nucleic acids in DNA. These long strings of information are then transcribed into shorter segments by a process known as transcription. The shorter strings are composed of a similar molecule called RNA that employs the same type of 4-base representation; and, each such RNA string represents a code for a specific molecule. In many cases the RNA molecules are not themselves end products, but merely an encoding of a different type of molecule called a protein. Proteins are also polymers composed of simpler components joined in sequence, but the building blocks of proteins are amino acids (instead of nucleic acids). As there are 20 different types of standard amino acids, it takes at 3 symbols in the 4-base RNA code to uniquely identify a single symbol in the 20-base protein code. In fact, there is a redundant encoding from the 64 possible, 4-base triples to the 20-base amino acids.

Since DNA is the carrier of heritability, this is a reasonable place to start our discussion. It is relatively easy to build a neural system that processes DNA. Typically,

Table 1 Nonlinear Model Results (pt.1)

Input data	Output	Method
DNA sequence	Promoter regions	Promoter region identification [10]
DNA sequence	RNA gene	Non-coding RNA gene finder [56]
DNA sequence	Functional RNA genes	Detection of functional RNA genes using feed-forward neural networks [15]
DNA sequence	Classifying rare events in human genome	Detection of rare event in unbalanced data using neural networks [16]
DNA sequence	Clustered gene expression patterns	Analyzing correlated gene expression patterns using unsupervised neural networks [31]
DNA sequence	DNA motifs	Identifying unknown DNA motifs on DNA sequences using unsupervised neural networks [4]
DNA sequence	Classification of DNA barcoding genes	Inferring species membership via DNA barcoding with back-propagation neural networks [68]
DNA sequence	mRNA's donor and acceptor sites	Predicting donor and acceptor location on human pre-mRNA with feed-forward neural networks [12]
AA sequence	Sequence classifications	Protein Sequence Classification using Bayesian neural networks [62]
AA sequence	Clustered sequences	Unsupervised Kohonen learning technique [26]
AA sequence	Coil locations	Coil prediction [30]
AA sequence	β-sheet locations	Predicting protein β-sheets using alignment, neural networks and graph algorithm [13]
AA sequence	β-turn locations	Prediction of protein β-turn structure using evolutionary information and neural networks [36]
AA sequence	Protein Structural domains	Decomposition of protein structures into structural domains using profile and ANN [28]
AA sequence	Protein domain boundaries	Predicting protein domain using bidirectional recurrent neural networks [60]
AA sequence	Disulphide bonds	Disulphide bond prediction with a 2D-recurrent network [59]
AA sequence	Prediction of residue contacts	2D-recurrent neural networks for Protein contact map prediction [58]

Table 2 Nonlinear Model Results (pt.2)

Input data	Output	Method
AA sequence	Secondary structure	Predicting the secondary structure of globular proteins using MLP [52]
AA sequence	Secondary structure	Prediction of protein secondary structure using sequence profiles and neural networks [53]
AA sequence	Secondary structure	Prediction of protein secondary structure using evolutionary information and neural networks [54]
AA sequence	Secondary structure	Prediction of protein secondary structure using Position Specific Scoring Matrix(PSSM) and neural networks [34]
AA sequence	Secondary structure	Prediction of protein secondary structure using hidden neural networks [47]
AA sequence	Secondary structure	Prediction of protein secondary structure using bidirectional recurrent neural networks [7]
AA sequence	Real values of the solvent accessibility	Feed-forward neural networks for predicting the real values of solvent accessibility of amino acid [2]
AA sequence	Real values of the solvent accessibility	Approximating the real-value relative solvent accessibility (RSA) of AA residues [1]
AA sequence	Protein binding sites	Binding site prediction with neural network [37]
AA sequence	Secondary structure, solvent accessibility, backbone structural motifs, and contact density	Predicting 1D structural properties using structural alignment method (SAMD) and recursive neural networks [50]
AA sequence	Signal peptides	Detection of signal peptides in proteins [51]
AA sequence	Detection of protein stability	Prediction of protein stability changes using statistical potentials and multilayer feed-forward neural networks [20]
AA sequence	Detection of protein disorders	Predicting protein disorder for N-, C- and internal regions [46]
AA sequence	Detection of motifs	Predicting proteasome cleavage motifs using artificial neural networks [38]
AA sequence	Detection of drug resistant factor	Predicting HIV drug resistance with neural networks [21]
AA sequence	Protein superfamilies	Classifications of protein sequences based on superfamily classes [66]

Table 3 Nonlinear Model Results (pt.3)

Input data	Output	Method
Mass spectrometry data	Diagnosis of tumours	Classifying human tumour and identification of biomarkers [8]
DNA microarrays	Diagnosis of cancers	Classification and prediction of cancers using gene expression profiling and artificial neural networks [39]
DNA microarrays	Diagnosis of breast cancers	Detecting breast cancer using artificial neural networks [45]
DNA microarrays	Classification of diseases	Classification of gene expression data using ensemble neural networks [48]

a sliding window of fixed length is applied to the sequence, and the nucleic acids that fall within the window are encoded in a one-hot fashion. That is, four input units are used to represent each nucleotide and exactly one of these units (corresponding to one of the four different nucleotides) is activated each time. In this section, we consider four different goals in analyzing DNA: (i) identifying RNA coding regions in the DNA (arbitrary and specific fRNA), (ii) identifying promoter regions in the DNA, (iii) detecting disease carriers, and (iv) DNA barcoding.

While the central dogma of molecular biology encompasses how DNA is transcribed into RNA and then translated into protein sequences, most DNA does not code for proteins. Originally, called "Junk DNA" these parts of the genome are beginning to be better understood. In some cases, DNA is transcribed into functional RNA (fRNA) that is never translated into a protein but rather performs a directly useful biological function. Such RNA can be referred to as "non-coding" and the DNA regions that prescribe it are called "non-coding genes". Non-coding RNA genes have been explored for their hidden and important roles in cells. A challenging task is the identification of non-coding RNA genes due to the diversity and the lack of consensus patterns for their genes. One avenue is to identify transcription factor binding sites: locations in the DNA where special molecules attach and begin the process of transcribing the DNA into RNA. A novel approach using fuzzy neural networks for non-coding RNA gene prediction was proposed in [56]. The hybrid approach has the advantages that give the nodes and parameters in the neural network physical meanings and provide a means to incorporate the qualitative prior knowledge by fuzzy set theory.

Another research area related to RNA is the detection of the gene encoding functional RNA (fRNA). In brief, fRNAs are the set of RNA genes which generate functional RNA products such as transfer RNA(tRNA) and microRNA(miRNA) without translation to protein. For instances, tRNA is involved in translation of the three-letter code in messenger RNA into the amino acids of proteins. In [15], a feed-forward neural network is employed for fRNA gene detection. Evolutionary computation is used to optimize the architecture of the neural networks. In other words, the neural network is evolved and optimized by deletions and insertions of nodes and connections and also adjusting the weights associated between two nodes.

Another type of pattern that can be found in DNA is the promoter region. These regions provide convenient places for the RNA polymerase proteins to attach to a DNA strand and begin the transcription process. In this fashion, these regions serve a regulatory role. Identifying promoter regions using artificial neural networks has been also studied in [10]. The traditional promoter prediction methods mainly search for motifs. However, recent studies in [35], [42] and [61] indicate that DNA structural features such as curvature, and stress-induced duplex destabilization (SIDD) also provide valuable information. In [10], SIDD profile data obtained from *E. coli* is used as the training data for the neural network.

One challenge faced by bioinformaticians is an usual sparsity of data. While there are often many long genetic sequences available, the most interesting phenomena are sometimes extremely rare. Therefore, a rare event leads to a variety of needle-in-a-haystack problems which have to be modelled and understood. Rare events are log normally distributed, so methods based on statistics that assume Gaussian distributions (e.g. arithmetic means) fail. However, sample stratification is a useful technique for rare event detection in unbalanced data especially in molecular biology. The technique makes each class in a sample data have equal weight in decision making. Using a neural network for sample stratification and detection of rare events was examined in [16]. The experiment was carried out on human genome DNA, and it showed significant improvement for rare event detection.

A common task with regard to the voluminous data in molecular biology is the detection of unique features from DNA sequences. In [4], an unsupervised learning class of ANNs, known as self-organizing map (SOM) [41], was studied in order to detect new motifs (domains) in DNA sequences. It was used to detect the signal peptide coding region on a dataset of human insulin receptor genes. SOMs are useful in pattern clustering and feature detection since this class of neural networks form internal representations that model the underlaying structures of input data. In the study, no prior knowledge, such as sequence alignment analysis, was embedded in the neural network. Yet, after the neural network training, the existence of minimal similarity patterns (MSPs) among the trained data was found by a statistical measure called "Tanimoto similarity" which is proportional to the difference between the input and weight vectors. The proposed method may potentially facilitate the identification of other DNA domains such as functional DNA patterns by performing further analysis on MSP clusters.

The final problem that we will discuss in this section stems from the field of taxonomy. Traditional taxonometric methods identify species by painstaking observation of morphological features—the physical characteristics of an organism. While this method has served scientists since before the days of Aristotle, it can be problematic. Many organisms are so small that observation of physical differences even using microscopy is difficult. In other cases, organisms have multiple life stages with very different forms that need to be individually identified, or significant differences among sexes. Sometimes the physical form of an organism is affected by its environment (including diet, habitat, etc.). In these cases, relying on the observation of physical traits is problematic. With the advent of genetic sequencing another approach is possible. By directly comparing the DNA of organisms it is possible to make species identifications [29]. Ideally, this is done by focusing on specific genetic traits that vary among species but not within species. A first approach might be to identify a specific gene with this property and then to measure differences among instances of this gene across organisms using a classical genetic distance measure (such as alignment scores). Current distance-based methods for species identifications using DNA barcoding sequences are frequently criticized for treating the nearest neighbour as the closest relative using a number of raw similarity scores. In [68], a feed-forward neural network is employed for the classification of DNA barcoding sequences. The results indicate a better performance compared to the previous methods such as basic local alignment search tool(BLAST) [3] which is a simple genetic distance-based method.

2.1 Example Application

In the following, we briefly explain an application of ANNs for the identification of donor and acceptor sites on messenger RNA (mRNA). In eukaryotic organisms an important stage before the translation of a messenger RNA molecule to a correct protein is the remove the introns (non-coding regions) and joining exons (coding regions) in a process known as RNA splicing. In other words, the DNA coding for a particular protein will often be discontinuous and interrupted by these introns. The splicing mechanism removes these introns and concatenates the exons to form a correct RNA molecule for the protein to be assembled.

An mRNA is a molecule that is copied from DNA during a process called transcription. An mRNA molecule carries a "blueprint" of the genetic information for synthesizing a protein. In vertebrates, small ribonucleoproteins recognize the splice sites by detecting the sequences around exon-intron transitions. In [12], a feed-forward neural network has been applied to predict splicing sites in human pre-mRNA. In this study, a joint prediction scheme for exon and intron regions was developed since the transitions between exon and intron regions control cutoff positions for the splicing process, and can therefore lead to the prediction of splicing sites. The dataset used for the training and test was obtained from GenBank Release 62.0.

A subset of the dataset was eliminated based on sequences with only one intron, no complete RNA transcript, or more than one gene. In total, 95 genes remained for training and testing after the eliminations based on the afore mentioned criteria. The dataset was divided into two parts in which 65 genes were used for the training of neural networks and the remaining 30 genes for testing. Since many genetic datasets that are collected in this way tend to have strong sister sequences that are nearly identical to each other and thus trivialize the problem, it is important to keep such sister sequences together (in either the training or testing datasets), rather than separating them (into training and testing). To avoid such high similarities among genes, the genes were alphabetically order prior to dividing the dataset.

Another challenge to measuring the predictive performance of neural networks arises since the non-donor/non-acceptor sites outnumber the donor/acceptor sites, resulting in largely imbalanced classes. To overcome the imbalanced classes, the correlation efficient method [49] was applied for the evaluation of the neural networks. The neural network inputs were prepared based on a sliding window that scanned the DNA sequence. The window length was the number of nucleotides within it. The nucleotides *A, G, C, T* and unknown nucleotides were represented using a one-hot encoding in 4-digit binary numbers as 1000, 0100, 0010, 0001 and 0000 respectively. A single output of the neural network predicted whether or not the nucleotide represented in the middle of the input window was a donor or acceptor site. The neural network had a single hidden layer, and its performance was evaluated with different numbers of neurons in that hidden layer and varying inputs neurons (equivalent to window size).

The experimental results indicated the optimal prediction of neural networks were achieved whit a window size of 15 nucleotides and 40 neurons in the hidden layer for donor sites as compared to a window size of 41 nucleotides and 20 neurons in hidden layer for acceptor sites. With the selected architectures for the neural networks, the percentage of positive prediction of donor and acceptor sites were 95%, whereas the percentage of false predictions for donor and acceptor sites were only 0.1% and 0.4% respectively.

2.2 Conclusion

In this section, we have surveyed some neural approaches to DNA analysis. The representation of a DNA sequence as an input vector to an artificial neural network via a sliding-window approach is straightforward and has been used to great effect in the methods described above.

3 Peptide Sequence Analysis

All biological functions are based on the interactions of proteins. Proteins serve as catalysts in many biochemical reactions and play critical roles in the structure and behaviour of all cells. Protein's chemical properties are determined by their amino acid constituents, as well as their shapes since the shape of a protein affects the accessibility of the amino acids. Since amino acid sequence generally determines the shape of a protein (prions are a notable exception), it should be possible to predict a protein's shape based on its amino acids. This quest has long been viewed as a "holy grail" of bioinformatics.

In order to tackle such an important and challenging problem, a number of sub-tasks to the problem of determining a protein's three-dimensional shape have been identified. Proteins exhibit regions of structural patterns whose shapes are well defined. They can be held together by bonds between non-neighbouring amino acids (in the protein's chain) called disulphide bonds. Moreover, there is a fairly strong structural homology (similar shapes) among homologous protein sequences (similar sequences).

Coding amino acid sequences as neural network inputs can be accomplished by a sliding window and a representation scheme for individual amino acids. Since there are twenty amino acids, it is possible to encode them by twenty input units with a one-hot encoding. This tends to result in a very large input vector which in turn leads to a large number of connections, and trainable parameters. Having too many trainable parameters can result in over fitting (especially if the sample space is small). Of course, twenty distinct values can be encoded in a 5-bit vector using a binary representation. This type of representation may not be ideal however, since certain patterns (00000 and 00001) are much closer in Euclidean or Hamming space than others (01100 and 10011). Thus, a binary representation can implement a bias in the network favouring output mappings which treat the similar input patterns similarly. By contrast, the 20-input one-hot approach places every amino acid at a point equidistant to every other amino acid using any metric (Euclidean, Hamming, etc.). A number of alternative amino acid encodings using physicochemical properties of amino acids have been proposed and employed in [44], [67].

Once an encoding of the input has been developed, the next task is to determine what the neural network should output or predict. The first stage in understanding a protein's structure (and thereby gleaning insight into its function) is often to recognize particular sequence patterns. In [62], a Bayesian neural network approach is used to classify protein sequences. The features selected from the protein sequences for the input of the neural networks are based on both the global and local similarities.

Organizing and searching for homologous sequences in DNA and protein databases are essential tasks. An interesting clustering result with an accuracy of 96.7% for protein sequences into families using ANNs was accomplished in [26]. The unsupervised Kohonen learning method [41] has been used to train the network and cluster protein sequences since the number of composition and protein families were not known in advance. The neural network clustering approach is different than the non-hierarchical statistical methods for clustering data that usually require the number of expected classes be defined in advance [5].

Fortunately, there are some constraints on the structures of proteins. Because of weak covalent bonds between the hydrogen atoms in the amino acids, the amino acids themselves are often drawn into tight, stable arrangements. Two common arrangements are helices (where covalent bonds form between nearby amino acids coiling the polymer), and sheets (where covalent bonds form between two or more long polymer strands that run parallel or anti-parallel to each other). A third structure called a *coil* is more of a catch-all category that covers irregular regions of proteins. These three structure categories are referred to as secondary structure—the primary structure being the linear sequence of amino acids, while the tertiary structure is the detailed three-dimensional shape of the protein.

ANNs have also been used to predict which parts of a protein form each type of secondary structure and the detailed arrangements within the secondary structures. The most challenging parts of the prediction are perhaps coils which are irregular structural patterns. In [30], neural networks were used to predict dihedral angle probability distributions of coils from protein sequences. The network inputs are organized in a predefined window of residues. The dihedral angle probability distribution is predicted for the middle residue. The results indicate improvements compared to those using statistical methods [43], [70].

In addition to the covalent hydrogen bonds between amino acids, there are also di-suplide bonds. These form between pairs of one particular type of amino acid, cysteine. These bonds are also known as bridges as they form connection between amino acids that would otherwise be separated larger regions of space. Having prior knowledge of disulphide bridge locations in a protein structure is very valuable for the prediction of the protein backbone conformation. A recurrent artificial neural network has been successfully applied for disulphide bonding prediction [59]. This approach creates a two-dimensional matrix of potential disulphide bridges. Each dimension represents one amino acid in the potential disulphide bridge. This matrix representation can then be used both to formulate the input and the output of a neural network. Additionally, recurrent connections can then be used to propagate information about amino acids that surround the potential disulphide bond. An important challenge that many bioinformatics approaches face is that the number of exemplars (actual proteins that exist in nature) tends to be relatively small compared to the input space (all possible amino acid chains). This can result in a high dimensional parameter space with only few points, and consequent overfitting (as noted previously). The work in this area uses alignment profiles and homologies to expand the input examples and thus mitigate this issue. In addition, the study shows that using multiple alignment profiles improves the prediction accuracy which emphasizes on the importance of evolutionary information.

A protein's secondary structure is defined based on its common 3D structural patterns. Protein secondary structures are grouped into three structural classes: the α-helix (a spiral conformation), the β-sheet (a twisted, pleated sheet) and the coil (the most irregular structural pattern). Most proteins are composed of sections of all three of these classes. It is possible to assign each amino acid in a protein to one of these categories based on the protein shape at that amino acid's location. Thus, an amino acid sequence can be converted into a structural class sequence called a secondary structure. Predicting secondary structure, in turn, can be used as initial information by methods using free energy minimization to study protein pathways leading ultimately to 3D structure predictions. The pioneering work of using neural networks for predicting the secondary structure of globular proteins was proposed in [52]. In the study, an MLP with one hidden layer was used. The main drawback of the method's architecture was the over-fitting problem. Several techniques were introduced to address the problem, such as ensemble averages by training independently different neural networks, different input information [53, 54] and alternate learning procedures [34]. Using multi-sequence alignments for the network input instead of raw amino acid sequence data has significantly improved the result because secondary structures tend to be preserved across homologous proteins.

In [54], evolutionary information obtained from protein sequence alignments is used as the neural network inputs, and in turn it increased the classification accuracy. The effectiveness of incorporating protein evolutionary information with a neural network has been also studied for predicting β-turn patterns [36]. A fast protein secondary structure prediction using MLP was also proposed in [47], where the outputs of the neural network are passed to a Hidden Markov Model (HMM) to optimize the predictions.

In addition, a bidirectional recurrent neural network was used for the protein secondary structure problem in [7] to overcome the limitations of the past methods that used fixed-size input windows. While a conventional network's input is limited by its fixed-size input layer, a recurrent network uses feedback to process information over time (this distinction is similar to that between combinatorial and sequential logic circuits). By processing information over time, the input is not limited to a fixed size. This results in an architectural parsimony whereby a network with fewer adaptable parameters is able to process large input patterns. This, in part, helps to overcome the overfitting problem. In a three-stage method proposed in [13] for predicting protein β-sheets, a recurrent neural networks has been used as a primary step to obtain the residue pairing probabilities of all pairs in β-sheets. The inputs of the neural network are generated from profile, secondary structure and solvent accessibility information.

In a protein, structural domains have important applications in protein folding, evolution, function and design. They are common structures that occur in multiple proteins. These native-like structures are independent of the rest of the protein in the sense that they remain folded if separated from the rest of the protein. Methods for protein domain decomposition, such as using graph theory, are not accurate when the number of structural domains in a protein is not known prior to partitioning. To effectively assess the quality of a partition, in [28] the structural information of a protein including the hydrophobic moment profile and a neural network were used to evaluate the quality of identified domains. Using neural networks contributed significantly to an increase in prediction accuracy from 74.5% to 81.9%. Bidirectional recurrent neural networks have been utilized to predict protein domain boundaries [60]. The performance of these neural networks relies on protein sequences, evolutionary information and protein structural features.

Detecting signal peptides (SP) in proteins using ANNs was studied in [51], and the Swiss-Prot protein database was used to evaluate the performance of the proposed method. Signal peptides are the short fragments of proteins that lead newly synthesized proteins to find their destinations. One advantage of the proposed method is its computational speed compared to those of other approaches in [22], [25].

Another method for describing the shape of a protein uses a similar technique described earlier for the disulphide bond prediction. Specifically, it aims to identify neighbouring amino acids (without the presence of covalent bonds, like those between sulphur atoms). These neighbouring amino acids are considered "contacts", and a complete 2D matrix showing the degree of proximity between all pairs of amino acids is called a "contact map". Protein contact maps have important applications in proteins such as inferring protein folding rates, evaluating protein models and improving protein 3D structure predictions. Protein contact maps are encoded as a matrices of residue-residue contacts within a distance threshold. The NNcon method proposed in [58] is a protein contact map prediction technique based on 2D-recurrent neural networks. It maps 2D input information into 2D output targets. NNcon has been ranked among the best contact map prediction methods in CASP8. NNcon can be used to predict both general residue-residue contacts and specific beta contacts in β-sheets.

In [2], a feed-forward neural network was used for predicting the real values of solvent accessibility of amino acid residues. Solvent accessibility identifies which parts of a protein are accessible on the surface of the three dimensional structure as opposed to lying in the interior hydrophobic core. Understanding accessibility sheds light on the chemical properties of the protein. The method in [2] predicts the real values of accessible surface area from the sequence of information without a prior classification of exposure states unlike the past techniques classifying residues into buried and exposed states with varying thresholds. The categorical nature of such methods reduces the amount of information. Surface accessibility values are regarded as features used to improve the techniques applied for protein structure prediction. In [1], a neural network-based regression method was used to approximate the real-value relative solvent accessibility (RSA) of amino acid residues. Unlike other methods, the approach is not based on a classification problem which needed arbitrary boundaries among the classes. Instead, the method employs several feed-forward and recurrent neural networks and eventually combines them into one consensus predictor.

Using the 1D structural properties of a protein is an alternate way of exploring the correlation between a protein sequence and its 3D structure. Predicting the structural properties is valuable for protein structure and function prediction. The automated structural alignment method, SAMD, proposed in [50] for protein 1D structure prediction employs a recursive neural network and uses remote homology information unlike most 1D prediction methods that do not incorporate the homology information into the prediction process. The method is able to predict four structural properties which are: secondary structure, solvent accessibility, backbone structural motifs and contact density. The structural information is coded into the templates of structural frequency profiles and used as additional inputs to the recurrent neural networks to predict 1D-structural properties of query sequences. The systems is capable of making predictions by relying on data of only remotely homologous sequences whose structures are known in the Protein Data Bank (PDB) [9].

Predicting protein function is important to understand protein folding mechanisms. Protein function information is also correlated to its 3D structure. There are a number of neural network applications in the protein function prediction area, in particular, the identification of protein binding sites such as in [37]. Also, designing proteins that are able to function robustly in unusual environments such as in extreme pH and temperatures is very important. An interesting exploration also would be to change protein properties with a number of substitutions, and then predicting whether the mutations affect the stabilities of the proteins. In [20], a fast and accurate method was proposed for predicting protein stability changes when amino acids are mutated in a protein sequence. Statistical potentials and a multilayer feed-forward neural network are the two main components used in this approach to predict protein stability changes.

Artificial neural networks have also been used to predict protein disorder for N-, C-termini and internal regions [46]. A polypeptide chain has two unbounded ends which are a carboxyl group (-*COOH*) called the C-terminus and an amino group (-*NH2*) called the N-terminus. The translation of a protein from a messenger RNA (mRNA) starts from the N-terminus and ends with the C-terminus. A protein's function depends on its 3D structure in native state when it is known to be completely folded. By contrast, it has been observed, e.g. in [27], [11], that some proteins are partially or completely unfolded in their native states. The so-called natively unfolded or disordered proteins were investigated and it was postulated that the disordered regions due to the lack of a fixed 3D structure could be the integral parts of a novel protein function [63], [65]. The experiment in [46] indicates a higher prediction accuracy for disordered regions compared to those of discriminant analysis and logistic regression methods [17], [24].

Predicting proteosome cleavage motifs has also been examined using artificial neural networks in [38]. The motif prediction is a crucial step to understanding a cellular process such as metabolic adaptation and regulation of immune responses. The artificial neural network application has been also examined for the prediction of HIV protease mutants [21]. Predicting the resistant factor helps current HIV therapies in developing more effective treatments.

3.1 Example Application

In the following, an ANN application for protein "superfamily" [19] classification is explained in summary. Although the term superfamily refers to a group of evolutionarily related proteins, it is also applied to a group of structurally or functionally related proteins. A common task with regards to amino acid sequences is the classification of these protein sequences into superfamilies which often possess a common origin, structure and function. An example of protein classification techniques at the superfamily level that employs artificial neural networks is ProCANS (Protein Classification Artificial Neural Networks System) [66]. ProCANS has been implemented on a Cray supercomputer and is used for classifying unknown proteins at the superfamily level by embedding the information of the Protein Identification Resource (PIR) database [55] which is organized according to the superfamily concept. The two main steps of the ProCANS system are the sequence encoding scheme, to extract information from sequences without knowing the importance of its features in the classification model, and modular network architecture, a collection of small feed-forward neural networks instead of one large neural network to increase the processing of large and complex databases [40].

The important part of the sequence encoding scheme used in the study is a hashing function called the *n*-gram extraction method [14]. Using the *n*-gram extraction method, all patterns of possible *n* consecutive residues (or an alphabetic set) in a sequence are extracted, and the total number of each pattern's occurrences is recorded. Then, the sequence is represented as an m-dimensional vector called a "count vector" where each element of the vector corresponds to each pattern's total count. For instances, with twenty amino acids there are four hundreds possible residue pairs (patterns) using a 2-gram (bigram) extraction method. In ProCANS, ten sequence encodings were used according to two alphabet sets: amino acids and exchange group sets consisting alphabets of size twenty and six respectively. The first five encodings are based on count vectors which combine the various patterns of amino acids and exchange group patterns. For example, the encodings "a2" and "a2e2" are the bigram amino acids and the concatenation of the bigram amino acids and exchange group patterns respectively. The rest of the five encodings are based on a "position vector". Position vectors are generated according to the n-gram method, however each element of the vector is the average of each pattern's position (or order) in the sequence.

The second step in the ProCANS system is database modularization. At this step, the PIR database is divided to multiple sets according to protein functional groups such as electron transfer proteins, growth factors, etc.. The described encoding schemes are applied on the sets, and each set is used for the training of a feedforward neural network called a "database module". In the study, the PIR database is divided into four sets and four neural networks called database modules are trained for each of the ten encoding schemes. All trained neural networks have a single hidden layer fixed with 200 neurons, but the number of inputs for each module depends on the selected encoding scheme. The total number of outputs for all four neural networks is 620 corresponding to 620 protein superfamilies. Moreover, each neural network is trained to classify a range of superfamilies. The number of superfamilies classified by the four database modules are 148, 157, 178 and 137. Each input neuron is fed with a real value, computed by the product of the n-gram count and the corresponding residue's frequency in nature. Then, the product is mapped to the range $[0,1]$.

To evaluate classification accuracy, a protein sequence is classified by all four database modules. Therefore, among 620 classification scores from 0 (no match) to 1.0 (perfect match), the superfamilies of the three highest scores are selected as the predicted superfamilies of the protein. The classification accuracy is expressed by the total number of correctly identified patterns (superfamilies), the total number of incorrectly identified patterns and the total number of unidentified patterns. A protein's superfamily is considered correctly classified if one of the best three scores is above a defined threshold value and matches the known superfamily number of the protein. The predictive accuracy is examined by comparing threshold values ranging from 0.01 to 0.9. By choosing a lower threshold value, more superfamilies (patterns) are identified which results a higher sensitivity (more true positives) and a lower specificity (more false positives). In contrast, a higher specificity and a lower sensitivity are achieved if a higher threshold value is chosen. The experimental results reached a 90% classification accuracy with 9% false positives.

At a threshold of 0.9 the classification accuracy decreases to 68% with zero percent false positive. Also, the results indicate that the best encoding scheme is the concatenation of 1-gram (single letter) amino acids and 2-gram protein exchange groups.

3.2 Conclusion

In this section, we have examined approaches that begin with amino acid sequences and aim to predicting protein structure and function. We have seen that there are many intermediate goals along the way to full 3D structure prediction. We have also noted the danger of overfitting, which is caused by a sparsity of exemplars in a high dimensional space and various methods that are effective at mitigating this problem.

4 Diagnostic Predictions

Even if we knew the functions of proteins or had a tool to predict them, we would still be at a loss to explain many biological processes. This is because most of these processes involve protein-protein interactions and the presence or absence of proteins are varied by specific and complex regulatory mechanisms. Many biological processes can be turned on or off by causing particular genes to be variably expressed under different conditions. The study of gene regulation seeks to understand this variable expression. Variable expression, in turn, can then be used to shed light on the processes going on in an organism. This can provide valuable diagnostic tools for medicine and other applications.

Modern microarray technology uses 2D arrays of short DNA or RNA sequences called "probes" that bind to specific complementary RNA strands found in cells. A microarray may contain several thousands such probes (or even millions with the newest technology). By providing florescently died RNA materials, it is possible to copy the RNA produced in a functioning cell, photograph such an array, and based on the brightness of specific grid points, measure the expression of specific RNA patterns. By building custom designed microarrays for specific genes or interesting gene segments, it is thus possible to capture a snap shot of the proteins being produced in a cell at a given point in time.

An example using this datatype is disease classification and the identification of indicative biomarkers for detecting the early onset of diseases. Deciphering the gene expression signatures for classification of cancerous diseases is challenging for such a high dimensional and complex dataset. In the seminal paper [39], a neural network was used to classify cancers into a number of diagnostic categories based on their gene expression signatures obtained by cDNA microarray analysis. In genetic engineering, a complementary DNA (cDNA) is a type of DNA synthesized from a messenger RNA template in which the introns are removed in order to be able to clone eukaryotic genes in prokaryotes. Selecting an optimal subset of gene markers is an extremely difficult task since there are a high number of gene markers, and these markers can reach over one million with new chip technology.

In [45], using ANNs resulted in reducing the gene signatures from 70 to 9 which accurately together predicted breast cancer from microarray data.

In [31], an unsupervised artificial neural network was used to analyze gene expression data obtained from DNA array experiments for correlated gene expression patterns. Clustering techniques are common tools applied to gene expression data. However, the proposed method uses a growing neural network which adopts the topology of a binary tree. The outcome is a hierarchical clustering that also has the robustness of a neural network. With regards to the rapidly growing DNA array technology and the huge amount of information, the proposed method is claimed to be faster and more accurate compared to the other hierarchical clustering techniques [64], [23], and [32].

In proteomics one of the pioneering applications of artificial neural networks was in the mining of the mass spectrometry data for the protein screening of cancer patients [8]. In the study, the ANN's weights were analyzed and the ions that had the highest contribution for the classification were identified. By further analyses, it was discovered that two ions in combination are able to predict the tumour grade with the highest accuracy.

4.1 Example Application

An application of ANNs for classification of gene expression data for disease diagnosis is explained in the following. DNA microarrays can be used to simultaneously measure the expression levels of large numbers of genes. As a result, gene expression data has become an effective tool in clinical purposes such as disease diagnosis. In order to develop a reliable technique to diagnose a disease, the bodies response can be measured via gene expression patterns in individual cells. Machine learning can then be used to identify and classify specific expression patterns and thereby label healthy and sick cells. Such patterns may include finding co-regulated genes. In this regard, the greatest challenge of this work is to build accurate classification models that are capable of processing the large amount of gene expression data consisting of thousands features (genes) and a few number of samples.

An essential component in mining such high dimensional data for key features is a robust feature selection technique. In [48], a combinatorial feature selection method and an ensemble neural network are used to classify gene expression data. As mentioned earlier, due to the limited samples of the gene expression data, the bootstrap method is used to resample the data 100 times. This increase of training data is necessary to ensure the accuracy, robustness and generalization of a classifier.

It has been verified that different feature selection techniques applied to a dataset result in different profiles for the dataset. Therefore, the combinatorial feature selection method provides more information for a classifier by employing three feature selection methods: ranksum test [69], principle component analysis(PCA) [57] and masked out clustering [33]. The ranksum test extracts and selects 30 top genes; the PCA method selects 15 principle components; and the masked out clustering clusters the data into 50 groups and then, using the t-test, selects the 30 top genes.

The proposed neural network ensemble consists of three feed-forward neural networks. Each competitive and cooperative neural network of the ensemble network has a single hidden layer with 10 units. The selected features produced by the ranksum, PCA and masked out clustering methods are separately fed as the inputs to each neural network in the ensemble network. Each neural network learns to classify based on the information extracted from the training data. The partial classification accuracies generated by all networks are aggregated according to a soft-voting scheme where the confidence of each network is considered as a voting value instead of the crisp values of 0 and 1.

An advantage of the proposed classifier is that its overall classification accuracies were higher or comparable to those of the other standard techniques on seven different types of gene expression datasets (Lung Cancer, DLBCL, etc.) due to combining the information extracted by the three different feature selection techniques.

4.2   Conclusion

In this section we have explored even higher levels of analysis, relating to the function and expression of genes. This analysis enables new diagnostic tools for diseases. Neural networks have been successfully used to identify markers and patterns in these cases, as well as feature selection.

5   Conclusion

In this chapter we have pursued the flow of information from the basic hereditary patterns in DNA to RNA to protein to functionality, and finally biological processes. We have reviewed a large number of neural network approaches to processing the data and making useful predictions. In completing any review such as this we have had to purposely omit some works. In general we have tried to give a broad overview of the scope of the field rather than focusing on variations on specific approaches. At the same time, there are constantly new developments in dealing with the rich and voluminous data being produced by modern molecular techniques, and these provide a great opportunity for future work. While we have tried our best to tie together in a logical way the diverse work covered in this survey using an information flow approach, there is one additional theme that resurfaces in our observations on the work. That is the challenge of working in high dimensional spaces, where the number of example patterns (although large and growing) is still fairly small compared to the dimensionality of the spaces to be considered. This creates a problem for any adaptive technique in the form of a danger of overfitting parameters to the data [6]. Regularization methods must be employed to manage this issue. Throughout this chapter we have pointed out some such methods which rely on input encoding, recurrent networks, alignment profiles, homologies and ensemble averages. We expect this to remain a challenge as datasets continue to become richer and be a prevalent theme of work in this area in the next decade.

Acknowledgement. Dr. Stefan C. Kremer is funded by the NSERC of Canada and uses that funding to support his students, including Mr. Masood Zamani.

References

1. Adamczak, R., Porollo, A., Meller, J.: Accurate prediction of solvent accessibility using neural networks-based regression. Proteins: Structure, Function, and Bioinformatics 56(4), 753–767 (2004)
2. Ahmad, S., Gromiha, M.M., Sarai, A.: Real value prediction of solvent accessibility from amino acid sequence. Proteins: Structure, Function, and Bioinformatics 50(4), 629–635 (2003)
3. Altschul, S.F., Gish, W., Miller, W., Myers, E.W., Lipman, D.J.: Basic local alignment search tool. Journal of Molecular Biology 215(3), 403–410 (1990)
4. Arrigo, P., Giuliano, F., Scalia, F., Rapallo, A., Damiani, G.: Identification of a new motif on nucleic acid sequence data using Kohonen's self-organizing map. Computer Applications in the Biosciences: CABIOS 7(3), 353 (1991)
5. Auray, J.P., Duru, G., Zighed, D.A.: Analyse des données multidimensionnelles: Les Méthodes de structuration. A. Lacassagne (1990)
6. Bai, B., Kremer, S.C.: In: Proceedings of the IEEE International Conference on Bioinformatics and Biomedicine Workshops (2011) (to appear)
7. Baldi, P., Brunak, S., Frasconi, P., Pollastri, G., Soda, G.: Bidirectional dynamics for protein secondary structure prediction. Sequence Learning, 80–104 (2001)
8. Ball, G., Mian, S., Holding, F., Allibone, R.O., Lowe, J., Ali, S., Li, G., McCardle, S., Ellis, I.O., Creaser, C., et al.: An integrated approach utilizing artificial neural networks and SELDI mass spectrometry for the classification of human tumours and rapid identification of potential biomarkers. Bioinformatics 18(3), 395–404 (2002)
9. Bernstein, F.C., Koetzle, T.F., Williams, G.J.B., Meyer, E.F.: et al. The protein data bank: A computer-based archival file for macromolecular structures. Journal of Molecular Biology 112(3), 535–542 (1977)
10. Bland, C., Newsome, A., Markovets, A.: Promoter prediction in e. coli based on SIDD profiles and artificial neural networks. BMC Bioinformatics 11(suppl. 6), S17 (2010)
11. Bloomer, A.C., Champness, J.N., Bricogne, G., Staden, R., Klug, A.: Protein disk of tobacco mosaic virus at 2.8 a resolution showing the interactions within and between subunits. Nature 276(5686), 362 (1978)
12. Brunak, S., Engelbrecht, J., Knudsen, S.: Prediction of human mRNA donor and acceptor sites from the DNA sequence. Journal of Molecular Biology 220(1), 49–65 (1991)
13. Cheng, J., Baldi, P.: Three-stage prediction of protein-sheets by neural networks, alignments and graph algorithms. Bioinformatics 21(suppl. 1), i75–i84 (2005)
14. Cherkassky, V., Vassilas, N.: Performance of back propagation networks for associative database retrieval. In: International Joint Conference on Neural Networks, IJCNN, pp. 77–84. IEEE (1989)
15. Cheung, M., Fogel, G.B.: Identification of functional RNA genes using evolved neural networks. In: Proceedings of the 2005 IEEE Symposium on Computational Intelligence in Bioinformatics and Computational Biology, CIBCB 2005, pp. 1–7. IEEE (2005)
16. Choe, W., Ersoy, O.K., Bina, M.: Neural network schemes for detecting rare events in human genomic DNA. Bioinformatics 16(12), 1062–1072 (2000)
17. Cox, D.R., Snell, E.J.: Analysis of binary data, vol. 32. Chapman & Hall/CRC (1989)
18. Crick, F.H.: On protein synthesis. In: Symposia of the Society for Experimental Biology, vol. 12, p. 138 (1958)

19. Dayhoff, M.O., McLaughlin, P.J., Barker, W.C., Hunt, L.T.: Evolution of sequences within protein superfamilies. Naturwissenschaften 62(4), 154–161 (1975)
20. Dehouck, Y., Grosfils, A., Folch, B., Gilis, D., Bogaerts, P., Rooman, M.: Fast and accurate predictions of protein stability changes upon mutations using statistical potentials and neural networks. Bioinformatics 25(19), 2537–2543 (2009)
21. Draghici, S., Potter, R.B.: Predicting HIV drug resistance with neural networks. Bioinformatics 19(1), 98–107 (2003)
22. Dyrløv Bendtsen, J., Nielsen, H., von Heijne, G., Brunak, S.: Improved prediction of signal peptides: Signalp 3.0. Journal of Molecular Biology 340(4), 783–795 (2004)
23. Eisen, M.B., Spellman, P.T., Brown, P.O., Botstein, D.: Cluster analysis and display of genome-wide expression patterns. Proceedings of the National Academy of Sciences 95(25), 14863 (1998)
24. Eisenbeis, R.A., Avery, R.B.: Discriminant analysis and classification procedures: theory and applications. Lexington Books (1972)
25. Fariselli, P., Finocchiaro, G., Casadio, R.: Speplip: the detection of signal peptide and lipoprotein cleavage sites. Bioinformatics 19(18), 2498 (2003)
26. Ferrán, E.A., Ferrara, P.: Clustering proteins into families using artificial neural networks. Computer Applications in the Biosciences: CABIOS 8(1), 39–44 (1992)
27. Fletcher, C.M., Wagner, G.: The interaction of eif4e with 4e-bp1 is an induced fit to a completely disordered protein. Protein Science 7(7), 1639–1642 (1998)
28. Guo, J., Xu, D., Kim, D., Xu, Y.: Improving the performance of domainparser for structural domain partition using neural network. Nucleic Acids Research 31(3), 944–952 (2003)
29. Hebert, P.D.N., Penton, E.H., Burns, J.M., Janzen, D.H., Hallwachs, W.: Ten species in one: Dna barcoding reveals cryptic species in the neotropical skipper butterfly astraptes fulgerator. Proceedings of the National Academy of Sciences of the United States of America 101(41), 14812 (2004)
30. Helles, G., Fonseca, R.: Predicting dihedral angle probability distributions for protein coil residues from primary sequence using neural networks. BMC Bioinformatics 10(1), 338 (2009)
31. Herrero, J., Valencia, A., Dopazo, J.: A hierarchical unsupervised growing neural network for clustering gene expression patterns. Bioinformatics 17(2), 126–136 (2001)
32. Iyer, V.R., Eisen, M.B., Ross, D.T., Schuler, G., Moore, T., Lee, J.C.F., Trent, J.M., Staudt, L.M., Hudson, J., Boguski, M.S., et al.: The transcriptional program in the response of human fibroblasts to serum. Science 283(5398), 83 (1999)
33. Jager, J., Sengupta, R., Ruzzo, W.L.: Improved gene selection for classification of microarrays. In: Pacific Symposium on Biocomputing 2003, Kauai, Hawaii, January 3-7, p. 53. World Scientific Pub. Co. Inc. (2002)
34. Jones, D.T.: Protein secondary structure prediction based on position-specific scoring matrices. Journal of Molecular Biology 292(2), 195–202 (1999)
35. Kanhere, A., Bansal, M.: Structural properties of promoters: similarities and differences between prokaryotes and eukaryotes. Nucleic Acids Research 33(10), 3165 (2005)
36. Kaur, H., Raghava, G.P.S.: A neural network method for prediction of β-turn types in proteins using evolutionary information. Bioinformatics 20(16), 2751–2758 (2004)
37. Keil, M., Exner, T.E., Brickmann, J.: Pattern recognition strategies for molecular surfaces: III. binding site prediction with a neural network. Journal of Computational Chemistry 25(6), 779–789 (2004)
38. Keşmir, C., Nussbaum, A.K., Schild, H., Detours, V., Brunak, S.: Prediction of proteasome cleavage motifs by neural networks. Protein Engineering 15(4), 287–296 (2002)

39. Khan, J., Wei, J.S., Ringner, M., Saal, L.H., Ladanyi, M., Westermann, F., Berthold, F., Schwab, M., Antonescu, C.R., Peterson, C., et al.: Classification and diagnostic prediction of cancers using gene expression profiling and artificial neural networks. Nature Medicine 7(6), 673–679 (2001)

40. Kimoto, T., Asakawa, K., Yoda, M., Takeoka, M.: Stock market prediction system with modular neural networks. In: International Joint Conference on Neural Networks, vol. 1, pp. 1–6. IEEE (1990)

41. Kohonen, T.: Self-organization and associative memory. In: Self-Organization and Associative Memory, 100 figs. XV, 312 pages. Springer Series in Information Sciences, vol. 8, p. 1. Springer, Heidelberg (1988)

42. Kozobay-Avraham, L., Hosid, S., Bolshoy, A.: Involvement of DNA curvature in intergenic regions of prokaryotes. Nucleic Acids Research 34(8), 2316 (2006)

43. Kuang, R., Leslie, C.S., Yang, A.S.: Protein backbone angle prediction with machine learning approaches. Bioinformatics 20(10), 1612 (2004)

44. Lac, H., Kremer, S.: Inducing fold dynamics from known protein structures using machine learning. PhD thesis, CIS, University of Guelph (April 2009)

45. Lancashire, L.J., Powe, D.G., Reis-Filho, J.S., Rakha, E., Lemetre, C., Weigelt, B., Abdel-Fatah, T.M., Green, A.R., Mukta, R., Blamey, R., et al.: A validated gene expression profile for detecting clinical outcome in breast cancer using artificial neural networks. Breast Cancer Research and Treatment 120(1), 83–93 (2010)

46. Li, X., Romero, P., Rani, M., Dunker, A.K., Obradovic, Z.: Predicting protein disorder for N-, C-, and internal regions. Genome Informatics Series, 30–40 (1999)

47. Lin, K., Simossis, V.A., Taylor, W.R., Heringa, J.: A simple and fast secondary structure prediction method using hidden neural networks. Bioinformatics 21(2), 152–159 (2005)

48. Liu, B., Cui, Q., Jiang, T., Ma, S.: A combinational feature selection and ensemble neural network method for classification of gene expression data. BMC Bioinformatics 5(1), 136 (2004)

49. Matthews, B.W.: Comparison of the predicted and observed secondary structure of T4 phage lysozyme. Biochimica et Biophysica Acta (BBA)-Protein Structure 405(2), 442–451 (1975)

50. Mooney, C., Pollastri, G.: Beyond the twilight zone: Automated prediction of structural properties of proteins by recursive neural networks and remote homology information. Proteins: Structure, Function, and Bioinformatics 77(1), 181–190 (2009)

51. Plewczynski, D., Slabinski, L., Ginalski, K., Rychlewski, L.: Prediction of signal peptides in protein sequences by neural networks. Acta Biochimica Polonica 55(2), 261–267 (2008)

52. Qian, N., Sejnowski, T.J.: Predicting the secondary structure of globular proteins using neural network models. Journal of Molecular Biology 202(4), 865–884 (1988)

53. Rost, B., Sander, C.: Improved prediction of protein secondary structure by use of sequence profiles and neural networks. Proceedings of the National Academy of Sciences of the United States of America 90(16), 7558–7562 (1993)

54. Rost, B., Sander, C.: Combining evolutionary information and neural networks to predict protein secondary structure. Proteins-Structure Function and Genetics 19(1), 55–72 (1994)

55. Sidman, K.E., George, D.G., Barker, W.C., Hunt, L.T.: The protein identification resource (PIR). Nucleic Acids Research 16(5), 1869 (1988)

56. Song, D., Deng, Z.: A novel ncRNA gene prediction approach based on fuzzy neural networks with structure learning. In: 2010 4th International Conference on Bioinformatics and Biomedical Engineering (iCBBE), pp. 1–5. IEEE (2010)

57. Speed, T.P.: Statistical analysis of gene expression microarray data. CRC Press (2003)

58. Tegge, A.N., Wang, Z., Eickholt, J., Cheng, J.: NNcon: improved protein contact map prediction using 2d-recursive neural networks. Nucleic Acids Research 37, w515–w518 (2009)
59. Vullo, A., Frasconi, P.: Disulfide connectivity prediction using recursive neural networks and evolutionary information. Bioinformatics 20(5), 653–659 (2004)
60. Walsh, I., Martin, A.J.M., Mooney, C., Rubagotti, E., Vullo, A., Pollastri, G.: Ab initio and homology based prediction of protein domains by recursive neural networks. BMC Bioinformatics 10(1), 195–214 (2009)
61. Wang, H., Noordewier, M., Benham, C.J.: Stress-induced DNA duplex destabilization (SIDD) in the e. coli genome: Sidd sites are closely associated with promoters. Genome Research 14(8), 1575 (2004)
62. Wang, J.T.L., Ma, Q., Shasha, D., Wu, C.H.: Application of neural networks to biological data mining: a case study in protein sequence classification. In: Proceedings of the Sixth ACM SIGKDD International Conference on Knowledge Discovery and Data Mining, pp. 305–309. ACM (2000)
63. Weinreb, P.H., Zhen, W., Poon, A.W., Conway, K.A., Lansbury Jr., P.T.: NACP, a protein implicated in alzheimer's disease and learning, is natively unfolded. Biochemistry 35(43), 13709–13715 (1996)
64. Wen, X., Fuhrman, S., Michaels, G.S., Carr, D.B., Smith, S., Barker, J.L., Somogyi, R.: Large-scale temporal gene expression mapping of central nervous system development. Proceedings of the National Academy of Sciences 95(1), 334 (1998)
65. Wright, P.E., Dyson, H.J.: Intrinsically unstructured proteins: re-assessing the protein structure-function paradigm. Journal of Molecular Biology 293(2), 321–331 (1999)
66. Wu, C., Whitson, G., Mclarty, J., Ermongkonchai, A., Chang, T.C.: Protein classification artificial neural system. Protein Science: A Publication of the Protein Society 1(5), 667 (1992)
67. Zamani, M., Chiu, D.: An evaluation of DNA barcoding using genetic programming-based process. Life System Modeling and Intelligent Computing, 298–306 (2010)
68. Zhang, A.B., Sikes, D.S., Muster, C., Li, S.Q.: Inferring species membership using DNA sequences with back-propagation neural networks. Systematic Biology 57(2), 202–215 (2008)
69. Ziegel, E.R.: Probability and statistics for engineering and the sciences. Technometrics 46(4), 497–498 (2004)
70. Zimmermann, O., Hansmann, U.H.E.: Support vector machines for prediction of dihedral angle regions. Bioinformatics 22(24), 3009 (2006)

Contributors

Nigel M. Allinson holds the Chair of Image Engineering at the University of Sheffield, UK, where he heads the Image Engineering Laboratory. He has published over 280 scientific papers, sits of several advisory boards of international research establishments, and helped found five spinout companies based research from his group.

Yoshua Bengio is Full Professor of the Department of Computer Science and Operations Research, head of the Machine Learning Laboratory (LISA), CIFAR Fellow in the Neural Computation and Adaptive Perception program, Canada Research Chair in Statistical Learning Algorithms, and he also holds an NSERC industrial chair. His main research ambition is to understand principles of learning that yield intelligence. He teaches a graduate course in Machine Learning, and supervises a large group of graduate students and post-docs. His research covers many areas of machine learning and is widely cited (over a 4300 citations found by Google Scholar in 2008).

Hendrik Blockeel holds a PhD in Computer Science from the Katholieke Universiteit Leuven, Belgium. He is currently a professor at the same university, and a part-time associate professor at Leiden University, The Netherlands. His research covers a variety of topics in the areas of machine learning, data mining, and applications of these in artificial intelligence, bio-informatics, and medical informatics. Particular topics of interest include relational data mining, statistical relational learning, prediction in structured output domains, inductive databases, query-based mining, and declarative experimentation. He has authored numerous publications in these areas, and authored or edited seven volumes. He is an action editor of Machine Learning and an editorial board member of several other peer-reviewed journals, is or was organizer or program chair of multiple international conferences, workshops and summer schools, and was a board member of the European Coordinating Committee for Artificial Intelligence from 2004 to 2010.

Willem Burgers studied applied physics at Fontys Hogescholen in Eindhoven (the Netherlands) where he received a BSc in 2001. The graduation project was done at Océ Technologies NV. He subsequently studied physics at the Radboud University Nijmegen and received his MSc in 2007 with a graduation project for Shell NV. Since then he has been working for SNN and SMART Research BV on various projects. His main expertise is development of algorithms and software for computational efficient implementations of machine learning methods for large scale problems.

Christopher Cameron is an undergraduate student in the Honours Biomedical Toxicology Baccalaureate programme at the University of Guelph. He has simultaneously completed a Minor in Computing Information and Science. His research focuses on the application, and comparison, of QSAR techniques to problems in toxicology.

Aaron Courville completed his PhD at Carnegie Mellon University in 2006. He is currently a research scientist in the LISA Lab at the University of Montreal. His research interests include deep learning methods, probabilistic models, nonparametric Bayesian methods and modelling animal learning.

Chrisina Draganova has an MSc in Computing Science, an MSc in Mathematics and Informatics, and a Ph.D. degree in Applied Mathematics. She has worked in several British universities including London Metropolitan University, South-Bank University, Kingston University, University College London, and at Veliko Turnovo University, Bulgaria. She was awarded UK National Teaching Fellowship in 2009. Her research includes the development of new methods in the area of approximation and interpolation with spline functions and innovative applications of neural networks. She has worked on developing methods for characterising the smoothest interpolation and optimal recovery of functions. In the last years her research has concentrated on applying effective neural network methods to a number of applications including: face interpretation, automatic age estimation, restoration of partial occluded shapes of faces, isolating sources of variation in multivariate distributions and the enhancement of learning and teaching. She has published her research results in numerous refereed conference proceedings and in peer-reviewed journals. She is also co-chair of the International Neural Network Society's Special Interest Group on Engineering Applications of Neural Networks

Francesco Gargiulo was born in Vico Equense, Naples, in 1981. He received the M.Sc. Degree in Telecommunication Engineering (*cum laude*) in 2006 and the Ph.D. degree in Computer and Control Engineering in 2009, both from the University of Naples Federico II. Since 2008 he is member of the International Association for Pattern Recognition (IAPR). His research interests involve pattern recognition, multiple classifier systems and computer security.

Liviu Goraş was born in Iasi, Romania, in 1948. He received the Diploma Engineer and the Ph.D. degree in Electrical Engineering from the "Gheorghe Asachi" Technical University (TU) Iasi, Iasi, Romania, in 1971 and 1978, respectively. Since 1973, he was successively Assistant, Lecturer, Associate Professor and, since 1994, he has been a Professor with the Faculty of Electronics and Telecommunications, TU Iasi. From September 1994 to May 1995, he was on leave, as a senior Fulbright Scholar, with the Department of Electrical Engineering and Computer Sciences, University of California at Berkeley. His main research interests include nonlinear circuit and system theory, cellular neural networks, and signal processing. He is the main organizer of the International Symposium on Signal, Circuits and Systems, ISSCS, held in Iasi every two years since 1993. Dr. Goras was the recipient of the IEEE Third Millennium Medal.

Mohamed Farouk Abdel Hady received his BSc. in 1999 from the Mathematics Department of the Faculty of Science at Cairo University. He received the MSc degree in 2005 from the same university and the thesis presented a new technique to extract fuzzy if-then rules from artificial neural networks. In February 2011, he received his doctoral degree from the Department of Neural Information Processing at Ulm University in Germany thanks to the doctoral scholarship that he got in October 2006 from the German Academic Exchange Service (DAAD) to pursue his study and research in Germany. Afterwards and until July, he was a post-doctoral researcher at the same department at the University of Ulm. Before joining Ulm University from 2001 to 2006, he was a researcher at IBM Cairo's "Technology Development Center", where he was involved in a variety of research and development (R&D) activities. From 2000 to 2005, he was a teaching assistant at the Department of Computer Science, Institute of Statistical Studies and Research, Cairo University. Since 2005, he has worked as an assistant lecturer at the same department. His research interests include machine learning, data mining, semi-supervised and active learning, multi-label learning, ensemble learning and pattern recognition, especially in learning from unlabeled data which is the main topic of his doctoral dissertation.

Markus Hagenbuchner holds a PhD (Computer Science, University of Wollongong, Australia). He is currently a senior lecturer in the School of Computer Science and Software Engineering at the University of Wollongong. He joint the machine learning research area in 1992, and commenced his research focus on Neural Networks for the graph structured domain in 1998. Dr. Hagenbuchner pioneered the development of Self-Organizing Maps for structured data. He is the team leader of the machine learning group at the University of Wollongong, and a deputy director of the Intelligent Systems Research Lab within the Information and Communication Technology Research Institute. He has been the chief investigator on a number of projects funded by the Australian government for which he received in excess of one million dollars in research funds. Dr. Hagenbuchner's work has resulted in breakthrough technologies in the machine learning area. His methods produced state-of-the-art performances on a range of benchmark problems in text mining, data mining, image processing, and document processing. His work has been recognized as a research strength of the University of Wollongong.

Bert Kappen studied particle physics in Groningen, the Netherlands and completed his PhD in this field in 1987 at the Rockefeller University in New York. Since 1989, he is conducting research on neural networks at the Radboud University Nijmegen, the Netherlands. Since 1997 he is associate professor and since 2004 full professor at this university. His group consists of 10 people and is involved in research on machine learning and computational neuroscience. He is director of the Dutch Foundation for Neural Networks (SNN). Since 2009, he is visiting professor at UCL's Gatsby Computational Neuroscience Unit in London.

Milly Kc is an early career researcher working in the field of machine learning and information retrieval. She graduated from Bachelor of Information and Communication Technology with a first class Honor in 2004. She then proceeded into research in the areas of machine learning and web-based information retrieval. In 2009, she earned a PhD in Computer Science and Software Engineering from University of Wollongong. Milly Kc worked as a research associate in University of Wollongong, in the Informatics Faculty's School of Computer Science and Software Engineering, from 2008 to 2010. Her research focuses on the distributed estimation or calculation of centralised ranking algorithms such as Pagerank. During her employment, she has collaborated and presented her work to research partners in Australia, Italy and Hong Kong. Milly Kc has also been actively involved in teaching subjects in her research area such as artificial intelligence, reasoning and learning, distributed computing and markup languages.

Stefan C. Kremer is an Associate Professor of Computer Science and Director of Bioinformatics at the University of Guelph. His research interests include computational intelligence and bioinformatics.

Paul C. Kainen received his Ph.D. in Algebraic Topology from Cornell University. In addition to neural networks, he has worked extensively in graph theory, is coauthor, with T. L. Saathy, of the *Four Color Problem* (McGraw-Hill, 1977; Dover, 1986), and is Director of the Laboratory for Visual Mathematics at Georgetown University.

Věra Kůrková received her Ph.D. in Topology from Charles University, Prague. Since 1990, she has worked as a scientist in the Institute of Computer Science of the Academy of Sciences of the Czech Republic. From 2002 to 2008 she was the Head of the Department of Theoretical Computer Science. She is a member of the Editorial Boards of the journals Neural Networks and Neural Processing Letters, and in 2008-2009 she was a member of the Editorial Board of the IEEE Transactions on Neural Networks. She is a member of the Board of the European Neural Networks Society and was the Chair of conferences ICANNGA 2001 and ICANN 2008. In 2010 she was awarded by the Czech Academy of Sciences the Bolzano

Medal for her contributions to mathematics. Her research interests include mathematical theory of neurocomputing and learning and nonlinear approximation theory.

Jing Li received the BEng degree from Nanchang University. She is currently a PhD student with the Department of Electronic and Electrical Engineering in the University of Sheffield, UK. Her research interests include content-based image retrieval, image data management, and pattern recognition. She has published in IEEE Transactions on Image Processing (TIP), Neurocomputing (Elsevier), International Journal of Imaging Systems and Technology (IJIST), amongst others.

Sajid A. Marhon is a PhD student in the School of Computer Science at the University of Guelph. His research focus is on gene finding techniques based on signal processing. Sajid is supervised by Stefan Kremer.

Claudio Mazzariello is currently a PostDoc researcher at the at the Department of Computer Science and Systems of the University of Naples Federico II; he received a Ph.D. degree in Computer Engineering from the University of Naples Federico II, Italy, in 2007. His research is mainly focused on multiple classifier systems, network security and attack detection, network user behavior modeling and critical infrastructure protection. Dr. Mazzariello is a member of the International Association for Pattern Recognition (IAPR) and a student member of the IEEE and ACM.

Dominic Palmer-Brown is a professor of neural computing and Dean of the Faculty of Computing at London Metropolitan University. His educational qualifications include a BSc (Hons) in Electrical and Electronic Engineering; an MSc in Intelligent Systems; and a PhD in Computer Science. In addition to working in universities, he has experience of R&D in the avionics industry, and as editor of Trends in Cognitive Sciences, Elsevier Science London. He is a Fellow of the British Computer Society. His research encompasses neural computing for data mining, pattern recognition, NLP, modelling user interaction, and enhancing virtual learning environments. He has supervised 12 PhDs to completion, and delivered invited keynote lectures at international conferences, most recently on the modal learning approach to neural computing. His publications include articles in Information Sciences, IEEE Transactions in Neural Networks, Neurocomputing, Connection Science, and Ecological Modelling. He was recently guest editor for a special issue of the journal Neurocomputing (Elsevier), and is currently guest editor for a special issue of the journal Neural Computing and Applications (Springer). He is co-chair of the International Neural Network Society's Special Interest Group on Engineering Applications of Neural Networks.

Andrea Passerini is Assistant Professor at the Department of Information Engineering and Computer Science of the University of Trento. His main research interests are in the area of machine learning, with a special emphasis on bioinformatics applications. In recent years he developed techniques aimed at combining statistical and symbolic approaches to learning, via the integration of inductive logic programming and kernel machines. He is also pursuing a deeper integration of machine learning approaches and complex optimization techniques. He co-authored more than forty scientific publications.

Marcello Sanguineti received the "Laurea" degree in Electronic Engineering in 1992 and the Ph.D. degree in Electronic Engineering and Computer Science in 1996 from the University of Genoa, were he is currently Assistant Professor. He is also Research Associate at Institute of Intelligent Systems for Automation of the National Research Council of Italy. He coordinated several international research projects on approximate solution of optimization problems. He is a Member of the Editorial Boards of the IEEE Transactions on Neural Networks, the International Mathematical Forum, and Mathematics in Engineering, Science and Aerospace. He was the Chair of the Organizing Committee of the conference ICNPAA 2008. His main research interests are: infinite programming, nonlinear programming in learning from data, network optimization, optimal control, and neural networks for optimization.

Carlo Sansone is currently Associate Professor of Computer Science at the Department of Computer Science and Systems of the University of Naples Federico II, where he leads the activities of the Artificial Vision and Intelligent System Lab. His research interests cover the areas of pattern recognition, information fusion, computer network security, biometrics and image forensics. He coordinated several projects in the areas of image analysis and recognition, network intrusion detection and IP traffic classification. Prof. Sansone co-authored more than 120 research papers in international journals and conference proceedings. He was co-editor of three books and two Special Issues. Prof. Sansone is a member of the International Association for Pattern Recognition (IAPR) and of the IEEE.

Friedhelm Schwenker received the PhD degree in mathematics from the University of Osnabrück in 1988. From 1989 to 1992 he was a scientists at the Vogt-Institute for Brain Research at the Heinrich Heine University Düsseldorf. Since 1992 he is a researcher and lecturer at the Institute of Neural Information Processing at the University of Ulm. Since 2008 he is the chair of the Technical Committee "Neural Networks and Computational Intelligence" of the International Association for pattern Recognition (IAPR). His main research interests are in artificial neural networks, machine learning, data mining, pattern recognition, applied statistics and approximation theory.

Ah-Chung Tsoi studied Electronic Engineering at the Hong Kong Technical College; Electronic Control Engineering and Control Engineering at University of Salford, England, where he obtained a MSc and a PhD degree in 1970 and 1972 respectively, and has served as Professor of Electrical Engineering at University of Queensland, Australia; Dean, Director of Information Technology Services, and then foundation Pro-Vice Chancellor (Information Technology and Communications) at University of Wollongong; Executive Director, Mathematics, Information and Communications Sciences Interdisciplinary cluster, Australian Research Council, Director, Monash e-Research Centre, Monash University, and Vice President (Research and Institutional Advancement), Hong Kong Baptist University, Hong Kong. In September 2010, he took up the position of Dean, Faculty of Information Technology, Macau University of Science and Technology. Professor Tsoi was instrumental and supervised a number of large software projects, e.g., in University of Wollongong, he supervised the implementation of a web based student management system and integrated it with the other enterprise systems, e.g., finance, human resources; in Australian Research Council, he was an in-house expert in assisting the implementation of a new online research grant application management system, called RMS and it is in use today by all researchers within Australia who are applying for grants with the Australian Research Council; in Monash University, he was instrumental in the establishment of a grid computing system, involving a compute cluster, and its connection to the national grid of computers in Australia; in Hong Kong Baptist University, he supervised the implementation of an in-house implementation of enterprise information system involving student management, finance, human resources, and research grant administration. Moreover, Professor Tsoi's research area is neural networks, which can be applied to text mining, data mining. In this aspect, he had provided new algorithms for modelling graphs with the inclusion of both the link information and content information of nodes. He had successfully applied these techniques to classifying the Wikipedia dataset, web spam detection, with state-of-the-art results.

Paul Ungureanu received the Diploma and the PhD degree in Electronic Engineering from the "Gheorghe Asachi" Technical University of Iasi, Romania in 2001 and 2010, respectively. His main areas of interest are pattern recognition in general and implementation of low level features extraction circuits. He has studied the CNN as features extractors and the implementation of Gabor filters. Since 2010 he is a member of the research team on "Algorithms and parallel architecture for signal acquisition, compression and processing" at the Faculty of Electronics, Telecommunications and Information Technology of Iasi.

Andrew Vogt has a doctorate in mathematics from the University of Washington and teaches mathematics and statistics at Georgetown University. In addition to co-authoring work with Paul Kainen and Věra Kůrková on neural networks, he does research in functional analysis, mathematical physics, and statistics.

Ion Vornicu is a Ph.D. candidate in Electronic Engineering at the "Gheorghe Asachi" Technical University of Iasi, Romania. During the doctoral studies he has carried out a research stage at the Institute of Microelectronics of Seville. He received his B.S. degree in Microelectronics in 2008 and master degree on Modern Signal Processing Techniques in 2009 from the Faculty of Electronics, Telecommunications and Information Technology of Iasi. His main areas of interest are the design and VLSI implementation of massively interconnected analog parallel architectures for image processing. He has received the "Best paper award 2009" at the

International Semiconductor Conference CAS, Sinaia. Since 2010 he is a member of the research team on "Algorithms and parallel architecture for signal acquisition, compression and processing" at the Faculty of Electronics, Telecommunications and Information Technology of Iasi.

Wim Wiegerinck studied theoretical physics at the University of Amsterdam, the Netherlands. He joined the SNN neural network group at the Radboud University Nijmegen (RU) as a PhD student. His thesis (1996) concerned stochastic learning in neural networks. He remained in the group as SNN researcher. He is assistant professor at the RU and vice-director/senior researcher at SMART Research BV, SNN's commercial outlet. His main expertise is in machine learning and Bayesian inference.

Masood Zamani received his B.Sc. in Computer and Mathematics from Amirkabir University of Technology in Iran. He completed MSc. degree in Computer Science from Ryerson University in Canada. He is currently studying for a PhD. in Computer Science at the University of Guelph in Canada. His research interests are, in general, the application of Machine Learning methods such as neural networks, support vector machines, uncertainty reasoning and evolutionary computation in bioinformatics. He currently works on protein tertiary structure prediction.

ShuJia Zhang is currently a final-year PhD student in University of Wollongong, Australia (UoW). She started her research work from participating a Summer Scholarship Program in UoW in 2006. She developed an early interest in Automatic Web Search Service Testing project, and earned Bachelor of Information Communication Technology Honor Degree from UoW, majored in Software Engineering, in 2007. She continued her research on Search engine testing area for one year when she worked as research assistant in UoW. She started PhD study since July 2008 and her thesis topic is Impact Sensitive Ranking of Structured Documents. The main research interest is to provide alternative ranking scheme by utilizing machine learning methods to encode structural information of the documents.

Author Index

Editors

Monica Bianchini received the Laurea degree (cum laude) in Applied Mathematics in 1989 and the PhD degree in Computer Science and Control Systems in 1995, from the University of Florence, Italy. She is currently an Associate Professor at the Department of Information Engineering and Mathematical Sciences of the University of Siena. Her research interests include machine learning (with a particular emphasis on theoretical aspects of learning in neural networks), optimization, approximation theory, and pattern recognition. She served/serves as an Associate Editor for IEEE TRANSACTIONS ON NEURAL NETWORKS (2003-09) and Neurocomputing (2002-present), and have been the editor of numerous books and special issue in international journals on neural networks/structural pattern recognition. She is a permanent member of the editorial board for IJCNN, ICANN, ICPR, ANNPR, ICPRAM and KES.

Marco Maggini received the laurea degree cum laude in Electronic Engineering in February 1991, and the PhD in Computer Science and Control Systems in 1995 from the University of Firenze. He is currently associate professor in Computer Engineering at the Department of Information Engineering and Mathematical Sciences of the University of Siena. His main research interests are on machine learning, with focus on neural networks and kernel methods, and related applications in Web technologies and Web search engines, Information Retrieval, Information Extraction, and Pattern Recognition. He is coauthor of more than one hundred scientific publications. He has served as an associate editor of the ACM Transactions on Internet Technology and he has been member of the program committees of several international conferences.

Dr. Lakhmi C. Jain serves as Adjunct Professor in the Faculty of Information Sciences and Engineering at the University of Canberra, Australia. Dr. Jain founded the KES International for providing a professional community the opportunities for publications, knowledge exchange, cooperation and teaming. Involving around 5000 researchers drawn from universities and companies world-wide, KES facilitates international cooperation and generate synergy in teaching and research. KES regularly provides networking opportunities for professional community through one of the largest conferences of its kind in the area of KES. www.kesinternational.org

His interests focus on the artificial intelligence paradigms and their applications in complex systems, art-science fusion, e-education, e-healthcare, unmanned air vehicles and intelligent agents.